T0271070

Steam Generators
and Waste Heat Boilers
For Process and Plant Engineers

MECHANICAL ENGINEERING
A Series of Textbooks and Reference Books

Founding Editor

L. L. Faulkner

*Columbus Division, Battelle Memorial Institute
and Department of Mechanical Engineering
The Ohio State University
Columbus, Ohio*

RECENTLY PUBLISHED TITLES

Steam Generators and Waste Heat Boilers: For Process and Plant Engineers
V. Ganapathy

Maintenance, Replacement, and Reliability: Theory and Applications, Second Edition
Andrew K.S. Jardine and Albert H.C. Tsang

Heat Exchanger Design Handbook, Second Edition,
Kuppan Thulukkanam

Vehicle Dynamics, Stability, and Control, Second Edition,
Dean Karnopp

*HVAC Water Chillers and Cooling Towers: Fundamentals, Application,
and Operation, Second Edition,*
Herbert W. Stanford III

Ultrasonics: Fundamentals, Technologies, and Applications, Third Edition,
Dale Ensminger and Leonard J. Bond

Mechanical Tolerance Stackup and Analysis, Second Edition,
Bryan R. Fischer

Asset Management Excellence,
John D. Campbell, Andrew K. S. Jardine, and Joel McGlynn

*Solid Fuels Combustion and Gasification: Modeling, Simulation, and Equipment
Operations, Second Edition, Third Edition,*
Marcio L. de Souza-Santos

Mechanical Vibration Analysis, Uncertainties, and Control, Third Edition,
Haym Benaroya and Mark L. Nagurka

Principles of Biomechanics,
Ronald L. Huston

Practical Stress Analysis in Engineering Design, Third Edition,
Ronald L. Huston and Harold Josephs

*Practical Guide to the Packaging of Electronics, Second Edition:
Thermal and Mechanical Design and Analysis,*
Ali Jamnia

Steam Generators
and Waste Heat Boilers
For Process and Plant Engineers

V. Ganapathy

CRC Press
Taylor & Francis Group
Boca Raton   London   New York

CRC Press is an imprint of the
Taylor & Francis Group, an **informa** business

CRC Press
Taylor & Francis Group
6000 Broken Sound Parkway NW, Suite 300
Boca Raton, FL 33487-2742

First issued in paperback 2017

© 2015 by Taylor & Francis Group, LLC
CRC Press is an imprint of Taylor & Francis Group, an Informa business

No claim to original U.S. Government works
Version Date: 20140331

ISBN 13: 978-1-138-07768-3 (pbk)
ISBN 13: 978-1-4822-4712-1 (hbk)

This book contains information obtained from authentic and highly regarded sources. Reasonable efforts have been made to publish reliable data and information, but the author and publisher cannot assume responsibility for the validity of all materials or the consequences of their use. The authors and publishers have attempted to trace the copyright holders of all material reproduced in this publication and apologize to copyright holders if permission to publish in this form has not been obtained. If any copyright material has not been acknowledged please write and let us know so we may rectify in any future reprint.

Except as permitted under U.S. Copyright Law, no part of this book may be reprinted, reproduced, transmitted, or utilized in any form by any electronic, mechanical, or other means, now known or hereafter invented, including photocopying, microfilming, and recording, or in any information storage or retrieval system, without written permission from the publishers.

For permission to photocopy or use material electronically from this work, please access www.copyright.com (http://www.copyright.com/) or contact the Copyright Clearance Center, Inc. (CCC), 222 Rosewood Drive, Danvers, MA 01923, 978-750-8400. CCC is a not-for-profit organization that provides licenses and registration for a variety of users. For organizations that have been granted a photocopy license by the CCC, a separate system of payment has been arranged.

Trademark Notice: Product or corporate names may be trademarks or registered trademarks, and are used only for identification and explanation without intent to infringe.

Library of Congress Cataloging-in-Publication Data

Ganapathy, V.
    Steam generators and waste heat boilers : for process and plant engineers / author, V. Ganapathy.
        pages cm. -- (Mechanical engineering)
    Includes bibliographical references and index.
    ISBN 978-1-4822-4712-1 (alk. paper)
    1. Boilers. I. Title.

TJ263.5.G36 2014
621.1'83--dc23   2014012069

Visit the Taylor & Francis Web site at
http://www.taylorandfrancis.com

and the CRC Press Web site at
http://www.crcpress.com

Dedicated to plant and process engineers who henceforth will be able to evaluate

boiler and HRSG performance without assistance from boiler suppliers!

Contents

Preface...xiii
Author..xxxi

1. **Combustion Calculations** 1
   Introduction...1
   Moisture in Air..1
   Combustion Calculations 2
      Excess Air from Flue Gas Analysis 8
      Simplified Combustion Calculations 10
   Estimation of Heating Values...12
   Burner Selection..13
   Combustion Temperatures 14
   Simplified Procedure for Estimating Combustion Temperatures 15
      Effect of FGR on Combustion Temperature 15
      Relating FGR and Oxygen in Windbox 16
   Gas Turbine Exhaust Combustion Calculations 17
      Relating Oxygen and Energy Input in Turbine Exhaust Gases 17
      Evaluating Fuel Quantity Required to Raise Turbine Exhaust Gas Temperature 19
      Firing Temperature versus Oxygen, Burner Duty, and Water Vapor.......21
   Boiler Efficiency 21
      Heat Loss Method 22
      Simplified Formulae for Boiler Efficiency 23
      Simplified Procedure for Obtaining Major Boiler Parameters............26
      Excess Air versus Efficiency 26
   Firing Fuels with Low Heating Values..27
   Firing of Multiple Fuels...29
   Emission Conversion Calculations: Steam Generators 29
   Converting ppmvd of NO_x to mg/Nm^3...32
   Converting Turbine Exhaust Emissions 34
   Low-Temperature Corrosion in Boilers 35
   Condensation of Acid Vapors in Low-Temperature Heat Sinks 38
      Acid Dew Point Temperature T_{dp}...38
   References 40

2. **Steam Generator Furnace Design**..41
   Advantages of Water-Cooled Furnaces 43
      Ring Header Design 44
   Heat Release Rates..46
      Steam Pressure 47
      Circulation Systems 52
   Furnace Exit Gas Temperature Evaluation..53
   Empirical Formula for Furnace Duty Estimation 57

Furnace Duty with Combination Firing...60
Distribution of Heat Flux around Tubes and Fins..61
Distribution of Radiation to Tube Banks...61
Relating Heat Flux from Furnace to inside Tubes...64
External Radiation to Heat Transfer Surfaces at Furnace Exit.....................65
        Direct Radiation to Superheater at Part Load...66
Terms Frequently Used in Furnace Performance..67
Estimating Boiling Heat Transfer Coefficient...67
Estimating Fin Tip Temperature...68
Boiling Process...69
Boiler Circulation..70
Thom's Method for Estimating Losses...75
Circulation Calculations...77
        Circulation Flow versus Load...81
Flow Stratification in Horizontal Tubes...82
Correlations for CHF (Critical Heat Flux) and Allowable Steam Quality.....82
Circulation Problems..85
Guidelines for Good Circulation System Design..85
References..86

3. **Steam Generators**...87
Introduction..87
Large Package Boilers..87
Water-Cooled Furnaces...90
Major Changes in Boiler Design..90
Absence of Air Heater..90
Emissions Affect Steam Generator Designs...92
Custom-Designed Boilers..97
Novel Ideas...97
        O-Type Package Boiler...98
        Dual-Pressure Design...99
        Small Changes, Big Benefits..99
Boiler Classification...103
Improving Boiler Efficiency...108
        Adding Condensate Heater to Improve Boiler Plant Efficiency...........109
        Understanding Boiler Performance...111
        Why Economizer Does Not Steam in Steam Generators.......................112
        Understanding Boiler Surface Areas...113
Steam Generators for Oil Sands Application..115
Superheaters...118
        Radiant versus Convective Superheaters...121
        How Emissions Impact Radiant and Convective Superheaters............124
        Arrangement of Superheaters...124
        Steam Temperature Control..125
Steam Generators with Import and Export Steam..129
Flow in Parallel Paths..129
Steam Inlet and Exit Nozzle Location...130
Off-Design Performance of Superheater...131

Case Study of a Superheater with Tube Failure Problems...135
    Problem at Low Loads with Inverted-Loop Superheaters 137
Steam Generator Design and Performance..138
    Data Required for Performing Steam Generator Analysis 139
    Checking Boiler Performance 142
    Evaluating Part Load Performance 146
Performance without Economizer...149
Tube Wall Temperature Estimation at Economizer Inlet...149
Methods to Minimize Low-Temperature Corrosion Problems......................................151
Precautions to Minimize Corrosion in Operation 154
Water Chemistry, Carryover, Steam Purity 155
Fire Tube Boilers...157
    Boiler Horse Power 160
    Sizing and Performance Calculations...160
References 163

4. Waste Heat Boilers...165
Introduction 165
    Gas Inlet Temperature and Analysis..168
    Flue Gas Composition and Gas Pressure 170
    Water Tube versus Fire Tube Boilers 173
    Circulation Systems 174
    Heat Recovery in Sulfur Plants..178
    Heat Recovery in Sulfuric Acid Plant 181
    Heat Recovery in Hydrogen Plants 182
    Combining Solar Energy with Heat Recovery Systems 184
    Gas Turbine HRSGs 184
    Natural, Forced Circulation, Once-Through Designs, and Special Applications 188
    Natural versus Forced Circulation HRSGs 191
    Emission Control in HRSGs 194
    Flow-Accelerated Corrosion 195
Water Tube Waste Heat Boilers: Sizing and Performance...197
    Optimizing Pinch and Approach Points in HRSGs 197
    HRSG Performance and Evaluating Field Data..199
    Efficiency of Waste Heat Boiler, HRSG 204
        Advantages of Supplementary Firing in HRSGs..204
    HRSG and Steam Generator Performance 205
    Optimum Utilization of Boilers 205
    Combined Cycle Plants and Fired HRSGs 206
    HRSGs with Split Superheaters 208
    Performance with and without Export Steam 208
    Providing Margins on Heating Surfaces 217
    Cement Plant Waste Heat Recovery 217
Kalina Cycle...223
    Advantages of Kalina Cycle 224
Fluid Heaters and Film Temperature..227
Design of Fire Tube Boilers...230
    Boiling Heat Transfer Coefficient h_o...231

Effect of Scale Formation..234
Estimation of Loss in Performance Due to Fouling 236
Tube Wall Thickness Calculations..236
Off-Design Performance with Addition of Economizer..........................238
Simulation of Fire Tube Boiler Performance 239
Effect of Gas Analysis on Performance..241
Critical Heat Flux q_c...242
Effect of Tube Size on Boiler Design 244
Estimating Tube Bundle Diameter 245
Simplified Approach to Evaluating Performance of Fire Tube Boilers 245
Estimating Tube Sheet Thickness..248
Estimating Tube Sheet Temperature with Refractory and Ferrules..........248
Calculation of Tube Sheet Temperature without Refractory on Tube Sheet 249
Air Heaters..250
Heat Pipes 251
Design of Tubular Air Heaters..252
Heat Transfer Inside and Outside Tubes 253
Part Load Performance..257
Specifying Waste Heat Boilers 258
Waste Heat Boiler Data (Thermal Design) 259
References 261

5. HRSG Simulation..263
Introduction..263
What Is HRSG Simulation? 263
Applications of HRSG Simulation..265
Understanding Pinch and Approach Points...266
Estimating Steam Generation and Gas–Steam Temperature Profiles.........267
Why Cannot We Arbitrarily Select the Pinch and Approach Points?.........270
Why Should Pinch and Approach Points Be Selected in Unfired Mode? 271
Simulation of HRSG Evaporator...272
Supplementary Firing 273
Off-Design Performance Evaluation 274
How Accurate Is Simulation?...279
Steaming in Economizer..281
Methods to Minimize Steaming in Economizer 283
Field Data Evaluation 284
Single- or Multiple-Pressure HRSG 286
Split Superheater Design...295
Fresh Air Firing..299
Efficiency of HRSG 300
Cogeneration Plant Application..301
Optimizing HRSG Arrangement 303
Application of Simulation to Understand the Effect of Ambient Conditions............305
Applying Margins on Exhaust Gas Flow and Temperature 306
Conclusion 308
References 308

6. Miscellaneous Boiler Calculations 309
   Condensing Economizers..309
      Water Dew Point of Flue Gases...310
      Energy Recoverable through Condensation 311
      Condensation Heat Transfer Calculations.......................................312
      Condensation over Finned Tubes 315
   Wall Temperature of Uninsulated Duct, Stack................................317
   Insulation Calculations 320
      Hot Casing Design...322
   Drum Coil Heater: Bath Heater Sizing 323
   Checking Heat Transfer Equipment for Noise and Vibration Problems 327
      Determining Natural Frequency of Vibration of Tubes 327
      Acoustic Frequency f_a...328
      Vortex Shedding Frequency f_e...328
      Damping Criterion..331
      Fluid Elastic Instability 332
   Steam Drum Calculations 333
      Steam Velocity in Drum..333
      Blowdown Calculations 335
      Drum Holdup Calculations 339
   Estimating Flow in Blowdown Lines...340
      Theory...340
   Flow Instability in Two-Phase Circuits 343
   Transient Calculations..346
      Analysis of an Evaporator...348
      Drum-Level Fluctuations..350
   Fan Calculations..351
      Density of Flue Gas, Air..351
      Pressure Drop in Ducts..352
      Fan Selection 353
      Why Should the Fan Capacity Be Reviewed at the Lowest Density Condition? 354
      Head Developed by Fan...355
      Fan Power Consumption..355
      Estimating Stack Effect 355
   References 356

Appendix A: Boiler Design and Performance Calculations 357
Appendix B: Tube-Side Heat Transfer Coefficients and Pressure Drop ...377
Appendix C: Heat Transfer Coefficients Outside Plain Tubes...................393
Appendix D: Nonluminous Heat Transfer Calculations...........................407
Appendix E: Calculations with Finned Tubes...413
Appendix F: Properties of Gases 447
Appendix G: Quiz on Boilers and HRSGs with Answers 477

Conversion Factors...485

Glossary...491

Nomenclature 493

Index...495

Preface

For over three decades, I was associated with boiler companies designing oil- and gas-fired custom package boilers and waste heat boilers that are in operation worldwide in chemical plants, refineries, cogeneration systems, and power plants. The boilers have been operating well and performing as predicted within the margin of measurement errors. As a specialist in thermal design of heat transfer equipment and out of interest, I would correlate field data from operating units with predicted performance to check the correction factors to be incorporated in heat transfer or gas pressure drop correlations that were used for designing the boilers. I would review operating data such as boiler exit gas temperature, steam generation, fuel input, gas pressure drop, evaporator, superheater performance, their tube wall temperatures, and other process-related parameters and compare the results with predicted data. Over the years this has helped me to fine-tune the calculation procedure and modify the correction factors used in the performance evaluation of various heat transfer surfaces. Through the publication of hundreds of articles during this period, I have also been sharing thermal design procedures for boilers and HRSGs with the engineering community. I have written a few books, the recent being *Industrial Boilers and HRSGs*, where the emphasis was on sizing and performance evaluation of steam generator and waste heat boiler components.

For the past decade, since the publication of the cited book, I have been playing the role of a boiler consultant. The transformation from a boiler designer to a boiler consultant has been gratifying as I can interact with plant engineers worldwide and learn from them about various boiler performance issues and suggest modifications to solve boiler-related problems or improve their performance (the word *boiler* refers to a steam generator as well as a waste heat recovery unit). Many plant engineers or even consultants have no clue as to how a boiler should be evaluated before it is bought or what information should be obtained from boiler suppliers regarding its thermal performance aspects. They rely entirely on the sales pitch from boiler suppliers and buy it without giving any critical thought to its design from a thermal performance viewpoint. Purchasing a boiler is thought of as purchasing a commodity. When purchasing a steam generator or HRSG, plant engineers have to evaluate several factors keeping in mind that the life of a boiler is about 30 to 40 years and that the plant will have to face the consequence of any wrong decision for a long time. Some of the considerations are

- Is the consultant engaged by the plant well versed in thermal or process engineering aspects?
- Is the boiler design reflecting the latest trends and improvements in technology?
- How can one rationalize the differences in surface areas among various proposals for the same boiler? Selecting the boiler with the maximum surface area is not the answer.
- Can the boiler handle future emission regulations on NO_x, CO, or UHC with minimum field modifications?
- Has operating cost due to fan power consumption or fuel consumption in boilers or impact of additional back pressure in HRSGs on gas turbine power output been considered? Life-cycle cost is more important than initial cost.

- Have provisions been made to deal with sulfuric acid dew point corrosion in heating surfaces while firing high-sulfur fuels or handling flue gas containing SO_2 and SO_3?
- While firing fuels containing ash or low-melting point solids, can the boiler front end withstand potential slagging and high-temperature corrosion?
- Has the HRSG supplier evaluated the economics of pinch point and approach point while arriving at the HRSG size and steam generation?
- Has the boiler supplier considered all heat sinks available in the plant and tried to maximize energy recovery? The plant engineer has a major role to play in this matter.
- Why is a dual-pressure HRSG offered? Can a single-pressure HRSG not do the same job?
- Why surface areas in finned heating surfaces vary so much between HRSG designs, and how to rationalize difference?
- Significance of heat flux in steam generators and its relation to circulation.
- Relation between superheater tube wall temperature and reduction in its life.
- Is the steam-side pressure drop reasonable at the design point considering all operating loads?
- Advantages of a given design over another, say insulated/refractory lined furnace wall over membrane wall furnace design or a convective superheater versus a radiant design in steam generator.
- Has the performance data at various loads and with different fuels been provided?
- Has complete geometric data of the boiler heating surfaces such as furnace, superheater, evaporator, economizer, and air heater been provided so that an independent evaluation of performance can be done before the boiler is purchased? Many boiler suppliers do not provide this information, and many plant engineers do not know that they should have this important data.
- Will the Kalina cycle be a better choice for their low-gas temperature heat recovery application rather than a steam system?
- Have design features to minimize the boiler or HRSG startup time been provided?

If issues such as those mentioned are not discussed or reviewed before the boiler is purchased, many problems related to thermal performance can surface during operation later on; these include issues such as

- Steam generation much lower than predicted (HRSGs and waste heat boilers)
- Superheater tube failure or overheating and sagging of tubes
- Steam temperature not achieved at rated load
- Flow stagnation in superheater tubes due to low-friction loss and high-gravity loss at low loads
- Too much spray water than envisaged while controlling steam temperature
- Circulation problems resulting in evaporator tube failures
- Steaming in economizer (in gas turbine HRSGs)
- Lower efficiency in steam generators than guaranteed

- Fuel consumption in supplementary fired burners higher than expected or predicted or higher firing temperature in HRSGs than predicted
- Longer startup time than predicted
- Operating costs too high and not envisaged

Plants also have to face several issues regarding boiler thermal performance:

- Unsure whether fouling on steam side or improper design caused tube overheating.
- Unsure whether fouling on gas or steam side caused lower steam output.
- Inability to understand the performance characteristics of steam generator and HRSG versus load and how to optimally load them to maximize plant steam generation and minimize fuel input.
- If operating parameters of a HRSG such as steam generation and exit gas temperature are different from those in the proposal, how to determine if it is due to variations in gas inlet conditions or if it is due to improper sizing of the boiler?
- If a steam generator is operating at say 70% capacity for which performance data such as efficiency or steam temperature or exit gas temperature was not provided by the boiler supplier, how does one find out if it is operating as it should?

When the plant engineer brings the operating problems to the attention of the boiler suppliers, naturally they defend their design and extraneous reasons are blamed for the problems. In my experience not many plant engineers have been familiar with heat transfer and thermal performance aspects, and as a result accept any report from boiler suppliers and as such, are unable to challenge the findings with calculations or studies they have done independently, at the same thing realizing that the design is really flawed. With numerous real world examples on steam generator and HRSG design and off-design performance aspects and calculation modules for major boiler components such as superheater, evaporator, air heater, economizer, and furnace (unfired and fired) given in this book, any plant or process engineer should be able to improve his knowledge in this field and become a competent thermal engineer or consultant and be able to question the boiler supplier if a compromised design is offered and correct the deficiencies *before* it is manufactured and not after the boiler is installed.

The book provides information on thermal design and performance aspects of steam generators and HRSGs, discusses recent developments in boiler technology, novel design ideas, recent trends in their design, and also how plants can save operating costs before the boiler is purchased. The emphasis is on thermal design, and not on manufacturing issues or mechanical design or stress analysis for which there are other excellent books. Several examples on typical performance-related problems faced by plant engineers with steam generators and HRSGs are completely worked out. At the request of many engineers, the metric and SI system has been followed in this book and wherever possible British units are shown in parentheses. Important heat transfer and pressure drop correlations are presented in all three units. Plant engineers with a background in process or heat transfer can go through them and develop calculation modules or computer codes to evaluate thermal performance of boilers and their components independently, without the aid of boiler suppliers. This I feel will be very helpful to plant engineers in the long run. After reviewing the numerous examples provided in the book, plant engineers can evaluate the performance of a steam generator starting from the furnace all the way to the economizer exit or the other way around; they can evaluate the off-design performance of a complete HRSG at any gas inlet condition or check the superheater tube wall temperature or its duty based on the geometric data provided, or an evaporator design considering the circulation system proposed, or check the fuel guarantees offered by boiler suppliers for steam generators or HRSGs. For off-design performance calculations, two methods have been provided and illustrated through several examples. Plant engineers can simulate the performance of their operating boilers and perform what if analysis and gather more information about existing units. The adage that knowledge is power is true in any field of endeavor and applies equally to the field of boiler design and performance evaluation. The problems that many plants face today with boilers and waste heat recovery systems are mainly due to the fact that the designs are accepted as such without being critically evaluated before they are purchased. If the plants do not have the resource to perform this evaluation, they may hire an independent consultant to do so. Remember that a plant has to use a boiler for over 30 or 40 years and time and money spent in the evaluation (before the boiler is purchased!) will pay dividends later in the form of low maintenance or repair costs or problem-free operation. When problems such as tube failures, overheating of tubes, lack of performance, and lower than predicted efficiency surface after a few months of operation, the boiler supplier always defends his design, and if the plant does not have knowledgeable thermal or process engineers to counter the supplier's claims with appropriate calculations and studies, the problems may never be fixed or resolved. I have seen some boilers, in some plants, which should have never been bought as supplied, because the plants spend a lot of time in shutdowns, repairs, and are unable to operate the boiler at the rated capacity.

Chemical plants, refineries, cement plants, cogeneration systems, and power plants invest a lot of their resources, time, money, and manpower in steam generators and waste heat boilers, as steam is the most important working fluid for power generation and process needs. The management expects these steam generators or waste heat boilers to perform well for several decades without problems so that plant engineers can concentrate on their main line of business—whether chemical manufacture or oil refining or power generation. However, this is not the case in many plants. I became an independent boiler consultant a decade ago, and I have had the opportunity to visit several plants worldwide facing severe boiler and HRSG problems. As a boiler designer, I can appreciate the good features in any boiler design offered at the proposal stage as well as spot poor design features or potential problem areas in design. However, in many cases, the boilers are purchased based on the specifications developed by the plant or a consulting firm engaged by them, which in a vast majority of cases does not have specialists in thermal design or performance aspects; the emphasis seems to be on hardware or nonpressure parts. Plant engineers and consultants are familiar with mechanical and electrical aspects of boilers such as instrumentation, controls, fans, pumps, insulation and refractory selection, materials of construction, welding, and erection issues. In fact, the boiler specification developed by their hired consultant will run into hundreds of pages on these topics with hardly a page or two on thermal design or performance aspects. Also, when a boiler is being purchased, engineers typically raise queries on these topics with which they are well versed in but not on thermal design as it is a subject that many are not familiar with! It is some form of Parkinson law perhaps! You discuss things you are familiar with and ignore topics such as heat transfer or process calculations with which you are not familiar. As a result of this, compromised designs are often purchased by plants, resulting in problems such as those discussed earlier. One may wonder if boiler suppliers are untrustworthy. The point is that a busy boiler company may have supplied an off-the-shelf boiler for an application without considering the special needs of the plant, they may have asked their apprentice engineers to design the boiler, or passed on the drawings of a similar boiler designed years ago without checking the suitability for the present situation, the plant engineer may not be aware that the boiler has very high operating costs, or the customer may not have provided all the information regarding plant parameters or their requirements for future operation such as changes in steam parameters or fuel data. Note that plant engineers have to be more cautious and responsible than boiler suppliers or their consultants before purchasing the boiler as they have to use the unit for 30 to 40 years, while for the consultants or boiler suppliers it is just one more boiler and they move on to other projects once the boiler or HRSG has been sold! Hiring a consultant who specializes in process heat transfer or thermal engineering can help a plant foresee problems in their boiler before it is purchased and save them a lot of money in the long run. Alternatively, plants should develop a team of in-house engineers who specialize only in boiler-related process calculations and who can check the design offered by the boiler supplier; the purpose of this book is exactly that, namely, to educate plant engineers and consultants in thermal design and performance calculations. Many reputable boiler suppliers while making a sales pitch say that they have thousands of such boilers and similar designs in operation and yet within a few months of operation, several problems cited earlier surface and the plant engineers run from pillar to post to fix the issues or they may never be fixed as the design modification would be as expensive as replacing the unit!

When I was a boiler designer in the United States, custom-designed boilers were offered by my company as these have several advantages, as discussed in Chapter 3, over standard or off-the-shelf designs. Standard or off-the-shelf boilers are predesigned for a particular steam output with tube geometry, tube spacing, height, width, length all fixed. In fact, they may have already been built and branded as, for example, 40 t/h or 70 t/h boilers. Note that the efficiency, steam temperature, gas pressure drop through the boiler, or fan power consumption is dependent on the quantity of flue gas flowing through the boiler. As explained in Chapter 3, with the introduction of flue gas recirculation (FGR) for NO_x control, one has to check the boiler performance based on actual flue gas flowing through the unit and not select a boiler from a catalog based on steam capacity. Local conditions such as site elevation and ambient temperature may also affect performance. A few plant engineers even today order a water tube boiler based on the model number. While such an approach may save time and money, it is suitable wherever emission regulations need not be complied with or when saturated steam is generated or when clean, standard fuels are fired. When the boiler has to comply with emission regulations or when special fuels are fired, about 15% excess air and 15%–30% FGR may have to be used depending on the NO_x levels desired at that particular location; the flue gas mass flow would be about 20%–40% more compared with a boiler without FGR and hence the back pressure will be at least 40% more, calling for a larger fan for the same steam output if the standard boiler is used. The boiler exit gas temperature will also be higher due to the higher flue gas mass flow, resulting in lower boiler efficiency as shown in Chapter 1, where the effect of excess air and exit gas temperature on boiler efficiency is discussed. Hence, based on actual excess air and flue gas flow, the furnace dimensions, superheater surfacing, convection bank tube spacing, boiler height, and economizer surface, to mention a few factors, have to be reviewed and adjusted to obtain a low back pressure and a high efficiency; in fact, I would recommend performing such a detailed evaluation for every boiler you purchase; Chapter 3 shows an example of how this is done. This re-engineering of the boiler on a case-by-case basis results in an efficient design with a smaller fan. So when a custom-designed steam generator was proposed for a project, the consultant who was evaluating various offers did not appreciate the fact that the boiler was custom-designed for the specific requirement and that it was not a standard model; he was questioning why our forced draft fan was smaller than that of the competition! He had also come to the conclusion that our fan will not work! I had to then explain to him about custom designing of boilers and the advantages accrued to its owner over a period of time and how our tube spacing and tube lengths were adjusted on a case-by-case basis to lower the gas pressure drop and maintain the efficiency considering the higher flue gas flow through the boiler arising out of local NO_x and CO regulations! Many large boiler suppliers do not want to change the boiler design or dimensions on a case-by-case basis as it is expensive, while a small boiler company can make these changes easily as this does not need approval from the corporate office.

To give another example, many plants even today purchase boilers or HRSGs based on surface area, which is a big mistake! Engineers tabulate surface areas offered by different vendors and if the prices are comparable, select the vendor with the largest surface area, and to add insult to injury, the customer thinks he is getting a bargain! As explained in Appendix E, surface area of a finned tube superheater or evaporator in a HRSG can vary by about 50%–100% for the same duty depending on the fin geometry chosen. Higher fin density leads to higher heat flux inside the superheater tubes, resulting in higher tube wall temperatures and shortened superheater life. Higher fin density also results in a lower overall heat transfer coefficient, necessitating a larger surface area for the same duty and higher gas pressure drop. The customer was under the impression that our superheater was too small as the surface area with our small fin density was about 60%–70% of that of the competition and hence would not achieve the steam temperature. Fortunately, the customer called and asked me this question and I sent him copies of the articles I had published on finned tubes explaining how the U values decrease with an increase in ratio of external to tube internal surface and the product of U × A (overall heat transfer coefficient × surface area) was the important thing to check and not A alone. (The problem is many boiler vendors will not provide information on U values but just the surface area, and unsuspecting engineers may evaluate the options based on surface area alone.) I explained to him about the increase in heat flux caused by poor selection of fins and how the tube wall temperature would increase and as a result how tube life would decrease. After this presentation, they were convinced and appreciated the importance of proper selection of fins in HRSGs. Plant engineers who are not process savvy will jump at the offer with the highest surface area and may have to deal with issues such as high heat flux inside tubes or overheating of tubes later! The book gives a few examples of the effect of fin geometry on the overall heat transfer coefficient and tube wall temperature in the evaporator and superheater so that plant engineers or consultants are not mislead by surface areas when buying a HRSG. Bottom line is "Don't go by surface area A alone. Ask for the U (overall heat transfer coefficient) for each heating surface!" Then check U × A, which should be the same for all vendors for any component with the same duty. The appendices also provide methods to evaluate heat transfer coefficients inside and outside plain and finned tubes so that U may be estimated independently for each heating surface and checked with data provided by the boiler vendor. There may be some differences in U values between boiler suppliers, but they should be small.

Many boilers are purchased based on initial cost without any consideration of operating costs arising from high gas pressure drop or lower efficiency or maintenance costs. For example, a refractory lined boiler will have higher maintenance costs associated with its startup and operation compared with a completely water-cooled furnace as discussed in Chapter 3. The cost of fuel and electricity has increased over the years and ignoring them in the evaluation would not be correct. Examples are provided in the book on evaluating gas pressure drop or selecting HRSGs with low pinch and approach points and low back pressure, which may be slightly more expensive but pay dividends in the long run in the form of lower operating costs and fuel consumption if the HRSG is fired. Methods to improve the efficiency of a steam generator or HRSG using secondary heating surfaces or by lowering the feed water temperature or by heating air using a closed loop glycol system are discussed.

When fuels with potential for acid dew point corrosion are fired in steam generators, some consultants suggest that the exit gas temperature be raised above the dew point. However, this does not prevent acid vapor condensing at the feed water inlet section of the economizer tubes, which may be operating below the acid dew point temperature. In fact, the temperature of the feed water should be raised to avoid condensation and consequent local corrosion and not the gas temperature alone as the tube side heat transfer coefficient is much higher than the gas side coefficient and hence governs the tube wall temperature. Raising the gas temperature above the dew point may prevent corrosion in the ducts downstream of the economizer but not in the tubes per se. One should also understand the fact that as the steam generator load decreases, the economizer exit gas temperature will also decrease and the ducts also can get corroded at part loads. This can be prevented by ensuring that the feed water temperature is above the acid dew point as discussed in Chapter 3, as a result of which the exit gas temperature will always be above the dew point at all loads! The book also discusses the option of using condensing economizers and its sizing procedure and evaluation of additional energy recovery and materials to be used in such applications. The importance of keeping a boiler warm during startup and shutdown while using sulfur-containing gases in boilers and HRSGs is also stressed. Examples show how one may estimate the cold end tube wall temperature in the air heater and economizer and also the effect of the counter flow versus parallel flow arrangement on cold end tube temperature with plain and finned tubes.

When specifying gas turbine HRSGs, consultants not familiar with process calculations or HRSG temperature profile analysis expect a certain exit gas temperature from the boiler or a certain amount of steam generation in the unfired mode. This may not be always attainable as the gas–steam temperature profile is dependent on the pinch and approach points chosen and the steam pressure and this is a thermodynamic limitation. The higher the steam pressure, the higher the exit gas temperature unless one adds a low-pressure steam system. Thus, one cannot predict a particular exit gas temperature without doing an analysis of temperature profiles. Also, many plant engineers are not aware of the fact that supplementary firing is the most efficient way of generating additional steam in a HRSG. If a refinery has both steam generators as well as gas turbine HRSGs, it is prudent to plan future HRSGs as fired units so that steam can be generated at more than 100% efficiency compared with about 93% (on lower heating value basis) in steam generators for gas firing. This point is explained in Chapter 5 on HRSG simulation. The cost of a HRSG is mainly a function of the exhaust gas flow through it, and since this remains the same whether the HRSG is unfired or fired, planning the HRSG for firing in the future is a good idea as the additional cost of burners and insulation will be minimal. On the other hand, the plant can obtain 100%–300% more steam through supplementary firing or furnace-fired HRSGs.

HRSG simulation is explained in Chapter 5. Without designing the HRSG per se, plant engineers can obtain valuable information on steam generation, burner duty, gas–steam temperature profiles, and off-design performance. Using this concept, it is also possible to obtain the efficiency of single-pressure versus dual-pressure HRSGs without designing the HRSG per se. One may also rearrange the boiler components to optimize energy recovery. Anyone familiar with energy balance calculations can perform this analysis. Often, consultants suggest a dual-pressure HRSG without doing a temperature profile analysis.

Dual-pressure HRSGs are not required for all cases involving HP and LP steam requirement. The ratio of LP steam flow to HP steam and the ratio of LP steam pressure to HP steam pressure impact the temperature profile, and sometimes it is more prudent to go with a single-pressure HRSG and take off the process steam from the drum and pressure-reduce it rather than install a dual-pressure HRSG that could be more expensive than a single-pressure unit as shown in Chapter 5. Unless the plant engineer or the consultant is familiar with such analysis, a dual-pressure HRSG may be specified by the plant or the consultant without any analysis, which can be expensive in the long run. It is my desire that a consultant or a plant engineer review any boiler or HRSG design offered to a plant and independently check if there is scope for improving the energy recovery.

Often HRSGs operate in the fired and unfired mode. Common practice is to locate the superheater downstream of the burner and design the superheater in a way that steam temperature is obtained in the unfired mode and then the stream is desuperheated in the fired mode to obtain the desired steam temperature using demineralized water spray. This design is not expensive as the superheaters are built as a single module. This approach, however, may increase the tube wall temperature of the superheaters in the fired mode; also some plants may not have demineralized water for attemperation. One novel approach discussed in the book is to split the superheater so that the final superheater is located beyond the gas turbine and the primary superheater is located downstream of the burner. This approach also lowers the tube wall temperature of the superheaters and eliminates the need for spray attemperation in many cases.

Low-temperature heat recovery as in cement plants offers many challenges. If steam is used as the working fluid, pinch point limits the heat recovery potential as the lower the inlet gas temperature, the higher the exit gas temperature unless there is multiple pressure steam generation. (The reason for this is explained in Chapter 5.) The Kalina cycle HRVG discussed in the book enables the exit gas temperature from the HRVG to be much lower than possible with steam systems due to the varying boiling point of the ammonia–water mixture working fluid. Gas inlet temperatures of 250°C–350°C are ideal heat sources for the Kalina cycle.

There are applications where space is a premium. In one project, a plant had to fit a steam generator of 50 t/h capacity inside a building of width 3.5 m. For example, in Chapter 3, a steam generator with a width of about 3 m is shown with finned tubes in the convection section. A conventional A- or D- or O-type boiler design would not have fitted into this space as its width would have exceeded 5 m and considering space around the boiler, it would have been impossible for a regular or conventional boiler to be installed. As a boiler designer, I came up with an idea of using a O-type boiler with finned tubes with a width of less than 3 m; to reduce the total boiler length, finned tubes were used in the convection bank. The plant engineers and consultants involved in the project appreciated what we were offering and hence the project was a success. If the plant engineers had not been open to this idea, they may have been without a boiler or destroyed the building and installed a standard boiler! It was explained to the plant personnel that finned tubes are used in HRSGs and the boiler offered simply borrowed that concept though it was the first time this was done in a steam generator!

Chapter 3 also discusses an elevated drum steam generator with external downcomers and risers. This concept helps one to shop assemble large capacity steam generators and minimize field work; only the drum with associated downcomer and riser piping is assembled in the field. This approach saves a lot of time and money for the end user. This concept is widely used in HRSG design, but it is new to steam generators. As plant engineers who could appreciate the circulation system were involved in the evaluation and purchase of this boiler, it was not difficult to sell the idea. Hence, it is important that engineers with a background in thermal design or process calculations are also involved in the purchasing decisions of a plant.

In package oil- and gas-fired steam generators, many boiler vendors offer a radiant super-heater even if the steam temperature required is only about 350°C–400°C. It is possible to attain about 500°C with a convective superheater, which has fewer tube failures or performance problems compared with the radiant design. The radiant superheater is more prone to failures at full as well as part loads compared with the convective design, as discussed in Chapter 3. Another common problem is that the flame may impinge on the superheater if the burner is not properly adjusted and there is always a dispute whether the burner is causing the tube failures or if the design of the superheater itself is flawed. Another error made by some consultants is to specify that the steam pressure drop in the superheater should be low; I have seen figures as low as 0.35 kg/cm^2 (5 psi) at full load. This is a very low value for a good design, particularly if the degree of superheat is large and the boiler has to operate over a wide range of steam flows. One has to review the steam velocity, the tube wall temperature, and the pressure drop at all loads. As a designer, I used to ask them about the flow distribution issues at lower loads with such a low pressure drop to start with. At 50% load, the pressure drop will be less than 0.1 kg/cm^2, leading to flow stagnation or reverse flow in some tubes with downward flow where the gravity head opposes the friction loss; hence, overheating or failure can occur if the gas temperature is high. The importance of tube side pressure drop in superheater and the significance of low flow in upward and downward paths are discussed in Chapter 3. Lack of understanding of process or thermal design is the reason for such ridiculous specifications. Also, if the superheater fails because of the low pressure drop suggested by the consultant, then who is at fault?

Some consultants while evaluating offers from different boiler suppliers do not even obtain the thermal or tube geometry data for the boiler surfaces, which is required to do a performance evaluation at a later date if need arises. The plant engineers feel that the consultant who has prepared the specifications will do a good job and do not interfere with the evaluation process. The consulting company may or may not have thermal or process engineers on their team. If they do not, then the plant may have to be at the mercy of the boiler supplier when a boiler problem arises later on, when they do not even have the minimum information to be able to evaluate the boiler's performance. As a boiler consultant, I have seen this problem in many plants and hence one of the services I provide is to evaluate boilers from thermal and performance viewpoints and ensure that all tube geometry data for steam generators and HRSGs are made available by the supplier before purchase. Chapters 3 and 4 present several examples of the format of the tube geometry data. Plant and process engineers should ensure that suppliers of boilers for steam generators and HRSGs provide information in this format so that future performance evaluation becomes easy. Presently many boiler suppliers get away with very sketchy information on their boiler tube geometry and performance details. Only when a problem arises in the boiler and the performance has to be checked do plant engineers run around looking for basic information about the boiler. The boiler supplier may not even be around when these data are required!

Consultants sometimes make the mistake of demanding certain margins over surface areas of various boiler components. This affects the performance of components located downstream. The misconception is that a 15% margin will provide about 10%–15% more duty! An example is worked out to prove that it is not so and this method of specifying boilers can cause problems in performance, particularly if several heat transfer surfaces are included in the design.

Many plant engineers are not aware of the differences in boiler efficiencies based on higher and lower heating values. It is the practice in the United States to state boiler efficiency on an HHV basis, whereas in Europe and Asia, it is based on LHV. The LHV efficiency is higher by a factor equal to the ratio of higher to lower heating value of the fuel. Hence, unless the basis of efficiency is mentioned, clearly evaluation can be skewed. Also, while specifying waste heat boilers, consultants sometimes provide the volumetric gas flow instead of mass flow. This practice should be avoided. One should provide the flue gas flow in mass units such as kg/h and the flue gas analysis as the gas properties and duty are affected by the gas analysis. Energy balance is done using mass of flue gas and not its volume. Variation in steam output can be even 5% or more if some boiler vendor assumes an analysis based on his experience while another vendor assumes a different gas analysis and computes the density. I have seen n different answers from n engineers for the density of flue gas at a given pressure and temperature!

Often performance information on steam generator or HRSG is not completely provided by the boiler supplier when a boiler is sold; little technical or process information is offered during the contract stage, but shrewd plant engineers should ask the boiler supplier to furnish more information on thermal performance such as part load performance and what happens if some steam parameters change such as steam pressure or feed water temperature or fuel analysis. Some plants may operate at say 25% load for most of the time and then go on to 100% load. The plant may need the specified steam temperature at 25% load also, which may not have been discussed. Once the boiler is started up and if it is found that the steam temperature is much lower, the boiler supplier cannot be blamed. Sometimes, a low BTU fuel may be fired after a few years of installation. This will increase the flue gas flow through the boiler and also affect the furnace absorption, steam temperature, backpressure, and efficiency. This scenario must be discussed with the boiler supplier before the boiler is purchased. The plant can make a decision whether to buy the larger fan now or later, when they switch to the low Btu fuel firing mode.

A steam plant was planning a large, new waste heat boiler; it was generating saturated steam in another waste heat boiler and wanted to superheat the saturated steam in the new boiler. This type of design calls for a large superheater and one has to see what happens to the steam temperature and the tube wall temperature when the import steam is absent. A large amount of spray would also be required to control the steam if import steam is absent. The superheater has to be designed considering both the cases. Refineries often encounter such situations as they have numerous process streams generating saturated steam and they would like to superheat this in another new boiler whether it is a steam generator or waste heat boiler. Separately fired superheaters may be inefficient and prone to problems compared with convective-type superheaters in steam generators. In solar plants that are combined with waste heat boilers or HRSGs, the saturated steam from the solar panels must be superheated in the waste heat boiler. Even when the solar panel does not generate steam, the superheater of the waste heat boiler should operate safely without overheating the tubes.

Another example of creativity in boiler design is discussed in Chapter 3. A plant wanted a steam generator to operate at a low pressure for the first few years and then at a higher pressure. Boiler vendors suggested superheater designs that had to be replaced in future as steam-specific volume decreases with pressure and the same design cannot be used while maintaining the steam capacity. Chapter 3 describes the concept used in the design offered, which made them choose the same superheater and operate the boiler at the same steam capacity at both steam pressure levels by simply changing the inlet and outlet steam connections. The customer understood our process calculations and went with our design concept. Other options would have been very expensive for the plant.

Often gas turbine HRSGs and waste heat boilers are designed for a particular gas inlet flow and temperature; however, in operation, the exhaust gas conditions may be different from the design. How can one tell if the HRSG performance is satisfactory or not? The HRSG supplier blames the drop in gas turbine exhaust gas flow or temperature as the reason for the poor HRSG performance, while the gas turbine supplier blames the HRSG design. The plant is caught between them. To minimize such issues, the HRSG supplier may be asked to provide data on the HRSG performance for various gas inlet conditions. With such a performance chart, the plant engineer knows what he can expect from the HRSG at different gas inlet conditions. Problems arise when such data are not sought before the HRSG is purchased. If an HRSG designed to generate 50 t/h of steam at 450°C in the unfired mode generates only 45 t/h of steam at a slightly lower steam temperature, plant engineers scramble around to find out why. The book gives an example on how to perform off-design performance calculations and ensure that the HRSG design is satisfactory at any operating point. Plant engineers may then challenge the HRSG supplier if they feel that the HRSG is not adequately designed.

Circulation is another important issue in boilers. If downcomers and riser pipes are not properly located or sized, circulation can be hampered. There can be reverse flows in evaporator tubes or stagnation if downcomer tubes are heated by flue gas. This point is discussed along with methods to estimate circulation in fire tube as well as water tube boilers in Chapter 2. Correlations for departure from nucleate boiling (DNB) are also given along with estimation of actual and allowable heat flux in evaporator tubes.

Thus, the book is aimed at plant engineers, process engineers, and consultants who want to understand and perform boiler thermal calculations and evaluate the performance of new or existing steam generators and waste heat boilers and become more knowledgeable buyers or plant engineers. Recent developments in boiler design and industry trends are addressed. It is hoped that this book will be a good reference book for plant engineers, consultants, and even boiler designers!

Problems have been worked out in metric units, and SI unit values are shown alongside in several examples; British units are also shown in parentheses for important data, formulae, and results; though SI units are used in college textbooks in the United States and in other parts of the world, engineers in the boiler industry in the United States still work with charts and data in British units; boiler companies in South America and the eastern part of the world such as Malaysia and Indonesia still work with metric units. Many boiler companies have yet to convert their design standards and charts to SI units and hence this approach. Important correlations throughout the book have been provided in tabular form in all three units so that plant engineers may quickly check the results using whichever system they are familiar with.

The starting point in the design of any heat transfer equipment is the analysis of thermal and transport properties of flue gas. This is obtained from combustion calculations and hence we start with this topic. Chapter 1 deals with basic combustion calculations given the fuel analysis and how one may obtain excess air value from oxygen readings in flue gas. Simplified combustion calculation procedures for various fuels help plant engineers arrive at quick estimates of air and flue gas quantities from plant operating data. For performing furnace performance calculations, the combustion temperature of products of combustion is required; estimation of this value for gaseous and oil fuels is illustrated with and without FGR. HRSGs are often supplementary fired, and consumption of oxygen in turbine exhaust as a function of fuel input should be known to process or plant engineers. The ASME heat loss method of efficiency calculation is explained with examples, and simplified equations for estimating boiler efficiency are provided. Conversion calculations for pollutants such as NO_x, CO, UHC, and CO from mass to volumetric units (SI, metric, British) are illustrated as plant engineers should have an idea of the emission levels of various pollutants and the different units used throughout the world. Acid dew point correlations for various acid vapors are also presented so that one can address the problem of low-temperature corrosion.

Chapter 2 describes calculations involving boiler furnace such as estimation of furnace exit gas temperature (FEGT) and furnace duty. The significance of heat release rates and heat flux inside tubes is discussed, as also the view factor determination based on tube diameter and spacing in the membrane wall enclosure, which is helpful for estimating local heat flux inside the tubes. FEGT is estimated by different methods. Information on FEGT enables one to check how downstream components such as superheaters, evaporators, and economizers perform. Distribution of external radiation to the heating surface at the furnace exit is critical if a radiant superheater is used. Recent developments in furnace design such as completely water-cooled furnaces are discussed. Correlations for critical heat flux are provided so that one may check if departure from nucleate boiling is likely. This is followed by circulation calculations.

Chapter 3 on steam generators describes trends in large steam generator designs such as multiple-module, elevated drum design types of boilers such as D, O, and A and forced circulation steam generators. The importance of custom designing and the advantages it offers to the end user are discussed with numerical examples. Pollution regulations limit emissions of NO_x and CO and methods used to meet the limits such as the use of SCR (selective catalytic reduction system) or FGR and how these methods impact boiler design and operating costs are explained. The effects of excess air, FGR on boiler performance, and methods of reducing operating costs using custom-designed boilers are discussed. Emission regulations have had a major impact on boiler designs during the last two decades, and these issues are addressed. The significance of the surface areas of various components and how it should not be the criterion for selecting boilers is explained. An efficiency improvement scheme using a glycol heat recovery system behind the economizer that is in operation in a large boiler plant is discussed. How efficiency may be improved using a lower feed water temperature is also explained with calculations for a natural gas-fired steam generator. Advantages of convective over radiant superheaters are explained with calculations of tube wall temperatures of superheaters. Novel designs of steam generators in operation are described so that plant and process engineers may apply these concepts in their plants if need arises. Various methods of steam temperature control are discussed. If a plant engineer goes through the many numerical examples and studies them, he will be able to do similar calculations for the boilers in his plant and even challenge the boiler supplier if some data are different or if the boiler is not performing well!

Performance calculations of a complete steam generator using field data or data provided by boiler suppliers are outlined step by step. Part load calculations are also illustrated. Plant engineers may go through these examples and see how they can apply these methods to their existing units and check if their calculations match the field data. Examples of tube wall temperature calculations for a superheater as well as cold end temperature of plain tube and finned tube economizers also will enable plant engineers to check their existing superheater for overheating issues or the economizer for low-temperature corrosion potential if any. Any plant engineer familiar with programming codes can convert these calculations into a program and evaluate the performance of their steam generator at any load without support from boiler suppliers, which is the objective of this book. Important data required for such an analysis are also listed so that plant engineers can obtain these data before buying the steam generator! Reliance on boiler suppliers is minimized by such practical examples!

Chapter 4 on waste heat boilers starts with the classification of waste heat boilers and addresses the importance of flue gas analysis; fire tube versus water tube boilers used in chemical plants, refineries, and cogeneration systems are described; heat recovery in sulfur plants, hydrogen plants, and cement plants and the effect of the fouling factor on performance are discussed. A description of a boiler in a cement plant using Kalina cycle is given. The advantage of this system over the use of steam water as the working fluid in low-temperature heat recovery systems is discussed. Features of unfired, fired, natural, and forced circulation gas turbine HRSGs are described. Combining the operation of steam generators with gas turbine HRSGs for better utilization of fuel is explained. Performance calculations for an operating HRSG are completely detailed so that plant engineers can check if the HRSG or waste heat boiler has been designed adequately based on operating parameters at any load! Consultants often demand a large margin on heating surfaces, and the impact of oversurfacing on steam generation and superheater temperature is shown with an example. Fluid heaters using waste flue gases are often used in chemical plants; heat flux inside tubes and film temperature estimation are an important aspect of this design and hence illustrated by an example. Cogeneration plants would like to take off saturated steam from the drum for process heating or other use and the balance will be superheated. An example of performance calculations when steam is exported from the drum of the waste heat boiler is illustrated. Plant engineers can obtain the steam temperature or superheater tube wall temperature when steam is exported. How emission regulations are impacting HRSG designs by increasing their operating costs as well as making the design more complicated is illustrated. Calculation for economizer and tubular air heater design and off-design performance and cold-end tube wall temperature estimation will help plant engineers foresee problems with low-temperature corrosion.

Calculations for design and off-design performance of fire tube boilers are detailed. Tube sheet temperature estimation with and without refractory and ferrules is included. Several examples illustrate the calculation of heat flux in fire tube and water tube boilers, tube wall temperatures, and film temperatures for heat transfer fluids. One of the important factors in sizing of boilers is the tube size. The impact of tube size on heat transfer coefficients in both fire tube and water tube boilers as well as on their weight and gas pressure drop is illustrated with examples. The advantages of using smaller diameter tubes in fire tube as well as in water tube boilers are illustrated with examples. Often plant engineers would add an economizer to their fire tube boiler to improve the efficiency. An example is provided showing how the revised duty and steam generation may be estimated when an economizer is added.

Information to be considered while developing specifications for a waste heat boiler is discussed at the end of the chapter, particularly the tube geometry data. Plant engineers should obtain a quotation with these data as it will help them analyze boiler performance when needed. Today, in several projects I see that this information is not readily available and plant engineers spend a lot of time trying to obtain this data. In one case, information on streams in a superheater was not clear and the boiler designer was not in business to help the plant engineer. It took a lot of time to inspect the boiler in operation and discuss with the fabricator and review header drawings before an estimate could be made. The number of streams in a superheater or economizer is an important piece of information, and process and plant engineers should have this if they want to evaluate the performance of the superheater. Appendix B discusses the significance of this term.

Chapter 5 describes the HRSG simulation process. This is an important tool for plant engineers planning cogeneration projects using gas turbine HRSGs. Using the concepts outlined, plant engineers can obtain so much information about their future or the operating HRSGs: How much steam can be generated given the exhaust gas flow conditions? How much fuel is required to generate additional steam? Will the economizer generate steam at low loads? How can one evaluate field data and compare it with the guarantee data? Is dual-pressure HRSG really required for given steam conditions or will a single-pressure HRSG do? How efficient is fresh air firing? If in an existing HRSG with a superheater the process steam is taken off the drum, what happens to the steam temperature? These and similar issues can be addressed by plant engineers without designing the HRSG or even without contacting the HRSG supplier. Hence, simulation is a great planning tool. Examples have been provided to illustrate these problems and results compared with physically designed HRSGs to show that the accuracy is reliable. They may use the results from simulation to check the HRSG performance or even its operating data.

Chapter 6 on miscellaneous calculations provides several procedures for calculations or sizing of components associated with boilers. Condensation heat recovery is explained in detail along with estimation of energy recoverable in water vapor condensation and sizing of plain or finned tube condensing economizers. Acoustic vibration analysis, flow in blow-down lines, simplified transient analysis, drum-level fluctuations, insulation and refractory performance calculations, pressure drop in air and flue gas ducts, wall temperature estimation in uninsulated stacks, and sizing of natural convection bath heaters and coils immersed in a boiler drum for cooling superheated steam are also illustrated. Simplified procedures for transient analysis will help plant engineers get an idea of the time to heat up boiler components or see how steam pressure or drum level fluctuates when there are upsets in flue gas flow or temperature or feed water flow.

The appendices provide calculation procedures for basic sizing and performance evaluation of heat transfer components and estimation of heat transfer coefficients inside and outside tubes. Procedures for evaluating flue gas properties are also given. Correlations for heat transfer and pressure drop outside and inside tubes are provided in SI, British, and metric units. The important concept of streams in boiler components is explained in Appendix B. Completely worked out examples on solid and serrated fins in Appendix E will enable plant engineers with programming skills to model economizer in steam generator or waste heat boiler components. Examples illustrate when finned tubes should be used and why they should not be used in some applications and how fin geometry should be selected. With several worked out real world examples for the evaluation of performance of steam generators or waste heat boilers, plant engineers have valuable resources to fall back on when such needs arise.

In conclusion, I can say that I have learnt a lot as a consultant during the last decade from numerous plant problems and performance issues and found that plant engineers depend solely on boiler suppliers when it comes to evaluating a boiler problem or finding a solution to it. It should not be so. Hence, this book has been written with the main purpose of educating plant engineers with a flair for process heat transfer on boiler design and performance calculations. It is my belief that if they study the appendices and go through the numerous practical and real world examples that have been worked out completely throughout the book, they will be able to evaluate the performance of boilers in their plants on their own, challenge boiler suppliers when poor or compromised designs are offered, suggest changes to proposed designs to lower operating costs, and analyze any boiler problem and arrive at the solution themselves or discuss intelligently with the boiler supplier if need be with their findings. Boiler designers will also find useful information throughout the book on design and performance aspects of various boiler components. Boiler suppliers will not be able to offer simple excuses for poor performance of their boilers as the plant engineers will now have backup calculations to prove their point and be well armed with information.

Worked Out Real World Problems of Interest to Plant Engineers

The book has several worked out examples in metric and SI units dealing with design and off-design performance aspects of steam generators, waste heat boilers (fire tube and water tube) and HRSGs, and their components such as furnace, evaporator, superheater, economizer, air heater, all based on real world examples. Any plant engineer with heat transfer or thermal sciences background at a collagen level should be able to follow these examples and apply the methodology to solve various boiler-related problems in their plant, check boiler designs offered during proposal stage, suggest improvements before the boiler is purchased, and challenge the boiler supplier if there are problems in operation. These examples provide plant engineers with tools and wherewithal to perform various types of boiler calculations, and they need not depend on boiler suppliers for help. They are now empowered to question boiler suppliers on design or performance issues and not accept any design or report from them as such. This is the main objective of the book.

A Few Typical Solved Problems

- A plant receives a quotation from a boiler vendor for a finned tube superheater. Information on process data and tube geometry details is given. Using heat transfer correlations presented in the Appendices, the U is estimated (overall heat transfer coefficient) and the adequacy of surface area is checked in addition to gas-steam side pressure drops and tube wall temperatures. To predict its off-design performance, two methods are discussed, one being the NTU method and the other the conventional method. (A similar analysis for evaporator, economizer, and air heater is given.) These examples will educate plant engineers and enable them to perform similar calculations and rectify any flaws in the component design before it is ordered.

- Using fuel data, furnace dimensions, and tube geometry details of an oil- and gas-fired steam generator, the complete performance of a steam generator is evaluated at full and part loads. One example starts from the economizer exit and works backwards all the way to the furnace, and another method starts from the furnace end and works through to the economizer exit. These examples will help plant engineers to check independently if the boiler supplier's surfacing of furnace, superheaters, boiler bank, economizer, fuel consumption and efficiency is reasonable and as guaranteed.

- An example shows how in a natural gas-fired steam generator the feed water temperature entering the economizer may be reduced to improve fuel utilization and boiler efficiency. The feed water entering the economizer is used to preheat the make-up water. One must ensure that the feed water is cooled to slightly above the water dew point of the flue gas.

- Performance of radiant and convective superheaters is evaluated to see which option results in flat steam temperature over load. The effects of excess air and flue gas recirculation on their performance and steam and tube wall temperatures are also evaluated. It may be noted that radiant superheaters are more prone to over-heating and tube failures and hence these examples will be informative and help plant engineers be careful when they evaluate a high-pressure, high-temperature steam generator design for possible purchase.

- An example shows the performance of plain and finned tube economizers in an oil-fired steam generator in both a counter flow and parallel flow arrangement and the resulting lowest tube wall temperature (to avoid acid dew point corrosion) and boiler efficiency. Plant engineers sometimes think of changing the economizer configuration to avoid acid corrosion problems and this example explains the implications.

- An example shows how one can evaluate the complete performance of a HRSG using field data and tube geometry details and check whether the original design is reasonable or not. Often HRSGs do not operate at design gas flow or steam flow conditions and hence this example will be helpful to plant engineers who can relate field data with design guarantees even if the steam generation or flue gas flows are significantly different in operation from those shown in the proposal.

- An example shows what happens to the superheated steam temperature and tube wall temperature when saturated steam in a waste heat boiler is taken off from the drum for process heating. The HRSG supplier may not have envisaged this mode of operation when supplying the boiler but the plant engineer may be required to check this option several years after the boiler has been in operation.

- Plant engineers can see how steam generator and HRSG characteristics vary with load; the effect of supplementary firing and savings in fuel input compared to steam generators may be evaluated. Thus, one may maximize steam generation with a minimum fuel input.

- Complete design and off-design performance calculations of fire tube boilers are explained. An example illustrates how boiler performance improves when an economizer is added. Plant engineers often think of adding an economizer to an existing boiler to improve efficiency and can carry out this exercise with minimal help from any boiler vendor or completely on their own.

- Calculations for tube sheet temperature in a fire tube boiler with and without tube sheet refractory are illustrated. Often tube sheet refractory falls off during operation, leading to overheating of tube sheet, and plant engineers can now investigate how high the tube sheet temperature can go without refractory.

- Design and off-design performance of tubular air heaters are illustrated with a few examples. Low-temperature corrosion is likely at low loads and plant engineers can check this.

- Significance of pinch and approach points is explained with several examples, including calculation of gas-steam temperature profiles and steam generation in single or multiple pressure HRSGs. Simulation of performance of complex HRSGs may be evaluated and optimized before approaching the HRSG supplier for a quotation.

- Examples compare results for a HRSG obtained from the simulation process with that obtained using actual tube geometry details to check accuracy of simulation techniques.

- Using a simulation method, examples illustrate if a single-pressure HRSG can be a better choice than a dual-pressure HRSG under certain circumstances. The effect of the ratio of HP (high pressure) to LP (low pressure) steam flow and steam pressure is illustrated. This helps one to arrive at the least-expensive HRSG configuration instead of simply accepting a design from a HRSG vendor.

- An example illustrates how a plant may optimize the configuration of a multi-module HRSG to maximize energy recovery using the simulation method. (HRSG suppliers may not have the time to study these options.)

- A simulation example illustrates how by splitting superheaters so that a portion is located upstream and another downstream of the burner a flat steam temperature may be obtained in both unfired and fired modes and over a wide load range without the need for a desuperheater!

- An example shows how the condensation duty may be estimated when flue gas is cooled beyond its water dew point. Examples also show how plain tube and finned tube condensing economizers may be sized.

- An example shows how a coil located inside the steam drum may be sized for a given duty. An example design of a steam desuperheater coil located inside the steam drum is given.

- Effects of sudden changes in process steam demand or feed water cut-off on drum level and steam pressure are illustrated with examples. Examples also show how one may estimate the time to heat up an evaporator or a superheater using flue gases.

Several more examples on basic and applied heat transfer calculations related to boiler, HRSG design, and performance are provided.

Articles on boilers published by the author can be downloaded from the website http://vganapathy.tripod.com/boilers.html. The author may be contacted at v_ganapathy@yahoo.com.

Author

V. Ganapathy is a consultant on steam generators and waste heat boilers based in Chennai, India. He has over 40 years of experience in the engineering of steam generators and waste heat boilers with emphasis on thermal design, performance, and heat transfer aspects. He also develops custom software on boiler design and performance. He holds a bachelor's degree in mechanical engineering from IIT Madras and an MSc (Engg.) from Madras University.

Ganapathy has published over 250 articles on steam generators and thermal design and has authored five books on boilers, the latest being *Industrial Boilers and HRSGs* (Taylor & Francis Group, Boca Raton, Florida). He also conducts intensive courses on boilers for chemical plants, refineries, and engineering consulting companies worldwide.

He has contributed several chapters to the *Handbook of Engineering Calculations* (McGraw Hill), *Encyclopedia of Chemical Processing and Design* (Marcel Dekker, New York), and recently a chapter on HRSG to the book *Power Plant Life Management and Performance Improvement* (Woodhouse Publishing, United Kingdom).

1

Combustion Calculations

Introduction

Boiler combustion and efficiency calculations are the starting point for boiler performance evaluation. These calculations enable the boiler designer or the plant engineer to estimate the boiler efficiency, air quantity required for combustion, and flue gas quantity generated in a boiler or heater; flue gas analysis that impacts convective and nonluminous heat transfer coefficients, adiabatic combustion temperature, flue gas specific heat, and enthalpy is also obtained from combustion calculations; these data in turn aid furnace performance evaluation, sizing, or performance evaluation of heat transfer equipment in the gas path, help evaluate air- and gas-side pressure drops, and also estimate the water and acid dew point temperatures. CO, NO_x, and CO_2 emission conversion calculations also require that results of combustion calculations are available. If NO_x emission is to be limited, then one has to estimate the amount of flue gas recirculation (FGR) to dampen the combustion temperature that impacts NO_x formation. Thus, basic and useful information for boiler performance evaluation can be generated from combustion calculations. The emphasis in this chapter is on oil and gaseous fuels. Sometimes, multiple fuels are fired in a boiler simultaneously, and this issue is also discussed.

Moisture in Air

Air is required for the combustion of fossil fuels. However, atmospheric air is never dry. It contains a certain amount of moisture due to local humidity and ambient temperature conditions. This adds to the volume of air to be handled by the forced draft fan and also increases the amount of water vapor in the flue gas. The amount of moisture in air may be obtained using psychometric chart (Figure 1.1) or estimated from the equation [1]:

$$M = 0.622P_w/(1.033 - P_w) \tag{1.1}$$

where P_w is the partial pressure of water vapor in air, kg/cm² a. For example, at 27°C, from steam tables, the saturated vapor pressure is 0.5069 psia = 0.0356 kg/cm² a, and if the relative humidity is say 65%, then

$$P_w = 0.65 \times 0.0356 = 0.0231 \text{ kg/cm}^2 \text{ a}$$

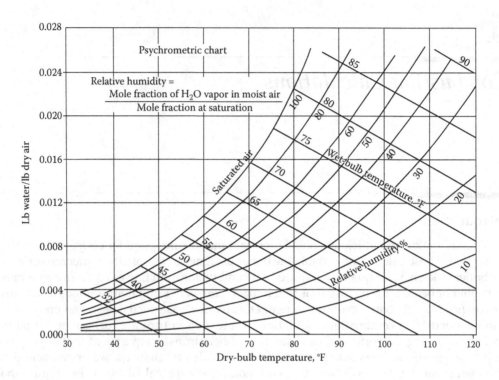

FIGURE 1.1
Moisture in air due to relative humidity.

Hence, M = 0.622 × 0.0231/(1.033 − 0.0231) = 0.0142 kg/kg air or 0.0142 lb/lb air. If 1000 kg of dry air is required for combustion in a boiler, the actual wet air that should be sent to the burner will be 1014.2 kg, and the boiler fan has to be sized for the volume of this amount of wet air.

Combustion Calculations

Knowing the fuel analysis, excess air, and ambient conditions, one may perform combustion calculations as shown in the following text.

Example 1.1

Natural gas having CH_4 = 83.4%, C_2H_6 = 15.8%, and N_2 = 0.8 by volume is fired in a boiler using 15% excess air. Ambient temperature is 20°C and relative humidity is 80%. Perform combustion calculations and determine the flue gas analysis.

Solution

From steam tables, the saturated vapor pressure at 20°C (68°F) is 0.34 psia = 0.0239 kg/cm² a. At 80% relative humidity, P_w = 0.8 × 0.0239 = 0.0191 kg/cm² a

M = .622 × 0.0191/(1.033 − 0.0191) = 0.012 (same information may be obtained from Figure 1.1). Combustion of methane may be expressed as $CH_4 + 2O_2 = CO_2 + 2H_2O$ or 1 mol of CH_4 requires 2 mol (volumes) of O_2 or 2 × 100/21 = 9.53 mol of air for combustion (air contains 21% volume of oxygen and rest nitrogen). Similarly,

$2C_2H_6 + 7O_2 = 4CO_2 + 6H_2O$ or 1 mol of ethane requires 3.5 mol of O_2 or 3.5 × 100/21 = 16.68 mol of dry air for combustion.

Tables 1.1 and 1.2, which give the air required for theoretical combustion of various fuel constituents, may also be used to arrive at these values.

Hence, 100 mol of fuel requires 83.4 × 9.53 + 15.8 × 16.68 = 1058.3 mol of theoretical dry air. Considering 15% excess air factor, actual dry air required = 1.15 × 1058.3 = 1217 mol. Excess air = 0.15 × 1058.3 = 158.7 mol; excess O_2 = 158.7 × 0.21 = 33.3 moles and N_2 formed = 0.79 × 1217 = 961 mol; air moisture = 1217 × 28.84 × 0.012/18 = 23.5 mol. (We multiplied moles by molecular weight [MW] to get the weight and then converted the moisture in air to volume basis by dividing by the MW of water vapor; here 28.84 is the MW of air and 18 that of water vapor.)

Tables 1.1 and 1.2 may also be used to get the amount of CO_2, H_2O, and N_2 formed. For example, 1 mol of methane forms one mole of CO_2. One mole of ethane forms two moles of CO_2. Similarly, 1 mol of CH_4 forms 2 mol of H_2O, and 1 mol of ethane forms 3 mol of H_2O. Hence, total amount of CO_2 and H_2O formed is

CO_2 = 83.4 × 1 + 2 × 15.8 = 115 mol;
H_2O = 2 × 83.4 + 3 × 15.8 + 23.5 = 237.7 mol (23.5 mol is the air moisture);
N_2 = 961 + 0.8 (fuel nitrogen) = 961.8 mol;
O_2 (excess) = 33.3 mol

Total moles of flue gas formed = 115 + 237.7 + 961.8 + 33.3 = 1347.8 mol

Flue Gas Analysis and Air–Flue Gas Quantities

% volume CO_2 in flue gases = (115/1347.8) × 100 = 8.5%
% volume H_2O = (237.7/1347.8) × 100 = 17.7%
N_2 = (961.8/1347.8) × 100 = 71.36%
% O_2 = (33.3/1347.8) × 100 = 2.47%.

This analysis is on wet basis. To convert to dry basis, one has to subtract the water content and recalculate the analysis. (Dry analysis is required as some instruments measure the oxygen content on dry basis from which excess air is computed.)

On dry basis, CO_2 = 8.5 × 100/(100 − 17.7) = 10.3%, O_2 = 2.47 × 100/(100 − 17.7) = 3%, and N_2 = 71.36 × 100/(100 − 17.7) = 86.7%. For efficiency calculations, one has to know the dry and wet air quantities and dry and wet flue gas formed per kg of fuel. See Table 1.3.

For nonluminous heat transfer calculations, one requires the partial pressures of CO_2 and H_2O. p_w = 237.7/1347.8 = 0.176 atm and p_c = 115/1347.8 = 0.085 atm.

MW of flue gas = 8.5 × 44 + 17.7 × 18 + 71.36 × 28 + 2.47 × 32 = 27.70.

CO_2 on mass basis in flue gas is required for emission calculations as it is considered a pollutant. % weight of CO_2 in flue gas = 8.5 × 44/27.7 = 13.5. One may also compute the emissions of CO_2/million J of heat input.

Amount of fuel fired per million J of energy input on higher heating value (HHV) basis = 106/(53.940 × 106) = 0.01854 kg fuel. (HHV of the fuel is 53,940 kJ/kg [or 53.94 × 106 J/kg] as shown later.) Hence, CO_2 formed = 20.4 × 0.01854 × 0.135 = 0.051 kg/MM J.

Per MM Btu (1054 MM J), the CO_2 emissions = 1054 × 0.051 = 53.75 kg (118.5 lb). (20.4 is the quantity of wet flue gas produced, kg/kg fuel, see Table 1.3.)

TABLE 1.1

Combustion Constants (Part 1)

No.	Substance	Formula	Mol. Wt[a]	Lb per cu ft[b]	Cu ft per lb[b]	Sp. gr. Air = 1000[b]	Heat of Combustion[c]			
							Btu/cu ft		Btu/lb	
							Gross	Net[d]	Gross	Net[d]
1	Carbon	C	12.01	—	—	—	—	—	14,093[g]	14,093
2	Hydrogen	H_2	2.016	0.005327	187.723	0.06959	325.0	275.0	61,100	51,623
3	Oxygen	O_2	32.000	0.08461	11.819	1.1053	—	—	—	—
4	Nitrogen (atm.)	N_2	28.016	0.07439[c]	13.443[c]	0.9718[e]	—	—	—	—
5	Carbon monoxide	C_2O	28.01	0.07404	13.506	0.9672	321.8	321.8	4,347	4,347
6	Carbon dioxide	CO_2	44.01	0.1170	8.548	1.5282	—	—	—	—
	Paraffin series C_nH_{2n+2}									
7	Methane	CH_4	16.041	0.04243	23.565	0.5543	1013.2	913.1	23,879	21,520
8	Ethane	C_2H_6	30.067	0.08029[c]	12.455[c]	1.04882[e]	1792	1641	22,320	20,432
9	Propane	C_3H_8	44.092	0.1196[c]	8.365[c]	1.5617[c]	2590	2385	21,661	19,944
10	n-Butane	C_4H_{10}	58.118	0.1582[c]	6.321[c]	2.06654[e]	3370	3113	21,308	19,680
11	Isobutane	C_4H_{10}	58.118	0.1582[e]	6.321[e]	2.06654[e]	3363	3105	21,257	19,629
12	n-Pentane	C_5H_{12}	72.144	0.1904[e]	5.252[e]	2.4872[c]	4016	3709	21,091	19,517
13	Isopentane	C_5H_{12}	72.144	0.1904[e]	5.252[e]	2.4872[e]	4008	3716	21,052	19,478
14	Neopentane	C_5H_{12}	72.144	0.1904[e]	5.252[e]	2.4872[e]	3993	3693	20,970	19,396
15	n-Hexane	C_6H_{14}	86.169	0.2274[e]	4.398[c]	2.9704[e]	4762	4412	20,940	19,403
	Olefin series C_nH_{2n}									
16	Ethylene	C_2H_4	28.051	0.07456	13.412	0.9740	1613.8	1513.2	21,644	20,295
17	Propylene	C_3H_6	42.077	0.1110[e]	9.007[e]	1.4504[e]	2336	2186	21,041	19,691
18	n-Butene (butylene)	C_4H_8	56.102	0.1480[e]	6.756[e]	1.9336[e]	3084	2885	20,840	19,496
19	Isobutene	C_4H_8	56.102	0.1480[e]	6.756[e]	1.9336[e]	3068	2869	20,730	19,382
20	n-Pentene	C_5H_{10}	70.128	0.1852[e]	5.400[e]	2.4190[e]	3836	3586	20,712	19,363

Aromatic series C_nH_{2n-6}										
21	Benzene	C_6H_6	76.107	0.2060ᶜ	4.852ᶜ	2.6920ᵉ	3751	3601	18,210	17,480
22	Toluene	C_7H_8	92.132	0.2431ᶜ	4.113ᵉ	3.1760ᵉ	4484	4284	18,440	17,620
23	Xylene	C_8H_{10}	106.158	0.2803ᵉ	3.567ᵉ	3.6618ᵉ	5230	4980	18,650	17,760
Miscellaneous gases										
24	Acetylene	C_2H_2	26.036	0.06971	14.344	0.9107	1499	1448	21,500	20,776
25	Naphthalene	$C_{10}H_8$	128.162	0.3384ᵉ	2.955ᵉ	4.4208ᵉ	5854ᶠ	5654ᶠ	17,298ᶠ	16,708ᶠ
26	Methyl alcohol	CH_3OH	32.041	0.0846ᵉ	11.820ᵉ	1.1052ᵉ	867.9	768.0	10,259	9,078
27	Ethyl alcohol	C_2H_5OH	46.067	0.1216ᵉ	8.221ᵉ	1.5890ᵉ	1600.3	1450.5	13,161	11,929
28	Ammonia	NH_3	17.031	0.0456ᵉ	21.914ᵉ	0.5961ᵉ	441.1	365.1	9,668	8,001
29	Sulfur	S	32.06	—	—	—	—	—	3,983	3,983
30	Hydrogen sulfide	H_2S	34.076	0.09109ᵉ	10.979ᵉ	1.1898ᵉ	647	596	7,100	6,545
31	Sulfur dioxide	SO_2	64.06	0.1733	5.770	2.264	—	—	—	—
32	Water vapor	H_2O	18.016	0.04758ᵉ	21.017ᵉ	0.6215ᵉ	—	—	—	—
33	Air	—	26.9	0.07655	13.063	1.0000	—	—	—	—

Source: Ganapathy, V., *Industrial Boilers and HRSGs*, CRC Press, Boca Raton, FL, 2003, p.238.

All gas volumes corrected to 60°F and 30 in. Hg dry. For gases saturated with water at 60°F, 1.73% of the Btu value must be deducted.

[a] Densities calculated from values given in the *Journal of the American Chemical Society*, February 1937.

[b] Calculated from atomic weights given in gL at 0°C and 760 mmHg in the International Critical Tables allowing for the known deviations from the gas laws. Where the coefficient of expansion was not available, the assumed value was taken as 0.0037 per 0°C. Compare this with 0.003662, which is the coefficient for a perfect gas. Where no densities were available, the volume of the mole was taken as 22.4115 L.

[c] Converted to mean Btu per lb (1/180 of the heat per lb of water from 32°F to 2120°F) from data by Frederick D. Rossini, National Bureau of Standards, letter of April 10, 1937, except as noted.

[d] Deduction from gross to net heating value determined by deducting 18,919 Btu/lb mol water in the products of combustion. Osborne, Stimson, and Ginnings, *Mechanical Engineering*, p. 163, March 1935, and Osborne, Stimson, and Flock, National Bureau of Standards Research Paper 209.

[e] Denotes that either the density or the coefficient of expansion has been assumed. Some of the materials cannot exist as gases at 60°F and 30 in. Hg pressure, in which case the values are theoretical ones given for ease of calculation of gas problems. Under the actual concentrations in which these materials are present, their partial pressure is low enough to keep them as gases.

[f] From third edition of Combustion.

TABLE 1.2

Combustion Constants (Part 2)

No.	Substance	Cu ft per cu ft of Combustible						Lb per lb of Combustible						Experimental Error in Heat of (±%)
		Required for Combustion			Flue Products			Required for Combustion			Flue Products			
		O_2	N_2	Air	CO_2	H_2O	N_2	O_2	N_2	Air	CO_2	H_2O	N_2	
1	Carbon	—	—	—	—	—	—	2.664	8.863	11.527	3.664	—	8.863	0.012
2	Hydrogen	0.5	1.882	2.382	—	1.0	1.882	7.937	26.407	34.344	—	8.937	26.407	0.015
3	Oxygen	—	—	—	—	—	—	—	—	—	—	—	—	—
4	Nitrogen (atm.)	—	—	—	—	—	—	—	—	—	—	—	—	—
5	Carbon monoxide	0.5	1.882	0.2382	1.0	—	1.882	0.571	1.900	2.471	1.571	—	1.900	0.045
6	Carbon dioxide	—	—	—	—	—	—	—	—	—	—	—	—	—
Paraffin series C_nH_{2n+2}														
7	Methane	2.0	7.528	9.528	1.0	2.0	7.528	3.990	13.275	17.265	2.744	2.246	13.275	0.033
8	Ethane	3.5	13.175	16.675	2.0	3.0	13.175	3.725	12.394	16.119	2.927	1.798	12.394	0.030
9	Propane	5.0	18.821	23.821	3.0	4.0	18.821	3.629	12.074	15.703	2.994	1.634	12.074	0.023
10	n-Butane	6.5	24.467	30.967	4.0	5.0	24.467	3.579	11.908	15.487	3.029	1.550	11.908	0.022
11	Isobutane	6.5	24.467	30.967	4.0	5.0	24.467	3.579	11.908	15.487	3.029	1.550	11.908	0.019
12	n-Pentane	8.0	30.114	38.114	5.0	6.0	30.114	3.548	11.805	15.353	3.050	1.498	11.805	0.025
13	Isopentane	8.0	30.114	38.114	5.0	6.0	30.114	3.548	11.805	15.353	3.050	1.498	11.805	0.071
14	Neopentane	8.0	30.114	38.114	5.0	6.0	30.114	3.548	11.805	15.353	3.050	1.498	11.805	0.11
15	n-Hexane	9.5	35.760	45.260	6.0	7.0	35.760	3.528	11.738	15.266	3.064	1.464	11.738	0.05
Olefin series C_nH_{2n}														
16	Ethylene	3.0	11.293	14.293	2.0	2.0	11.293	3.422	11.385	14.807	3.138	1.285	11.385	0.021
17	Propylene	4.5	16.939	21.439	3.0	3.0	16.939	3.422	11.385	14.807	3.138	1.285	11.385	0.031
18	n-Butene (butylene)	6.0	22.585	28.585	4.0	4.0	22.585	3.422	11.385	14.807	3.138	1.285	11.385	0.031
19	Isobutene	6.0	22.585	28.585	4.0	4.0	22.585	3.422	11.385	14.807	3.138	1.285	11.385	0.031
20	n-Pentene	7.5	28.232	35.732	5.0	5.0	28.232	3.422	11.385	14.807	3.138	1.285	11.385	0.037

Aromatic series C_nH_{2n-6}

21 Benzene	7.5	28.232	35.732	6.0	3.0	28.232	3.073	13.297	10.224	3.381	0.692	10.224	0.12
22 Toluene	9.0	33.878	32.878	7.0	4.0	33.878	3.126	13.527	10.401	3.344	0.782	10.401	0.21
23 Xylene	10.5	39.524	50.024	8.0	5.0	39.524	3.165	13.695	10.530	3.317	0.849	10.530	0.36
Miscellaneous gases													
24 Acetylene	2.5	9.411	11.911	2.0	1.0	9.411	3.073	13.297	10.224	3.381	0.692	10.224	0.16
25 Naphthalene	12.0	45.170	57.170	10.0	4.0	45.170	2.996	12.964	9.968	3.434	0.562	9.968	—
26 Methyl alcohol	1.5	5.646	7.146	1.0	2.0	5.646	1.498	6.482	4.984	1.374	1.125	4.984	0.027
27 Ethyl alcohol	3.0	11.293	14.293	2.0	3.0	11.293	2.084	9.018	6.934	1.922	1.170	6.934	0.030
28 Ammonia	0.75	2.823	3.573	—	1.5	3.323	1.409	6.097	4.688	—	1.587	5.511	0.088
29 Sulfur	—	—	—	SO_2	—	—	0.998	4.285	3.287	1.9928 SO_2	—	3.287	0.071
30 Hydrogen sulfide	1.5	5.646	7.146	1.0	1.0	5.646	1.409	6.097	4.688	1.880	0.529	4.688	0.30
31 Sulfur dioxide	—	—	—	—	—	—	—	—	—	—	—	—	—
32 Water vapor	—	—	—	—	—	—	—	—	—	—	—	—	—
33 Air	—	—	—	—	—	—	—	—	—	—	—	—	—

Source: Ganapathy, V., *Industrial Boilers and HRSGs*, CRC Press, Boca Raton, FL, 2003, p238.

All gas volumes corrected to 60°F and 30 in. Hg dry. For gases saturated with water at 60°F, 1.73% of the Btu value must be deducted.

[a] Calculated from atomic weights given in the *Journal of the American Chemical Society*, February 1937.

[b] Densities calculated from values given in gL at 0°C and 760 mmHg in the International Critical Tables allowing for the known deviations from the gas laws. Where the coefficient of expansion was not available, the assumed value was taken as 0.0037 per 0°C. Compare this with 0.003662, which is the coefficient for a perfect gas. Where no densities were available, the volume of the mole was taken as 22.4115 L.

[c] Converted to mean Btu per lb (1/180 of the heat per lb of water from 32°F to 2120°F) from data by Frederick D. Rossini, National Bureau of Standards, letter of April 10, 1937, except as noted.

[d] Deduction from gross to net heating value determined by deducting 18,919 Btu/lb mol water in the products of combustion. Osborne, Stimson, and Ginnings, *Mechanical Engineering*, p. 163, March 1935, and Osborne, Stimson, and Flock, National Bureau of Standards Research Paper 209.

[e] Denotes that either the density or the coefficient of expansion has been assumed. Some of the materials cannot exist as gases at 60°F and 30 in. Hg pressure, in which case the values are theoretical ones given for ease of calculation of gas problems. Under the actual concentrations in which these materials are present, their partial pressure is low enough to keep them as gases.

[f] From third edition of Combustion.

TABLE 1.3

Dry and Wet Flue Gas Analysis

Basis	Wet	Dry
% volume CO_2	8.5	10.3
H_2O	17.64	0
N_2	71.36	86.7
O_2	2.47	3.0

Molecular weight of the fuel = $(83.4 \times 16 + 15.8 \times 30 + 0.8 \times 28)/100 = 18.3$.
w_{da} = kg dry air/kg fuel = $1217 \times 28.84/1830 = 19.18$.
w_{wa} = kg wet air/kg fuel = $19.18 + 23.5 \times 18/1830 = 19.41$.
w_{dg} = kg dry flue gas/kg fuel = $(115 \times 44+33.3 \times 32 + 961 \times 28)/1830 = 18$.
w_{wg} = kg wet flue gas/kg fuel = $(115 \times 44+33.3 \times 32 + 961.8 \times 28 + 237.7$
$\times 18)/1830 = 20.4$.

Water Formed per kg of Fuel

This is an important piece of information as it gives an idea of how much water can be condensed in a condensing economizer if used. In the earlier example, the MW of flue gas = 27.7. The % weight of H_2O in the flue gas = $17.64 \times 18/27.7 = 11.46$. It was shown earlier that w_{wg} = amount of wet flue gas formed per kg of fuel = 20.4. Hence, the amount of water vapor formed per kg fuel fired = $0.1146 \times 20.4 = 2.34$ kg. Typically, for natural gas, it varies from 2.15 to 2.4. Similar calculations may be carried out for fuel oil. For no. 2 fuel oil at 15% excess air, it can be shown that % volume CO_2 = 11.57, H_2O = 12.29, N_2 = 73.63, O_2 = 2.51. MW of flue gas = $11.57 \times 44 + 12.29 \times 18 + 73.63 \times 28 + 2.51 \times 32 = 28.72$.% weight of H_2O in flue gas = $12.29 \times 18/28.72 = 7.7\%$. The amount of flue gas produced per kg fuel is about 18. Hence, $0.077 \times 18 = 1.39$ kg of water vapor is produced per kg of fuel fired at 15% excess air. On an average, it varies from 1.3 to 1.5 kg/kg fuel, much smaller than that for natural gas. Hence, more energy can be recovered in latent heat form with flue gas from combustion of natural gas than with fuel oils.

Excess Air from Flue Gas Analysis

In operating plants, data on flue gas analysis will be available using which the plant engineer may arrive at the excess air used. This will help the plant engineer evaluate the boiler efficiency, and air and flue gas quantities. A formula that is widely used to obtain excess air E in % is [2]

$$E = 100 \times (O_2 - CO/2)/[0.264N_2 - (O_2 - CO/2)] \tag{1.2}$$

where O_2, CO, and N_2 are % volume of oxygen, carbon monoxide, and nitrogen on dry flue gas basis. Another formula is used when an Orsat-type analyzer is used for analyzing the flue gases; SO_2 will be absorbed with CO_2. The oxygen is on dry volumetric basis.

$$E = K_1O_2/(21 - O_2) \tag{1.3}$$

$$\text{Constant } K_1 = (100C + 237H_2 + 37.5S + 9N_2 - 29.6O_2)/(C + 3H_2 + 0.375S - 0.375O_2) \tag{1.4}$$

where C, H_2, N_2, S, and O_2 are fraction by weight of carbon, hydrogen, nitrogen, sulfur, and oxygen in fuel.

Let us check the value of constant K_1 for Example 1.1.
% weight of CH_4 in fuel = 83.4 × 16/(83.4 × 16 + 15.8 × 30 + 0.8 × 28) = 0.729.
% weight of C_2H_6 in fuel = 15.8 × 30/(83.4 × 16 + 15.8 × 30 + 0.8 × 28) = 0.259
% N_2 by weight in fuel = 0.012
Carbon C in CH_4 = 0.75 × 0.729 = 0.5467 and fraction hydrogen = 0.1823
Carbon in C_2H_6 = (24/30) × 0.259 = 0.2072 and fraction hydrogen = 0.0518

Total C fraction by weight = 0.5467 + 0.2072 = 0.7539, and total hydrogen by weight = 0.0518 + 0.1823 = 0.2341. Hence,

$$K_1 = [100 × 0.7539 + 237 × 0.2341 + 9 × 0.012]/[0.7539 + 3 × 0.2341] = 89.9$$

Let us compute the heating value of the fuel from the weight fractions. Lower heating value (LHV) = 0.729 × 21,529 + 0.259 × 20,432 = 20,980 Btu/lb = 11,655 kcal/kg = 48,800 kJ/kg (where 21,529 and 20,432 in Btu/lb are the LHV of methane and ethane from Tables 1.1 and 1.2).

$$HHV = 0.729 × 23,879 + 0.259 × 22,320 = 23,188 \text{ Btu/lb} = 12,882 \text{ kcal/kg}$$
$$= 53,940 \text{ kJ/kg} = 53.93 × 10^6 \text{ J/kg}.$$

K_1 may also be obtained from Table 1.4 for various fuels. If oxygen is measured on true volume basis (wet basis), then one uses constant K_2 for excess air evaluation as shown:

$$K_2 = (100C + 363H_2 + 37.5S + 9N_2 - 29.6O_2)/(C + 3H_2 + 0.375S - 0.375O_2) \qquad (1.5)$$

Many modern analytical techniques, such as those employing infrared or paramagnetic principles, also measure on a dry gas basis because they require moisture-free samples to avoid damage to the detection cells. These analyzers are set up with a sample conditioning system that removes moisture from the gas sample. However, some analyzers, such as in situ oxygen detectors employing a zirconium oxide cell, measure on the wet gas basis. Results from such equipment need to be corrected to a dry gas basis before they are used in the ASME equations.

These values of K_1 and K_2 have been arrived at after performing calculations on several fuels with different fuel analysis and hence give a good working average value.

TABLE 1.4

Constants K_1 and K_2 for Excess Air Evaluation

Fuel	K_1	K_2
Carbon	100	100
Hydrogen	80	121
Carbon monoxide	121	121
Sulfur	100	100
Methane	90	110.5
Oil	94.5	105.9
Coal	97	103.3
Blast furnace gas	223	225
Coke over gas	89.3	114.2

TABLE 1.5

Combustion Constants A (Air Required for 1 MM kJ fuel), kg

Fuel	A, kg/GJ	Max CO_2 (Dry Flue Gas)	A, lb/MM Btu
Blast furnace gas	247	24.6–25.3	575
Bagasse	279	20	650
Coke oven gas	288	9.23–10.6	670
Refinery and oil gas	309	13.3	720
Natural gas	312–314	11.6–12.7	727–733
Fuel oils, furnace oil, and lignite	318–323	14.25–16.35	740–750
Bituminous coals	327	17.8–18.4	760
Coke	344	20.5	800

Using Table 1.4, for methane (natural gas), $K_1 = 90$. O_2 on dry flue gas basis is 3%. Hence, excess air = $90 \times 3/(21 - 3) = 15\%$. If wet basis is used, then $K_2 = 110.5$. Then, excess air = $110.5 \times 2.47/(21 - 2.47) = 14.7$—close to 15%. These constants are good estimates for a type of fuel, and so some minor variations may be expected depending on actual fuel analysis. There is another approximate method to get the excess air from CO_2 values, but the accuracy is not good; see Table 1.5.

E = max CO_2 on dry flue gas basis/%CO_2 measured. In our example, from Table 1.5, max $CO_2 = 12\%$, while actual is 10.3%. Hence, E = 12/10.3 = 1.165 or 16.5% excess air. This is only an estimate. The O_2 basis is more accurate.

Simplified Combustion Calculations

One may develop a suitable computer code for performing detailed combustion calculations using the procedure given earlier. However, when one is interested only in an estimate of air or flue gas quantities generated in a boiler or heater, the simplified combustion calculation procedure described in the following text may be used. Each fuel such as natural gas, coal, or oil requires a certain amount of dry stoichiometric air for combustion per million joules of energy fired on HHV basis (or per MM Btu fired). This quantity does not vary much with fuel analysis for a type of fuel and therefore becomes a valuable method of estimating air for combustion when fuel analysis is not readily available.

For solid fuels and oil, the dry air w_{da} in kg/kg fuel can be obtained from fundamentals:

$$w_{da} = 11.53C + 34.34 \times (H_2 - O_2/8) + 4.29S \tag{1.6}$$

where C, H_2, and S are carbon, hydrogen, and sulfur in the fuel by fraction weight.

$$\text{For gaseous fuels, } w_{da} = 2.47 \times CO + 34.34 \times H_2 + 17.27 \times CH_4 + 13.3 \times C_2H_2$$
$$+ 14.81 \times C_2H_4 + 16.12 \times C_2H_6 - 4.32O_2 \tag{1.7}$$

(CO, H_2, and CH_4 are in fraction by weight. Only some constituents are shown here; if there are other combustibles, one can include those by using values from Tables 1.1 and 1.2. For example, CO requires 2.47 kg air/kg fuel and H_2 requires 34.34 kg air/kg fuel.

Example 1.2

Estimate the amount of air required per million kJ of fuel oil fired. $C = 0.875$, $H_2 = 0.125$, and $°API = 28$.

Solution

The HHV of fuel oils may be written as

$$HHV = 17{,}887 + 57.5°API\text{-}102.2\%S \ (Btu/lb) \tag{1.8a}$$

or

$$HHV = 41{,}605 + 133.74°API\text{-}237.7\%S \ (kJ/kg) \tag{1.8b}$$

Using this, $HHV = 45{,}350 \ kJ/kg$.

Using (1.6), the amount of dry stoichiometric air required $= 11.53 \times 0.875 + 34.34 \times 0.125 = 14.38 \ kg/kg$ fuel.

Here, 1 million kJ of energy on HHV basis has $10^6/45{,}350 = 22.05$ kg fuel; hence, 22.05 kg fuel requires $14.38 \times 22.05 = 317$ kg dry air.

(1 MM Btu of energy has $10^6/19{,}500 = 51.3$ lb fuel. The HHV, in Btu/lb, is 19,500. Hence, 1 MM Btu fuel requires $51.3 \times 14.38 = 738$ lb of dry air.)

Example 1.3

Check the combustion air required for million kJ of natural gas having methane = 83.4%, ethane = 15.8%, and nitrogen = 0.8% by volume.

Solution

MW of the fuel $= 0.834 \times 16 + 0.158 \times 30 + 0.008 \times 28 = 18.3$.

% weight of methane $= 83.4 \times 16/18.3 = 72.9$; % weight of ethane $= 15.8 \times 30/18.3 = 25.9$

$w_{da} = 17.27 \times 0.729 + 16.12 \times 0.259 = 16.76 \ kg/kg$ fuel (17.27 and 16.12 were taken from Tables 1.1 and 1.2).

$HHV = (0.729 \times 23{,}876 + 0.256 \times 22{,}320) \times 2.326 = 53{,}776 \ kJ/kg$ (convert from Btu/lb to kJ/kg using a factor of 2.326). Here, 1 million kJ has $10^6/53{,}776 = 18.60$ kg fuel; hence, air required for firing 1 million kJ $= 18.6 \times 16.76 = 312$ kg.

Let us take 100% methane.

$HHV = 23{,}879 \times 2.326 = 55{,}543 \ kJ/kg$; methane requires 17.265 kg air/kg (from Tables 1.1 and 1.2).

Hence, 1 million kJ of energy has $10^6/55{,}543 = 18$ kg, and this requires $18 \times 17.265 = 310.8$ kg dry stoichiometric air. Thus, for a variety of fuels, we can perform this calculation and show that within a small % variation, the dry stoichiometric air required for combustion of a type of fuel is a constant (see Table 1.5).

Example 1.4

A fired heater is firing natural gas at 15% excess air in a furnace; energy input is 50 million kJ/h on an HHV basis. Assume $HHV = 52{,}500 \ kJ/kg$. Determine the air and flue gas produced.

Solution

We don't have the fuel analysis and hence use the simplified procedure. Air required $= 50 \times 313 \times 1.15 \times 1.013 = 18{,}231 \ kg/h$ (1.013 refers to the moisture in air at the ambient temperature of 27°C and relative humidity of 60%. Factor A = 313). Fuel required = 50 × $10^6/52{,}500 = 952 \ kg/h$; hence flue gas produced $= 18{,}231 + 952 = 19{,}183 \ kg/h$.

Working the same problem in British units, (50 × $10^6 \ kJ/h = 50 \times 10^9/1055 \ Btu/h = 47.39$ MM Btu/h. 1 MM Btu requires 728 lb dry theoretical amount of air. Hence, actual air required $= 47.39 \times 728 \times 1.15 \times 1.013 = 40{,}190 \ lb/h$).

Gaseous fuels are often fired in steam generators, and Table 1.6 gives the fuel gas analysis for a few typical gaseous fuels, and Table 1.7 shows the ultimate analysis for typical fuel oils.

TABLE 1.6

Analysis of Some Gaseous Fuels

Gas	Coke Oven	Blast Furnace	Natural Gas	Natural Gas	Refinery Gas
% Volume hydrogen	47.9	2.4			45
Methane	33.9	0.1	83.4	97	12.5
Ethylene	5.2	—			
Ethane			15.8	1	7
Propane					23
Butane					11
Carbon monoxide	6.1	23.3			
Carbon dioxide	2.6	14.4		0.5	
Nitrogen	3.7	56.4		1.5	
Oxygen	0.6		0.8		
Water vapor	—	3.4			
Hydrogen sulfide					1.5
Specific gravity (relative to air)	0.413	1.015	0.636	0.569	0.76
HHV, MJ/m^3 (Btu/ft^3)	21.98 (590)	3.12 (83.8)	42.06 (1129)	37.2 (1000)	49.85 (1340)

TABLE 1.7

Analysis of Typical Fuel Oils

Grade or No.	1	2	4	5	6
Carbon, % weight	85.9–86.7	86.1–88.2	86.5–89.2	86.5–89.2	86.5–89.2
Hydrogen	13.3–14.1	11.8–13.9	10.6–13.0	10.5–12	9.5–12
Nitrogen	0–0.1	0–0.1			
Sulfur	0.01–0.5	0.05–1.0	0.2–2.0	0.5–3.0	0.7–3.5
°API	40–44	28–40	15–30	14–22	7–22
Ash	—	—	0–0.1	0–0.1	0.01–0.5
HHV, kJ/kg	45,764	45,264	43,821	43,170	42,333
HHV, Btu/lb (avg)	19,675	19,460	18,840	18,560	18,200
HHV, kcal/kg	10,930	10,811	10,467	10,311	10,111

Estimation of Heating Values

When the ultimate analysis of fuels is known, the heating values can be estimated using

$$HHV = 14{,}500C + 62{,}000(H_2 - O_2/8) + 4000S \tag{1.9a}$$

$$LHV = HHV - 9720H_2 - 1110W \tag{1.9b}$$

Some texts use 9200 instead of 9720. Difference in LHV is small.

Where W is the weight fraction of moisture in fuel, and C, H_2, O_2, and S are the fractions by weight of carbon, hydrogen, oxygen, and sulfur. LHV and HHV are in Btu/lb. Multiply by 2.326 to obtain values in kJ/kg.

If fuel oil has 87.5% by weight carbon and 12.5% by weight hydrogen, HHV = 14,500 × 0.875 + 62,000 × 0.125 = 20,437 Btu/lb = 47,538 kJ/kg = 11,353 kcal/kg.

LHV = 20,427 – 9,200 × 0.125 = 19,277 Btu lb = 44,838 kJ/kg = 10,709 kcal/kg.

If ultimate analysis is not known, one may estimate the heating values of fuel oils using [3]

$$HHV = 17{,}887 + 57.5 \times {}^{\circ}API - 102.2S \tag{1.10a}$$

$$LHV = HHV - 92 \times H_2 \text{ (HHV and LHV in Btu/lb)} \tag{1.10b}$$

$$\% \ H_2 = F - 2122.5/({}^{\circ}API + 131.5) \tag{1.11}$$

$$\text{where } F = 24.5 \quad \text{for } 0 <= {}^{\circ}API <= 9 \tag{1.12}$$

$$F = 25 \quad \text{for } 9 <= {}^{\circ}API <= 20$$

$$F = 25.2 \quad \text{for } 20 <= {}^{\circ}API <= 30$$

$$F = 25.45 <= {}^{\circ}API <= 40$$

Burner Selection

The burner is an important component of any boiler. Burner designs have undergone several iterations during the last decade. Burner suppliers are offering burners with single-digit NO_x emissions with low CO levels competing with the selective catalytic reduction (SCR) system. However, these burners require a large amount of FGR on the order of 35%, and flame stability is a concern at low loads. Fuel or air staging or steam injection are the other methods used to limit NO_x. Today, single burners are used for capacities up to 300–350 MM Btu/h (316.5–369 GJ/h). Figure 1.2 shows a Natcom burner firing natural gas. It is a maintenance-free (refractory-less) burner. Burners are also available for 9 ppm NO_x on natural gas. More on emissions and NO_x control is discussed in Chapter 3. Table 1.8 shows typical burner emissions.

FIGURE 1.2
Low NOx burner. (Courtesy of Natcom Burner, Division of Cleaver Brooks, Thomasville, GA.)

TABLE 1.8

Typical Burner Emissions

"Standard" Burner Emissions Values (in ppm, Ref. at 3% O_2, Dry)	Natural Gas	No. 2 Oil	No. 6 Oil
NO_x	83	96[1]	340[2]
CO	50	100	100
SO_2*	Nil	28	30
VOC	9.58	11.98	14.37
PM_{10}	7	50	100
Low-NO$_x$ Burner Emissions Values (in ppm, ref. at 3% O$_2$, Dry)			
NO_x	30	78	300
CO	50	100	100
SO_2*	Nil	28	30
VOC	9.58	11.98	14.37
PM_{10}	7	50	100
Ultra Low-NO$_x$ Burner Emissions Values (in ppm, ref. at 3% O$_2$, Dry)			
NO_x	9	78	NA
CO	50	100	NA
SO_2*	Nil	28	NA
VOC	9.58	11.98	NA
PM_{10}	7	50	NA

Source: Courtesy of Cleaver Brooks, Thomasville, GA.

Note: * SO_2 depends on sulfur content.

Combustion Temperatures

Combustion temperature is an important parameter in combustion calculations. Burner suppliers can estimate the NO_x formation from the combustion temperature of the fuel, which depends on the fuel analysis and excess air. If FGR is introduced to lower NO_x, then the combustion temperature is reduced. NO_x reduces significantly as the combustion temperature is reduced, and hence, FGR is often a good and less expensive way to control NO_x. Figure 1.3 shows the effect of combustion temperature on NO_x for natural gas.

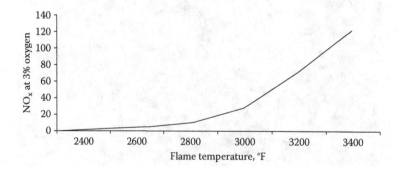

FIGURE 1.3

Typical NO_x formation versus flame temperature for natural gas.

Example 1.5

Let us determine the combustion temperature for the fuel gas in Example 1.1. The LHV is 11,655 kcal/kg (20,980 Btu/lb or 48,800 kJ/kg). Fraction methane by weight = 0.729 and that of ethane = 0.259.

Solution

Amount of flue gas produced = 20.4 kg/kg fuel. Assume air is not heated and is at ambient conditions. Hence, the enthalpy of the products of combustion = 48,800/20.4 = 2392.1 kJ/kg = 572.2 kcal/kg (dividing LHV of fuel by the amount of flue gas produced). From Table F.10 showing enthalpy of products of combustion, T_{ad} = 1807°C (3285°F). This is the adiabatic combustion temperature, and the actual flame temperature will be slightly less due to dissociation losses of about 1%. Hence, 1788°C (3250°F) is more likely.

Let us say 15% FGR at 150°C from economizer exit is introduced into the boiler. From Example 1.1, we see that the ratio of flue gas to fuel is about 20.4 at 15% excess for natural gas. The LHV of the fuel is 11,655 kcal/kg as shown earlier.

FGR flow = 0.15 × 20.4 = 3.06 kg flue gas. At 150°C, the enthalpy is 36 kcal/kg (reference 15°C). Air at 27°C has enthalpy = 2.7 kcal/kg.

Hence, enthalpy of air, fuel, and flue gas mixture = (19.4 × 2.7 + 11,655 + 3.06 × 36)/20.4/1.15 = 503.7 kcal/kg or combustion temperature = 1600°C (2912°F); enthalpy obtained from Table F.10 by interpolation using the flue gas analysis shown in Example 1.1.

Simplified Procedure for Estimating Combustion Temperatures

An estimate of adiabatic combustion temperature may be made using the following formula. The air and flue gas produced are estimated using the simplified combustion calculation procedure.

$$T_{ad} = [LHV + A\alpha \times HHV \times C_{pa} \times (t_a - t_{amb})/10^6]/[(1 + A\alpha \times HHV/10^6) \times C_{pg}] \quad (1.13)$$

Basically, we estimate the air and flue gas produced using the A values discussed earlier and then use the specific heats of air and flue gas to obtain T_{ad}. α is the excess air factor. If 20% excess air is used, then α = 1.2. T_{ad}, t_a, and t_{amb} are the adiabatic combustion temperature, combustion air temperature, and ambient temperature respectively.

Example 1.6

Fuel oil is fired at 15% excess air in a boiler; estimate the combustion temperature.

Solution

LHV = 18,430 Btu/lb = 42,872 kJ/kg, HHV = 19,636 Btu/lb = 45,673 kJ/kg. There is no air heater. Hence $t_a = t_{amb}$. Hence, 1 million kJ on HHV basis has 10^6/45,673 = 21.89 kg oil.

Theoretical air required for 10^6 kJ = 320 kg from Table 1.5; hence, actual air = 1.15 × 320 = 368 kg; flue gas produced = (368 +2 1.89) = 390 kg or ratio of flue gas to fuel = (390/21.89) = 17.8.

T_{ad} = 42,872/(17.8 × 1.328) = 1813°C (3295°F). (1.328 is the flue gas specific heat in kJ/kg °K.)

Effect of FGR on Combustion Temperature

Introducing flue gas at the point of combustion reduces the flame temperature significantly, which in turn lowers NO_x levels.

TABLE 1.9

Combustion Temperatures with and without FGR

Excess Air, %	15	15	15	15	15	15
FGR, %	0	15	25	0	15	25
Fuel	N. gas	N. gas	N. gas	No. 2 oil	No. 2 oil	No. 2 oil
Tad, °C	1775	1589	1485	1866	1662	1559
Tad, °F	3227	2892	2705	3390	3023	2838
W_g/W_f	20.97	24.12	26.22	17.83	20.5	22.29

FGR may be introduced at the FD fan suction as shown in Figure 3.8, or it may be introduced at the burner using a separate FGR fan. The suction system or induced system does not require an additional fan, but the FD fan must be sized to handle the higher flow at higher temperature of the flue gas and air mixture. Table 1.9 shows the effect of FGR on combustion temperatures.

(*Natural gas*: 97% methane, 2% ethane, and 1% propane. Flue gas analysis% volume $CO_2 = 8.29$, $H_2O = 18.17$, $N_2 = 71.07$, $O_2 = 2.46$. LHV = 49,864 (21,437 Btu/lb), HHV = 55,272 kJ/kg (23,762 Btu/lb). *Fuel oil*: carbon = 87.5% by weight, hydrogen = 12.5%,°API = 32. Flue gas analysis: % volume $CO_2 = 11.57$, $H_2O = 12.29$, $N_2 = 73.63$, $O_2 = 2.51$. LHV = 43,059 kJ/kg (18,512 Btu/lb), HHV = 45,885 kJ/kg (19,726 Btu/lb)).

Relating FGR and Oxygen in Windbox

FGR affects the oxygen in the windbox by diluting it. One may measure the oxygen values in the windbox and relate it to the FGR rate used. This may be used for relating actual FGR rates used with NO_x emissions.

Example 1.7

A boiler firing natural gas at 15% excess air uses 54,118 kg/h of combustion air, and about 6,352 kg/h of flue gases is recirculated. Flue gas analysis is as follows: % volume $CO_2 = 8.29$, $H_2O = 18.17$, $N_2 = 71.07$, $O_2 = 2.47$. Determine the oxygen levels in the windbox. Assume air is dry and has 77% nitrogen and 23% oxygen by weight.

Solution

The amount of nitrogen in air = $0.77 \times 54{,}118 = 41{,}671$ kg/h and oxygen = $(54{,}118 - 41{,}671) = 12{,}447$ kg/h. MW of flue gas = $(8.29 \times 44 + 18.17 \times 18 + 71.07 \times 28 + 2.47 \times 32)/100 = 27.61$.

% CO_2 by weight in flue gas is $8.29 \times 44/27.61 = 13.21$. Similarly, % weight of $H_2O = 11.84$, $N_2 = 72.07$, $O_2 = 2.88$. The mixture has a flow of $6352 + 54{,}118 = 60{,}470$ kg/h

CO_2 in the mixture = $0.1321 \times 6352 = 839$ kg/h

$H_2O = 0.1184 \times 6352 = 750$ kg/h (neglect air moisture)

$N_2 = 0.7207 \times 6352 + 41{,}671 = 46{,}249$ kg/h

$O_2 = 12{,}447 + 0.0288 \times 6{,}352 = 12{,}630$ kg/h

Converting this to% volume, we have

$$CO_2 = (839/44)/[(839/44) + (750/18) + (46{,}249/28) + (12{,}630/32)]$$
$$= 100 \times (19.07/2107) = 0.9\%$$

Similarly, $H_2O = 1.98\%$, $N_2 = 78.37\%$, and $O_2 = 18.75\%$. We see that the % volume of oxygen has come down to 18.75 due to FGR. For finding out the % FGR for a desired NO_x level, this reading of oxygen may be used as a reference.

Gas Turbine Exhaust Combustion Calculations

Supplementary firing or furnace firing is often done in HRSGs to increase the HRSG efficiency while generating additional steam on the order of 200%–400%. HRSGs that recover energy from gas turbine exhaust are often fired using distillate fuel or natural gas to raise the gas temperature entering the HRSG. Heavy oils are rarely used as emissions of NO_x, and particulates and oxides of sulfur formed could affect the finned surfaces of the HRSG and the SCR catalyst if used. Depending upon the firing temperature, the HRSGs may be called supplementary-fired or furnace-fired HRSGs. This is discussed in Chapter 4. Here we will see how the combustion calculations are performed and how these calculations are different from those for a steam generator. Typically, turbine exhaust contains over 13%–15% by volume of oxygen in the exhaust, and one can increase the firing temperature by addition of fuel alone till an oxygen level of about 3%–3.5% by volume is reached without using additional air. Only fuel is introduced in the burner, and the oxygen available in the exhaust is consumed. A typical duct burner is shown in Figure 1.4. Arrangement of an HRSG with duct burner is shown in Figure 1.5.

Relating Oxygen and Energy Input in Turbine Exhaust Gases

Gas turbine exhaust gases typically contain 13%–16% oxygen by volume compared to 21% in atmospheric air. If steam is injected in the gas turbine for NO_x control, the oxygen content will be further reduced. Still there is enough oxygen to raise the exhaust gases to about 1600°C (see Figure 1.6). Sometimes, augmenting air is introduced at the burner to ensure a stable combustion process.

The energy Q in GJ/h required to raise the temperature of exhaust gases from t_1 to t_2 °C is given by an energy balance around the burner, but approximately it is $Q = 10^{-6} \times W_g \times (h_2 - h_1)$

FIGURE 1.4
Burner for turbine exhaust. (Courtesy of Natcom Burner, Division of Cleaver Brooks, Thomasville, GA.)

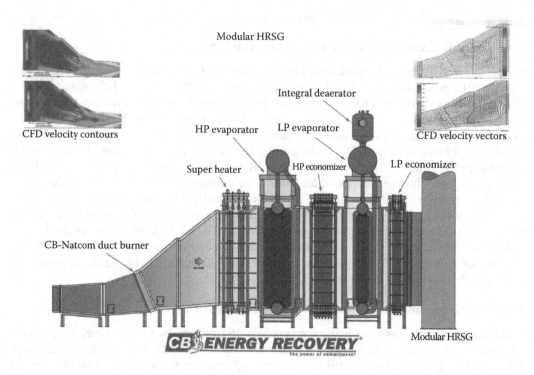

FIGURE 1.5
Duct burner in a HRSG. (Courtesy of Natcom Burner, Division of Cleaver Brooks, Thomasville, GA.)

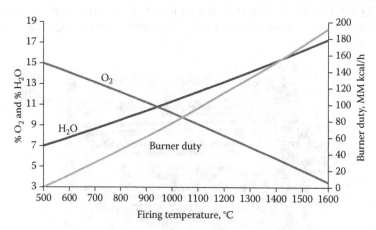

FIGURE 1.6
HRSG firing temperature, burner duty, and exhaust gas analysis.

where h_1, h_2 are the enthalpies of the gas before and after combustion in kJ/kg, and W_g is the exhaust flow in kg/h. The fuel quantity is small and can be neglected when compared to the exhaust gas flow. A more accurate expression will be $(W_g + W_f)h_2 - W_g h_1 = 10^6 \times Q$ where W_f = fuel consumption in kg/h.

$W_f = 10^6 \times Q/LHV$ where LHV is the lower heating value in kJ/kg. If O% volume of oxygen is available in the exhaust gas, the equivalent amount of air W_a in W_g kg/h of exhaust gases may be shown to be

$W_a = (100/23) \times W_g \times O \times 32/100/28.4 = 0.049\ W_g \times O$ kg/h air where the oxygen in % volume is converted to mass basis by multiplying with its MW of 32 and dividing by the exhaust gas MW of 28.4 (typical gas turbine exhaust analysis is used. % volume $CO_2 = 3$, $H_2O = 7$, $N_2 = 75$, and $O_2 = 15$). The (100/23) is the conversion from oxygen to air on mass basis.

The energy input on HHV basis = $10^6 \times HHV \times (Q/LHV)$. Now 1 GJ of energy input requires A kg of air for combustion as shown in Table 1.5. Hence, $10^6 \times HHV \times (Q/LHV)$ requires $10^6 \times HHV \times (Q/LHV) \times A$ kg/h air = $W_a = 0.049 \times W_g \times O$. Simplifying, $Q = 0.049 \times 10^{-6} \times W_g \times O \times LHV/(A \times HHV)$. For typical natural gas and fuel oils, (LHV/A/HHV) may be approximated as 0.00287. Hence within ± 3% margin,

$$Q = 140 \times 10^{-6} \times W_g \times O = 0.000140\ W_g \times O;\ Q\ \text{is in GJ/h},\ W_g\ \text{in kg/h} \qquad (1.14a)$$

$$Q = 60\ W_g\ O\ \text{in British units}.\ Q\ \text{in Btu/h},\ W_g\ \text{in lb/h} \qquad (1.14b)$$

For example, with a fuel input of 30 GJ/h (28.44 MM Btu/h), the % volume of oxygen consumed with 70,000 kg/h (154,000 lb/h) of exhaust gases will be O = 30/(0.000140 × 70,000) = 3%. This can raise the temperature of 70,000 kg/h gas by about 360°C.

Evaluating Fuel Quantity Required to Raise Turbine Exhaust Gas Temperature

The following example shows how this is done. A computer program is ideal for this exercise as one has to obtain a rough estimate of the fuel quantity and fine-tune it using the enthalpy of the exhaust gas obtained after combustion. The gas analysis will vary with the firing temperature assumed, and hence several iterations may be required. However, the following manual calculation shows the procedure.

Example 1.8

Let us compute the fuel quantity required to raise the temperature of 500,000 kg/h of gas turbine exhaust gases from 500°C to 800°C and the final exhaust gas analysis. Exhaust gas analysis entering the burner is % volume $CO_2 = 3$, $H_2O = 7$, $N_2 = 75$, and $O_2 = 15$. Fuel analysis is: methane = 97%, ethane = 2%, and propane = 1%. HHV = 55,335 kJ/kg and LHV = 49,867 kJ/kg.

Solution

This is a trial and error process. One has to assume the fuel input, perform combustion calculations and arrive at the exhaust gas analysis, calculate the gas enthalpies at inlet and exit of the burner, and do the following energy balance:

$$W_1 h_{g1} + Q_f = (W_1 + W_f) h_{g2}\ (Q_f\ \text{in kJ/h},\ h_{g1},\ h_{g2}\ \text{in kJ/kg})$$

where

W_1, W_2 are the exhaust gas flows entering and leaving the burner, kg/h
h_{g1}, h_{g2} are the enthalpies of the exhaust gas before and after the burner

Note that the gas analysis will be different after combustion, and hence a few iterations are required to obtain the gas enthalpy, which is again a function of gas analysis and temperature. Also,

$$W_2 = W_1 + W_f$$

where W_f is the fuel flow, kg/h.

Let us first convert the incoming exhaust gas analysis from volumetric to mass basis.

$$MW = 0.03 \times 44 + 0.07 \times 18 + 0.75 \times 28 + 0.15 \times 32 = 28.38$$

Fraction weight of $CO_2 = 0.03 \times 44/28.28 = 0.0465$, $H_2O = 0.07 \times 18/28.28 = 0.044$. $N_2 = 0.75 \times 28/28.38 = 0.74$ and $O_2 = 0.15 \times 32/28.38 = 0.169$.

Enthalpy at 500°C: From Table F.11, $h_{g1} = 0.0465 \times 118.2 + 0.044 \times 231 + 0.74 \times 124.5 + 0.169 \times 113.8 = 127$ kcal/kg $= 531.75$ kJ/kg.

Mass of CO_2 in incoming exhaust gases $= 0.0465 \times 500,000 = 23,250$ kg/h.

Mass of $H_2O = 0.044 \times 500,000 = 22,000$ kg/h.

Mass of $N_2 = 0.74 \times 500,000 = 370,000$ kg/h.

Mass of $O_2 = 84,750$ kg/h by difference.

Now convert the fuel analysis to mass basis. $MW_f = 97 \times 0.16 + 2 \times 0.3 + 1 \times 0.44 = 16.56$.

Fraction weight of $CH_4 = 97 \times 0.16/16.56 = 0.937$. $C_2H_6 = 2 \times 0.3/16.56 = 0.036$ and $C_3H_8 = 1 \times 0.44/16.56 = 0.027$. Let the burner duty $= 160$ GJ/h on LHV basis. (One may estimate the burner duty as $W_g \times 1300 \times (800 - 500)$ GJ/h where 1300 refers to the approximate specific heat of the flue gas in J/kg°C. Here Q = 195 GJ/h). But let us continue with our assumed value of 160 GJ/h and see what the final temperature is.

$W_f = 160 \times 10^6/49,867 = 3,208$ kg/h. CH_4 in fuel $= 0.937 \times 3208 = 3007$ kg/h. C_2H_6 in fuel $= 0.036 \times 3208 = 116$ kg/h and $C_3H_8 = 0.027 \times 3208 = 86.6$ kg/h.

CH_4 of 1 kg requires 3.99 kg of oxygen for combustion from Tables 1.1 and 1.2. So 3007 kg/h requires $= 11,998$ kg/h. Similarly 1 kg of C_2H_6 requires 3.725 kg oxygen and so 116 kg/h requires $= 116 \times 3.725 = 432$ kg/h and C_3H_8 requires $= 86.6 \times 3.629 = 314$ kg/h oxygen. Hence, oxygen in exhaust gas after combustion will be reduced and will be $84,750 - 11,998 - 432 - 314 = 72,006$ kg/h.

Similarly, CO_2 formed due to combustion of the fuel $= 3007 \times 2.744 + 116 \times 2.927 + 86.6 \times 2.994 = 8850$ kg/h. After the burner $= 23,250 + 8,850 = 32,100$ kg/h.

H_2O after combustion $= 3007 \times 2.246 + 116 \times 1.798 + 86.6 \times 1.634 + 22,000 = 29,105$ kg/h

Hence products of combustion contain $CO_2 = 32,100$ kg/h, $H_2O = 29,105$ kg/h, $N_2 = 370,000$ kg/h and $O_2 = 72,006$ kg/h. Total $= 503,211$ kg/h. Converting to mass fractions, $CO_2 = 32,100/503,211 = 0.0637$, $H_2O = 29,105/503,211 = 0.0578$, $N_2 = 370,000/503,211 = 0.7353$, and $O_2 = 72,006/503,211 = 0.143$.

From energy balance around the burner, neglecting heat losses,

$$500,000 \times 531.75 + 160 \times 10^6 = 503,208 \times h_{g2} \quad \text{or} \quad h_{g2} = 846.32 \text{ kJ/kg} = 202.1 \text{ kcal/kg}.$$

h_{g2} from the earlier flue gas analysis at 800°C $= 0.0637 \times 205.4 + 0.0578 \times 391.6 + 0.7353 \times 207 + 0.143 \times 190.7 = 215.2$ kcal/kg.

Hence exhaust gas temperature after firing is short of 800°C. Hence another iteration is required. From computer program, one can show that the fuel input required is 189 GJ/h on LHV basis.

$$W_f = 189 \times 10^6/49,867 = 3,790 \text{ kg/h}; \quad CH_4 = 0.937 \times 3,790 = 3,551 \text{ kg/h},$$
$$C_2H_6 = 0.036 \times 3,790 = 136 \text{ kg/h} \quad \text{and} \quad C_3H_8 = 102.3 \text{ kg/h}$$

O_2 required $= 3551 \times 3.99 + 136 \times 3.725 + 3.629 \times 102.3 = 15,046$.

O_2 after burner $= 84,750 - 15,046 = 69,704$ kg/h.

CO_2 after burner $= 23,250 + 2.744 \times 3551 + 2.927 \times 136 + 102.3 \times 2.994 = 33,698$ kg/h.

$H_2O = 22,000 + 3551 \times 2.246 + 136 \times 1.798 + 102.3 \times 1.634 = 30,387$ kg/h.

$N_2 = 370,000$ kg/h. Total flue gas after burner $= 503,789$ kg/h.

Converting to mass fractions, $CO_2 = 33,685/503,789 = 0.067$, $H_2O = 30,380/503,786 = 0.06$ $N_2 = 370,000/503,786 = 0.734$, $O_2 = 69,704/503,789 = 0.138$.

Enthalpy h_{g2} after burner $= (500,000 \times 531.75 + 189 \times 10^6)/503,789 = 902.9$ kJ/kg.

h_{g2} at 800°C $= 0.067 \times 205.4 + 0.06 \times 391.6 + 0.734 \times 207 + 0.138 \times 190.7 = 215.7$ kcal/kg $= 903$ kJ/kg. This agrees with the fuel input of 189 GJ/h (LHV basis). Hence, 189 GJ/h is the fuel consumption required to raise the exhaust gas to 800°C.

Moles of flue gas: $(33,698/44) + (30,387/18) + (370,000/28) + (69,704/32) = 17,846$

% volume $CO_2 = 100 \times (33,698/44)/17,845 = 4.29\%$. $H_2O = (30,387/18)/17,846 = 9.46\%$ $N_2 = (370,000/28)/17,846 = 74\%$ $O_2 = (69,704/32)/17,846 = 12.2\%$ or by difference 12.25%.

TABLE 1.10

Flue Gas Analysis at Burner Inlet and Exit

Data	Entering	Leaving
Gas flow, kg/h	500,000	503,770
% volume CO_2	3	4.29
H_2O	7	9.46
N_2	75	74.00
O_2	15	12.25
Temperature, °C	500	800

A summary of the burner performance is shown in Table 1.10.

Using the simplified formula $Q = 0.0001400\ W_g \times O$, we have $Q = 0.000140 \times 500,000 \times (15 - 12.25) = 192.5$ GJ/h (46 MM kcal/h). In British units, $Q = 60 \times 500,000 \times 2.204 \times 2.75 = (182$ MM Btu/h).

Note that it is easy to get a good estimate of fuel consumption simply based on oxygen difference in the turbine exhaust gas. One may also obtain the firing temperature once the fuel analysis is known. Since measurement of gas temperature after the burner is not very accurate due to measurement errors and variation gas temperatures across the cross section of the duct, either the detailed combustion calculation for the fuel input based on exhaust gas oxygen content or the simplified procedure as described earlier will be an added check.

Firing Temperature versus Oxygen, Burner Duty, and Water Vapor

Plant engineers should have an idea of how the various components of the flue gas change after firing as the flue gas analysis affects the HRSG heat transfer calculations.

Figure 1.6 shows the O_2, H_2O in % volume after combustion in a duct burner and the fuel input on LHV basis with natural gas firing for various firing temperatures. Gas turbine exhaust temperature = 500°C with the following analysis: % volume CO_2 = 3, H_2O = 7, N_2 = 75, O_2 = 15. Exhaust gas flow = 500,000 kg/h. Duct burners with insulated casing may be used for the HRSG design up to 900°C, and beyond this, membrane wall furnaces with duct burner or register burners are used depending on burner supplier. Typically, duct burners are used up to 5%–6% oxygen after combustion. Beyond this, register burners are used as the flame temperature increases significantly and mixing of fuel and air is better.

Supplementary firing in gas turbine HRSGs increases the efficiency of fuel utilization. Efficiency of energy recovery increases as firing is increased. This is explained in Chapter 4 on waste heat boilers.

Boiler Efficiency

There are basically two methods of determining boiler efficiency.

1. Direct method in which the energy output and fuel input are measured and efficiency = output/input

   Fuel input estimation requires accurate information on flow of fuel and its heating value. While with gaseous fuels a fairly accurate measurement of flow may be obtained, with oil fuels due to its viscosity and variations with temperature, accurate fuel consumption is difficult to measure. With solid fuels, it is more so difficult. With modern instruments, steam flow and temperatures can be measured to determine the energy output. Boiler blowdown is avoided during the testing as the amount of blowdown cannot be determined accurately due to the two-phase critical flow in the blowdown line. See Chapter 6 on miscellaneous calculations for the estimation of blowdown flow.

2. The indirect method simply estimates the various losses in the boiler as described here. An advantage of this method is that errors in measurements do not impact the efficiency much. For example, if the fuel input measurement to a boiler has an error of 1% in the direct method, the error will be 92 ± 0.9% and the efficiency can vary from 91.1 to 92.9%.

   A 1% error in, say, exit gas temperature measurement (which is the major heat loss) will result in a variation of 8 ± 0.08 or 8.08–7.92%. (92% is typical boiler efficiency for gaseous fuels with about 15% excess air and about 150°C exit gas temperature); hence efficiency will be 100 − 8.08 = 91.92 to 100 − 7.92 = 92.08%. One may feel that the indirect method favors the boiler supplier rather than the boiler owner; however, if the exit gas temperature (for the same feed water inlet temperature in the economizer, steam parameters, and ambient conditions) is higher by, say, 10°C, the efficiency decreases by 0.45%, a significant deviation. Hence this method that requires lesser manpower and time for testing is widely accepted in industrial practice.

There are two ways of stating efficiency, one based on HHV and the other on LHV. As discussed earlier, $\eta_{HHV} \times HHV = \eta_{LHV} \times LHV$.

The European and Asian countries generally follow the LHV method while in the United States, efficiency is often stated based on HHV. Hence one should be aware of this practice and not be misled by the values of efficiency and investigate further. With oil firing, the ratio of the heating values is about 6.5%, while for natural gas, it is about 9.5%. So one should question the heating value basis used whether it is on HHV or LHV before comparing different boiler proposals or performance. The exit gas temperature from the boiler also is an indicator of efficiency.

Heat Loss Method

The various heat losses are [4]

1. Dry gas loss L_1,%

$$L_1 = 24w_{dg}(t_g - t_a)/HHV \tag{1.15a}$$

2. Loss due to combustion of hydrogen and moisture in fuel, L_2.

$$L_2 = 100 \times (9H_2 + W) \times (584 + 0.46t_g - t_a)/HHV \tag{1.15b}$$

where 584 is the latent heat of water vapor in kcal/kg and 0.46 the specific heat of water vapor in kcal/kg °C

3. Loss due to moisture in air L_3

$$L_3 = 46Mw_{da}(t_g - t_a)/HHV \tag{1.15c}$$

4. Loss due to CO formation

$$L_4 = 100 \times [CO/(CO + CO_2)] \times C \times 5644/HHV \qquad (1.15d)$$

5. Casing or radiation loss L_5

A more accurate way to compute this loss is to compute the heat losses at various sections based on the calculated or measured casing temperature, which depends on the insulation thickness used and ambient wind and temperature conditions as illustrated in Chapter 6 on miscellaneous calculations. Then the heat loss in kW/m^2 may be estimated for the entire casing and then obtain the % heat loss as shown in the following. It is generally very small in oil- and gas-fired package boilers, in the range of 0.2%–0.5%. For smaller units (up to 20 t/h of steam), it may be high, and for larger units it is smaller. We are not dealing with solid fuel-fired boilers, which can have refractory-lined casing with higher heat losses.

$$\text{A quick estimate of } L_5 = 10^{0.62 - 0.42 \log(Q)} \qquad (1.15e)$$

where Q is the boiler duty in MM Btu/h. If duty of a boiler is 200 MM kJ/h (189.7 MM Btu/h), the heat loss = $L_5 = 10^{[(0.62 - 0.42 \times \log(189.7)]} = 0.46\%$.

Simplified Formulae for Boiler Efficiency

Boiler efficiency mainly depends upon the excess air and exit gas temperature and reference temperature. The following equations provide a quick solution based on 1% casing and other unaccounted losses.

$$\text{For natural gas: } \eta_{HHV} = 89.4 - (0.002021 + 0.0351 \times EA) \times \Delta T \qquad (1.16a)$$

$$\eta_{LHV} = 99 - (0.00202 + 0.0389 \times EA) \times \Delta T \qquad (1.16b)$$

For fuel oils (no 2, diesel; may be used for heavy oil with lesser accuracy)

$$\eta_{HHV} = 92.9 - (0.002336 + 0.0351 \times EA) \times \Delta T \qquad (1.16c)$$

$$\eta_{LHV} = 99 - (0.00249 + 0.03654 \times EA) \times \Delta T \qquad (1.16d)$$

ΔT is the difference between exit gas temperature and reference temperature in °C.
EA is the excess air factor. If excess air is 15%, then EA = 1.15.

Example 1.9

Determine the efficiency of a boiler firing the natural gas fuel as in Example 1.2. Assume exit gas temperature = 204°C and ambient temperature is 20°C. Relative humidity is 80%. Use a casing loss of 1%. HHV = 12,882 kcal/kg and LHV = 11,655 kcal/kg.

Solution

Combustion calculations have already been done in Example 1.1. Let us use the results.
Dry flue gas w_{dg} = 18 kg/kg fuel. Moisture in air = $w_{wa} - w_{da}$ = 19.52 – 19.29 = 0.23 kg/kg fuel.

Water vapor formed due to combustion of fuel = 20.4 − 0.23 − 18 = 2.17 kg/kg fuel

1. Dry gas loss = 100 × 18 × 0.24 × (204 − 20)/12,882 = 6.17%
2. Loss due to combustion of hydrogen and moisture in fuel = 100 × 2.17 × (584 + 0.46 × 204 − 20)/12,882 = 11.1%
3. Loss due to moisture in air = 100 × 0.23 × 0.46 × (204 − 20)/12,882 = 0.15%
4. If CO in flue gases = 100 ppm, then CO = 0.01% volume in flue gases. CO_2 = 8.5% from earlier combustion calculations. Carbon content in fuel was estimated as 0.7539 kg/kg fuel earlier.
   L_4 = 100 × (0.01/8.5) × 0.7539 × 5,644/12,882 = 0.039%
5. Casing loss = 1.0%
   Total losses = 6.17 + 11.1 + 0.15 + 1.0 + 0.038 = 18.458%. Hence, efficiency on HHV basis = 100 − 18.458 = 81.54%. Efficiency on LHV basis = 81.54 × 12,882/11,655 = 90.12%.
   Using the simplified formula,

$$\eta_{HHV} = 89.4 − (0.002021 + 0.0351 × 1.15) × 184 = 81.6\%$$

$$\eta_{LHV} = 99 − (0.00202 + 0.0389 × 1.15) × 184 = 90.39\%$$

Example 1.10

Casing temperature of a boiler was measured as 82°C when the ambient temperature was 29°C. Wind velocity = 134 m/min. Determine the casing loss if the surface area of the boiler is 231 m². Casing emissivity may be taken as 0.1.

Solution

From Chapter 6, one may determine the casing loss q.

$$q = 5.67 × 10^{−11} × 0.1 × [(273 + 82)^4 − (273 + 29)^4] + 0.00195 × (82 − 29)^{1.25} × [(134 + 21)/21]^{0.5}$$
$$= 0.043 + 0.755 = 0.798 \text{ kW/m}^2 \text{ (252 Btu/ft}^2\text{h)}$$

The total heat loss = 231 × 0.798 = 184.3 kW (231 m² is the total surface area of the boiler casing). If boiler duty is 50 MW, the heat loss = 184/50,000 = 0.0037 or 0.37%.

Note that as boiler load decreases, the heat loss in kW will not diminish (as it depends only on ambient conditions and wind velocity) while the duty decreases. Hence, casing loss as a % will increase. At 50% load, the casing loss will be about 0.75% in our example. While the flue gas heat losses decrease as the load decreases, the casing loss increases in indirect proportion to the load. Thus, a parabolic trend is seen for the efficiency versus load curve. Table 1.11 shows how the boiler efficiency and various losses vary with load.

Example 1.11

If fuel oil is fired at 20% excess air in a boiler and ΔT = 150°C, then

$$\eta_{HHV} = 92.9 − (0.002336 + 0.0351 × 1.2) × 150 = 86.23\%$$

$$\eta_{LHV} = 99 − (0.00249 + 0.03654 × 1.2) × 150 = 92\%$$

Note that these equations assume a total heat loss of 1%. At part loads, the heat loss will be higher as explained earlier and in Chapter 6. One has to adjust the efficiency to account for the higher heat loss at lower loads.

TABLE 1.11

Typical Boiler Performance versus Load

Boiler Load, %	100	75	50	25	
Boiler duty	72.81	54.61	36.41	18.20	MM kcal/h
Ambient temp.	26.7	26.7	26.7	26.7	°C
Relative humidity	60	60	60	60	%
Excess air	17	17	17	17	%
Flue gas recirculation	22	22	22	22	%
Fuel input (HHV)	88.61	66.42	44.34	22.34	MM kcal/h
Heat release rate (HHV)	743,453	557,291	372,038	187,434	kcal/m^3 h
Heat release rate (HHV)	520,692	390,312	260,562	131,275	kcal/m^2 h
Steam flow	149,728	112,296	74,864	37,432	kg/h
Process steam	0	0	0	0	kg/h
Steam pressure	76.5	76.5	76.5	76.5	kg/cm^2 g
Steam temp.	293	293	293	293	±5°C
Feed water temp.	172	172	172	172	±5°C
Water temp. lvg eco	233	226	219	212	±5°C
Blowdown %	2	2	2	2	%
Boiler exit gas temp.	386	359	332	307	±5°C
Eco exit gas temp.	184	181	178	176	±5°C
Air flow	136,420	102,261	68,267	34,394	kg/h
Flue gas to stack	143,310	107,426	71,715	36,131	kg/h
Flue gas through boiler	174,838	131,059	87,492	44,080	kg/h
Flue gas analysis, losses, efficiency, %					
Dry gas loss	5.42	5.29	5.19	5.13	%
Air moisture	0.15	0.14	0.14	0.14	%
Fuel moisture	11.04	11.01	10.99	10.98	%
Casing loss	0.35	0.47	0.70	1.40	%
Unacc/margin	0.50	0.50	0.50	0.50	%
Efficiency, lhv	91.58	91.63	91.50	90.81	%
Efficiency, hhv	82.55	82.59	82.48	81.85	%
Furnace back pr.	369.53	205.82	91.18	23.20	mm wc
% vol CO$_2$	8.13	8.13	8.13	8.13	
H$_2$O	18.02	18.02	18.02	18.02	
N$_2$	71.11	71.11	71.11	71.11	
O$_2$	2.74	2.74	2.74	2.74	
SO$_2$	0.00	0.00	0.00	0.00	
Fuel flow	6888	5163	3448	1736	kg/h

Gas-% volume

Methane	98.0	Boiler surface areas—m^2	
Ethane	0.5	Furnace volume	119
Propane	0.1	Furnace proj. area	171
Butane	0.0	Evaporator	1372
Hydrogen	0.0	Economizer	6296
Carbon dioxide	0.6		
Nitrogen	0.8		
LHV-kcal/kg	11,594.0		
HHV-kcal/kg	12,863.0		

Simplified Procedure for Obtaining Major Boiler Parameters

Major parameters like fuel, air, and flue gas flow, boiler efficiency, and excess air may be obtained if the boiler duty and excess air are known.

Example 1.12

Determine the efficiency, fuel flow, excess air, and air and flue gas quantity produced when no. 2 fuel oil with HHV = 45,737 kJ/kg is fired in a boiler. Oxygen is measured as 3.2% dry in the flue gas. Boiler duty (energy absorbed by steam) = 84.66 MW. Boiler exit gas temperature = 184°C and ambient is at 27°C. Relative humidity = 60%.

Solution

This is the type of situation a plant engineer often faces, working with a few field data and sketchy information on fuel. First estimate the excess air using the following expression:

$E = K_1 O_2/(21 - O_2)$ where K_1 from Table 1.4 is 94.5. $E = 94.5 \times 3.2/(21 - 3.2) = 17\%$

Using the simplified equation for efficiency, $\eta_{HHV} = 92.9 - (0.002336 + 0.0351 \times EA) \times \Delta T$

We have $\eta_{HHV} = 92.9 - (0.002336 + 0.0351 \times 1.17) \times (184 - 27) = 86\%$
Fuel input = 84.66/0.86 = 98.44 MW = 98,440 kW = 98,440 kJ/s
Fuel flow = 98,440/45,737 = 2.152 kg/s = 7748 kg/h

From Table 1.5, 320 kg of air is required per 10^6 kJ heat input. Hence, air required = 0.09844 × 320 × 1.17 × 1.013 × 3600 = 134,400 kg/h (1.013 is the conversion from dry to wet air at 60% relative humidity (Figure 1.1). Flue gas produced = 134,400 + 7,748 = 142,148 kg/h. To obtain the flue gas analysis, refer to Table 1.15, later in the chapter, and interpolate for 17% excess air. To obtain more detailed results, one may use a computer program or perform detailed combustion calculations.

Excess Air versus Efficiency

Figure 1.7 shows the effect of excess air and exit gas temperature on boiler efficiency with natural gas firing on LHV basis. With distillate oil firing, it will be about 0.5% higher. To obtain efficiency on HHV basis, multiply the values by 0.904 for natural gas and by 0.94 for distillate fuel oil and by 0.95 for heavy fuel oil.

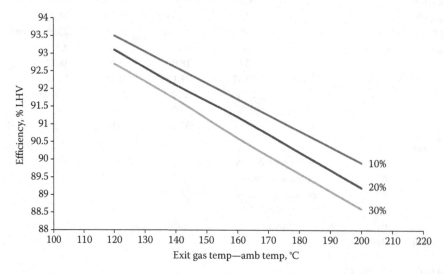

FIGURE 1.7
Excess air versus boiler efficiency on LHV basis for natural gas.

Firing Fuels with Low Heating Values

Table 1.6 shows a few gaseous fuels with low heating values. When fuels with low heating values are fired in a boiler, one has to be concerned about the following:

1. Low combustion temperature resulting in a low gas temperature at the furnace exit. Sometimes a supplementary fuel such as oil or natural gas is fired to a small extent to maintain a stable flame. If a superheater is located directly at the furnace exit, the steam temperature may have to be checked as the LMTD will be significantly lower.

2. Large amount of flue gases will be generated for the same steam generation while firing the low Btu fuel. This increases the convective heat transfer and also the gas velocities in various heating surfaces. If a boiler is operated on natural gas as well as a low Btu fuel, then the turndown will be lower due to low gas velocities and flow distribution concerns while firing natural gas at low loads.

3. The boiler exit gas temperature will be higher due to the higher flue gas flow, and this decreases the boiler efficiency.

4. The back pressure will also be higher due to the higher gas pressure drop with larger flue gas flow. Fans may have to be checked for suitability on low-Btu fuel firing. Tables 1.12 through 1.14 show the performance of a boiler with two different fuels.

Example 1.13

A boiler generating 100 t/h of saturated steam at 28.7 kg/cm² g is fired with two fuels as shown in Table 1.12. Column 1 shows the low heating value fuel, and the other is the typical natural gas with a high heating value and the fuel analysis.

Table 1.13 shows the flue gas analysis and the various losses and boiler efficiency. It may be seen that the % volume water vapor is much lower with the low BTU fuel. The convective heat transfer coefficient will be high, while the nonluminous coefficient will be much smaller. The back pressure is nearly double with the low BTU fuel. Fan power consumption will be very high due to the high back pressure. The exit gas temperature is about 40°C higher due to the higher mass flow of flue gas with low-BTU fuel. If a superheater is present, depending on its location (convective or radiant), the steam temperature could also be significantly different.

TABLE 1.12

Fuels Fired in the Boiler

Gas, % Volume	Low Heating Value Fuel	Nat Gas
Methane	0.1	97
Ethane	0	1
Hydrogen	2.4	0
Carbon monoxide	23.3	0
Carbon dioxide	14.4	0.5
Nitrogen	56.4	1.5
Water vapor	3.4	0
LHV, kcal/kg	590	11,480
HHV, kcal/kg	600	12,735

TABLE 1.13

Losses and Efficiency

	Low Heating Value Fuel	Nat Gas
Dry gas loss	8.88	3.59 %
Air moisture	0.10	0.10 %
Fuel moisture	1.84	10.63 %
Casing loss	0.30	0.30 %
Unacc/margin	0.50	0.50 %
Efficiency, lhv	89.79	94.17 %
Efficiency, hhv	88.39	84.89 %
Furnace back pr	344.00	162.00 mm wc
% vol CO_2	23.60	8.25 %
H_2O	4.70	18.25 %
N_2	70.48	71.04 %
O_2	1.22	2.46 %
SO_2	0.00	0.00
Fuel flow	105,726	5187 kg/h

TABLE 1.14

Boiler Performance on Natural Gas and Low-Btu Fuel

Boiler Load, %	100	100	
Boiler duty	55.82	55.81	MM kcal/h
Ambient temp.	26.7	26.67	°C
Relative hum.	60	60.0	%
Excess air	15	15	%
FGR	0	0	
Fuel input (hhv)	66.05	63.43	MM kcal/h
Heat rel. rate (HHV)	659,686	633,529	kcal/m³ h
Heat rel. rate (HHV)	441,478	423,975	kcal/m² h
Steam flow	100,000	100,000	kg/h
Steam pressure	38.7	38.7	kg/cm² g
Steam temp.	248	248	±5°C
Feed water temp.	110	110	±5°C
Water temp. lvg eco	175	219	±5°C
Blowdown	2	2	%
Boiler exit gas temp.	365	405	±5°C
Eco exit gas temp.	133	159	±5°C
Air flow	99,957	75,706	kg/h
Flue gas to stack	105,144	181,432	kg/h
Fuel	Nat. gas	Low BTU	
Max gas velocity—boiler	34	44	m/s
Economizer	9	14	m/s

Firing of Multiple Fuels

Often, two or even three fuels may be fired in a boiler simultaneously. A low heating value waste gaseous fuel along with oil and natural gas may be fired to generate steam in boilers. In such cases, the % contribution of each fuel should be known to arrive at the efficiency and flow calculations. One may also manipulate the % heat input of each fuel to match the oxygen in the flue gas. Presented in the following text are results for a case from a computer program in which natural gas and fuel oil are fired simultaneously with about 70% being the contribution from natural gas. One may calculate the efficiency based on each fuel separately and then on weighted fuel input basis. The flue gas analysis on individual fuel basis and weighted basis also may be estimated as shown later. One may adjust the fuel contribution of each fuel to match the oxygen in flue gas and find out the contribution of each fuel.

Example 1.14

A steam generator is fired with natural gas and fuel oil. The exit gas temperature is measured as 404°F. If the fuel gas contribution is about 71% and that of oil 29%, determine the air, flue gas flows, analysis, and overall boiler efficiency.

Solution

The results from a computer program are shown in Figure 1.8. The efficiency on individual fuel basis and weighted overall basis is shown as also the air and mixed flue gas flow and analysis. Combustion temperature of each fuel is also obtained. Plant engineers may develop a computer code for this type of calculation and vary the ratio among fuels till fuel flow data matches with the estimated value. If a particular fuel is measured with more accuracy, we may change the ratio of fuels to reflect that data. If natural gas fuel measurement works out to about 4000 lb/h in this case, then we know that the ratio is as shown. Else we can repeat the calculations till we hit the fuel flow measured and obtain a set of all pertinent data shown.

Emission Conversion Calculations: Steam Generators

From basic combustion calculations and flue gas analysis, one can perform several emission-related calculations that are shown in the following text. The burner suppler can give an idea of emissions of NO_x, CO, and SO_x based on fuel input in kg/GJ (lb/MM Btu), Table 1.8. Conversion to parts per million volume dry basis (ppmvd) requires some effort and is dependent on flue gas analysis.

Emissions of NO_x, UHC (unburned hydrocarbons, SO_x, and CO are often stated in ppmvd at a reference level of 3% oxygen or on g/GJ basis (lb/MM Btu) or in mg/N m³. The relation between the terms is explained later for natural gas and no. 2 fuel oils.

(Natural gas: $C_1 = 97\%$, $C_2 = 2$, $C_3 = 1$; HHV and LHV = 55,263 and 49,920 kJ/kg (23,759 and 21,462 Btu/lb respectively. No 2 oil: C = 87.5%, $H_2 = 12.5\%$, °API = 32 (HHV = 45,885 and 43,058 kJ/kg (19,727 and 18,512 Btu/lb respectively). Relative humidity 60%, and 70°F ambient temperature. Note that changes in relative humidity and ambient temperature will impact the flue gas analysis very slightly.]

For converting mass basis to moles, divide by MWs. MW of NO_x is taken as 46, while that for the flue gas may be obtained from combustion calculations, which require information on fuel analysis and excess air. See typical data in Table 1.15.

Boiler duty and efficiency calculations

Combination firing—two fuels fired

PROJECT sample units—British   Boiler Parameters

1. Boiler duty—MM Btu/h = 109.93
2. Excess air—% = 18
3. Amb. temp.—°F = 80
4. Exit gas temp.—°F = 404
5. Ref. humidity—% = 60
   Flue Gas Analysis—% vol (wet/dry)
6. CO_2 8.99/10.73
7. H_2O 16.19
8. N_2 71.92/85.81
9. O_2 2.9/3.46
10. SO_2 —
11. Total air—lb/h = 115,402
12. Total flue gas—lb/h = 121,312
13. Efficiency—% HHV = 82.7
14. Efficiency—% LHV = 82.7
15. Adiab. comb temp.—°F = 3207

1. Steam press—psig = 400
2. Steam temp.—°F = 750
3. Steam flow—lb/h = 100,000
4. Feed water temp.—°F = 320
5. Blowdown—% = 1
6. Process steam—Lb/h =
   (gas flow, analysis, air flow, efficiencies shown here pertain to effective average conditions if multiple fuels are fired.)

Air heated by boiler flue gas or no air heater

Fuel no. 1 gas—% volume [% duty = 71]
Methane = 97, ethane = 2, propane = 1
Fuel and air input HHV—MM Btu/h = 95.54 fuel fred—lb/h = 4021 ad. comb temp.—°F = 3164
Air/fuel ratio = 20.5 gas/fuel = 21.5 eff-lhv—% = 90.55 eff-hhv—% = 81.69
Losses: dry gas—% = 6.22 air moisture = 0.17 fuel moisture = 11.12 radiation = 0.3 unaccounted = 0.5
Flue gas analysis:% vol CO_2 = 8.1 H_2O = 17.8 N_2 = 71.22 O_2 = 2.89 SO_2 =.
LHV—Btu/lb = 21,439 HHV—Btu/lb = 23,764 fuel temp. °F = 80 air temp. °F = 80
NO_x—ppmv/0.1 lb per MM Btu HHV = 83.1 CO—ppmv/0.1 lb CO per MM Btu HHV = 136.6

Fuel no. 2 oil—% weight [% duty = 29]
Carbon = 87. Hydrogen = 13. Sulfur =. Nitrogen =. Oxygen =. °API = 32 moisture =.
Fuel AND air input HHV—MM Btu/h = 37.38 fuel fired—Lb/h = 1895 ad comb temp.—°F = 3288
Air/fuel ratio = 17.41 gas/fuel = 18.41 eff-lhv—% = 91.12 eff-hhv—% = 85.28
Losses: dry gas—% = 6.71 air moisture = 0.17 fuel moisture = 7.04 radiation = 0.3 unaccounted = 0.5
Flue gas analysis:% vol CO_2 = 11.29 H_2O = 12.04 N_2 = 73.73 O_2 = 2.94 SO_2 =.
LHV—Btu/lb = 18,463 HHV—Btu/lb = 19,727 fuel temp. °F = 80 air temp. F = 80
NO_x—ppmv/0.1 lb per MM Btu HHV = 77.5 CO—ppmv/0.1 lb CO per MM Btu HHV = 127.4

FIGURE 1.8
Results from combination firing of fuels.

Emissions of NO_x in mass units such as g/GJ may be converted to ppmvd from the following equation:

$$V_n = 10^6 \times Y \times (N/46) \times (MW/W_{gm}) \times (21 - 3)/(21 - O_2 \times Y) \tag{1.17}$$

where
   MW = molecular weight of wet flue gases
   V_n is the amount of NO_x equivalent of 1 g/GJ in ppmvd
   O_2 = % volume of oxygen in wet flue gases
   Y = 100/(100 − %H_2O) where %H_2O is the volume of water vapor in wet flue gases
   N = NO_x produced in g/GJ (we may replace this by CO or SO_x as shown later)
   W_{gm} = flue gas produced per GJ fuel fired, g, obtained as shown later (lb/MM Btu values shown in brackets)

TABLE 1.15

Conversion from g/GJ to ppmvd of NO_x, CO, UHC

Excess Air, %	0	10	20	30	0	10	20	30
Fuel	Gas	Gas	Gas	Gas	Oil	Oil	Oil	Oil
CO_2	9.47	8.68	8.02	7.45	13.49	12.33	11.35	10.51
H_2O	19.91	18.38	17.08	15.96	12.88	11.90	11.07	10.36
N_2	70.62	71.22	71.73	72.16	73.63	74.02	74.34	74.62
O_2	0	1.72	3.18	4.43	0	1.76	3.24	4.5
MW	27.52	27.62	27.68	27.77	28.87	28.85	28.84	28.82
W_{gm} (g/GJ)	331,300	362,600	393,900	425,200	339,700	371,400	403,300	435,000
W_{gm} (lb/MM Btu)	768	841	914	966	790	864	938	1011
V_n, ppmvd-g/GJ	1.933	1.933	1.933	1.9335	1.818	1.816	1.813	1.810
V_n, ppmvd-lb/MM Btu	831	831	831	831	782	782	782	782
V_c, ppmvd-g/GJ	3.1756	3.1756	3.1756	3.176	2.987	2.986	2.978	2.974
V_c, ppmvd-lb/MM Btu	1365	1365	1365	1365	1284	1284	1281	1279
V_h, ppmvd-g/GJ	5.557	5.557	5.557	5.558	5.226	5.226	5.212	5.210
V_h, ppmvd-lb/Mm Btu	2390	2390	2390	2390	2247	2247	2241	2240
V_s, ppmvd-g/GJ	1.389	1.389	1.389	1.389	1.3067	1.305	1.303	1.300
V_s, ppmvd-lb/MM Btu	597	597	597	597	562	561	560	560

Let us see how V_n is obtained. For the case of zero excess air, air required per GJ is obtained using the million GJ method discussed earlier or from detailed combustion calculations. From the A values in Table 1.5 and the HHV of fuel for the natural gas case.

Flue gas per GJ = $314 + 10^6/55,263 = 332$ kg/GJ = 332,000 g/GJ. It may also be computed more accurately from a computer program using the fuel analysis, results of which are shown in Table 1.15. Fuel consumed per GJ = $10^6/55,263 = 18.1$ kg/GJ. MW = 27.53; Y = 100/(100 − 19.91) = 1.2486; $O_2 = 0$

Let us convert 1 g/GJ to obtain V_n, the NO_x in ppmvd.

$$V_n = 10^6 \times Y \times (N/46) \times (MW/W_{gm}) \times (21-3)/(21-O_2 \times Y)$$
$$= 10^6 \times 1.2486 \times (1/46) \times (27.53/331,300) \times 18/(21-0)$$
$$= 1.933 \text{ ppmvd (1 lb/MM Btu} = 430 \text{ g/GJ} = 831 \text{ ppmvd)}.$$

Let us check at 30% excess air for natural gas:

$$\text{Flue gas per GJ} = 314 \times 1.3 + 10^6/55,263 = 426.3 \text{ kg/GJ} = 426,300 \text{ g/GJ}$$

approximately or more accurately 425,200 from Table 1.15.

$$Y = 100/(100 - 15.96) = 1.19.$$

$$V_n = 10^6 \times 1.19 \times (1/46) \times (27.77/425,200) \times 18/(21 - 4.43 \times 1.19) = 1.9335 \text{ ppmvd}$$

Hence 1 g/GJ of NO_x = 1.933 ppmvd or 1 lb/MM Btu NO_x = (453.7/1.05485) × 1.933 = 831 ppmvd. (453.7 is the number of g in 1 lb and 1.05485 is the conversion from MM Btu to GJ). Similarly, for CO, one would use 28 in the denominator of this equation (1.17) instead of

46 and obtain 1 g/GJ = 3.176 ppmvd (1 lb/MM Btu = 1362 ppmvd) (to convert to lb/MM Btu from g/GJ multiply by 430).

For UHC, insert MW of 16 in the earlier equation (UHC is considered as methane, which has 16 as its MW). Then, $V_u = 5.543U$, where U is the UHC in g/GJ and V_u is the ppmvd of UHC. 1 lb/MM Btu UHC = 5.543 × 430 = 2383 ppmvd.

For SO_x, using an MW of 64, we have $V_s = 1.389S$, where S is the emissions in g/GJ and V_s is in ppmvd or 1lb/MM Btu of SO_x = 1.389 × 430 = 597 ppmvd.

For no. 2 oil, using values from Table 1.15, for the case of zero excess air, for 1 g/GJ NO_x, $W_{gm} = 320 + 10^6/45,885 = 341.79$ kg/GJ = 341,790 g/GJ or, more accurately, from combustion calculations, 339,700. Y = 100/(100 − 12.88) = 1.1478

$V_n = 10^6 \times 1.1478 \times (1/46) \times (28.87/339,700) \times 18/(21 − 0) = 1.818$ ppmvd or 1 lb/MM Btu NO_x = 1.818 × 430 = 781 ppmvd. It can be shown that at 30% excess also, the value in ppmvd is the same as shown in Table 1.15.

Similarly, 1 g/GJ of CO is equivalent to $V_c = 2.987$ ppmvd. Or 1 lb/MM Btu CO = 1285 ppmvd. 1 g/GJ of UHC or $V_u = 5.22$ ppmvd and 1 lb/MM Btu UHC = 2245 ppmvd.

1 g/GJ of SO_x = 1.303 ppmvd and 1 lb/MM Btu = 1.303 × 430 = 560 ppmvd

---

Converting ppmvd of NO_x to mg/Nm^3

Let 1 N m^3 of flue gas contain 1 mg of NO_x. Hence, volume of NO_x = 22.4 × (1/46) × 10^{-6} = 0.487 ppmvd. Hence 1 ppmvd = 1/0.487 = 2.053 mg/N m^3 of NO_x.

Similarly, 1 ppmvd of CO = 1.25 mg/N m^3 (as MW of CO = 28 versus 46 for NO_x) and (22.4/64) mg/N m^3 of SO_x = 1 ppmvd or 2.857 mg/N m^3.

From Table 1.16, one can convert emissions of pollutants to any unit. For example, 1 lb/MM Btu of NO_x in oil firing is 1/0.000128 = 781 ppmvd. CO of 1 g/GJ in gas firing will be 1/0.315 = 3.174 ppmvd.

Example 1.15

Fuel oil containing 1.5% sulfur is fired in a furnace. What is the emission of SO_2 in g/GJ if the HHV = 46,000 kJ/kg (19,776 Btu/lb)? What is the ppmw if 15% excess air is used for firing this oil in a boiler?

TABLE 1.16

Summary of Emission Conversions

Item	ppmvd	mg/Nm³	g/GJ	lb/MM Btu
NO_x gas	1	2.053	0.516	0.00120
NO_x oil	1	2.053	0.551	0.00128
CO gas	1	1.25	0.315	0.000732
CO oil	1	1.25	0.335	0.000786
UHC gas	1	0.714	0.180	0.000419
UHC oil	1	0.714	0.193	0.000445
SO_x gas	1	2.857	0.720	0.001675
SO_x oil	1	2.857	0.767	0.001786

1 g of S gives rise to 2 g of SO_2. Hence, SO_2 emission = 0.015 × 2 × 1,000 × 10^6/46,000 = 652 g/GJ (1.516 lb/MM Btu)

Fuel of 1 GJ requires 320 kg air for combustion (Table 1.5). With 15% excess air, flue gas generated = 1.15 × 320 = 10^6/46,000 = 389.7 kg/GJ

Fuel input of 1GJ generates 0.03 × 21.7 = 0.651 kg SO_2. Hence, % volume wet of SO_2 in flue gas = (0.651/64)/(389.7/28.5) = 744 ppmv wet., where 28.5 is the assumed MW of flue gas and 64 that of SO_2. If 3% of SO_2 converts to SO_3, the amount of SO_3 in ppmv in flue gas = 744 × 0.03 × 64/80 = 17.8 ppmv (80 is the MW of SO_3).

These conversion calculations are important when the acid dew point needs to be estimated.

Example 1.16

Natural gas with C_1 = 97% volume, C_2 = 2%, C_3 = 1% is fired with 15% excess air in a boiler. If fuel contains 700 ppm H_2S, what is the ppmv of SO_2 formed? If 2% of this converts to SO_3, what is the ppmv SO_3 formed? Let ambient temperature = 27°C and relative humidity = 60% so that the amount of moisture in air = 0.013 kg/kg.

One comes across this type of problem often in a boiler containing H_2S. It is necessary to find out the SO_2 and then the SO_3 formed so that the acid dew point may be computed.

Amount of theoretical dry air required for combustion of 100 moles of fuel = 97 × 9.528 + 2 × 16.675 + 1 × 23.821 = 981.4 mol. Hence, actual air = 1.15 × 981.4 = 1128.6 mol. Moisture in air = 1128.6 × 0.013 × 27.5/18 = 22.4 mol.

CO_2 = 97 + 3.5 × 2 + 5 × 1 = 109 mol

H_2O = 2 × 97 + 3 × 2 + 4 × 1 + 22.4 = 226.4 mol

N_2 = 0.79 × 1128.6 = 891.6 mol

O_2 = (1128.6 – 981.4) × 0.21 = 30.9 mol

Total = 109 + 226.4 + 891.6 + 30.9 = 1257.9 mol

% volume CO_2 = 100 × 109/1257.9 = 8.66. H_2O = 100 × (226.4/1257.9) = 18.00% N_2 = 100 × (891.6/1257.9) = 0.70.88% and O_2 = 100 × 30.9/1257.9 = 2.46%

MW flue gas = 8.66 × 44+18x18 + 70.88 × 28 + 2.46 × 32 = 27.68

700 ppm H_2S in fuel gives rise to 700/12.579 = 55.6 ppmv SO_2 in flue gas

Assuming 3% conversion of SO_2 to SO_3, the amount of SO_3 in flue gas = 0.03 × 55.6 × 64/80 = 1.33 ppmv SO_3. (First convert the SO_2 to weight basis and then to SO_3 on volume basis. This value is used in acid dew point estimation.)

As discussed earlier, higher combustion temperatures will result in higher NO_x. When fuels containing hydrogen are fired, the NO_x levels will be higher. Table 1.17 gives an idea of the NO_x and CO with various fuels.

TABLE 1.17

Typical Emissions from Various Fuels

Gas	NO_x, lb/MM Btu (g/GJ)	CO (lb/MM Btu) (g/GJ)
Natural gas	0.1 (43)	0.08 (34.4)
Hydrogen gas	0.15 (64.5)	0
Refinery gas	0.1–0.15 (43–64.5)	0.03–0.08 (12.9–34.4)
Blast furnace gas	0.03–0.05 (12.9–21.5)	0.12 (51.6)
Producer gas	0.05–0.1 (21.5–43)	0.08 (34.4)

Converting Turbine Exhaust Emissions

Gas turbine exhaust gases have limits on NO_x and CO, and often it is required to check their values. If a duct burner is introduced, it adds to the NO_x and CO. Emissions of these gases are often referred to 15% oxygen in the exhaust, unlike boiler emissions that have 3% oxygen as the basis. The conversion is done as follows:

$$V_n = 100 \times (w/46)/(W_g/MW)/(100 - \% \ H_2O) \times (21 - 15)$$
$$\times \ 10^6/[21 - 100 \times \%O_2/(100 - \%H_2O)], \ ppmvd$$

where
O_2 is the % volume of oxygen present in the wet exhaust gases
w is the flow rate of NO_x (reported as NO_2), kg/h
W_g is the exhaust gas flow, kg/h
MW is the molecular weight of flue gases

Example 1.17

Determine the NO_x and CO emissions in ppmvd, 15% oxygen dry basis if 550,000 kg/h of exhaust gases contain 25 kg/h NO_x and 15 kg/h of CO. Exhaust gas analysis: % volume $CO_2 = 3.5$, $H_2O = 10$, $N_2 = 75$, $O_2 = 11.5$

$$MW = 0.01 \times (3.5 \times 44 + 10 \times 18 + 75 \times 28 + 11.5 \times 32) = 28$$

$$V_n = 10^6 \times [100 \times (25/46)/(550,000/28)/(100 - 10)] \times (21 - 15)/[21 - 100 \times 11.5/(100 - 10)]$$
$$= 22.4 \ ppmvd$$

V_c = emissions of CO = $(15/25) \times 22.4 \times 46/28 = 22.0$ ppmvd (ratio of mass and MWs of CO to NO_x.

Example 1.18

Let us compute the emissions in a natural gas supplementary-fired HRSG. Exhaust gases of 500,000 kg/h have to be raised to 800°C from 500°C. Exhaust gas analysis is % volume $CO_2 = 3$, $H_2O = 7$, $N_2 = 75$, and $O_2 = 15$, and contains 8 kg/h of NO_x and CO each. Fuel input is 45 MM kcal/h = 188.37 GJ/h. Fuel HHV = 55,335 kJ/kg and LHV = 49,867 kJ/kg. The burner contributes 21.5 g/GJ NO_x and CO each (on LHV basis). Flue gas analysis after combustion is % volume $CO_2 = 4.28$, $H_2O = 9.51$, $N_2 = 74.02$, and $O_2 = 12.17$. Determine the NO_x and CO in ppmvd before and after the burner (see Example 1.8).

Solution

MW of incoming exhaust gases = $3 \times 44 + 7 \times 18 + 75 \times 28 + 15 \times 32 = 28.38$. At exit of burner, flue gas quantity = 503,778 kg/h. MW = $4.28 \times 44 + 9.51 \times 18 + 74 \times 28 + 12.17 \times 32 = 28.2$

Before the burner, $V_n = 10^6 \times [100/(100 - 7)] \times (8/46) \times (28.38/500,000) \times (21 - 15)/(21 - 15 \times 100/93) = 13.0$ ppmvd

$V_c = 13 \times 46/28 = 21.4$ ppmvd

Burner duty on LHV basis = 45 MM kcal/h = 188.4 GJ/h. At burner exit, amount of NO_x and CO = $8 + 188.4 \times 21.5/1000 = 12.05$ kg/h.

$$V_n = 10^6 \times (100/90.49) \times (12.05/46) \times (28.2/503,778) \times (21 - 15)/(21 - 12.17 \times 100/90.49)$$
$$= 12.9 \ ppmvd. \quad V_c = 12.9 \times 46/28 = 19.7 \ ppmvd$$

TABLE 1.18

Typical Allowable Emission Rates for a Combined Cycle Plant in California

Unit	Pollutant	lb/h (kg/h)	lb/MM Btu (g/GJ) or ppmvd
CTG/HRSG with firing	PM	28.2 (12.79)	0.012 (5.16)
	SO_x	5.7 (2.59)	0.0023 (0.99)
	NO_x	28.6 (12.98)	3 ppmvd at 15% O_2
	VOCs	35.2 (15.97)	0.015 (6.45)
	CO	98.5 (44.69)	20 ppmvd at 15% O_2
	Formaldehyde	5 (2.27)	0.002 (0.86)
Auxiliary boiler	PM	0.19 (0.086)	0.005 (2.15)
	SO_x	0.09 (0.04)	0.0024 (1.032)
	NO_x	3.5 (1.59)	0.092 (39.6)
	VOCs	0.49 (0.22)	0.013 (5.59)
	CO	2.1 (0.95)	0.055 (23.65)

Note: PM, particulate matter; VOC, volatile organic chemical.

It is interesting to note that the NO_x and CO in ppmvd are lower after combustion than before. This should be carefully noted, and one should not assume that NO_x and CO have decreased after firing. This is due to the change in flue gas analysis and basis of oxygen level used. One should note that NO_x and CO in kg/h has increased after combustion due to contribution from the burner.

Typical limits for a combined cycle plant in California are shown in Table 1.18.

Low-Temperature Corrosion in Boilers

During the process of combustion of fuels such as oil, gas, or coal containing sulfur or hydrogen sulfide, SO_2 is formed. Some of it (1%–3%) is converted to SO_3, which can combine with water vapor to form sulfuric acid. The amount of SO_2 converted to SO_3 depends on the presence of catalysts such as vanadium and ferric oxide commonly found on tube surfaces in oil-fired boilers due to the ash present in the fuel or on oxidized alloy steel superheater tubes. Excess air available in the flue gas plays a role in increasing the formation of SO_3 (Figure 1.9), which in turn increases the acid dew point temperature. The corrosion rate increases as the tube wall temperature decreases, and again as the wall temperature approaches the water dew point range, it again increases (Figure 1.10). Studies conducted by Land Combustion have shown similar trends. The amount of water vapor in the flue gas also increases the acid dew point. Economizers and air heaters are prone to this corrosion attack. Table 1.19 gives an idea of SO_3 in oil-fired boilers in ppm as a function of excess air.

There are two ways of handling corrosion problems. Allow acid condensation to happen by using a low-temperature water or heat sink while protecting the equipment using corrosion-resistant materials. The second method is to avoid condensation of acid vapor by increasing the lowest surface or tube wall temperature. Boiler and air heater suppliers suggest charts such as those shown in Figures 1.11 and 1.12 to avoid low-temperature

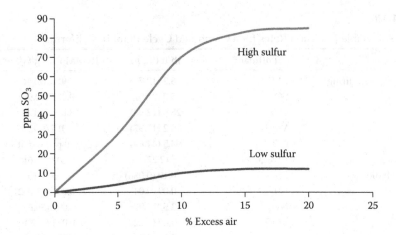

FIGURE 1.9

Effect of excess air on SO_3 formed in oil-fired boilers.

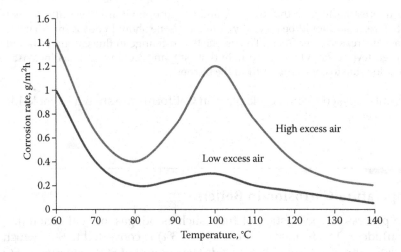

FIGURE 1.10

Typical corrosion rate versus cold-end temperature.

TABLE 1.19

SO_3 in Flue Gas, ppm

	Sulfur, %	0.5	1.0	2	3.0	4.0	5.0
Fuel	Excess air,%						
Oil	5	2	3	3	4	5	6
	11	6	7	8	10	12	14

Source: Ganapathy, V., *Oil Gas J.*, April 1978.

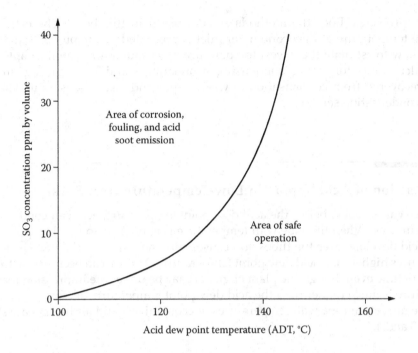

FIGURE 1.11
The relationship between SO_3 and acid dew point temperature. (Courtesy of Land Combustion Inc., Derbyshire, United Kingdom.)

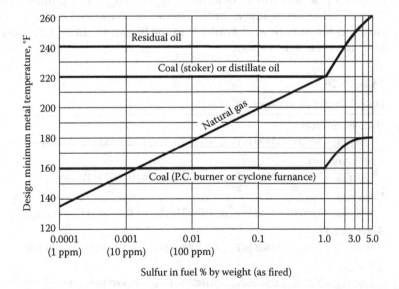

FIGURE 1.12
Limiting tube metal temperatures to avoid external corrosion in economizers and air heaters when burning fuels containing sulfur.

corrosion problems. Both the methods are discussed in this book. The estimation of tube wall temperature at the economizer inlet is presented in Chapter 3, while Chapter 4 shows how to estimate the lowest temperature in a tubular air heater. Chapter 6 illustrates calculations for sizing condensing economizers and determining amount of energy recovered from condensation of water vapor and also discusses materials suitable for condensation service.

Condensation of Acid Vapors in Low-Temperature Heat Sinks

When flue gas is cooled below the acid dew point temperature, acid can condense on the heating surfaces. When the tube wall temperature is at or close to the dew point, a small film of acid develops over the tubes and causes corrosion, though the flue gas may be at temperatures higher than acid dew point temperature. Damage can occur to the tubes over a period of time even though the plant engineer may be under the impression that the gas temperature is much higher than the acid dew point temperature [8]. Examples for calculating the minimum tube wall temperatures in economizers and air heaters are shown in Chapter 3 and 4.

Acid Dew Point Temperature T_{dp}

First one should have an idea of the acid dew point temperature T_{dp} to find out if corrosion is a possibility. A practical way to determine T_{dp} is using a dew point meter. Another method is to use correlations available in the literature. Presented in the following text are a few correlations for sulfuric acid available in the literature. Table 1.20 shows the dew point correlations for a few acid vapors.

TABLE 1.20

Dew Points of Acid Gases[a]

Hydrobromic acid

$1000/T_{dp} = 3.5639 - 0.1350 \ln P_{H_2O} - 0.0398 \ln P_{HBr} + 0.00235 \ln P_{H_2O} \ln P_{HBr}$

Hydrochloric acid

$1000/T_{dp} = 3.7368 - 0.1591 \ln P_{H_2O} - 0.0326 \ln P_{HCl} + 0.00269 \ln P_{H_2O} \ln P_{HCL}$

Nitric acid

$1000/T_{dp} = 3.6614 - 0.1446 \ln P_{H_2O} - 0.0827 \ln P_{HNO_3} + 0.00756 \ln P_{H_2O} \ln P_{HNO_3}$

Sulfurous acid

$1000/T_{dp} = 3.9526 - 0.1863 \ln P_{H_3O} + 0.000867 \ln P_{SO_2} - 0.000913 \ln P_{H_2O} \ln P_{SO_2}$

Sulfuric acid

$1000/T_{dp} = 2.276 - 0.0294 \ln P_{H_2O} - 0.0858 \ln P_{H_3SO_4} + 0.0062 \ln P_{H_2O} \ln P_{H_2SO_4}$

Source: Okkes, A.G., *Hydrocarbon Process.*, July, 1987; Hsiung, K.Y., *Chem. Engineer.*, 1981, 127; Ganapathy, V., *Oil Gas J.*, 1978, 105.

[a] T_{dp} is dew point temperature (K), and P is the partial pressure (mmHg). Compared with published data, the predicted dew points are within about 6 K of actual values except for H_2SO_4, which is within about 9 K.

Example 1.19

Flue gas analysis in a boiler is as follows: % H_2O = 12, SO_2 = 0.02, HCl = 0.0015, and rest oxygen and nitrogen. Gas pressure = 250 mm wc. Compute the dew points of sulfuric and hydrochloric acids given that 2% of SO_2 converts to SO_3.

To use the correlations in Table 1.20, one must convert the gas pressure to mmHg (mm mercury).

250 mm wc = 250/13.6 = 18.38 mm Hg. Absolute pressure = 760 + 18.38 = 778.38 mm Hg or 778.38/760 = 1.02457 atm abs.

p_w = 0.12 × 1.02457 × 760 = 93.44 mm Hg. ln p_w = 4.537

p_{HCL} = 0.0015 × 778.38 = 0.1168 mmHg. ln p_{HCl} = −2.1473

p_{SO3} = 0.02 × 0.0002 × 778.38 = 0.0031 mmHg. ln p_{SO3} = −5.7716

For HCl, T_{dp} = 3.7368 − 0.1591 × 4.537 + 0.0326 × 2.1473 − 0.00269 × 4.537 × 2.1473 = 3.0588 or T_{dp} = 327 K = 54°C = 129°F. Note that HCl dew point is low and close to water dew point, while sulfuric acid dew point is much higher.

For sulfuric acid (Verhoff correlation),

$1000/T_{dp}$ = 2.276 − 0.0294 × 4.537 + 0.0858 × 5.7716 − 0.0062 × 4.537 × 5.7716 = 2.4755 or T_{dp} = 404 K = 131°C = 268°F.

The following correlation for sulfuric acid was proposed by Okkes [5]:

$$T_{dp} = 203.25 + 27.6 \log p_w + 10.83 \log p_{SO3} + 1.6 (\log p_{SO3} + 8)^{2.19}$$

$$p_{SO3} = 0.0031 \text{ mmHg} = 4.1 \times 10^{-6} \text{ atm. Log } p_{SO3} = -5.3872$$

$$p_w = 93.44 \text{ mmHg} = 0.1229 \text{ atm. Log } p_w = -0.9104$$

$$T_{dp} = 203.25 - 27.6 \times 0.9104 - 10.83 \times 5.3872 + 1.6(2.6128)^{2.19} = 128.4°C = 263°F$$

Haase and Borgman proposed the following correlation based on experimental data [3]:

T_{dp} = 255 + 27.6 log (PH_2O)+18.7 log (PSO_3) where T_{dp} is in °C, gas partial pressures in atm.

Using this, we get T_{dp} = 255 − 27.6 × 0.910 − 18.7 × 5.387 = 120°C = 264°F.

Once an idea of the acid dew point temperature is obtained, one may proceed to compute the water dew point also as discussed later and evaluate the energy recoverable. Suitable materials such as alloy 22 steels or 304 stainless steels may be selected for the exchanger depending on the acid concentration. Calculation of tube wall temperatures at the cold end of an economizer is shown in Chapter 3 using which method one may establish how many rows of tubes are prone to condensation of acid or water vapor. Similarly, calculation of cold-end temperatures at normal and part loads is shown for a tubular air heater in Chapter 4. Plant engineers may perform similar calculations and ensure that the lowest tube wall temperatures are above the dew point of acid or water vapors to ensure that no corrosion occurs. The other option is to select materials to minimize corrosion for such equipment. This is discussed in Chapter 6.

References

1. V. Ganapathy, *Applied Heat Transfer*, Pennwell Books, Tulsa, OK, 1982, p13.
2. Babcock and Wilcox, *Steam: Its Generation and Use*, 38th edn., New York, 1978.
3. V. Ganapathy, *Industrial Boilers and HRSGs*, CRC Press, Boca Raton, FL, 2003, p238.
4. ASME, Power Test Code. Performance test code for steam generating units, PTC 4.1, New York, ASME 1974.
5. AG Okkes, Get acid dew point of fluegas, *Hydrocarbon Processing*, July 1987.
6. K.Y. Hsiung, Predicting dew point of acid gases, *Chemical Engineering*, Feb. 9,1981, p127.
7. V. Ganapathy, Estimate combustion gas dew point, *Oil and Gas Journal*, April 1978, p105.
8. V. Ganapathy, Cold end corrosion—Causes and cures, *Hydrocarbon Processing*, Jan. 1989, p57.

2

Steam Generator Furnace Design

The furnace of a modern high-pressure boiler is the most important part of the steam generator and may well be compared to the heart of the human body. The performance of the furnace affects the performance of each and every surface behind it such as superheater, evaporator, economizer, or air heater. When one talks about boiler furnace, the burner also has to be considered as an integral part of it. If combustion is incomplete or poor, NO_x and CO formation may increase, adding to the cost of emission control equipment; CO formation also decreases boiler efficiency. Fires are likely in oil- and coal-fired boilers if combustion in the furnace is poor. There have been instances when oil burners in the furnace did not operate properly, resulting in oil droplets being carried over by the flue gases and deposited in the economizer or air heater, leading to fires or explosions. The discussions that follow pertain to package steam generators firing oil or gaseous fuels with a capacity ranging from 15 to 250 t/h of steam.

If the furnace is undersized, the temperature of the flue gas leaving the furnace will be high, and superheaters located downstream facing the furnace will receive a large amount of direct radiation, resulting in high tube wall temperatures and even in tube failures. If the furnace is oversized, then the exit gas temperature will be lower, and steam temperatures may not be achieved at desired loads. In coal-fired boilers or in heavy oil–fired boilers where ash is present in the fuel, the furnace exit gas temperature must be lower than the ash melting temperature; else, molten ash can deposit on heating surfaces at the furnace exit, and removing them could be a challenge as rock-like substances can form on the screen or superheater tubes blocking the flue gas flow passage, thus affecting the performance of downstream equipment. This adds to the boiler operating as well as maintenance costs. Flue gas pressure drop will increase, adding to the cost of operation. Deposits on tubes also cause high-temperature corrosion problems.

In boilers firing solid fuels, the furnace is maintained at near-zero gas pressure so that the products of combustion do not leak to the atmosphere. Forced and induced draft fans are used to achieve this. However, due to the negative pressure in the furnace all the way to the boiler exit, there can be ingress of atmospheric air in poorly designed furnaces, through openings in furnace walls and ducts; this causes heat losses and lowering of flue gas temperatures and steam temperatures. Oil- and gas-fired boilers are always of the pressurized furnace type and use only forced draft fans. Leakage of flue gas from the furnace or boiler to the atmosphere is likely if there are openings that are not sealed properly such as the intersection of a membrane wall section and a refractory-lined wall.

The lesser the refractory usage in a furnace, the better the design. With refractory-lined pressurized furnace (Figure 2.1), it is difficult to maintain a leak-proof enclosure between the refractory walls and the water-cooled tubes as a result of which flue gases can leak to the atmosphere leading to inefficiency as well as corrosion of the casing, particularly when the fuel contains sulfur compounds. The tangent tube design is a slight improvement

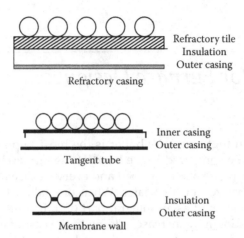

FIGURE 2.1
Furnace construction—membrane wall, tangent tube, and refractory.

over the refractory casing design, but it has the potential for leakage across the partition wall. During operation, the tubes are likely to bend or flex due to thermal expansion paving the way for leakage of flue gases from the furnace to the convection bank. This results in higher CO emissions and loss of energy; the difference in gas pressure between the furnace and convection exit can be as high as 250–400 mm wc, which can cause a lot of bypassing of flue gas from the furnace to the convection section, affecting the furnace duty.

If catalysts are used for NO_x or CO control, the pressure differential between the furnace and the convection pass can be even higher. The furnace exit gas temperature also may not be close to that predicted leading to underperformance of the superheater. Hence, present-day designs use completely water-cooled furnace with membrane walls (Figure 2.2).

FIGURE 2.2
Water-cooled membrane wall furnace showing burner openings. (Courtesy of Cleaver Brooks Inc, Engineered Boiler Systems, Thomasville, GA.)

Furnace dimensions must be such that the flame does not impinge on the walls resulting in overheating. Hence, boiler designers often discuss the furnace geometry with burner suppliers before designing the boiler and ensure that the fuels can be fired safely and efficiently with the least amount of emission of CO and NO_x. Plant and process engineers should also be involved in this process of burner selection if the plant is likely to fire a different fuel at a later date.

Advantages of Water-Cooled Furnaces

The furnace is completely water cooled except for the opening for manholes and burners as shown. The header system of feeding the front wall from the lower drum with water and taking the steam water mixture to the steam drum and the O-ring for accommodating the burner is a recent development. This design developed by Nebraska Boiler Company has been in use in several hundred boilers worldwide. The O-ring concept is also an elegant way of providing an opening for the burner compared to the complex bent tube design. In order to ensure that steam bubbles are pushed upward, twisted steel strips are placed in the O-ring. This concept is similar to the rifled tubes used in high-pressure boilers to improve the wetting of the tube periphery by water and hence increasing the limit of critical heat flux (CHF). To improve circulation in the front wall, an orifice of calculated dimension is placed in the D header. Circulation calculation for boilers with such a design is shown elsewhere.

The advantages of the completely water-cooled furnace are as follows:

1. The front, rear, and side walls are completely water cooled and are of membrane wall construction, resulting in a leak-proof enclosure for the flame. The entire furnace expands and contracts uniformly, thus avoiding casing expansion concerns. When refractory is used on the front, side, or rear wall, the sealing between the hotter membrane wall and the refractory casing is difficult, and hot gases can sometimes leak from the furnace to the outside as the furnace is pressurized. This can cause corrosion of the casing, particularly if oil fuels containing sulfur are fired.

2. Problems associated with refractory maintenance are eliminated. Also there is no need for annual shutdown for maintenance of the boiler plant to inspect or repair the refractory, thus lowering the cost of owning the boiler.

3. Fast start-up and shutdown rates are difficult with refractory-lined boiler because of the possibility of causing cracks in the refractory. However, with the completely water-cooled furnace, start-up rates are limited only by the thermal stresses in the drums and are much quicker; the evaporator and furnace tubes may be welded to the drums instead of being rolled if start-ups are frequent. With boilers maintained in hot standby conditions using steam-heated coils located in the mud drum, even 5–10 min start-ups are feasible. With a separate burner whose capacity is about 5%–7% of the main burner, the boiler can be maintained hot or at pressure. This concept has been used in a few *quick start* boilers in cogeneration plants where the boiler is required to come on line to full load within 2–3 min of firing. Hence, maintenance costs and fuel costs during each start-up are reduced, benefitting the end user in the long run.

4. Heat release rates on area basis will be lower by about 7%–15% for the water-cooled furnace for a given volume compared to the refractory-lined boiler due to the absence of refractory on front or rear walls and floor. Some oil- and gas-fired boilers designed several decades ago still have refractory on the floor, making the furnace less effective. Modifying the boiler to replace refractory with membrane wall will make the furnace more effective and lower the heat flux marginally. A lower area heat release rate (AHRR) also helps lower NO_x as can be seen from the correlations developed by a few burner suppliers.

5. Reradiation from the refractory on the front wall, side walls, and the floor increases the flame temperatures locally, which results in higher NO_x formation. Of the total NO_x generated by the burner, a significant amount is formed in the burner flame base; hence, providing a cooler environment for the flame near the flame helps lower NO_x.

6. Circulation of steam water mixture in all the sections of the boiler, particularly the floor tubes, was a concern decades ago, which prompted many boiler suppliers to place refractory over the floor tubes. With better furnace designs, operational experience, and advanced burner designs, refractory is not used nowadays.

   Heat fluxes in packaged boilers are generally low compared to those of utility boilers. To further protect the floor and roof tubes, tubes are inclined about 3°–5° to the horizontal. Also considering the low steam pressure (less than 110 barg), circulation has never been an issue as evidenced by the operation of hundreds of boilers firing oil and gas and generating up to 200 t/h of steam. The tube-side velocities are also reasonable to ensure that stagnation does not occur in the tubes, and hence, refractory on the floor is not required. Besides, burner suppliers have improved their design significantly in the past decade, and concerns about oil spilling on the floor causing local burning have been addressed. Furnace dimensions are also chosen such that the flame does not impinge on the furnace enclosure.

7. Package boilers today do not use air heaters for heat recovery. Economizers alone are used. Use of ambient air versus hot air lowers the combustion temperature in the furnace and thus lowers NO_x. The air- and gas-side pressure drops are also lower without the air heater, thus reducing the fan size and operating cost.

Ring Header Design

The ring header concept for burner opening has been developed by Nebraska Boiler Company (Figure 2.3). There are several hundred units operating with this design worldwide for the past three decades. Engineers and plant operators who have been exposed to this design sometimes wonder about the circulation and reliability of the front wall.

The O-ring header design for burner opening in fact offers lesser resistance to the flow of the steam–water mixture in the front wall compared to the complex tube bending configuration offered by several boiler manufacturers. Lesser resistance to flow means better circulation.

The following discussion is only on the losses in the ring header and not the inlet–outlet header system; the aforementioned system has also simplified the front and rear wall design by using D-header pipes that are custom sized; other designs have much more complicated way of feeding front wall and taking out the steam–water mixture back to the drum and hence offer more resistance to circulation.

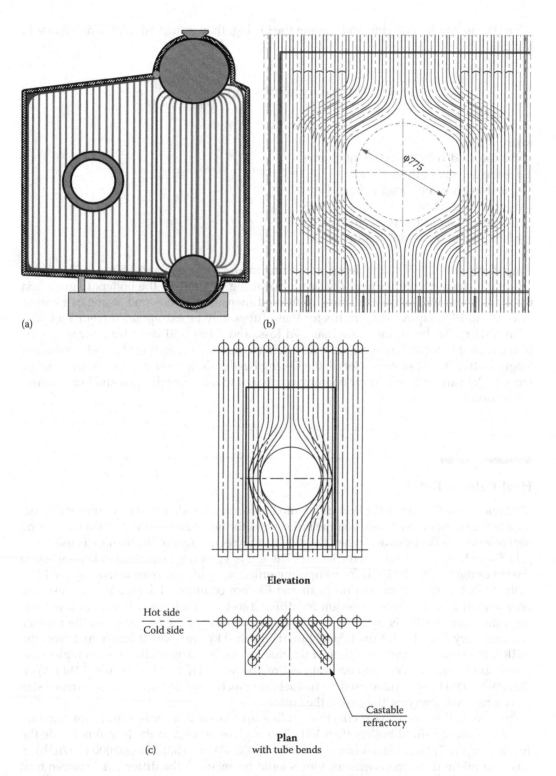

FIGURE 2.3
(a–c) O-ring and typical tube bending for burner opening. (Courtesy Thermodyne Technologies, Chennai, India.)

For the riser tubes entering and leaving the O-ring, there is just an inlet and exit loss L, which is

$$L = 1.5\ VH\ (velocity\ head) = 1.5\ V^2 \frac{\rho}{(2g)} = \frac{fL_e\rho V^2}{(2gd)}$$

where
   V is the fluid velocity in the tubes, m/s
   ρ is the density of fluid, kg/m^3
   L_e is the equivalent length of bend, m
   d is the tube ID, m
   f is the friction factor, typically 0.02–0.023
   g is the acceleration due to gravity, m/s^2

Simplifying this, L_e = 1.5 d/f. Hence, the O-ring incurs a loss of about 1.5 × d/0.02 = 75d.

If one looks at the tubes near the burner opening in some of the boilers (Figure 2.3a through c), you will see a lot of two- to three-dimensional bends and at angles ranging from 60° to 180°. Equivalent lengths for these tubes may be extrapolated from Table B.7. Even a sharp 90° bend has more than 57d loss, and there will be at least three to four of them in the critical tube region near their burner, far exceeding the 75d equivalent length in the O-ring system. Hence, the O-ring header design offers a much lower resistance to flow around the burner opening compared to the other designs and hence better circulation.

Heat Release Rates

Heat release rate is simply the heat input into the furnace divided by the volume of the furnace or the surface area. When volume is the basis, the heat release rate is called volumetric heat release rate (VHRR), and when the effective projected area of the furnace is used, it is called area heat release rate. In the United States, it is a practice to use higher heating value (HHV) of the fuel as the basis for computing efficiency and heat release rates, while LHV is the basis for these terms in European and Eastern countries. Furnace heat release rate on volumetric basis is more relevant for difficult-to-burn fuels such as coal or low heating value fuels. VHRR is indicative of the residence time of the flue gases in the furnace and may vary from 155 kW/m^3 (15,000 Btu/ft^3 h) to 310 kW/m^3 (30,000 Btu/ft^3 h). Lower the VHRR, the more the residence time of the flue gas in the furnace, the more complete the combustion process. In oil- and natural gas-fired boilers, VHRR as high as 620–1240 kW/m^3 (60,000–120,000 Btu/ft^3 h) are used as the fuels are much easier to burn and the furnace size can be reduced, lowering the cost of the boiler.

Furnace AHRR on furnace effective cooling area basis is a more significant parameter in oil- and gas-fired boilers than VHRR as it gives an idea of the heat flux inside the furnace tubes. Typical values range from 316 to 632 kW/m^2 (100,000–200,000 Btu/ft^2 h) in gas- and oil-fired steam generators. One should be aware of the difference between heat release rates and furnace duty. Furnace duty is the energy absorbed by the furnace, while heat release rate is the energy inputted into the furnace on unit area or volume basis.

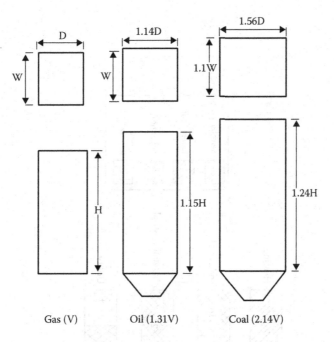

FIGURE 2.4
Impact of fuel on furnace size.

Only about 30%–40% of the net heat input is absorbed in the furnace, and the balance is absorbed in the boiler convective heating surfaces downstream of the furnace. The furnace duty also gives an idea of the heat flux inside the furnace tubes. A high AHRR also increases NO_x formation and the furnace exit gas temperature. Figure 2.4 compares the furnace sizes for different fuels. Due to the larger residence time required for combustion, solid fuel–fired furnaces will be larger than those firing natural or oil. Oil-fired furnaces will be slightly larger than that of natural gas–fired furnaces due to the higher heat flux with oil combustion process.

Steam Pressure

Steam pressure also affects the furnace size and, of course, the boiler arrangement. If the pressure is low, then the latent heat of steam is high, and hence, the furnace and evaporator heating surface will be larger compared to the case where steam pressure is very high as in utility boilers. Figure 2.5a shows the distribution of energy as a function of steam pressure. Since the total sensible or liquid plus latent heat of steam is absorbed in the economizer, evaporator tubes, and furnace, proper proportioning of duty is a must among these surfaces. Very high-pressure boilers may have simply a furnace followed by superheaters and an economizer as the latent heat is small (Figure 2.5b), while low-pressure boilers will have a furnace, a large evaporator (sometimes called a boiler bank), and an economizer (Figure 2.5c and d). To minimize furnace size, the evaporator surface is sometimes added in the furnace itself as water platens.

Figure 2.5b shows a steam generator for low- to medium-pressure steam generation (1500–5000 kPa). A large evaporator in the form of boiler bank tubes is used. The latent heat is absorbed in the furnace and in the bank tubes, which is a three-pass evaporator.

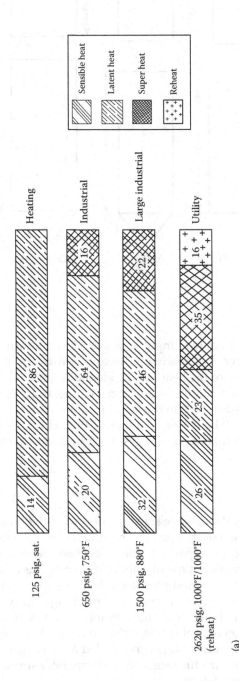

(a)

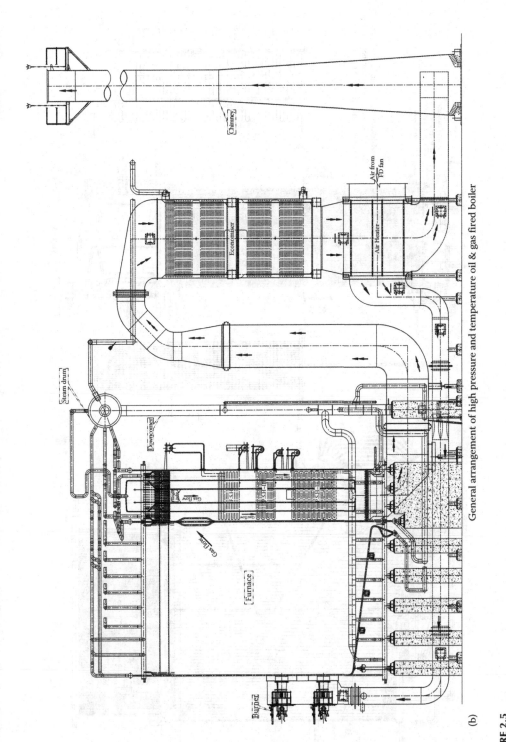

(b)

General arrangement of high pressure and temperature oil & gas fired boiler

FIGURE 2.5

(a) Distribution of energy in boilers as a function of steam pressure. (b) High-pressure steam generator. (Courtesy of ISGEC Heavy Engineering Ltd, India.)

(Continued)

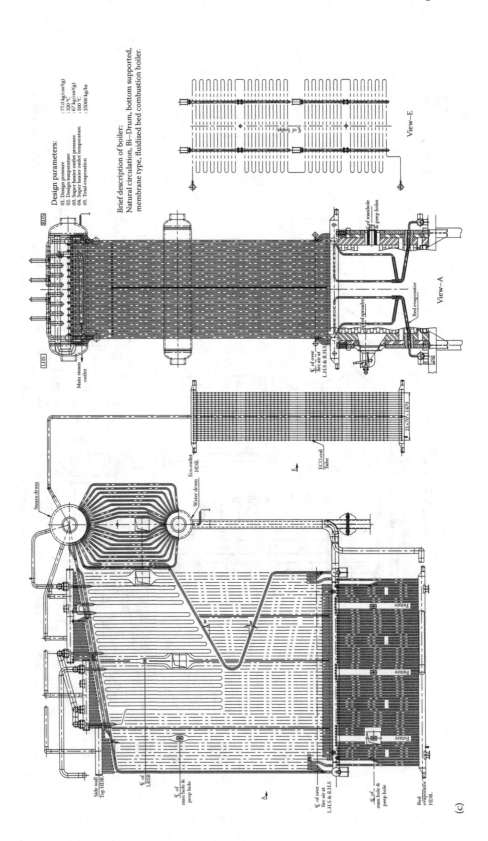

Design parameters:

01. Design pressure : 77.0 kg/cm²(g)
02. Design temperature : 320 °C
03. Super heater outlet pressure : 67 kg/cm²(g)
04. Super heater outlet temperature : 500 °C
05. Total evaporation : 25000 kg/hr

Brief description of boiler:
Natural circulation, Bi–Drum, bottom supported, membrane type, fluidised bed combustion boiler.

(c)

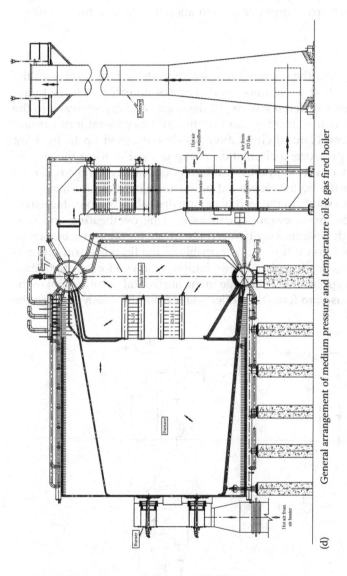

(d)   General arrangement of medium pressure and temperature oil & gas fired boiler

FIGURE 2.5 (*Continued*)
(c) Steam generator for medium-pressure steam. (Courtesy of Thermodyne Technologies, Chennai, India.) (d) Low-pressure steam generator with large evaporator. (Courtesy ISGEC Heavy Engineering Ltd, India.)

The liquid heat or the energy absorbed in the economizer is not much compared to high-pressure boilers, and hence, it is small.

Figure 2.5c shows a steam generator for a high-pressure boiler (11,000 kPa). The liquid heat (or the energy absorbed in the economizer) is large, and hence, we see a large economizer. The latent heat is small, and hence, we have only the furnace to absorb the latent heat and a large superheater for the superheat. Since the design pressure is rather high, an external steam drum is used with high ligament efficiency in order to reduce its thickness. Locating the drum up above also helps in improving circulation through the furnace tubes.

Circulation Systems

Furnaces may be classified based on the circulation of steam–water mixture through the evaporator tubes. Natural circulation is common as it requires no external force to aid circulation (Figure 2.6). The density difference between the colder and denser water in the downcomers and the hotter less dense steam–water mixture in the evaporator or furnace tubes aids the circulation process. Natural circulation systems are used up to 165 barg (2400 psig) steam. Beyond this pressure, a circulation pump is used to force the steam–water mixture through the furnace tubes. In once-through boilers, there is no circulation system, and the water entering the boiler comes out as steam. Flow velocity inside the tubes of about 1–1.5 m/s and zero solids in the feed water are desirable to avoid stagnation of flow and deposition of solids inside evaporator tubes causing overheating and possible tube failures. Once-through designs may be used at any steam pressure, but above the critical pressure of steam, only once-through is possible as the liquid–steam phase is absent. Forced circulation designs are described later. Chapter 3 shows a forced circulation steam generator used in oil field application. Here due to the horizontal arrangement of the furnace, circulation pumps are used to force the boiler water from the drum through the furnace tubes back to the drum.

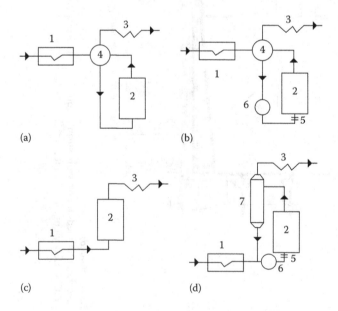

FIGURE 2.6
Boiler circulation methods. (a) Natural; (b) forced circulation; (c) once-through; and (d) once-through with superimposed circulation. *Note:* (1) Economizer; (2) furnace; (3) superheater; (4) drum; (5) orifice; (6) circulating pumps.

Furnace Exit Gas Temperature Evaluation

Furnace has been often described as the heart of the boiler. Hence, its sizing assumes paramount importance. Too large a furnace would be expensive. The furnace exit gas temperature will also be low resulting in larger surface area requirements for superheaters and evaporators downstream of the furnace. Too small a furnace can cause overheating of superheater tubes, and if ash-containing oil fuels are fired, slagging or melting of ash can occur causing high-temperature corrosion problems in the superheaters. Hence, the furnace duty or energy absorbed by steam in the furnace, which also fixes the furnace exit gas temperature, should be evaluated with care. Note that the determination of furnace exit gas temperature is not an accurate science and has to be linked with the type of furnace, excess air, flue gas recirculation (FGR) rate used, location of burners, and boiler load. Hence, there is no universally accepted procedure for this calculation.

For the process and plant engineers, a simplified procedure for estimating the furnace performance is given here. Energy balance yields the following equation:

$$Q_f = A_p \in_w \in_f \sigma(T_g^4 - T_o^4) = W_f \times LHV \times (1 - \text{casing loss-unaccounted loss}) - W_g \times h_e \quad (2.1)$$

where
Q_f is the energy transferred to the boiler furnace, kW
W_f, W_g is the fuel, flue gas flow, kg/s
LHV is the lower heating value of fuel, kJ/kg
h_e is the enthalpy of flue gas at furnace exit, kJ/kg
A_p is the effective projected area of the furnace cooling surface, m^2
$\in w, \in f$ is the emissivity of wall and flame
σ is the Steffan–Boltzman constant = $5.67 \times 10^{-8} \, W/m^2 \, K^4$
T_g, T_o is the furnace gas temperature, and furnace tube wall temperature, K

T_g is defined in a few ways. Some use the furnace exit gas temperature itself for T_g, while some others use the adiabatic combustion temperature. A reasonable agreement between measured and predicted values of the furnace exit gas temperature prevails when T_g is taken as the furnace exit gas temperature plus 160°C–175°C. Using the procedure and example discussed in Appendix A, one may arrive at the furnace exit gas temperature from economizer exit gas temperature. A correlation may be developed based on heat input and the furnace exit gas temperature for a given type of furnace.

The emissivity of wall $\in_w$ varies depending on whether the furnace surfaces are clean or covered with slag. It can vary from 0.6 to 0.9. Soot blowing also changes this value considerably. Flame emissivity $\in_f$ is given by

$$\in_f = \beta(1 - e^{-KPL}) \quad (2.2)$$

where
P is the gas pressure in atmospheres
L is the beam length of the furnace, m
K is the attenuation factor that depends on the fuel type and presence of ash and its concentration. For a nonluminous flame,

$$K = (0.8 + 1.6P_w) \times (1 - 0.00038T_g) \times \frac{(P_c + P_w)}{\left[(P_c + P_w)L\right]^{0.5}} \quad (2.3a)$$

P_c and P_w are partial pressures of CO_2 and H_2O in atm. For a semiluminous flame, the ash particle size and concentration enter into the calculations.

$$K = \left(0.8 + 1.6P_w\right) \times \left(1 - 0.00038T_e\right) \times \frac{\left(P_c + P_w\right)}{\left[\left(P_c + P_w\right)L\right]^{0.5}} + \frac{7\mu}{\left(d_m^2 T_e^2\right)^{0.33}} \qquad (2.3b)$$

where
    d_m is the mean effective diameter of ash particles in microns
    μ_m = 13 for coals ground in ball mills, 16 for coals ground in medium- and high-speed
        mills, and 20 for coals ground in hammer mills
    μ is the ash concentration in g/Nm^3

β = flame filling factor = 1.0 for nonluminous flames, 0.75 for luminous sooty flames of liquid fuels, and 0.65 for luminous and semiluminous flames of solid fuels. For a luminous oil or gas flame, $K = 0.0016T_e - 0.5$, where T_e, the furnace exit gas temperature, is in K.

These equations give only a trend or an approximation for the evaluation of furnace exit gas temperature. A wide variation between measured and estimated furnace exit gas temperature can exist due to the combustion process, location of the burners, type of firing (corner or front wall), how large the furnace is, and how much of it is filled with the flame whether there are soot particles in the fuel to mention a few. The unfilled portions are subject to gas radiation only, and the emissivity (0.15–0.3) can be far less than that estimated for a luminous flame.

In practice, many flames are a combination of luminous and nonluminous portions. The emissivity then is obtained by combining the two.

$\in_f = m\in_l + (1 - m)\in_n$ where $\in_f, \in_l, \in_n$ are emissivity of flame, luminous, and nonluminous portions and m is a coefficient. For natural gas, m = 0.1, while for fuel oils, it is 0.55. For solid fuels, since the flame is bright all over the furnace, m = 1.

Charts are available to estimate the furnace exit gas temperature directly based on furnace net heat input and its effective projected cooling area as shown later. However, the charts are developed based on furnace type and burner arrangement and evaluation of field data as discussed in Appendix A. Hence, each boiler manufacturer may have his chart or correlations for furnace performance evaluation, and no universal procedure exists as in the case of estimation of convective heat transfer coefficients. Furnace performance evaluation is described by a few as *black magic* and an art. The following chart shows how the furnace exit gas temperature varies with different fuels, and Figures 2.7 and 2.8 show how the furnace exit gas temperature varies with net heat input for different fuels. Corrections have to be made for various factors such as excess air, FGR, and burner location based on field data and experience.

Example 2.1

Determine the furnace exit gas temperature when natural gas (methane = 97%, ethane 3% by volume) is fired in a boiler furnace at 15% excess air. The net heat release rate per effective area is 250 kW/m^2. Furnace dimensions are approximately 2.44 m wide, 3.35 m high, and 9.75 m long (EPRS = 127 m^2). Flue gas analysis % volume CO_2 = 8, H_2O = 18, N_2 = 71.5, O_2 = 2.5. The tube walls may be assumed to be at 660 K. Its method of estimation is discussed later.

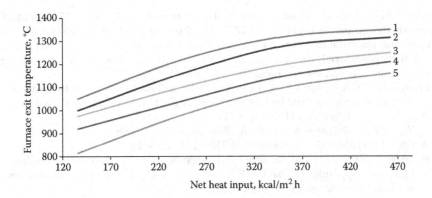

FIGURE 2.7
Typical furnace exit gas temperatures for some fuels: (1) pulverized coal, (2) natural gas, (3) fuel oils, (4) grate fired bagasse, (5) grate fired coal. (1 is the topmost curve and 5 is the bottommost curve.)

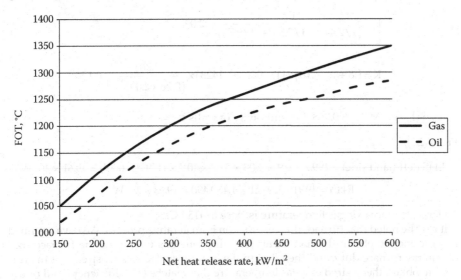

FIGURE 2.8
Furnace exit gas temperature for oil and natural gas. (From Ganapathy, V., *Industrial Boilers and HRSGs*, CRC Press, Boca Raton, FL, 2001, p112.)

The calculation of net heat release rate itself involves the following steps:

- Calculation of boiler duty (energy absorbed by steam) in kW.
- Calculation of boiler efficiency (see Chapter 1), from which the fuel flow is obtained.
- Net heat input (fuel flow × LHV).
- Net heat input divided by furnace effective projected radiant surface area (water-cooled surfaces). (For surfaces covered with refractory, a multiplication factor of 0.1–0.2 is used.)

Assume one has done all the calculations given earlier. From combustion calculations (Chapter 1), it can be shown that the HHV of the fuel gas = 55,345 kJ/kg and its LHV = 49,915 kJ/kg.

It may be shown that each GJ of natural gas fired on HHV basis requires a theoretical amount of 314 kg of dry air (Table 1.2). This amount of heat input requires

$1 \times 10^6/55,345 = 18.07$ kg of fuel. So ratio of dry air to fuel is $314/18.07 = 17.37$, or at 15% excess air, dry air per kg fuel is $1.15 \times 17.37 = 19.97$ or wet air about 20 kg/kg fuel as air always contains some moisture. Hence, the ratio of flue gas to fuel is $19.97 + 1 = 20.97$. If wet air is considered, the ratio of flue gas to fuel W_g/W_f is about 21. From energy balance in the furnace, we can write the following:

From (2.1), $Q_f = A_p \epsilon_w \epsilon_f \sigma(T_g^4 - T_o^4) = W_f$ LHV $- W_g h_e$ (heat losses are on the order of 0.5% in oil- and gas-fired boilers and hence neglected). Dividing by W_f, we have

$A_p \epsilon_w \epsilon_f \sigma (T_g^4 - T_o^4)/W_f =$ LHV $- W_g h_e/W_f$ or

$Q_f/A_p = 250,000$ W/m$^2 = W_f \times$ LHV$/A_p$. Rearranging the terms,

$A_p/W_f =$ LHV$/250,000 = 49,915,029/250,000 = 199.7$ m^2 s/kg

$P_c = 0.08$, $P_w = 0.18$. Assume furnace exit gas temperature $T_e = 1150°C$. $h_g = 1430.49$ kJ/kg (see Appendix F). Use $T_g = 1150 + 170 = 1320°C = 1593$ K (170°C more than the FEGT [furnace exit gas temperature])

$$K = (0.8 + 1.6P_w) \times (1 - 0.00038Te) \times \frac{(P_c + P_w)}{\left[(P_c + P_w)L\right]^{0.5}}$$

$$L = \frac{1.7}{(1/2.44 \ + \ 1/3.35 \ + \ 1/9.75)} = 2.09 \text{ m}$$

$$K = (.8 + 1.6 \times .18) \times (1 - .00038 \times 1423) \times \frac{(0.26)}{(0.26 \times 2.09)^{0.5}} = 0.176$$

$\epsilon_f = (1 - e^{-.176 \times 2.09}) = 0.308$. Let emissivity of walls $\epsilon_w = 0.9$

From (2.3),

LHS (left-hand side) $= 199.7 \times 0.9 \times .308 \times 5.67 \times 10^{-8} \times (1593^4 - 660^4) = 19.61 \times 10^6$ W

RHS $= 49,915,029 - 21 \times 1,430,490 = 19.93 \times 10^6$ W

Hence, furnace exit gas temperature is close to 1150°C.

It may be noted that furnace duty estimation is an iterative process. We have assumed an efficiency to obtain the fuel input. In reality, one has to assume the furnace size, obtain the furnace duty, and then design the heating surfaces downstream of the furnace to obtain the desired exit gas temperature that matches the efficiency used or use the efficiency obtained in the earlier calculations. A well-written computer program is the solution. The earlier example is to explain the methodology only.

The furnace net heat input $= 250 \times 127 = 31,750$ kW. Fuel input $= 31,750/49,915 = 0.636$ kg/s. Flue gas generated $= 0.0636 \times 21 = 13.357$ kg/s

The furnace duty $= 31,750 - 13.357 \times 1430.49 = 12,642$ kW. The average heat flux $= 12,642/127 = 99.5$ kW/m^2

Example 2.2

In the same furnace if no. 2 oil at 15% excess air is fired, then determine the exit gas temperature. Flue gas analysis is $CO_2 = 12$, $H_2O = 12$, $N_2 = 73.5$, $O_2 = 2.5$. Fuel oil LHV $= 43,050$ kJ/kg and HHV $= 45,770$ kJ/kg. Net heat input is the same as before, namely, 250 kW/m^2.

Solution

$A_p/W_f =$ LHV$/250,000 = 43,050 \times 10^3/250,000 = 172.2$ m^2 s/kg

Assume furnace exit gas temperature is 1050°C $= 1323$ K.

$$K = (0.8 + 1.6 \times 0.12) \times (1 - 0.00038 \times 1323) \times 0.24/(0.24 \times 2.09)^{0.5} = 0.167$$

For theoretical combustion of fuel oil, 1 GJ heat put requires 320 kg of dry air (see Chapter 1). Or $1 \times 10^6/45{,}770 = 21.85$ kg of fuel requires 320 kg dry air. Or ratio of flue gas to fuel at 15% excess air $= 1.15 \times 320/21.85 + 1 = 17.84$ or say 18 kg/kg wet air.

For a luminous flame, the emissivity consists partly of luminous and a nonluminous portion. Then $\epsilon_f = 0.55 \times \epsilon_l + .45 \times \epsilon_{nl}$

$$K = .0016 \times 1323 - .5 = 1.616. \; \epsilon_l = 0.75 \times \left(1 - e^{-1.616 \times 2.09}\right) = 0.724$$

$\epsilon_{nl} = (0.8 + 1.6 \times .12) \times (1 - .00038 \times 1323) \times .24 / (.24 \times 2.09)0.5 = 0.167.$
Hence, $\epsilon_f = 0.55 \times 0.724 + .45 \times 0.167 = 0.473$

$$\text{LHS} = 172.2 \times 0.473 \times 0.9 \times 5.67 \times (14.93^4 - 6.6^4) = 19.86 \times 10^6 \text{ W}$$

(Here again, we added 170°C to the assume FEGT.)

$$\text{RHS} = 43.05 \times 10^6 - 18 \times 1{,}296{,}280 = 19.72 \times 10^6 \text{ W}$$

Hence, furnace exit gas temperature is close to 1050°C.

This is only an estimate. It is preferable to use charts developed based on field data and experience as shown in Figure 2.8. Furnace exit gas temperature is a function of so many variables as mentioned earlier. Hence, experience and field testing can provide more reliable data for a particular furnace configuration. One may also work backward from economizer exit and use the field data to establish T_e as discussed in Appendix A.

Once the furnace exit gas temperature is estimated, the furnace duty has to be computed. This is discussed later.

Empirical Formula for Furnace Duty Estimation

In European and Russian boiler calculation practice, the Gurvich method has found wide acceptance. The following equation relates the furnace exit gas temperature T_e with combustion temperature T_c [2].

$$\frac{T_e}{T_c} = \frac{B_o^{0.6}}{\left(Bo^{0.6} + Ma_f^{0.6}\right)} \tag{2.4}$$

where
T_e, T_c are the furnace exit gas temperature and adiabatic combustion temperatures
M is an empirical factor impacted by fuel and gas temperature profile in the furnace
B_o is the Boltzman number

$$B_o = \frac{\varphi W_f V_c}{(\sigma \psi A p T_e)} \tag{2.5}$$

where
V_c is the average heat capacity of the flue gas in the furnace in the range of $(T_c - T_e)$ formed by 1 kg of fuel burned, kJ/kg K
φ is the furnace efficiency or (1-% heat loss/100) and is typically 0.995
σ is the Stefan–Boltzman constant $= 5.67 \times 10^{-11}$ kW/(m^2 K^4)
ψ is an empirical factor $= x\zeta$
ζ is 0.55 for fuel oil and 0.65 for gas

For coal firing, it varies from 0.35 to 0.55. For refractory-covered surfaces, it is 0.1–0.2. ζ is corrected by a factor y dependent on the furnace wall tube spacing to diameter. For pitch to diameter S_T/d of 1.25, it is 0.97, and for S_T/d of 2, it is 0.82. If the furnace has different sections with different tube spacing or partly covered with refractory, one may use the following weighted average value of ψ.

$$\psi = \Sigma yi \ Api \ \zeta i/Ap \tag{2.6}$$

$$a_f = \text{coefficient of thermal radiation} = 1/[1 + \{1/\in_f - 1\}\psi] \tag{2.7}$$

Flame emissivity $\in_f$ has been discussed earlier for nonluminous and luminous flames.

$$M = \text{a coefficient} = A - BX \tag{2.8}$$

where A and B are empirical coefficients depending on fuel fired. For gas and oil, A = 0.54 and B = 0.2. For coal, A = 0.59 and B = 0.5. X is the relative position of the highest temperature zone in the furnace (Figure 2.9a). If x_1 is the distance of burner 1 from bottom of

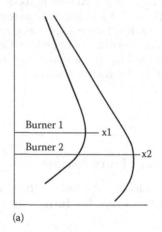

(a)

(b)

FIGURE 2.9
(a) Temperature profiles along furnace length or height (x1, x2 measured from furnace bottom). (b) CFD modeling of furnace temperature profile.

furnace and x is the distance from bottom of furnace to furnace exit, then X is the relative position = x_1/x. If the burner is in the front wall as in a D-type boiler, one may use X = 0 as x_1 is 0. Chapter 1 shows how one may estimate the adiabatic combustion temperature T_c for various fuels from excess air.

$$\frac{T_e}{T_c} = \frac{1}{[M\{5.67 \cdot 10^{-11} \psi A_p a_f T_c^3/(\varphi W_f V_c)\}^{0.6} + 1]} \tag{2.9}$$

T_c is obtained through combustion calculations as shown in Chapter 1. It is a function of excess air, combustion air temperature, and fuel analysis. W_f = fuel input kg/s.

Example 2.3

Solve Example 2.1 using Gurvich method. The boiler generates 45,372 kg/h (100,000 lb/h) steam at 25 kg/cm² g (355 psig) using 105°C (221°F) feed water. Natural gas at 15% excess air is fired as in Example 2.1. Compute also the energy absorbed in the furnace. D-type boiler with burner at front walls is used. The furnace is completely water cooled. No refractory is used. Pitch-to-diameter ratio of tubes is 1.75. LHV = 49,884 kJ/kg. Ambient air enters the furnace, and hence, additional heat input by air is not considered.

Solution

Data such as flue gas and combustion temperature and ratio of flue gas to fuel are obtained from combustion calculations (Chapter 1). Assume boiler heat losses = .5%. Hence, φ = 0.995. Furnace projected area based on using 253 kW/m² net heat input is 127 m². Adiabatic combustion temperature = 1780°C = 2053 K. Fuel quantity = 2344 kg/h = 0.651 kg/s.

Flue gas quantity = 49,216 kg/h = 13.67 kg/s. Ratio of flue gas to fuel = 49,216/234 = 21. ψ = 0.88 × 0.65 = 0.572. ϵ_f = .308 from Example 2.1. a_f = 1/[1 + {1/0.308 – 1}]0.572] = 0.437.

Assume T_e = 1150°C. From the properties of flue gases, enthalpy of gases (from Appendix F) at 1780°C and 1150°C = 565 kcal/kg and 342 kcal/kg. Hence, V_c = 21 × (565 – 342)/(1780 – 1150) = 7.43 kcal/kg K = 31.1 kJ/kg K.

M = 0.54 as the burners are on front wall.

Hence, T_e/T_c = 1/[M{5.67 × 10⁻¹¹ ψA_p a_f T_c^3/(φW_f V_c)}⁰·⁶ + 1] = 1/[0.54{5.67 × 10⁻¹¹ × 0.572 × 127 × 0.437 × 2053³/(0.995 × 0.651 × 31.1)}⁰·⁶ + 1] = 0.683 or T_e = 2053 × 0.683 = 1402 K = 1129°C. Enthalpy of flue gas at 1129°C from Appendix F is 335.5 kcal/kg = 1404.7 kJ/kg.

Hence, furnace duty Q_f = 0.651 × 49,915–21 × .651 × 1404.7 = 13,291 kW (45.36 MM Btu/h). The average heat flux is then Q_f/A_p = 13,291/127 = 104.6 kW/m² (33,055 Btu/ft² h). To estimate the fin tip temperature or CHF-related issues, the maximum heat flux considering the variation in temperature profile of the flue gas is considered.

One has to increase the furnace area in case the furnace exit gas temperature is high and is likely to cause slagging problems at the furnace exit. However, the example is shown to illustrate the methodology.

Example 2.4

Estimate the furnace exit gas temperature using the Gurvich method when fuel oil is fired in the same furnace. LHV of fuel oil = 43,059 kJ/kg. Fuel quantity = 0.755 kg/s. Use the data from Chapter 1. Fifteen percent excess air is used. Air enters at ambient temperature.

Adiabatic combustion temperature = 1866°C = 2139 K and ratio of flue gas to fuel = 17.83 from Chapter 1. Fuel quantity = 2718 kg/h = 0.755 kg/s.

ψ = 0.88 × 0.55 = 0.484. ϵ_f = 0.473 from Example 2.2. a_f = 1/[1 + {1/0.473 – 1}]0.484] = 0.65

Assume T_e = 1100°C. From the properties of flue gases, enthalpy of gases (from Appendix F) at 1866°C and 1100°C = 572 kcal/kg and 314.7 kcal/kg. Hence, V_c = 17.83 × (572 – 314.7)/(1866 – 1100) = 5.99 Kcal/kg K = 25.07 kJ/kg K. M = 0.54 as the burners are on front wall.

Hence, $T_e/T_c = 1/[M\{5.67 \times 10^{-11}\psi A_p\, a_f\, T_c^3/(\phi W_f\, V_c)\}^{0.6} + 1] = 1/[0.54\{5.67 \times 10^{-11} \times 0.484 \times 127 \times 0.65 \times 2139^3/(0.995 \times 0.755 \times 25.07)\}^{0.6} + 1]$ = 0.6267 or T_e = 2139 × 0.6267 = 1341 K = 1068°C. Enthalpy of gas at 1068°C is 304.4 kcal/kg = 1274.5 kJ/kg.

Furnace duty Q_f = 0.755 × 43,059–17.83 × 0.755 × 1274.5 = 15,352 kW (52.4 MM Btu/h).

The heat flux and furnace duty are higher on oil firing. Hence, the circulation is checked based on heat flux obtained in oil-fired case. The furnace exit gas temperature is also lower on oil firing compared to natural gas. When a boiler is fired with both fuel oil and natural gas, it is common practice to design the boiler and superheater surfaces on oil firing to ensure the desired steam temperature is obtained and, when firing natural gas, the superheater steam temperature will only be higher, which can be controlled by various methods as discussed in Chapter 3.

(Note that the terms furnace heat input and furnace duty are different. The duty is used to compute the heat flux inside the tubes and to check for departure from nucleate boiling [DNB]. This is explained later).

It may be noted that the furnace calculations give an idea of the exit gas temperature, and accuracy can vary depending on burner type and location. Boiler designers obtain field data on the performance and work backward the furnace exit gas temperature and arrive at correcting curves. Unlike convective heat exchanger calculations where the correlations for heat transfer are well established and more reliable, furnace performance evaluation is still a gray area. The error in estimation of the furnace exit gas temperature could be as high as ±100°C (180°F).

Furnace Duty with Combination Firing

Example 1.14 showed how the boiler efficiency and other pertinent data may be obtained when multiple fuels are fired in a boiler using field data. To compute boiler efficiency from field data, input the steam parameters, exit gas temperature and % oxygen values (dry or wet), fuel analysis and vary the ratio among fuels fired till the field measured oxygen of 3.46% dry is obtained. It may be possible to measure one of the fuels consumed accurately, and hence, we may cross-check the field data using this information.

The plant engineer may arrive at the furnace exit gas temperature by working backward from the economizer as shown in Chapter 3. The computation of furnace exit gas temperature may be done as follows:

1. Obtain the fuel input on LHV basis for natural gas and oil.

2. Based on the total net heat input of both fuels and the furnace projected area, obtain the furnace exit gas temperature for natural gas and oil. Since the ratio is about 70:30, the furnace exit gas temperature may be approximated as 70% of the value for natural gas and 30% of the value for oil.

3. Then the furnace duty may be estimated by the methods discussed earlier. The mixture flue gas analysis and enthalpy data are used. One may check if the furnace exit gas temperature matches with the value obtained by back calculation and if the total steam duty matches with the sum of the duties of all components. Any difference may be attributed to the estimation of furnace exit gas temperature as it has the largest uncertainty in the estimation process, more so with multiple-fuel firing.

Distribution of Heat Flux around Tubes and Fins

Furnace wall is typically of membrane wall design (Figure 2.10). The tubes are of outer diameter d with tube spacing of S_T. Furnace calculations and estimation of duty give some idea of the heat flux in the furnace based on projected area. However, this is an average value, and based on burner location, flame shape, excess air used, and field data on similar units, the boiler designer has some idea of the maximum heat flux. The ratio of maximum to average heat flux may vary from 1.2 to 1.5. Once the maximum flux is determined, one has to see how it varies around the tube periphery and along the fins. The actual flux at a location has to be obtained by multiplying the maximum flux by the view factor. Figure 2.10a shows how the view factor is estimated at any point on the tube and also along the membrane wall. Let us find out the view factor at a point on the tube defined by angle θ (Table 2.1).

Draw a tangent at the point on the circumference at angle θ to the adjacent tube as shown in Figure 2.10a. Draw a tangent at the point on the circumference on the tube itself. Now the angle φ between the two tangents is calculated.

Then view factor corresponding to angle θ is (1 + cos φ)/2.

Similarly, at any point on the fin, as shown in Figure 2.10a, draw two tangents to the adjacent tubes. If φ1 and φ2 are the angles, then

View factor on fin = (cos φ1 + cos φ2)/2

It is clearly seen that at the top of the tube facing the furnace radiation, the angle θ = 90. φ = 0. Hence, view factor = 1. At the angle θ = 0, if the fin spacing is 101 mm and tube OD = 50 mm, then the angle θ = 90 + sin⁻¹(25.4/76.2) = 109°. Then cos109° = −0.326 and view factor = (1 − 0.326)/2 = 0.337 as shown in the earlier table.

At the fin midpoint, sin φ1 = sin φ2 = 25.4/50.8 = 0.5. cos 30° = 0.866. Hence, view factor = 0.866. Using the view factors, one may estimate the local heat flux.

Distribution of Radiation to Tube Banks

When boiler screen section or superheater is located at the furnace exit, direct radiation from the furnace is absorbed by these heating surfaces and increases the heat flux inside the tubes and hence the tube wall temperature. This radiation is generally absorbed completely within the first four to five rows of tubes depending on the tube spacing. Figure 2.11 shows the distribution of external radiation to different rows depending on tube spacing. Direct radiation adds to the energy absorbed by the tubes and increases the metal temperature and may cause overheating and thermal expansion problems. Hence, one should be careful while locating superheaters at the furnace exit. When a screen section is used at the furnace exit, the tubes will not be overheated as these tubes operate at slightly above saturation temperature of steam, while a superheater operates at much higher temperature.

The following formula predicts the radiation to the tubes:

$$a = 3.14\left(\frac{d}{2S_T}\right) - \frac{d}{S_T}\left[\sin^{-1}\left(\frac{d}{S_T}\right) + \left\{\left(\frac{S_T}{d}\right)^2 - 1\right\}^{0.5} - \frac{S_T}{d}\right] \quad (2.10)$$

where a = fraction of energy absorbed by row 1. The second row absorbs (1 − a)a, the third row, [1 − {a + (1 − a)a}a], and so on.

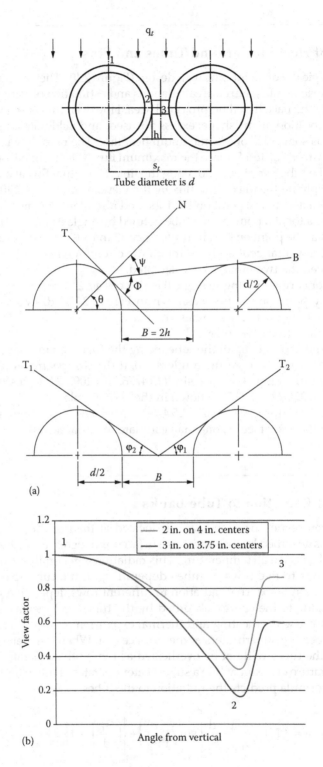

FIGURE 2.10
(a) View factor estimation. (b) View factors for 2 in OD tube on 4 in and 3 in OD tube on 3.75 in. spacing around tube periphery.

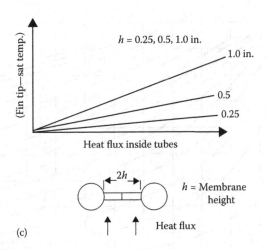

(c)

FIGURE 2.10 (*Continued*)
(c) Relating fin tip temperature to heat flux.

TABLE 2.1

View Factors for Typical Configurations of Membranes

Angle from Horizontal θ	View Factor—Case 1	View Factor—Case 2
0	0.33	0.17
15	0.51	0.37
30	0.68	0.60
45	0.82	0.79
60	0.92	0.92
75	0.98	0.98
90	1.00	1.00
Quarter point on fin	0.83	0.57
Midpoint of fin	0.86	0.60

Note: Case 1:2 in. OD tubes on 4 in. spacing. Case 2:3 in. OD tube on 3.75 in. spacing.

Example 2.5

A superheater with 50 mm OD tubes at 200 mm spacing is exposed to direct furnace radiation. Estimate the energy absorbed by the first four rows. Assume Q_r the direct radiation is 1 MW. (The radiation to the superheater may be estimated based on the heat flux and opening area of the exit plane as discussed earlier.)

Solution

$d = 50$ mm, $S_T = 200$ mm.
  $a = 3.14 \times 50/2/200 - 50/200[\sin^{-1}(50/200) + \{(200/50)^2 - 1\}^{0.5} - 200/50] = 0.3925 - 0.25(0.2526 + 15^{0.5} - 4) = 0.361$. Hence, first row absorbs 0.361 MW. (See Figure 2.11, which gives the total energy absorbed from external radiation for a certain number of rows. For two rows, the total absorbed is 0.6 MW as seen from the figure.)
  The second row absorbs $(1 - 0.361) \times 0.361 = 0.2306$ MW (total of two rows absorb nearly 0.6 MW). The third row receives $[1 - (0.361 + 0.2306)] \times 0.361 = 0.147$ MW.
  The fourth row $= [1 - (0.361 + .2306 + 0.147)] \times 0.361 = 0.094$ MW and so on.
  Typically, a minimum of four rows will absorb the external radiation completely. If the tube spacing were smaller, then a large amount of radiation is absorbed within the first few rows resulting in high heat flux to these tubes; hence, it is better to use a wider

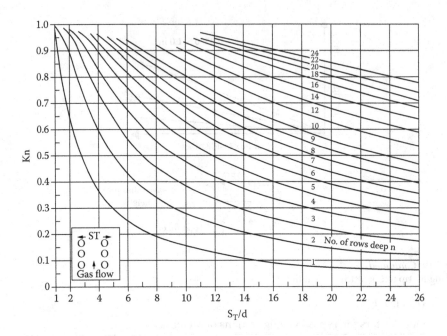

FIGURE 2.11
Distribution of external radiation to tubes.

spacing when the external radiation is large so that the radiation is spread over more rows, and heat flux is not intense in the first two or three rows. Screen tubes in boilers and heaters perform this function. Good boiler designs shield the superheater from furnace radiation by using more than six to eight rows of screen tubes. This is discussed in Chapter 3 on steam generators.

Relating Heat Flux from Furnace to Inside Tubes

The average or maximum heat flux in the furnace may be estimated using the methods suggested earlier. However, in order to check for DNB and also to estimate the boiling heat transfer coefficients, the heat flux inside the tubes is required.

From Figure 2.10,

$$q_p S_T = q_c(\pi d/2 + 2h) \tag{2.11}$$

where q_c is the average heat flux referred to the developed surface or periphery.

Then,

$$q_c \, d/d_i = q_i$$

where
d, d_i refer to tube OD and ID, mm
h is the membrane height, mm
S_T is the tube spacing, mm
q_p is the heat flux on projected area basis, kW/m^2
q_c is the circumferential heat flux, kW/m^2
q_i is the heat flux inside the tubes, kW/m^2

Simplifying,

$$q_i = q_p \, S_T d / \{(\pi d/2 + 2h)d_i\}$$

Example 2.6

If $q_p = 87.5$ kW/m^2, d = 51 mm, d_i = 44 mm, h = 15 mm, S_T = 81 mm, then

$$q_i = 87.5 \times 81 \times 51/(3.14 \times 51/2 + 30)/44 = 74.6 \text{ kW/m}^2$$

One may correct for the heat flux distribution along the flame length (see Figure 2.9 for vertical and horizontal furnaces) and use these values to check for DNB as discussed later.

External Radiation to Heat Transfer Surfaces at Furnace Exit

The calculations provided earlier give an estimate of energy absorbed by the furnace. However, there will be direct radiation to heat transfer surfaces such as superheaters or boiler screen section at furnace exit. One may estimate this as follows:

The energy absorbed by furnace from Example 1 = 12,642 kW. The furnace area = 127 m^2. The average heat flux = 99.5 kW/m^2. Let the opening at furnace exit = 3.35 × 1.2 = 4 m^2. Then approximately 99.5 × 4 = 398 kW energy is radiated to the tube section at furnace exit. However, one has to correct for the actual emissivity at the furnace exit and the smaller beam length and the distribution of heat flux along the flame length. Equations and charts are available in Ref. [7] for estimating the correction factor, and it may be seen that about 0.5–0.6 times the average flux is radiated to the surfaces beyond the furnace. Hence, the direct radiation may be taken as 250 kW, and the furnace absorbs 12,642 − 250 = 12,392 kW. If the screen section is located at the furnace exit, then this radiant energy is added to the evaporator duty; thus the external radiation does not cause any concern when the screen section absorbs this radiation as the boiling coefficient inside tubes is high and does not increase the tube wall temperature much; however, when a superheater is located at the furnace exit, then this has to be considered as contributing to the superheater duty. The first few rows of tubes depending on tube spacing will receive this external radiation and can overheat the superheater tubes. The distribution of external radiation to tube bank was discussed earlier. The first four to five rows typically absorb this radiation completely with the first row bearing the maximum brunt.

If we estimate the gas emissivity at the furnace exit and compute the external radiation, we get a different value for external radiation. From Example 2.1, P_c = 0.08, P_w = 0.18. At the furnace exit and opening to the convection bank, the beam length is different. Using a width of 2.44 m, height = 3.35 m, and length = 1.2 m,

$$L = 1.7/(1/2.44 + 1/3.35 + 1/1.2) = 1.1 \text{ m.}$$

$$K = (.8 + 1.6 \times .18) \times (1 - 0.00038 \times 1423) \times (0.26)/(0.26 \times 1.1)^{0.5} = 0.24$$

$$\epsilon_f = (1 - e^{-.24 \times 1.1}) = 0.23. \quad \text{Let } \epsilon_w = 0.9$$

Direct radiation to furnace opening to tubes at, say, 660 K = 5.67 × 10^{-11} × 0.23 × 0.9 × (1423^4 − 660^4) × 4 = 184 kW. (Opening to furnace = 3.35 × 1.2 = 4 m^2)

Let us say that we have located the radiant superheater at the furnace exit (not a good idea). Let there be 12 tubes of diameter 50.8 and length 3.35 m facing the furnace at a spacing of 100 mm. One can estimate the average heat flux in the furnace and the energy radiated at the furnace exit as discussed earlier and compute the energy absorbed row by row. The fraction a absorbed by the first row is given by

$$a = 3.14 \, (d/2S) - d/S[\sin^{-1}(d/S) + \{(S/d)^2 - 1\}^{0.5} - S/d]$$

For a $S_T/d = 2$, $a = 3.14/4 - 0.5 \times [\sin^{-1}(0.5) + 30^{0.5} - 2] = .785 - 0.5 \times [0.523 + 1.732 - 2] = 0.66$ [the value for $\sin^{-1}(0.5)$ is in radians].

The first row absorbs the maximum of 66% of the external radiation. Let 0.66 × 184 = 121 kW of energy be absorbed in the first row. Then the additional heat flux due to direct radiation is 121 × 860 = 104,438 kcal/h. If the steam generation is 40,000 kg/h, and there are 20 streams in 5 tubes/row (on each header) and 4 along gas path with 3 passes in 12 rows (see Figure B.4h for explanation of streams), flow through each tube will be about 2,000 kg/h. With a baffle after every 4 rows of tubes, the additional enthalpy absorbed by steam is (104,438/12/2,000) × 3 = 13 kcal/kg. Assuming steam-specific heat of 0.65 kcal/kg °C, the additional pickup in steam temperature and wall temperature can be at least 20°C. This can decrease the life of the superheater by several years if you check using the LMP charts for the material in consideration (see Appendix E). The tubes may be even operating well above their limits and can fail. Hence, locating the superheater at the furnace exit is not good practice as discussed in Chapter 3. Advantages of convective superheater design have been elaborated in Chapter 3. Due to mixing with slightly cooler steam in subsequent rows away from the furnace, the enthalpy total pickup by the first row could be slightly less but still a factor to be considered. If, for argument, we assume the first row facing the furnace starts out at 350°C and gets heated to 400°C, then this row can pick up an additional 20°C as shown and can cause overheating. Several superheaters located at furnace exit have failed for this reason. Also we estimate a particular furnace exit gas temperature. What if it is about 50°C higher? The radiation will be about (1473/1423)^4 or 15% more, or the temperature increase is higher by 23°C.

Direct Radiation to Superheater at Part Load

Let us see what happens to the direct radiation at part loads. Taking proportionate values, at 60% load, the NHI/EPRS value may be taken as 0.6 × 250 = 150 kW/m^2. The furnace exit gas temperature is about 1050°C from Figure 2.8. The external radiation to the first row is = 0.66 × 5.67 × 10^{-11} × 0.23 × 0.9 × (1323^4 − 660^4) × 4. = 99 kW = 85,140 kcal/h. The steam flow is 24,000 kg/h. The additional enthalpy absorbed by row 1 facing the furnace will be (85,140/12/1,200) × 3 = 18 kcal/kg or a temperature rise of 27°C (compared to 20°C at 100% load). This characteristic of radiant superheaters is shown in Figure 3.28.

That is why plant engineers should be concerned about purchasing boilers with a radiant superheater. Operating it at low loads is also a risk as discussed in Chapter 3. Unless a good screen section is used with a minimum of four to five rows, the external radiation can play havoc with the tube temperature and the life of the superheater. The other issues with radiant superheater design are discussed in Chapter 3.

Terms Frequently Used in Furnace Performance

Area heat release rate: This is the heat input on HHV basis divided by the furnace effective projected radiant surface area or also called EPRS, m². If air-heating is absent, AHRR will be fuel fired (kg/s) × HHV(J/kg)/EPRS(m²). The units are in kW/m² (Btu/ft² h)(kcal/m²h).

EPRS: Effective projected radiant surface is the total projected area of the furnace walls, which are effective (water-cooled membrane walls) less the openings for burners and man-ways. If refractory is used on some surfaces, the effectiveness is reduced by using appropriate multiplication factor of 0.1–0.2 discussed elsewhere. The opening to the convection pass is also included in this estimation of EPRS.

Volumetric heat release rate: This is fuel input on HHV basis divided by the volume of furnace (kW/m³)(Btu/ft³h)(kcal/m³h).

Furnace duty: Energy absorbed by furnace, kW (MM kcal/h)(MM Btu/h).

Net heat input: This is the heat input to the furnace based on LHV of the fuel; unaccounted and casing loss are considered in its estimation and also heat input from air if any.

Heat flux: This is the energy converted to tube side divided by inner surface area of the tubes in the furnace (kW/m²) (kcal/m²h) (Btu/ft²h). Table 2.2 gives an idea of the various terms and the units.

Estimating Boiling Heat Transfer Coefficient

An idea of boiling heat transfer coefficient helps in evaluating the tube wall temperature and the membrane tip temperature. Tube wall temperature is also required to estimate the interchange of radiation energy between the hot flue gas and the membrane walls. Subcooled boiling heat transfer coefficient inside tubes for water may be estimated by the following equations:

According to Collier

$$\Delta T = 0.0225 \; e^{(-P/86.9)} q^{0.5} \qquad (2.12)$$

TABLE 2.2

Furnace Parameters

Term	SI	British	Metric
Area heat release rate	316 kW/m²	100,000 Btu/ft²h	272,000 kcal/m²h
Volumetric heat release rate	1,034 kW/m³	100,000 Btu/ft³h	889,600 kcal/m²h
Heat flux	316 kW/m²	100,000 Btu/ft²h	272,000 kcal/m²h
Furnace duty	1,000 kW	3.413 MM Btu/h	0.860 MM kcal/h
Heating value of fuel	46,520 kJ/kg	20,000 Btu/lb	11,111 kcal/kg
Boiling coefficient	11,372 W/m²K	2,000 Btu/ft²h °F	9,780, kcal/m²h °C
Steffan–Boltzman constant	5.67×10^{-8} W/m²K⁴	0.1713×10^{-8} Btu/ft²R⁴	4.87×10^{-8} kcal/m²K⁴

Note: How to use the table: Heating value of 10,000 kcal/kg = 18,000 Btu/lb = 41,868 kJ/kg. Heat transfer coefficient of 10,000 W/m² K = 1,759 Btu/ft² h °F. Volumetric heat release rate of 70,000 Btu/ft³ h = 724 kW/m³.

and from Jens and Lottes,

$$\Delta T = 0.792 \ e^{(-P/62)} q^{0.25} \tag{2.13}$$

P is the steam pressure, bara
q is the heat flux, W/m^2

ΔT is the difference between saturation temperature and tube wall temperature, K
The boiling heat transfer coefficient is then $h_i = q/\Delta T$, W/m^2 K

Example 2.7

Determine the boiling heat transfer coefficient for steam when the heat flux inside the tubes is 189,270 W/m^2 (60,000 Btu/ft^2 h) and steam pressure = 83 bara (1,203 psia).
Using Collier,

$$\Delta T = 0.0225 \ e^{(-P/86.9)} q^{0.5} = 0.0225 \ e^{(-83/86.9)} \times (189{,}270)^{0.25} = 3.7°C$$

Using Jens and Lottes,

$$\Delta T = 0.792 \ e^{(-P/62)} q^{0.25} = 0.792 \times e^{(-83/62)} \times 189{,}270^{0.25} = 4.3°C$$

h_i varies from 189,270/3.7 = 51,154 W/m^2 K to 189,270/4.3 = 44,016 W/m^2 K (7,746 Btu/ft^2 h °F). Since boiling coefficients are so large compared to the gas-side heat transfer coefficient, the error in furnace tube wall temperature estimation or furnace performance will be marginal or negligible.

Estimating Fin Tip Temperature

Fin tip temperatures in membrane wall furnace depend on several factors such as cleanliness of water or tube-side fouling, fin geometry, heat flux tube diameter and spacing, and view factor as discussed earlier.

Assuming membranes are longitudinal fins heated from one side, the following equation may be used to determine the fin tip temperature:

$$(t_g - t_b)/\cosh (mh) = t_g - t_t \tag{2.14}$$

where t_g, t_b, and t_t are the gas temperature, fin base temperature, and fin tip temperatures, respectively, °C.

Due to the high tube-side boiling coefficients, on the order of 15,000–50,000 W/m^2 K, fin base temperatures will be a few °C higher than the saturation steam temperature, assuming fouling is minimal.

h is the membrane height, m

$$m = (h_g C/KA)^{0.5} \tag{2.15}$$

where
h_g is the gas-side heat transfer coefficient, W/m^2 K
C is the 2b + L where b = fin thickness and L = fin length, m
K is the fin thermal conductivity, W/m K
A is the cross section of fin = bL
C/A for long fins = (2b + L)/bL = 1/b

Example 2.8

In a boiler furnace, gas temperature at a location is 1200°C. Gas-side coefficient is assumed to be 170 W/m² K. Fin height = 12.7 mm, and thickness = 6.25 mm. Fin base is at 315°C. Thermal conductivity of fin = 35 W/m K. Determine the fin tip temperature.

$$mh = 0.0127 \times (170/.00635/35)^{0.5} = 0.352. \cosh(0.352) = 1.063 \quad \text{or}$$

$$t_t = 1200 - (1200 - 315)/1.063 = 367°C$$

As steam pressure increases, the ratio S_T/d is reduced to lower the fin tip temperature. It may be 2 at low pressures (<40 bara) and about 1.5 or less at higher steam pressures.

Low-pressure (1000–5000 kPa) boiler furnaces typically use membrane panels with 50.8 mm tubes at 88–101 mm pitch. The fin height will be smaller as the pressure increases due to increase in saturation temperatures and hence fin tip temperatures. High-pressure boilers (above 9000 kPa) may even use 50.8 mm tubes at 63.5 mm spacing. Figure 2.10c relates heat flux with fin tip temperatures in membrane walls.

Boiling Process

When thermal energy is applied to furnace tubes, the process of boiling is initiated. The fluid leaving the furnace tubes and going back to the steam drum is not 100% steam but is a mixture of water and steam. The ratio of the mixture flow to steam generated is known as the circulation ratio (CR). Typically, the steam quality in the furnace tubes is 5%–8%, which means that the mixture is mostly water with CR ranging from 20 to 12. As discussed later in natural circulation units, CR depends on the resistance of the various circuits and energy absorbed in each of them. In a D-type boiler, for example, the convection bank tubes and side walls have shorter tubes, while the side wall forming the D has a much longer length, and hence, the CR will be different in each parallel path.

Nucleate boiling is the process preferred in boilers. In this process, the steam bubbles generated by the energy input are removed by the mixture at the same rate so that the tubes are kept cool. The steam bubbles are contained within the water film, which keeps the tube inner surface wet and hence cool. Boiling heat transfer coefficients are very high on the order of 25,000–40,000 W/m² K (about 7,900–12,650 Btu/ft² h °F) as shown earlier. When the intensity of heat flux exceeds a value known as the CHF, then the process of nucleate boiling is disrupted. The bubbles formed inside the tubes are not removed adequately by the mixture flowing inside. The bubble formation interferes with the flow of mixture and forms a superheated steam film inside the tubes, which has a lower heat transfer coefficient and hence can increase the tube wall temperatures significantly as shown in Figure 2.12. It is the boiler designer's job to ensure that heat fluxes are kept within limits.

Generally, heat fluxes in package boilers are quite small compared to utility boilers as the furnace sizes are quite large relative to the energy input. Besides, they operate at much lower pressures, and hence, rarely does one hear of tube failures due to high heat flux or an associated problem called DNB. The actual heat fluxes in package boilers range from 126 to 222 kW/m² (40,000–70,000 Btu/ft² h), while CHF levels could be in excess of 790 kW/m² (250,000 Btu/ft² h) at the CR and steam pressures seen in package boilers. Rifled tubes are used in high-pressure boilers to ensure that tube inner surfaces are wetted by the water film. They have spiral grooves cut into their inner wall surface. The swirl flow induced by

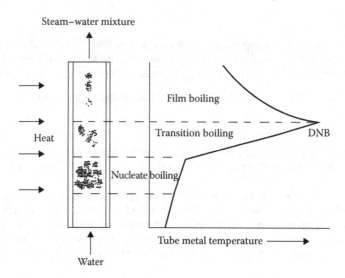

FIGURE 2.12
Boiling process and DNB in boiler tubes.

the rifled tubes not only forces more water outward on to the wall surfaces but also promotes mixing between the phases to counteract the gravitational stratification effects in a nonvertical tube. Horizontal tubes have a lower limit for CHF values as stratification of flow is a possibility compared to vertical tubes. Rifled tubes can handle a much higher heat flux about 50% more than plain tubes. As they are expensive, large utility high-pressure steam generators use them in high heat flux areas only.

At DNB conditions, the bubbles of steam forming on the hot tube surface begin to interfere with the flow of water to the tube surface and eventually coalesce to form a film of steam that blankets the hot tube inner surface. The transition from nucleate boiling to steady-state film boiling is unstable because of the sweeping away of the coalescing bubbles. As this unsteady transition approaches full film boiling, the tube wall temperature fluctuates significantly, and large temperature increases on the order of 30°C–60°C are seen as shown in Figure 2.12.

Boiler Circulation

A large percentage of boilers operating today are of natural circulation design (Figure 2.13). The difference in density between the colder water entering the boiler through the downcomers and the hotter mixture of steam and water flowing through the riser tubes drives the circulation process. Figure 2.13a shows a typical evaporator and the downcomers, risers associated with it. The cold feed water from the economizer or from a deaerator enters the drum, mixes with the riser steam and water flow, and part of it flows out of the drum as saturated steam while the rest returns through the downcomers to restart the process of circulation. CR is the term used to indicate the ratio of the mixture flowing through the downcomer evaporator system (or riser system) to that of saturated steam generated. A CR (circulation ratio) of 10 in a boiler of capacity 12 t/h of steam means that 120 t/h of steam water mixture is circulating through the downcomer and evaporator–riser system. This is only an average value. Depending on the energy absorbed in a particular circuit, the CR may vary row by row or section by section.

In elevated drum packaged boiler (Figure 2.13b), downcomers are connected to the bottom drum, and steam is generated in several parallel circuits such as the front wall, D-tubes, rear wall tubes, and bank tubes and finally through the riser pipes goes back to the steam drum. Each evaporator path may have a different tube size and equivalent length, which determines its resistance to flow and hence CR. When engineers talk about CR, they mean an average value, but one has to do the calculations to find out the CR in each parallel path. The first few rows or, say, 15% of the bank tubes, which face hot flue

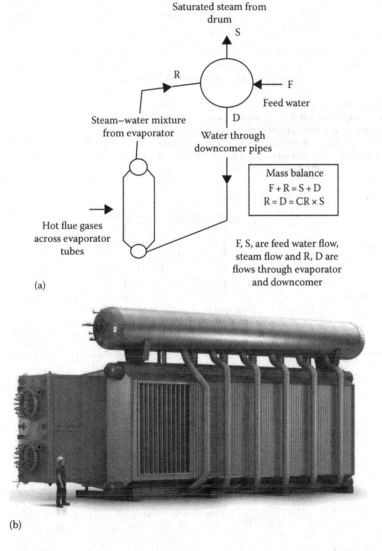

(a)

(b)

FIGURE 2.13
(a) A natural circulation system. (b) Steam generator with elevated drum and external downcomer–riser system. (Courtesy of Cleaver Brooks, Thomasville, GA.) *(Continued)*

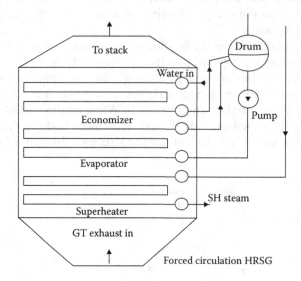

This is a forced circulation system generally seen with horizontal evaporator tubes. A circulation pump is used to suck water from the drum and ensure its flow through the evaporator tubes back into the drum. The pump capacity determines the CR, which can vary from three to eight depending on the designer's choice.

(c)

(d)

FIGURE 2.13 (Continued)
(c) Forced circulation system. (d) High-pressure boiler with external downcomers and risers.

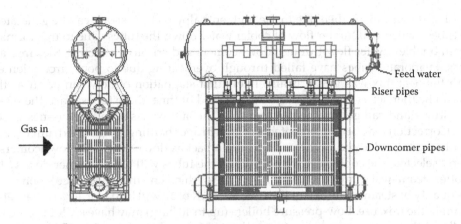

Gas in

Feed water

Riser pipes

Downcomer pipes

Elevated drum water tube boiler

Feed water enters the drum from the boiler feed pump or from an economizer. This mixes with the water from the evaporator tubes and the mixture flows down the downcomer pipes to the bottom of the evaporator tubes. Steam–water mixture is generated in the evporator tubes which is separated into water and steam and then the saturated steam leaves the drum. The amount of steam–water mixture circulated in the system divided by the steam generated is called the circulation ratio. CR is established by an iterative process. It can vary from 5 to 40 depending upon so many variables discussed.

(e)

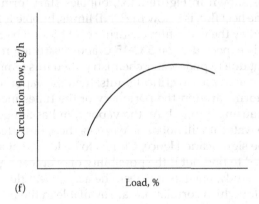

(f)

FIGURE 2.13 (*Continued*)
(e) Cross-flow boiler with elevated drum and external downcomers and risers. (f) Load versus circulation rate.

gases, may be generating more than 70% of the steam in the boiler banks, and hence, there is a need to do a section-by-section analysis.

In many package boilers, the drum is integral as shown in Figure 2.3a. In these boilers, external downcomers are not used. Portion of the bank tubes act as downcomers. These are typically located at the cooler end of the bank tubes, where the energy transfer to the tubes is minimal, and hence, the water enthalpy pickup is not much. The number of heated downcomers is obtained through circulation calculations to ensure that the enthalpy pickup is not significant. If tubes that are in very high gas temperature zone are selected as downcomers, then due to the high enthalpy pickup, steam can be generated in these tubes, which will hinder flow of cooler water down the tubes. Due to this formation of steam bubbles, water flow through downcomers is disrupted, affecting the circulation process. Evaporator tubes have failed through overheating due to poor circulation even though the heat flux was low. It is also likely that stagnation of flow can occur as these tubes can neither act as risers. Hence, over a period of time, these tubes reach the flue gas temperatures and can get overheated and fail. The author has seen these issues in some plants. Correction may be implemented through proper baffling inside the drum to ensure that only the tubes located in the cool gas region act as downcomers. If downcomers are properly selected, the cool water flowing down the tubes will have a higher density than the hotter steam–water mixture in the riser tubes, which ensures good circulation.

The quality of steam at the exit of the riser for a CR of 10 will be 0.1, or 10% is the wetness of steam in the mixture. Low-pressure boilers (up to 40 barg) may have a CR ranging from 15 to 40, while higher-pressure units may have a CR ranging from 5 to 15. This is determined by the balance between the available head and the losses in the circulation system. In a forced circulation system, a circulating pump takes water from the drum and forces it through the evaporator tubes. In these types of boilers (Figure 2.13c), one may predetermine the CR, depending on experience. Since pumps assure that circulation will occur, a low value of 3–8 may be used depending on the cost of pump and electricity. Note that the operating costs are higher if circulating pumps are used.

Circulation calculations or determination of CR in each circuit is not the end in itself. A CR of 10 or 15 does not mean much unless factors such as heat flux inside the tubes, mass flow inside the tubes, steam pressure, and orientation of evaporator tubes whether horizontal or vertical are considered.

In evaporator tube as shown in Figure 2.13a, bubbles start forming as the water gets heated in the risers. If the heat flux is below the DNB limits, bubble formation is not intense, and tubes will be wetted by the water film ensuring cool tubes. The tube wall temperature in a normal riser tube is expected to be 5°C–15°C above saturation temperature. If tube-side deposits are present due to poor water chemistry, then this temperature can shoot up. If the heat flux suddenly increases to critical limits, then the vapor formation is so intense that the film of vapor forms around the periphery of the tubes increasing the tube temperature significantly causing it to fail. As the vapor film has a much lower heat transfer coefficient compared to water, it will not be able to cool the tubes effectively, and hence, the temperature rise will be significant. Hence, CR has to be looked at along with the several other variables discussed to find out if the possibility of overheating exists.

As part of circulation study, one has to check the actual heat flux inside the tubes with the allowable heat flux for which correlations are available in the literature. There must be sufficient margin between the two to avoid what is called DNB condition. The heat flux inside the tubes at DNB conditions is called the CHF. DNB checks are initiated when heat fluxes exceed 600–700 kW/m^2 (190–212,000 Btu/ft^2 h) for steam pressures less than140 barg (2,030 psig). There are correlations in the literature for CHF depending tube diameter, mass flow inside the tubes, pressure, and tube orientation. Note that heat flux in a plain tube is much lower than in a finned tube. Hence, one should be careful in designing evaporators or once-through steam generators in a fired HRSG; evaporators are designed with varying fin density in each row so the high gas temperature regions have low heat flux and vice versa. The fin density varies from, say, 2 in high gas temperature region to, say, 5 in a much cooler gas temperature region.

It may be noted that a circulation pump is not always required with horizontal tube evaporators. Natural circulation system as discussed in Chapter 4 for a steel mill waste heat boiler was designed without circulating pumps. This evaporator had plain tubes, and hence, the heat flux inside the tubes was not high. The streams and the location of the drum should be selected with care, and circulations should be done to ensure that the fluid velocity inside the tubes is reasonably high and will not cause stagnation or separation of flow. The heat flux also is an important variable. With finned tubes, the heat flux inside the tubes particularly if the boiler is fired will be high, and hence, circulation pumps provide some safety factor.

While designing a steam generator, the sizing calculations are first performed followed by off-design calculations as discussed in Chapter 3 and Appendix A. The maximum load conditions or the highest furnace heat flux cases are generally chosen for circulation study. The energy absorbed in each section of the boiler such as furnace front and rear walls, side walls, and bank tubes is evaluated along with steam generation in each section. For example, the furnace duty may be estimated as shown earlier. Then based on the average heat flux in the furnace and considering variations in heat flux along the flame length, one can assign duty to each section from which the steam flow is estimated. Oil-fired boilers have a higher operating heat flux in the furnace compared to gas-fired boilers as discussed earlier.

Figure 2.9 shows how the furnace gas temperature (and hence the heat flux) varies along the furnace length. The maximum point occurs a short distance away from the burner location and cooling occurs along the furnace length. This fact is also considered while arriving at the estimation of steam generation in each section of the furnace. As one can see, the distribution is subject to many variables, and no clear formula or empirical equation can provide this information. It is generally based on the experience of the boiler supplier and field data of operating units. Losses from two-phase flow must be estimated for performing circulation studies. The next section deals with estimation of various losses.

Thom's Method for Estimating Losses [3,4]

The circulation calculations balance the static head available with the losses occurring in the steam loop. The downcomer flow is single phase, and hence, the following equation may be used for estimating the pressure loss as discussed in Appendix B.

$$\Delta P = 810 \times 10^{-6} \, f \, L_e v \, w^2 / d_i^5 \qquad (2.16)$$

where, w is the flow in each pipe in kg/s.

The *friction loss* ΔP_f in kPa in the heated riser tube is estimated by the following equation:

$$\Delta P_f = 38 \times 10^{-12} \, f \, v_f \, L \, G_i^2 \, r_3 / d_i \qquad (2.17)$$

where
   f is 4× Fanning's friction factor = Moody's friction factor
   L is the total effective length over which two-phase friction loss occurs, m
   v_f is the specific volume of saturated water, m^3/kg
   G_i is the tube-side mass velocity, kg/m^{2}h
   d_i is the tube ID, m
   r_3 is Thom's friction loss factor (Figure 2.14b)

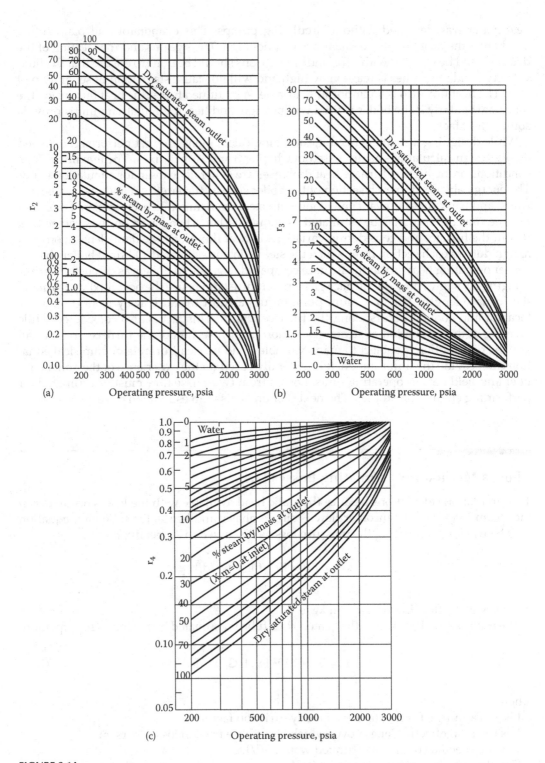

FIGURE 2.14
Thom's multiplication factor for (a) acceleration loss and (b) friction loss. (c) Thom's two-phase multiplication factor for gravity loss.

TABLE 2.3

Two-Phase Pressure Drop Correlations

SI	British	Metric
$\Delta P_a = 7.65 \times 10^{-11} \, v \, Gi^2 r_2$	$1.664 \times 10^{-11} \, v_f G^2 r_2$	$\Delta P_a = 7.8 \times 10^{-13} \, v_f Gi^2 r_2$
$\Delta P_f = 38 \times 10^{-12} \, fv_f LG_i^2 r_3/d_i$	$4 \times 10^{-10} \, fv_f LG^2 r_3/d_i$	$\Delta P_f = 0.388 \times 10^{-12} \, fv_f LG_i^2 r_3/d_i$
$\Delta P_g = 0.00981 \, Lr_4/v_f$	$0.00695 \, Lr_4/v_f$	$\Delta P_g = 0.0001 \, Lr_4/v_f$
G, kg/m² h	lb/ft² h	G, kg/m² h
w, kg/s	lb/h	w, kg/s
L, m	ft	L, m
f, Moody's friction factor	Fanning friction factor = Moody's friction factor/4	f, Moody's friction factor
di, m	In	di, m
v_f, m³/kg	ft³/lb	m³/kg
ΔP, pressure loss, kPa	psi	kg/cm²

The *gravity loss* in the heated riser section in kPa is given by

$$\Delta P_g = 0.00981 \, Lr_4/v_f \tag{2.18}$$

where

L is the length of the boiling section in m

v_f is the specific volume of saturated water, m³/kg

r_4 is the multiplication factor for gravity loss (Figure 2.14c)

ΔP_a is the *acceleration loss* in kPa due to change of phase, which is significant at low steam pressures and high mass velocities and is given by

$$\Delta P_a = 7.65 \times 10^{-11} \, v_f Gi^2 r_2 \tag{2.19}$$

r_2 is the acceleration loss multiplication factor as shown in Figure 2.14a.

The correlations are shown in Table 2.3 in all three systems of units.

Circulation Calculations

Circulation calculation in a natural circulation boiler is an iterative process.

1. First, a CR is assumed based on experience and type of steam generator and pressure. For low-pressure boilers (<70 barg), CR could range from 12 to 35, while for higher-pressure units, it can range from 5 to 15. CR = 1/x where x is the steam quality, fraction. Flow through the downcomers, evaporators, and risers is CR x steam generated.

2. The feed water temperature entering the drum should be known from thermal calculations, which should have already been done. This includes the complete boiler performance as illustrated in Chapter 3, furnace calculations, and estimation of energy absorbed in each parallel path such as front wall, rear wall, and side

walls. The resistance of each path must also be known. Mixture enthalpy entering the downcomers is calculated through an energy balance around the drum, Figure 2.13a.

$$h_{fw} + CR \times h_e = h_g + CR \times h_m \tag{2.20}$$

where h_{fw}, h_e, h_m, and h_g are the enthalpy of feed water, steam–water mixture leaving the evaporator, water entering the downcomers, and saturated steam, respectively.

It should be noted that cool water entering the downcomer aids better circulation. If h_m is close to or at saturation, then bubbles will be formed in the region near the entry of downcomer pipes, and circulation will be hindered. There may be several downcomers in a boiler each with a different length and pipe size. A flow and pressure balance calculation is done to arrive at the flow in each downcomer and riser.

As the downcomer flow enters the evaporator, it gets heated, and boiling starts after a distance from the bottom of the furnace. This is called the boiling height, L_b.

$$L_b = L \times CR \times W_s (h_f - h_m)/Q \tag{2.21}$$

(All in consistent units)
h_f is the enthalpy of saturated liquid
Q is the energy absorbed in the furnace
L is the furnace height
W_s is the steam generated by the boiler

1. The static head available is first estimated. It is typically the distance from the drum water level to the bottom of the furnace; then the following losses are estimated:

2. Losses in the downcomers including entry and exit losses.

3. Losses in the boiling height and two-phase losses such as friction, acceleration, and gravity using Thom's method or similar well-known two-phase correlations.

4. If there are unheated risers, their losses are estimated. Again there may be several parallel paths with different pipe sizes and lengths. Flow in each parallel path is evaluated by estimating the pressure drop in each path for different flows and ensuring that the pressure drop is the same. This is an iterative process. These calculations are preferably done using a computer program.

5. Loss in drum internals. This is impacted by the presence of cyclones inside the drum for steam separation.

The static head is balanced with various losses, and if they match, the CR assumed is fine; else, another CR is assumed and the calculations are repeated. If there are multiple parallel paths (a D-type boiler has front wall, rear wall, side wall, and boiler bank tubes), then flow and pressure balance calculations are carried out to ensure that the total losses in each parallel path are the same. CR in each parallel path may differ depending on its energy absorbed, tube size used, and resistance to flow.

Once the circulation calculations are done, the flow in each evaporator tube will be known along with the steam quality at its exit. Checks for heat flux in each path and DNB based on exit quality are then done. The variation of heat flux along flame length and view factors based on tube spacing enter into these calculations. If there is a potential for DNB, then efforts are taken to revise the downcomer and riser system pipe sizes to improve the CR in the boiler; else, the heat flux in the furnace has to be reduced, and the boiler design is redone. Based on experience and field data, one should be able to arrive at a good system considering these concerns.

Example 2.9

A waste heat boiler similar to the system shown in Figure 2.13a has the following data: steam pressure = 55.1 barg (800 psig). Static head available (from drum centerline to bottom of evaporator bundle) = 8 m. Steam drum is located far away from the evaporator, which is generating 58 t/h of steam; feed water temperature is 120°C.

Two downcomers of ID 285 mm and length 14 m and with six 90° bends are feeding the evaporator with two risers of ID 317 mm and of length 11 m with five bends are feeding drum. The evaporator bundle is 6 m high and has 420-44 mm ID tubes. Determine the approximate CR in the evaporator. Drum internals loss = 352 kg/m².

Solution

First, the energy balance is performed in the drum assuming a CR. Say CR = 11.

At 55.1 bar, from steam tables, $v_g = 0.035$ and $v_f = 0.0013$ m³/kg. Enthalpy of feed water $h_f = 121.2$ kcal/kg, enthalpy of saturated steam $h_g = 666.1$ kcal/kg, and saturated water $h_l = 284.27$ kcal/kg. Saturated steam temperature = 271°C.

Assume CR = 11. Mixture enthalpy leaving the drum $h_e = 0.0909 \times 666.1 + 0.9091 \times 284.27 = 319$ kcal/kg. From energy balance in the drum,

$11 \times 319 + 121.2 = 666.1 + 11h_m$ or $h_m = 269.5$ kcal/kg. From steam tables, this corresponds to 259°C and $v_m = 0.001271$ m³/kg. Hence, the water will start boiling after some distance from the bottom as the saturated steam temperature is 271°C.

1. Static head available = 8/.001271 = 6294 kg/m²
2. To compute the losses in downcomers, the velocity of water in downcomer must be known. $V_{dc} = 1.246wv_m/d_i^2$ (see Appendix B for formula for velocity inside tubes). Flow in each downcomer w = 11 × 58,000/2/3,600 = 88.6 kg/s. Then $V_{dc} = 1.246 \times 88.6 \times 0.001271/0.285/0.285 = 1.73$ m/s. Velocity head VH = $V^2/(2gv_m) = 1.73 \times 1.73/2/9.8/0.001271 = 120$ kg/m² (g = 9.8 m/s²). The entrance plus exit loss in downcomer = 1.5VH = 180 kg/m².

Equivalent length of downcomer (see Table B.7 for equivalent length of 90° bends) = 11 + 32 × 0.285 × 6 = 54.7 m. Total equivalent length = 11 + 54.7 = 65.7 m

Friction loss in downcomer = $810 \times 10^{-6}fL_e v_m w^2/d_i^5 = 810 \times 10^{-6} \times 0.012 \times 66 \times 0.001271 \times 88.6 \times 88.6/0.285^5 = 3.4$ kPa = 347 kg/m²

Boiling height is then determined. $L_b = CR \times L \times (h_l - h_m)/(h_v - h_f) = 11 \times 6(284.27 - 269.5)/(666.1 - 121.2) = 1.79$ m. Note that a large boiling length reduces the CR as the net thermal head for forcing the two-phase flow through the evaporator tubes reduces. A higher feed water temperature entering the drum helps but not close to saturation as that will hinder flow through downcomer tubes in case the downcomer tubes are partly heated.

The gravity loss in the boiling length before start of boiling = 1.79/0.001285 = 1393 kg/m² (an average specific volume between that of saturated liquid and the colder water entering the evaporator is taken).

The gravity loss in the evaporator section is $\Delta P_g = 0.00981Lr_4/v_f \cdot r_4$ from Figure 2.14c = 0.65. $\Delta P_g = 0.00981 \times (6 - 1.79) \times 0.65/.0013 = 20.62$ kPa = 2104 kg/m².

The friction loss may be shown to be very small due to the large number of tubes used. $r_3 = 2.3$ from Figure 2.14b. $\Delta P_f = 38 \times 10^{-12} f v_1 L G_i^2 r3/d_i$

$G = 58,000 \times 11 \times 4/(3.14 \times 420 \times .044^2) = 999,531$ kg/m²h

$\Delta P_f = 38 \times 10^{-12} \times 0.02 \times 0.0013 \times 4.21 \times (999,531)^2 \times 2.3/.044 = 0.217$ kPa $= 22$ kg/m²

Average mass flow in each tube $= 58,000 \times 11/420/3,600 = 0.42$ kg/s

Friction loss in the single-phase heated section $= \Delta P_f = 810 \times 10^{-6} \times 0.02 \times 1.79 \times 0.001285 \times 0.42^2/0.044^5 = 0.04$ kPa $= 4$ kg/m²

At entrance to evaporator bundle, velocity $= 1.246 \times 0.42 \times 0.001271/0.044^2 = 0.34$ m/s

Specific volume of steam mixture at evaporator exit $= 0.0909 \times 0.035 + (1 - 0.0909) \times 0.0013 = 0.004363$ m³/kg (1/CR = 0.0909)

Entrance loss $= 0.5$VH $= 0.5 \times 0.34 \times 0.34/2/9.8/.001271 = 2.32$ kg/m²

Exit velocity $= 1.246 \times 0.42 \times 0.004363/.044/.044 = 1.179$ m/s. Exit loss $= 1$ VH $= 1.179 \times 1.179/2/9.8/.004363 = 16.3$ kg/m²

$$\text{Acceleration loss } \Delta P_a = 7.65 \times 10^{-11} v_f G_i^2 r_2$$

$r_2 = 1.45$ from Figure 2.14a. $\Delta P_a = 7.65 \times 10^{-11} \times 0.0013 \times (999,531)^2 \times 1.45 = 0.14$ kPa $= 14$ kg/m²

Velocity of steam in riser pipe $= 1.246 \times 88.6 \times 0.004363/0.317^2 = 4.79$ m/s

Inlet and exit loss $= 1.5$VH $= 1.5 \times 4.79^2 \times 4.79/(2 \times 9.8 \times 0.004363) = 402$ kg/m²

$L_e =$ equivalent length $= 11 + 5 \times 32 \times 0.317 = 61.7$ m

Friction loss $= 810 \times 10^{-6} f L_e v_e W^2/d_i^5 = 810 \times 10^{-6} \times 0.011 \times 61.7 \times 0.004363 \times 88.6 \times 88.6/0.317^5 = 0.095$ kPa $= 952$ kg/m² (using homogeneous model)

The difference in height between the drum centerline and the evaporator header is about 2 m. The gravity head in the riser pipe $= 2/0.004363 = 458$ kg/m². (Using the relation between slip factor, void fraction, and quality, one can come up with more accurate estimation of gravity loss.)

Summary of Losses

1. Downcomer inlet and exit loss $= 180$ kg/m²
2. Downcomer friction loss $= 347$ kg/m²
3. Gravity loss in evaporator section $= 1393 + 2104 = 3497$ kg/m²
4. Friction loss in evaporator length $= 22 + 4 + 2 + 16 = 44$ kg/m²
5. Acceleration loss $= 14$ kg/m²
6. Riser inlet and exit loss $= 402$ kg/m²
7. Riser friction loss $= 952$ kg/m²
8. Unheated riser pipe gravity head $= 458$ kg/m²
9. Drum internals loss $= 352$ kg/m² (0.5 psi is a conservative estimate)

Total losses $= 6246$ kg/m². Available head $= 6294$ kg/m². Hence CR is in the range of 11. A computer program is helpful in performing these calculations more accurately. One can consider variation in steam generation in individual paths, differences in downcomer and riser sizes and lengths of parallel paths, effect of feed water temperature, and so on. Also this is a tedious calculation as we have to assume a CR and perform all these calculations and check if available head matches the losses. If not, another CR is assumed, and these calculations are repeated. For the earlier illustration, a CR close to the computer-calculated value was chosen. The manual calculation showed a simplistic model for illustration. The first few rows of tubes of evaporator will be generating more steam, and hence, the CR in these tubes could be slightly lower. If risers or downcomers with different lengths are used, the flow in each could vary. One should then check the heat flux in each section and ensure that the CHF levels are not reached.

Example 2.10

Pressure drop in two-phase flow may be estimated using the average specific volume as well as by using Thom's method. Let us compare the difference. In a once-through boiler, 15,000 kg/h of two-phase mixture at 105.5 kg/cm² a absolute pressure with exit quality of 80% flows inside a pipe of inner diameter 50 mm. For an effective length of 50 m, compute the friction loss using Thom's method and homogeneous model using average specific volume of mixture.

Solution

From steam tables, $v_g = 0.01728$ m³/kg, $v_f = 0.001469$ m³/kg

Mixture-specific volume at exit of boiler = $0.8 \times 0.01728 + 0.2 \times 0.001469 = 0.0141$ m³/kg

Average specific volume in the boiler = $(0.001469 + 0.0141)/2 = 0.007785$ m³/kg

w = 15,000/3,600 = 4.17 kg/s. r_3 from Figure 2.14b = 5.7

$\Delta P = 810 \times 10^{-6} \times 0.02 \times 50 \times 4.17^2 \times 0.007785/0.05^5 = 354$ kPa = 3.6 kg/cm² [51 psi]

$\Delta P(\text{Thom}) = 38 \times 10^{-12} \times 0.02 \times 0.001469 \times (15,000 \times 4/3.14 \times 0.05 \times 0.05)^2 \times 5.7 \times 50/0.05 = 371$ kPa = 3.79 kg/cm² [54 psi]

Hence, in the absence of Thom's curves, one may estimate the two-phase friction loss using an average specific volume in the boiling section.

Example 2.11

A simplified version of circulation calculations for the boiler in Figure 2.13b is shown. The boiler is a natural gas–fired D-type boiler with external drum, downcomers, and risers and generates 190 t/h of saturated steam at 40 kg/cm² a. Energy balance calculations were done, and the energy transferred in each section of furnace such as D tubes, front wall, rear wall, and the convection bank tubes was estimated using a thermal performance program. The steam generation in each section was estimated. Then, the geometric data for the circulation system, namely, the number of external downcomers, risers, and lengths of each section were inputted in a circulation program, results of which are shown in Table 2.4.

The front and rear walls have a CR of about 20, while the D-tubes have a CR of about 12 and the bank tubes about 19 giving an overall average CR of 16.8.The calculations can be more detailed with split-up of the bank tubes into several sections. However, this is only for illustrating that CR varies in each circuit depending on its resistance to flow and steam generation.

Natural circulation is adopted in small and large boilers up to a pressure of about 165 barg. Beyond this, forced circulation is required as the density difference between steam and water decreases.

Circulation Flow versus Load

The amount of flow circulated through a steam generator typically shows a trend shown in Figure 2.13f. As the load increases, the heat flux in various parts of the evaporator increases, and hence, the difference in density between the cooler downcomer water mixture and the flow in the evaporator tubes increases increasing the circulation flow through the system. However, beyond the maximum load, the resistance offered by the riser circuits and drum internals dominates and reduces the flow circulating through the system. If the CR at 100% load is, say, 15, then at 80% load, it could be 17, and at 120% load, it could be 11. At 100% load, the mixture flow = 15 × 100 = 1500. At 80% load, 80 × 17 = 1360, and at 120% load, 120 × 11 = 1320. Overloading a steam generator is not recommended for this reason, and critical circuits must be evaluated for DNB concerns.

TABLE 2.4

Results of Circulation Calculations

Steam Pressure				40		Section		Steam kg/h		CR
Sat temperature				249		Front wall		5,300		20
Feed water temp				194		Rear wall		2,830		20.5
Sp vol steam				0.0508		D tubes		58,300		12.1
Sp vol water				0.0012		Bank tubes		127,000		19.1
Average CR				**16.8**		**Units**				
Total static head				7.80		Flow		kg/h		
Total static head				0.63		Length		m		
Downcomer losses				0.10		Pressure		kg/cm² a		
Riser losses				0.10		Press loss		kg/cm²		
Riser gravity loss				0.11		Tube dia		mm		
Drum internal loss				0.04		Velocity		m/s		
Mixture temp				246		Sp volume		m³/kg		

Path	Level	Sort	Dia	Length	Bends	Flow	Velocity	No tubes	Pr. drop	Item
1	1	1	289	8	3	3,246,264	2.81	6	0.10	dc
1	1	1	289	1.22	0	3,246,264	7.08	8	0.10	riser

Steam	Quality	Accln	Gravity	Friction	Total flow	Tubes	Tube ID	Boil ht.	Length	CR
5,300	0.05	0.002	0.29	0.01	106,000	30	56.00	4.9	4.90	20.0
2,630	0.049	0.001	0.292	0.002	53,893	30	56.00	4.9	4.90	20.5
58,300	0.082	0.01	0.246	0.046	707,524	120	56.00	4.9	8.50	12.1
127,000	0.052	0.001	0.286	0.007	2,423,664	1320	45.00	4.9	4.90	19.1

Flow Stratification in Horizontal Tubes

With horizontal tubes in two-phase flow, one has to also check if flow stagnation is likely. Froude number is an indicator of inertial forces to gravity forces, and if it is above 0.04, flow separation is unlikely. Froude number $F = G^2/\rho^2 gd$ where G = mass flow inside the tubes (liquid only), kg/m² s. ρ = density of liquid (807 kg/m³), g = acceleration due to gravity (9.8 m/s²), and d = tube inner diameter (0.026 m).

Correlations for CHF (Critical Heat Flux) and Allowable Steam Quality

One may obtain the heat flux given the steam quality, mass flow inside tubes, steam pressure, and tube inner diameter or conversely obtain the allowable steam quality given a heat flux. There are several correlations for CHF, and the Kastner correlation [4] is one of them. It gives the allowable quality at any mass flow and heat flux and steam pressure.

For 0.49 < P < 2.94 MPa [5]

$$x_c = 25.6(1000q)^{-0.125} \, m^{(-0.33)}(1000d_i)^{-0.07} \, e^{0.1715P} \tag{2.22}$$

For 2.94 < P < 9.8 MPa

$$x_c = 46(1000q)^{-0.125} \, m^{(-0.33)}(1000d_i)^{-0.07} \, e^{-0.0255P} \tag{2.23}$$

For 9.8 < P < 19.6 MPa

$$x_c = 76.6(1000q)^{-0.125} m^{(-0.33)}(1000d_i)^{-0.07} e^{-0.0795P} \tag{2.24}$$

(Steam pressure P in MPa, d_i in m, q in kW/m^2, m in kg/m^2s)

Example 2.12

Steam is generated at 50 bara in a boiler using 40 mm tubes. Heat flux is 300 kW/m^2. Mass velocity of steam water mixture inside tubes is 750 kg/m^2 s. Determine the maximum allowable quality or CR to avoid DNB conditions.

$$x_c = 46(1000q)^{-0.125} m^{(-0.33)}(1000d_i)^{-0.07} e^{-0.0255P}$$

$$P = 50 \text{ bara} = 5 \text{ MPa}, \quad m = 750 \text{ kg/m}^2 \text{ s}, \quad d_i = .040 \text{ m}, \quad q = 300 \text{ kW/m}^2.$$

$x_c = 46 \times (300,000)^{-0.125} \ 750^{-0.33} \ 40^{-0.07} \ e^{-0.0255 \times 5} = 0.73$ or 73% allowable quality or a CR of 1.37 minimum. CR in practice will be much higher than this, and hence, this heat flux is acceptable.

Macbeth correlation gives the CHF, given the other variables, and takes the following form:

$$q_c = 0.5025 \ h_{fg} \ d_i^{-0.1} \ (G_i)^{0.51} (1 - x) \tag{2.25}$$

where
q_c = CHF, kW/m^2
x is the steam quality, fraction
d_i is the tube ID, m
G_i is the mass velocity, kg/m^2 s

Example 2.13

If G = 816.7 kg/m^2s, x = 0.2, d_i = 0.0381 m, and P = 6.9 MPa (h_{fg} = 361.1 kcal/kg), then

$$q_c = 0.5025 \times 361.1 \times (0.0381)^{(-0.1)} \times 816.7^{0.51} \times 0.8 = 6150 \text{ kW/m}^2$$

This being a theoretical correlation does not account for tube-side fouling, and the actual value of CHF could be 20%–30% of this.

Groeneveld's lookup tables are also used to check for CHF. Table 2.5 shows an extract from 1996 tables.

TABLE 2.5

Groeneveld's Lookup Tables for CHF in $kW/m2$ for 8 mm Tubes

Pressure, kPa	G, kg/m^2 s	X = 0.2	0.4	0.6	0.8
3000	500	5660	3392	2745	1320
	1000	5620	3079	1925	830
	1500	5043	2691	1080	499
	2000	4507	2279	608	330
5000	500	5178	3975	3040	1769
	1000	4957	3447	2066	1034
	1500	4530	2983	1194	899
	2000	3984	2557	668	650

Source: IAEA April 2001 report, Thermo-hydraulic relationships for advanced water cooled reactors, IAEA-TECDOC-1203.

Note: P in kPa, G in kg/m^2 s, CHF in kW/m^2. Use a correction factor of 0.79 for tube ID > 16 mm. This table is based on a tube ID of 8 mm. For diameter above 16 mm, use a correction factor of 0.79.

Example 2.14

At 50 barg and with 1000 kg/m^2 s mass velocity and 40 mm tube ID and quality = 0.2, the CHF = 4957 × 0.79 = 3916 kW/m^2

Though there are several correlations for CHF, many have been developed in laboratories under controlled conditions. They may show different CHF values for the same steam parameters and tube geometry. Hence, charts such as Figure 2.15 developed by boiler firms have more practical value as the results are backed by operation of steam generators [6].

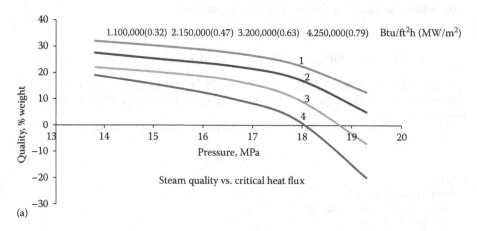

(a)

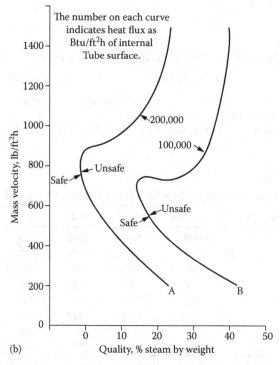

(b)

FIGURE 2.15
(a) Allowable steam quality as a function of heat flux Btu/ft^{2}h (MW/m^2) and steam pressure. (b) Allowable quality for nucleate boiling at 2700 psia. (From Babcock and Wilcox, *Steam: Its Generation and Use*, 38th edn., B&W, Barlerton, OH, 1992.)

Circulation Problems

The common problem in boiler circulation system is that the downcomer tubes are not properly located or shielded from hot flue gas. If located in the hot gas zone, steam bubbles are likely to be formed at the downcomer inlet preventing the flow of cold water to the bottom of the boiler for circulation. This generally happens through improper baffling inside the drum. The boiler designer may think that the tubes are acting as downcomers, while due to improper baffling, they may be acting as risers. Sometimes, the head is inadequate to ensure flow as downcomers, and hence, stratification occurs with steam bubble formation at downcomer inlet resulting in the stagnation of flow. The tubes can then reach the flue gas temperature and fail even though they may be in 500°C–600°C gas temperature region and the average heat flux in the convection bank is very low, well below the DNB limits. So DNB is not the problem but vapor formation, stratification, and overheating of tubes, which are neither capable of acting as downcomers or risers.

High feed water temperature close to saturation temperature or steaming water entering the drum from the economizer is also a concern. If one computes the mixture enthalpy entering the downcomers, he may find it above saturated water enthalpy or above saturation temperature. The bubble formation at the downcomer inlet prevents water flow to the bottom of the boiler for creating a circulation pattern.

Guidelines for Good Circulation System Design

The following are the guidelines used while designing the downcomer system in a package boiler.

1. The designated downcomer tubes should be located at the coolest gas temperature region; that is where the flue gases make a turn to the economizer. If they are located in a high gas temperature region, the enthalpy absorbed by these tubes can be high, resulting in steam bubble formation inside the tubes, which can hinder downward flow of water and hence the circulation process. Formation of steam does two negative things. It decreases the density of the water in the downcomer, which in turn reduces the available head for circulation and also physically prevents the free flow of downcomer water. So locating downcomers in hot gas zones should be avoided. If inevitable, then the downcomers must be insulated.

2. The gas temperature entering the downcomers should be as low as possible to ensure that even if stagnation occurs, the tubes will not be overheated. A good value is less than 450°C at full load.

3. Belly pans are used to collect steam from risers, and all the water for circulation is allowed to flow from drum normal level through the downcomer tubes. Design velocity chosen for downcomer flow is generally in the range of 1–3 m/s at full load.

4. Proper baffling to be done inside the drum to ensure that tubes, which are supposed to be risers, are inside the belly pan area (see Figure 6.10). This provides these tubes an additional head for circulation. (Density of steam–water mixture in the baffle space of belly pan is lower than the water in the drum; also the static head equivalent to the radius of the drum less the belly pan height is available for

circulation process.) If, say, the average CR is 16 (steam quality $x = 1/CR = 0.055$) in the boiler, the specific volume of the mixture $= 0.055 \times 0.0435 + 0.945 \times 0.00125 = 0.00356$ m³/lb, where 0.0435 and 0.00125 are the specific volumes of saturated steam and water, respectively, at 46 bara drum pressure. Hence, the density of the mixture $= 1/0.00356 = 281$ kg/m³. Assume that the normal water level is at 600 mm in the drum and the belly pan height is 125 mm, then the extra head available for the baffled tubes is $(0.600 - 0.125) \times 281 = 133$ kg/m², which helps a lot considering that the net head for circulation is in the order of about 3500 kg/m². So riser tubes should be under the belly pan. Tubes acting as risers in the water-filled region will have difficulty in circulating as this additional head is not available.

5. It is very important that downcomers do not take suction from locations in drum where vaporization or heat flux is intense resulting in sucking of bubbles into downcomers. This will interfere with circulation and also prevent a normal downcomer from acting as a downcomer. In large waste heat boilers, a vortex breaker is provided at the suction line to break up these bubbles.

6. Swaging of tubes inside drums is better avoided as it adds to the flow resistance and impacts circulation. Also, it adds to two-phase flow instabilities. Resistance to flow at the riser end should be avoided, while resistance at downcomer inlet or feed water inlet in once-through boilers improves the stability of two-phase flow as discussed in Chapter 6.

References

1. V. Ganapathy, *Industrial Boilers and HRSGs*, CRC Press, Boca Raton, FL, 2001, p112.
2. V. Ganapathy, *Applied Heat Transfer*, Pennwell Books, Tulsa, OK, 1982, p60.
3. V. Ganapathy, Boiler circulation calculations, *Hydrocarbon Processing*, January 1998, p100.
4. J.R.S. Thom, Prediction of pressure drop during forced circulation boiling of water, *International Journal of Heat Transfer*, 7, 709–724, 1964.
5. D. Tucakovic et al., Thermal hydraulics of evaporating tubes in the forced circulation loop of a steam boiler, *FME Transaction*, 36, 1, 2008.
6. Babcock and Wilcox, *Steam: Its Generation and Use*, 38th edn., Published by B&W, Barlerton, Ohio, 1992.

3

Steam Generators

Introduction

Oil- and gas-fired steam generators or boilers as they are often called form an essential part of any power plant or cogeneration system. The steam-based Rankine cycle has been synonymous with power generation for centuries. Though the steam parameters such as pressure and temperature have been steadily increasing during the last several decades, the function of the boiler remains the same, namely, to generate steam at the desired conditions efficiently and with low operating costs while meeting the local emission regulations. Low-pressure steam is used in cogeneration plants for heating or process applications while high-pressure steam is used for generating power via steam turbines. Boiler capacities have undergone significant changes during the past 50 years. Today oil- or gas-fired shop-assembled package boiler generating 100–140 t/h of steam at 100 barg, 500°C, is not uncommon (Figure 3.1). Modular design or elevated steam drum design (Figures 3.2 and 3.3) is also evolving to handle yet larger capacities of about 200 t/h steam, while large forced circulation boilers using treated water from oil wells are finding application in oil field applications (Figure 3.4). Another factor that has introduced significant changes in the design of boilers and associated systems is the stringent emission regulations in various parts of the world requiring CO and NO_x control systems such as flue gas recirculation (FGR) or selective catalytic reduction system (SCR) adding to the complexity of the boiler design and layout. High efficiency also remains a goal coupled with low fan power consumption; hence, boiler suppliers are offering both custom-designed boilers, wherever required, and standard boilers. This chapter deals with oil- and gas-fired steam generators only and not with solid fuel–fired boilers.

Large Package Boilers

Large boilers in the range of 100–200 t/h capacity were once field erected considering the limitations of local rail or road shipping dimensions. Considering the high cost of site erection, during the last three decades, boiler suppliers have come up with modular designs to minimize the labor cost, field erection, and delivery time by maximizing the work done in the shop.

For larger capacities, in the range of 150–200 t/h modular design, Figure 3.2 is an interesting option. Since shipping dimensions are not limiting the size, the furnace dimensions

FIGURE 3.1
Large package steam generator. (Courtesy of Cleaver Brooks Inc., Engineered Boiler Systems, Thomasville, GA.)

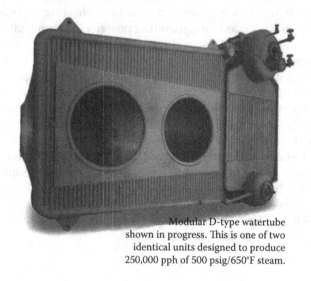

Modular D-type watertube
shown in progress. This is one of two
identical units designed to produce
250,000 pph of 500 psig/650°F steam.

FIGURE 3.2
Modular steam generator design. (Courtesy of Cleaver Brooks Inc., Engineered Boiler Systems, Thomasville, GA.)

can be chosen independently based on liberal furnace heat release rates; the width of the convection bank can also be large as it is also shipped separately, and hence, the bank tube spacing or that of the superheater if used in the convection bank can have longer lengths or wider tube spacing to minimize the gas velocity and pressure drop, as a result of which the fan can be made smaller and fan power consumption lower. The steam drum can be integral with the boiler bank tubes or can be shipped separately and assembled in the field.

Another option for large units is the elevated drum unit (Figure 3.3), which shows a 193 t/h boiler generating saturated steam for oil field application. The concept of external

FIGURE 3.3
Elevated drum steam generator for large capacity. (Courtesy of Cleaver Brooks Inc., Engineered Boiler Systems, Thomasville, GA.)

FIGURE 3.4
Large-capacity high-pressure forced circulation steam generator for oil sands application. (Courtesy of Cleaver Brooks Inc., Engineered Boiler Systems, Thomasville, GA.)

drum and circulation system has been widely used in waste heat boilers, and the package boiler industry has started adapting these concepts for the benefit of the end users. Considering the high cost of fuel and electricity, these boilers have been designed with a closed-loop glycol heating system, which offers about 2.5% more efficiency over a typical steam generator with an economizer, which is discussed later. Combining the modular and elevated drum boiler concepts, oil- and gas-fired steam generators of capacity 200–300 t/h are viable. With minimum field installation time and lower cost of erection, these recent designs are economically attractive to the end user.

Water-Cooled Furnaces

Refractory-lined boilers are becoming a thing of the past, and today's package boiler has completely water-cooled furnaces as shown in Figure 3.2; the water-cooled furnace design has several advantages over refractory-lined units as discussed in Chapter 2. One advantage of the completely water-cooled furnace design is that it enables quick starting of boilers. In combined cycle plants, one important application of steam generators is to provide steam to the steam turbine glands when steam from the HRSGs is not available. A package steam generator is used as a standby boiler, whose important purpose is to supply steam at a short notice, say, when the gas turbine trips. They are called quick start boilers or standby boilers. A steam-heated coil is inserted in the mud drum, which keeps the boiler always warm; sometimes a small additional burner of capacity 3%–5% of the total burner capacity is kept operating all the time, and the boiler makes a small amount of steam at the operating pressure or close to it. When required, it can be brought on line to the required capacity in less than 1–2 min. A refractory-lined boiler will not be suitable for this type of application.

Major Changes in Boiler Design

Major changes have been introduced in the design of oil- and gas-fired steam generators of today. Decades ago, boilers were concerned with only two issues, namely, efficiency and operating cost. Low boiler efficiency meant higher fuel cost, and a large gas pressure drop across heating surfaces meant higher fan power consumption. Each addition of 25 mm wc (1 in wc) pressure drop in a boiler of 45 t/h (100,000 lb/h) capacity results in about 5 kW of additional fan power consumption. The difference in efficiency between 5 and 15% excess air operation on natural gas is about 0.4%. Every 22°C drop in flue gas exit temperature adds 1% to the boiler efficiency. Therefore, steam generators were operating at the lowest possible excess air, say, 5%. The effect of excess air and exit gas temperature on boiler efficiency is shown in Figure 3.5 for natural gas firing. With limits on CO and NO_x emissions today, this is no longer possible, and higher excess air rates are used as explained later. The expression, "no NO_x, no SO_x, and no rocks" aptly describes the direction where boiler systems are headed, namely, emission-governed equipment [1].

Absence of Air Heater

The steam generator configuration has changed in the last few decades. The tubular or regenerative air heater, which was once an integral part of the steam generator, is now replaced by the economizer. Air heaters are rarely used in steam generators where NO_x emissions have to be limited. Air heater increases the combustion temperature in the furnace and thus increases the NO_x levels adding to the cost of pollution control equipment. With natural gas at 15% excess air, each 55°C (100°F) increase in combustion air temperature increases the flame temperature by about 36°C (65°F). In addition, air heater increases

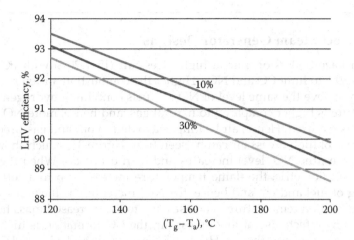

FIGURE 3.5
Effect of excess air and exit gas temperature on boiler efficiency for natural gas firing.

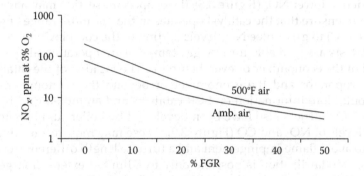

FIGURE 3.6
Effect of air temperature on NO_x.

the operating cost as it has a larger gas- and air-side pressure drop of about 100–120 mm wc compared to the economizer, which has a nominal pressure drop of only 25–30 mm wc. Figure 3.6 shows the effect of air temperature and FGR on NO_x emissions. High air temperature increases NO_x significantly, and hence, air temperature is kept at ambient or in some high-efficiency systems at less than 60°C–70°C. Air heaters find application in steam generators burning low heating value fuels or solid fuels where flame stability is improved by using hot air but not in natural gas– or fuel oil–fired steam generators. However, several heat recovery systems use them, and their design and performance aspects are discussed in Chapter 4 on waste heat boilers.

There are a few options for increasing the boiler efficiency without using the conventional air heaters. Higher efficiency is achieved in a package boiler today by using a condensing economizer or a closed-loop glycol or thermic fluid heating and cooling system or a scavenger system or by using a low feed water temperature, which are discussed later.

Emissions Affect Steam Generator Designs

Present-day package boilers operate at high excess air (15%–20%) with FGR rates ranging from 0% to 30% to limit CO and NO_x. Higher the combustion air temperature, higher the FGR rate to achieve the same level of NO_x. If fuels containing hydrogen are fired, the flame temperature is higher compared to natural gas, and hence, more NO_x is generated and more FGR is required. Hence, air heating is avoided in oil- and gas-fired boilers. The reason for the use of high excess air can be seen from Figure 3.7, which shows that as the excess air increases, the NO_x level increases and then drops off. When the excess air is a little above stoichiometric, the flame temperature increases up to a certain level due to good mixing of fuel and air, and hence, the NO_x increases; as the excess air increases further, the combustion temperature decreases due to the increased mass flow of air and consequent cooling effect; also at low excess air, the CO formation is high due to poor mixing between the fuel and the air. Hence, 15% excess air and 15–25% FGR are widely used in all gas- and oil-fired package boilers instead of 5% excess air a few decades ago. This increases the boiler size by 15%–25% if gas pressure drop has to be limited. A larger economizer is also required to maintain the same stack flue gas temperature and efficiency as the design without FGR.

If SCR is added to lower NO_x (Figure 3.8), the evaporator section may have to use a gas bypass system to ensure that the catalyst operates in the optimum range of gas temperatures (350°C–420°C) to guarantee NO_x levels adding to the complexity of the boiler construction. Catalysts are available for low gas temperature operation also, which can be located ahead of the economizer or even behind it. The location of the catalyst, whether between the evaporator and the economizer or beyond the economizer, also affects the boiler layout. More information on SCR catalysts and ammonia grid is provided in Chapter 4. Lo-NO_x burners also have been developed by boiler suppliers to handle the low emission levels of NO_x and CO (Figure 3.9). These may require a minimum furnace cross section to avoid flame impingement and a furnace length different from a standard furnace. Thus, standardization is possible only to a limited extent in large water tube boiler designs. While standard boiler applications find a place in the industry, there are numerous applications where a custom-designed boiler would fit the needs of the end

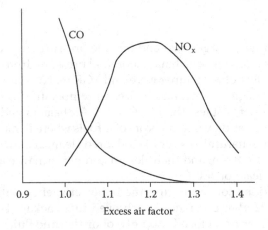

FIGURE 3.7
Typical NO_x and CO levels versus excess air.

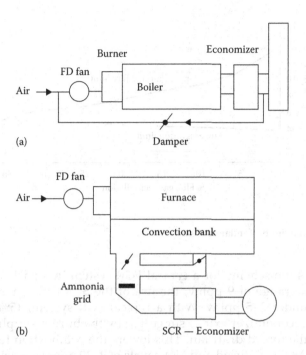

FIGURE 3.8
Arrangement of (a) FGR and (b) SCR system in package boiler for NO_x control.

Dual burner, 600 MMBtu/h

FIGURE 3.9
Low NO_x burner. (Courtesy of Cleaver Brooks Inc., Natcom Burners, Thomasville, GA.)

user better and lower the user's cost of owning the boiler. Plant engineers prefer custom-designed boilers as it lowers the operating costs and saves them money in the long run.

Operators must also consider the risk of operating a boiler near the limits of inflammability when using high FGR rate. Figure 3.10 shows the narrowing between the upper flammability limit and lower ignition limit as FGR increases. Integrating control systems to maintain fuel/air ratios at high FGR rates is difficult because FGR dampens the combustion process to the ragged edges of flammability-flameouts and flame instability. Full metering combustion control systems with good safety measures are necessary in such cases (Figure 3.11a and b).

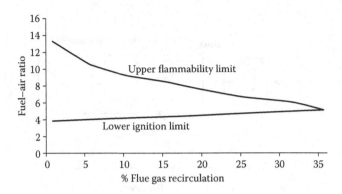

FIGURE 3.10
Flue gas recirculation and limits of inflammability.

Figure 3.8 shows the scheme for a typical FGR system in a package boiler that can achieve NO_x in the range of 9 ppmv; the use of the SCR can achieve 90%–95% NO_x removal or NO_x under 3–5 ppmv. With a typical FGR system, 15%–30% of flue gas quantity from the economizer exit as suggested by the burner supplier is recirculated to the suction of the forced draft fan. This lowers the combustion temperature of the flame, and hence, NO_x is reduced and NO_x levels of 9–30 ppmv are achieved for natural gas firing. FGR also increases the combustion air temperature entering the forced draft fan. With 26°C combustion air (80°F) and 15% FGR at 160°C (320°F) flue gas temperature, the mixed air temperature is about 44°C (112°F). The air density at the fan also decreases adding to the volume of air to be handled by the fan. FGR also affects the gas temperature profile through the boiler and affects the furnace heat absorption and heat flux.

The superheater performance is also impacted by FGR depending on where it is located, whether radiant or convective. This is discussed later. Custom designing of boiler evaporator section, which is the source of major gas-side pressure drop, can help lower the gas pressure drop arising out of higher mass flow through the boiler due to FGR. Tube length can be increased or tube spacing increased to reduce the gas velocity and pressure drop.

Catalyst operates efficiently in a range of gas temperatures, typically between 340°C and 420°C. If one looks at the gas temperature profile along the flue gas path, one may note that the gas temperature decreases as the load decreases (Figure 3.12). Hence, a gas bypass system is often used to maintain the gas temperature at the catalyst at low loads. The flue gas should be taken from a reasonably high gas temperature zone and mixed with the flue gas at the evaporator exit as shown; the ammonia injection system should be located upstream of the SCR and should have sufficient mixing length so that the flue gas can react with ammonia. The SCR catalyst pressure drop is typically in the range of 50–75 mm wc. One can thus see how complex a steam generator has become due to NO_x and CO emission regulations, not to speak of the operating and installation costs! Hence, a standard off-the-shelf boiler cannot be used for applications involving emission controls, particularly those using the SCR system. Figure 3.13 shows the scheme of ammonia injection system and the SCR. The ammonia grid is installed ahead of the catalyst box, and one should ensure good mixing of ammonia with the flue gas before it passes through the catalyst.

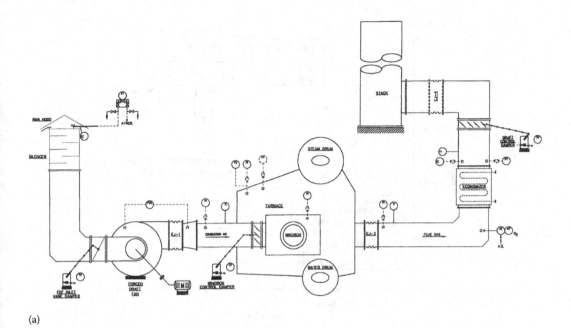

(a)

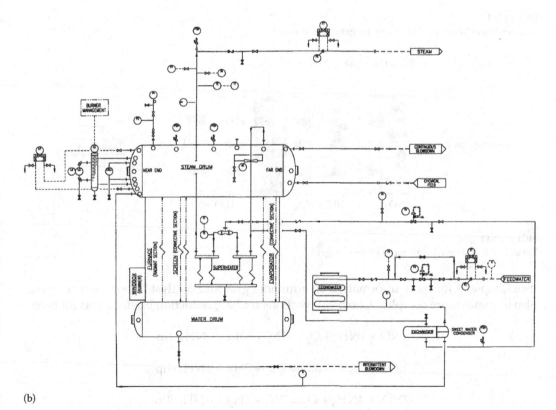

(b)

FIGURE 3.11
(a) Scheme of boiler controls—gas side. (b) Scheme of controls—steam side. (Courtesy of Cleaver Brooks Inc., Engineered Boiler Systems, Thomasville, GA.)

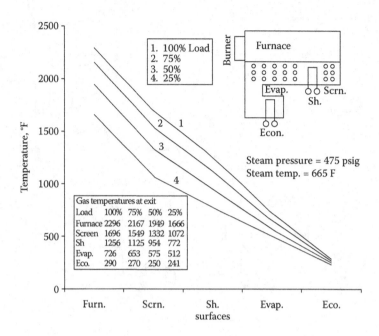

FIGURE 3.12
Flue gas temperature profile in a steam generator versus load.

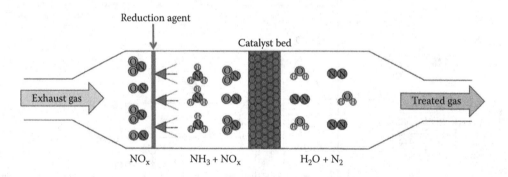

FIGURE 3.13
Typical SCR system with ammonia injection grid.

(See Chapter 4 for more information on ammonia grid and catalyst.) Hence, boiler layout also becomes more complex. Chemical reactions for NO_x reduction briefly are as follows:

$$4NO + 4NH_3 + O_2 \Rightarrow 4N_2 + 6H_2O + NH_3 \text{ slip}$$

$$NO + NO_2 + 2NH_3 \Rightarrow 2N_2 + 3H_2O + NH_3 \text{ slip}$$

$$2NO_2 + 4NH_3 + O_2 \Rightarrow 3N_2 + 6H_2O + NH_3 \text{ slip}$$

$$6NO + 4NH_3 \Rightarrow 5N_2 + 6H_2O + NH_3 \text{ slip}$$

$$6NO + 4NH_3 \Rightarrow 5N_2 + 6H_2O + NH_3 \text{ slip}$$

The idea is to convert the NO_x to N_2 and H_2O using the ammonia. Excess ammonia called ammonia slip should be as low as possible as it is also considered a pollutant. Controls and instrumentation are used to ensure that the slip is within levels permitted by local regulations.

Sulfur-containing fuels present problems for the catalyst. Vanadium present in the catalyst convert SO_2 to SO_3, which can react with excess ammonia to form ammonium sulfate or with water vapor to form sulfuric acid causing problems such as fouling and plugging of tubes downstream of the gas path. One method is to limit the number of hours of operation on fuels containing sulfur. Lowering ammonia slip helps, but this can affect the NO_x reduction efficiency. Environmentally, ammonium sulfate and bisulfate are particulates that contribute to visible haze and acidify lakes and ground areas where they settle out of the air. Sulfates are formed according to the following equations:

$$SO_3 + NH_3 + H_2O \rightarrow NH_4HSO_4$$

$$SO_2 + 2NH_3 + H_2O \rightarrow (NH_4)_2SO_4$$

Ammonium sulfate is a sticky substance that can be deposited on heat transfer surfaces such as finned economizer. If the ammonia slip is less than 10 ppm and SO_3 is less than 5 ppm, experts say that the probability of ammonium sulfate formation is nil unless the gas temperature is low on the order of 200°C at the catalyst. One may keep the boiler warm during shutdown or in standby conditions to prevent condensation of acid vapors.

Custom-Designed Boilers

Emission regulations and the associated operating costs have forced boiler designers to move away from the concept of standard designs to custom-designed boilers, particularly with large steam generators with high-pressure superheated steam. Designs are optimized to meet local emission regulations with high efficiency, while the operating costs are reduced. Allowable NO_x can vary from <5 ppmv in some locations in the United States to even as high as 76 ppmv in some other parts such as Canada. The boiler configuration and size will be different even though the steam parameters may be the same. FGR of 20% may be necessary to meet the emission limits in a particular area, while FGR may not be required in another area; the boiler cross section should increase by about 20% to maintain the same back pressure for the same steam generation; else the operating cost increases! One option is to manipulate the tube spacing or boiler tube lengths or widths to obtain the desired performance with the same gas pressure drop, in other words, custom designing. To quote a boiler salesman, "A gas- or oil-fired water tube boiler for California will not be the same as the boiler for China or South America even if the steam parameters are identical!" Examples of custom-designed boilers follow.

Novel Ideas

Novel ideas have been incorporated in package boilers by the author when he was a boiler designer to benefit the end user. Finned tubes in evaporators were unheard of in steam generators three decades ago, while they are inevitably used in gas turbine heat recovery steam generator (HRSGs) as they make the HRSGs compact. The author toyed with the idea of using finned tubes in the cooler region of the convection bank to achieve the desired performance with a lower gas pressure drop and a shorter length of convection section and started implementing the idea in a few boilers. Consultants who were well informed in process and thermal issues had no objection while conservative engineers even wondered if the concept will work! In applications involving a large degree of superheat (low steam pressure with a high degree of superheat), portion of the multistage superheaters have also been designed and built using finned tubes in order to make the convection bank compact; else, a large amount of space will be required for the superheater as also for the evaporator section. Hence, custom designing has become a way of life, and the end user is ultimately benefitted. Creativity is helpful in any business, and in package boiler industry, the author has been associated with a few interesting applications, which may be considered untraditional or novel!

O-Type Package Boiler

A natural gas–fired package boiler of capacity 50 t/h was required by a plant. However, the building in which the boiler was to be located was hardly 2.75 m wide, and it was difficult to install a conventional A- or O- or D-type boiler as the width would definitely exceed 3.5 m for this capacity. Borrowing the concepts used in the design of HRSGs, the author came up with the design shown in Figure 3.14. An O-type boiler was designed with the furnace dimensions required by the burner supplier to meet the emission limits. The burner was mounted on the front wall. There was no second pass! The convection section was an extension of the furnace with a combination of plain and finned tubes. The length of the convection bank would typically be the same as the furnace length if plain tubes were used; however, here by using finned tubes, the length of the convection bank was cut short to about 3 m (instead of the usual 8 m) so that the total boiler length was about 11 m only. The labor cost was also reduced as fewer tubes were used in the convection bank. If it were a plain tube convection section, it would have 75 rows of tubes deep, while with finned tubes, only 16 rows deep were required. Nine plain tubes were used ahead of the finned tubes to ensure that the tube wall and fin tip temperatures of the finned tubes were not exceeding carbon steel limits. The furnace back pressure was also slightly less with the finned tube design as fewer rows were required. Some data regarding the boiler and a plain tube equivalent boiler are given in Table 3.1. Several natural gas–fired package boilers designed by the author with finned tube superheaters and evaporators are in operation for the past several decades.

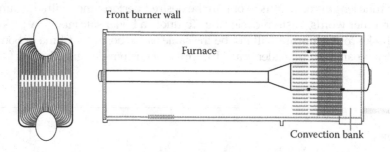

FIGURE 3.14
An O-type steam generator using a short finned convection. (Courtesy of Cleaver Brooks Inc., Engineered Boiler Systems, Thomasville, GA.)

TABLE 3.1

Comparison of Finned Tube Convection Bank versus Plain Tube Design

Item	Plain Tube (A, D, O Types)	Finned Tube (O Type)
Steam generation, kg/h	55	55
Steam temperature, °C	231	231
Steam pressure, kg/cm² g	28	28
Furnace width × height	1.9 × 2.75	1.9 × 2.75
Furnace length, m	7.6	7.6
Tubes/row convection	16	16
Number of rows deep	75	9 + 16 (9 plain, 16 finned)
Fin details	—	4 × 75 × 0.05 × 0.172
Total boiler width, m	3.7	1.85
Total boiler length, m	8	11
Gas temp to convection, °C	1303	1303
Gas temp. to eco, °C	313	351
Exit gas temp., °C	149	149
Efficiency, % LHV	93	93
Furnace back press, mm wc	116	100

Note: Economizer is made 5% larger in finned tube design as evaporator exit gas temperature is more.

Dual-Pressure Design

Another application where custom designing and creativity helped is shown in Figure 3.15. An 80 t/h boiler was required to generate steam at 10 barg and 360°C for the first few years and later on generate 80 t/h of steam at 45 kg/cm² g at about 400°C.

Designing a boiler to operate at two different pressures at the same capacity is a challenging task. Due to the lower specific volume at higher pressure and vice versa, the steam velocity and the steam pressure drop will be about four times more in the low-pressure case if the same number of streams were used in the superheater (for the meaning of streams refer to Appendix B), but pressure drop has to be reasonable in both high- and low-pressure operation. While many boiler suppliers came up with a superheater design that had to be replaced for the higher-pressure operation later, the concept shown in Figure 3.15 was developed by the author's team, which enabled the use of the same superheater at both operating conditions.

The suggested concept enabled the plant not to make any changes to the superheater inside the boiler when the steam pressure changed. In the initial low-pressure mode of operation, there would be two inlets to the superheater from the opposite direction as shown. The steam-specific volume would be higher, but as the path traveled by steam is shorter with smaller (about 50%) steam flow, the pressure drop will not be that much more. In the high-pressure case, steam inlet would come from one end and leave at the other end as shown. Thus, only the intermediate nozzle and one of the piping to the superheater from the drum had to be blanked; since these were accessible from outside, there was no need to pull out the superheater for the modification later, and hence, the plant preferred this concept.

Small Changes, Big Benefits

Custom designing process involves teamwork and understanding among design and fabrication groups in a boiler company. While many boiler suppliers like to keep on doing

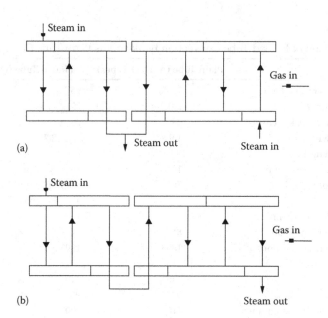

FIGURE 3.15
Superheater arrangement for (a) low-pressure and (b) high-pressure operations.

what they have been doing for decades, many small boiler companies have found ways to offer better value to the end user if they modified the existing designs whenever necessary. In the following example, a standard boiler design for an application involving FGR was found to have a very high back pressure, though the initial cost was attractive. The back pressure was reduced by making a minor change to the convection bank as a result of which the operating cost of the boiler was significantly reduced. The boiler cost did not increase, while the fan size and power consumption were drastically reduced. A strong thermal engineering team is required in a boiler company to perform this type of analysis. Plant engineers should review the thermal performance and question the boiler suppliers regarding operating costs and back pressure. Since many boiler companies do not want to spend time reengineering their boiler and want to push existing designs, plant engineers will not get a satisfactory answer! However, by becoming more knowledgeable, they can easily see how a particular boiler design can be improved. The following example illustrates the point.

Example 3.1

A boiler supplier proposed a steam generator firing distillate oil with a capacity of 90,750 kg/h (200,000 lb/h) of saturated steam at 27.6 barg (400 psig) using 110°C (230°F) feed water to a chemical plant; to meet low emissions of NO_x, the boiler supplier suggested the need for FGR at 17% and an excess air of 15%. On reviewing the boiler performance of the standard boiler, the plant engineer felt that the back pressure or flue gas pressure drop in the convection bank was high. The author also felt that the operating cost could be significantly reduced by making minor changes to the convection bank design without impacting the performance. The results are shown in Table 3.2 for the standard and custom designs showing how a small change can bring in so much relief in operating cost! The economizer exit gas temperature is the same in both cases,

TABLE 3.2

Comparison of Standard and Custom-Designed Boiler

Item	Standard Boiler		Custom Boiler	
Flue gas flow, kg/h	110,940		110,950	
Furnace exit temperature, °C	1,130		1,130	
Leaving evaporator, °C	386		422	
Leaving economizer, °C	148		149	
Evaporator surface, m²	552		552	
Economizer surface, m²	2,370		2,495	
	Evaporator	**Economizer**	**Evaporator**	**Economizer**
Tubes/row	12	20	12	20
Number of rows deep	105	19	105	20
Tube length, m	2.743	3.66	2.743	3.66
Transverse pitch, mm	133	114	159	114
Gas press drop, mm wc	286	33	150	36

Notes: Gas analysis: % volume CO_2 = 11.57, H_2O = 12.29, N_2 = 73.63, O_2 = 2.51. Fouling factor gas = 0.0006, and tube side = 0.0002 m²h °C/kcal. Tube size = 50.8 × 44.7 mm. Fins configuration of economizer: 197 fins/m, 19 mm high, 1.27 mm thick, 4.36 mm serration.

and hence, the efficiency is the same. The boiler furnace and convection bank surface area remained the same. However, the evaporator tube spacing was manipulated in the custom boiler to lower the gas pressure drop. To compensate for the slightly higher gas temperature entering the economizer, one more row was added to the economizer. Hence, boiler cost did not see any increase! (A large company in my personal opinion may not be amenable to such design modifications as it involves several layers of bureaucracy to make these small changes or they may say that they have thousands of such standard boilers in operation worldwide, and if the plant engineers are not process-savvy, they may go along with the poor design losing money in operating costs year after year!)

The boiler practically is the same in both cases except for the tube spacing of the evaporator section and an additional row of tube in the economizer. However, the operating cost of the custom-designed boiler reduces by over 130 mm wc. The fan power saved is $2.7 \times 10^{-6} \times 90,000 \times 130/(1.16 \times 0.7)$ = 39 kW (1.16 kg/m³ is the density of air and 0.7 is the efficiency of fan motor combination and air flow = 90,000 kg/h). At 15 cents/kWh, the annual savings in operating cost due to this slight modification is = 39 × 0.15 × 8,000 = $46,800. Hence, the custom design pays for itself in a short period. It is not sufficient if the plant engineer reviews the painting or insulation or the welding aspects or weight of the boiler. The objective of the book is to make plant engineers process-savvy so that they can review boiler performance from boiler suppliers and suggest improvements or question their data!

Plant engineers can also verify the performance data for the standard and custom boilers as follows. Appendix A shows the procedure.

Solution

Calculations for Evaporator Performance in Standard Boiler

Gas mass velocity G = 110,940/[12 × 2.743 × (0.133 − 0.0508) = 41,000 kg/m²h = 11.39 kg/m²s.

Average gas temperature in the bank tubes = (1130 + 386)/2 = 758°C = 1031 K. A good estimate of gas film temperature = 0.5 × (758 + 231) = 494°C or close to 500°C. Gas properties for heat transfer are evaluated at this temperature. From Appendix F,

$C_p = 0.2863$ kcal/kg °C, $\mu = 0.1209$ kg/m h, $k = 0.046$ kcal/m h °C. Tube wall temperature may be assumed to be 237°C, a little more than saturation due to the high tube-side boiling heat transfer coefficient of about 10,000 kcal/m² h °C (assumed). As discussed in Appendix A, the tube-side coefficient will not impact U.

Let us use Grimson's correlations for h_c. For $S_T/d = 133/50.8 = 2.62$ and $S_L/d = 101.6/50.8 = 2$, from Table C.2, $B = 0.213$, $N = 0.64$.

Reynolds number Re $= 41,000 \times 0.0508/0.1209 = 17,227$. Nu $= 0.213 \times 17,227^{0.64} = 109.5 = h_c \times 0.0508/0.046$ or $h_c = 99.15$ kcal/m² h °C. A correction factor of 10% is used with Grimson's correlations for convection bank with membrane panel end walls (based on the author's experience). Hence, use $h_c = 109$ kcal/m² h °C.

Calculation of nonluminous heat transfer coefficient h_n. Using methods discussed in Appendix D, beam length L $= 1.08 \times (0.133 \times 0.1016 - 0.785 \times 0.05058 \times 0.0508)/0.0508 = 0.244$ m. K $= (0.8 + 0.12 \times 1.6) \times (1 - 0.00038 \times 1031) \times 0.238/(0.238 \times .244)^{0.5} = 0.596$.

$$\epsilon_g = 0.9 \times [1\text{-}exp(.244 \times 0.5960)] = 0.122$$

$h_n = 5.67 \times 10^{-8} \times 0.122 \times [1031^4 - 510^4]/(1031 - 510) = 14.1$ W/m² K $= 12.12$ kcal/m²h°C (1031°K is the average gas temperature and 510 K is the average tube wall temperature.)

$1/U_o = 1/(109 + 12.12) + 0.0006 + 0.0505 \times ln(50.8/44.7)/(2 \times 35) + 50.8/(44.7 \times 9800) + 0.0002 \times 50.8/44.7$ or $U_o = 107.6$ kcal/m² h °C (125.1 W/m² K)

Let us check the evaporator exit gas temperature. One may use the following equation for evaporator performance as shown in Appendix A.

Ln[(1130–231)/(T – 231)] $= 107.6 \times 552/(110,941 \times 0.303)$ or T $= 385$°C agrees with the data given by the boiler supplier (0.303 is the specific heat of flue gas at the average gas temperature).

Flue Gas Pressure Drop

MW of flue gas $= 44 \times 0.1157 + 18 \times 0.1229 + 28 \times 0.7363 + 32 \times 0.0251 = 28.72$. $S_T/d = 2.625$. $S_L/d = 2$.

Density at average gas temperature of 758°C $= 12.17 \times 28.72/(273 + 758) = 0.339$ kg/m³.

Reynolds number at 758°C $= 41,000 \times 0.0508/0.1469 = 14,178$. See Appendix C for flue gas pressure drop calculations.

Friction factor f $= 14,178^{-0.15}[0.044 + 0.08 \times 2/\{1.625^{(0.43 + 1.13/2)}\}] = 0.034$

$\Delta P = 0.204 f G^2 N_d/\rho_g = 0.204 \times 11.39^2 \times 0.034 \times 105/0.339 = 279$ mm wc, which is close to the boiler supplier's data.

Calculations for Evaporator Performance in Custom Boiler

Now let us look at the option of $S_T = 159$ mm. Average gas temperature is (1130 + 422)/2 = 776°C. G $= 110,940/[12 \times 2.743 \times (0.159 - 0.0508) = 31,150$ kg/m²h $= 8.65$ kg/m² s.

Approximate gas film temperature $= 0.5 \times (776 + 231) = 503$°C, which is not much different from the earlier case. So let us use the same flue gas properties.

For $S_T/d = 3.125$, $S_L/d = 2$, $B = 0.198$, and $N = 0.648$ from Appendix C. Re $= 31,150 \times 0.0508/0.1209 = 13,088$. Nu $= 0.198 \times 130,880.648 = 92.1$ or $h_c = 92.1 \times 0.046/0.0508 = 84.3$ kcal/m² h °C. Using 10% margin as before, $h_c = 91.8$ kcal/m² h °C.

Beam length L $= 1.08 \times (0.159 \times 0.1016 - 0.785 \times 0.05058 \times 0.0508)/0.0508 = 0.300$ m

K $= (0.8 + 0.12 \times 1.6) \times (1 - 0.00038 \times 1049) \times 0.238/(0.238 \times .300)^{0.5} = 0.531$

$$\epsilon_g = 0.9 \times [1 - exp(-.0300 \times 0.531] = 0.132$$

$h_n = 5.67 \times 10^{-8} \times 0.132 \times [1049^4 - 510^4]/(1049 - 510) = 15.87$ W/m^2 K $= 13.65$ kcal/m^{2}h °C

$1/U_o = 1/(91.8 + 13.65) + 0.0006 + 0.0505 \times \ln(50.8/44.7)/(2 \times 35) + 50.8/(44.7 \times 9800) + 0.0002 \times 50.8/44.7$ or $U_o = 95$ kcal/m^{2}h °C (110.5 W/m^2 K).

Let us compute the exit gas temperature T. Ln$[(1130 - 231)/(T - 231)] = 95 \times 552/(110{,}941 \times 0.303)$ or T $= 420$°C agrees with the data given by the boiler supplier.

Flue Gas Pressure Drop

Reynolds number at average gas temperature of 776°C $= 31{,}150 \times 0.0508/0.1486 = 10{,}649$.
Friction factor $f = 10{,}649^{-0.15}[0.044 + 0.08 \times 2/\{2.125^{(0.43 + 1.13/2)}\}] = 0.0297$.

$\Delta P = 0.204fG^2N_d/\rho_g = 0.204 \times 8.65^2 \times 0.0297 \times 105/0.333 = 143$ mm wc, which is close to the value shown. Density of flue gas at average gas temperature is 0.333 kg/m^3.

Economizer Performance

A simple way of checking the economizer duty will be explained as the economizer design or cross section is unchanged. Calculate the LMTDs for both cases and the duty that is proportional to the temperature rise of the water. The ratio of the duty for custom to standard economizer is $Q_c/Q_s = (U_c \times \Delta T_c \times 2495)/(U_s \times \Delta T_s \times 2370) = 87/76$ (87°C and 76°C refer to water temperature rise, which is a proxy for the duty as water flow is the same in both cases and the surface areas are, respectively, 2495 and 2370 m^2).

$$\Delta T_c = [(386 - 186) - (148 - 110)]/\ln[(386 - 186)/(148 - 110)] = 97.5°C.$$

$$\Delta T_s = [(422 - 197) - (149 - 110)]/\ln[(422 - 197)/(149 - 110)] = 106°C$$

Hence, $(U_c \times 106 \times 2495)/(U_s \times 97.5 \times 2370) = 87/76$ or $U_c/U_s > 1$. That is, it is sufficient if U_c is equal to U_s to transfer the desired duty in the economizer. However, note that U_c will be higher than U_s as the average gas temperature in the economizer is now higher. Hence, the duty Q_c can be easily achieved. *Hence, the custom design is a far better offering than the initial standard design.* Only by challenging the boiler supplier, a better design emerged. Thus, plant engineers should have process-oriented background, and this book will help them become a little more process-savvy.

Boiler Classification

Boilers may be classified into several categories depending on the following:

1. By application as utility, marine, or industrial boiler. Utility boilers are the large steam generators used in power plants generating 500–1000 MW of electricity. They are generally fired with solid fuels, pulverized coal, or fluidized bed furnaces. Utility boilers generate high-pressure, high-temperature superheated and reheated steam; typical steam parameters are 16,500 kPa (165 barg), 540°C/540°C. A few utility boilers generate supercritical steam at 240 barg, 590°C/590°C/590°C with double reheat systems. Industrial boilers used in cogeneration plants generate steam at varying steam pressures of 10–100 barg at steam temperatures ranging

from saturation to 500°C. They are generally fired on oil and gas. Industrial boilers may import saturated steam from other boilers to be superheated or export saturated steam with the balance being superheated. Hence, their designs are more challenging.

2. By circulation methods: Figure 2.6 illustrates the concepts of natural, forced, or controlled and once-through steam generators (OTSGs). Natural circulation designs are used up to 165 barg steam pressure. Beyond this, a controlled circulation system using circulating pumps is preferred. A small operating cost is involved for moving the steam water mixture through the evaporator tubes. However, the circulation ratio (CR) may be kept small based on the design and experience of the boiler supplier. CR of 4–7 is common. The once-through design concept may be implemented at any pressure of steam. However it requires zero solids feed water; else deposits inside tubes while evaporation occurs would increase the evaporator tube wall temperature and damage the tubes particularly in the high heat flux region.

3. By fuel type: Type of fuel-fired influences the boiler size significantly. Solid fuel requires low heat release rates as combustion is difficult, and more residence time is required in the furnace. Oil- and gas-fired furnaces can be smaller. The furnace absorption with oil firing is more due to the higher intensity of flame radiation, and hence, the furnace for oil fuels is slightly larger. This is discussed in Chapter 2.

    Steam pressure also affects the boiler furnace size. As can be seen from Chapter 2, the latent heat of steam is much higher at lower steam pressure compared to higher steam pressure. The liquid heat or the energy pickup from feed water temperature at economizer inlet to the saturated steam point is much higher as the steam pressure increases. Hence, very high-pressure boilers may have a small furnace and no evaporator or boiler bank to speak of and a large superheater and a large economizer, while a low steam pressure boiler will have a large furnace and a boiler bank or evaporator and a small superheater and a small economizer.

4. Type of boiler: package boilers are classified as A, D, or O type depending on their construction as shown in Figures 3.16 and 3.17.

O-type boilers: This boiler has a steam drum and a bottom drum and has a cross section like the letter O. It is symmetrical in design and hence can be shipped easily, which is why this type of design is favored in rental boiler industry where a boiler may be used at various locations during its lifetime. The gas outlet is at the burner end and at the top. An economizer may be located in the gas outlet duct. If a convective superheater is required, it has to be split up as shown in the A-type boiler. If two-stage superheater with interstage attemperation is required, it adds to the complexity of piping, and two desuperheaters are required. A radiant superheater as shown in Figure 3.16a may be used, but unless the furnace is properly sized, flame can impinge on the superheater tubes, causing it to sag or even overheat and fail.

The A type is similar in features to the O-type boiler and being symmetrical in construction is well suited for frequent movements as a rental boiler. The convective superheater has to be split and located in the two parallel sections of the convection bank, while the radiant superheater is similar to that used in the O type. Using a single-stage radiant section is not advisable as temperature control is difficult. Hence, a combination of convection and radiant superheaters is preferred. This is discussed later.

The D type is well suited for large boiler capacities with high controlled steam temperatures (Figures 3.16d and 3.17a, b). Since all the flue gas goes through the convection bank, there is no question of maldistribution between the two convection passes as in A- or O-type boilers.

When steam temperature control is required, it is desirable to use a screen section followed by a two-stage superheater with a desuperheater in between. The economizer can be located in the horizontal gas flow direction as shown in Figure 3.17a to make the arrangement simple. More space is required for the D-type boiler as the gas exit is at the side and not at the boiler front as in O- and A-type boilers. Figure 3.17b shows the economizer of a D-type boiler in the vertical gas flow direction. The stack can be mounted on top of the economizer.

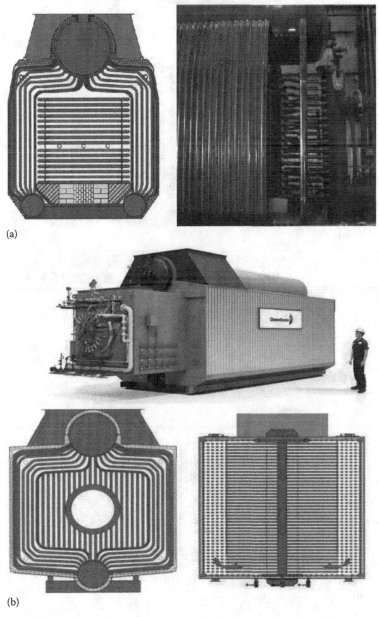

(a)

(b)

FIGURE 3.16
(a) A-type boiler, (b) O-type boiler,

(*Continued*)

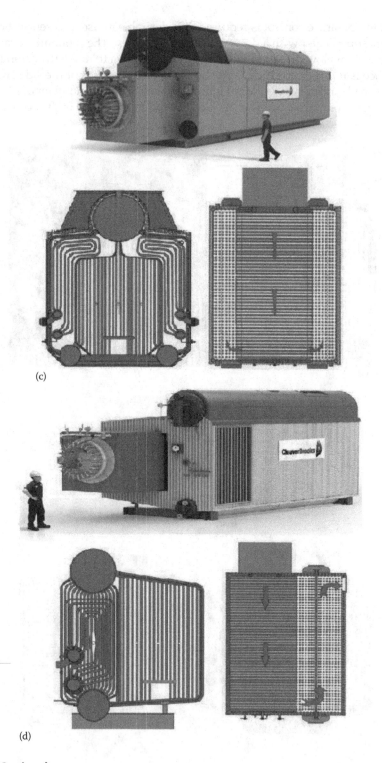

FIGURE 3.16 (*Continued*)
(c) A-type boiler with superheater, and (d) D-type boiler with superheater. (Courtesy of Cleaver Brooks Inc., Engineered Boiler Systems, Thomasville, GA.)

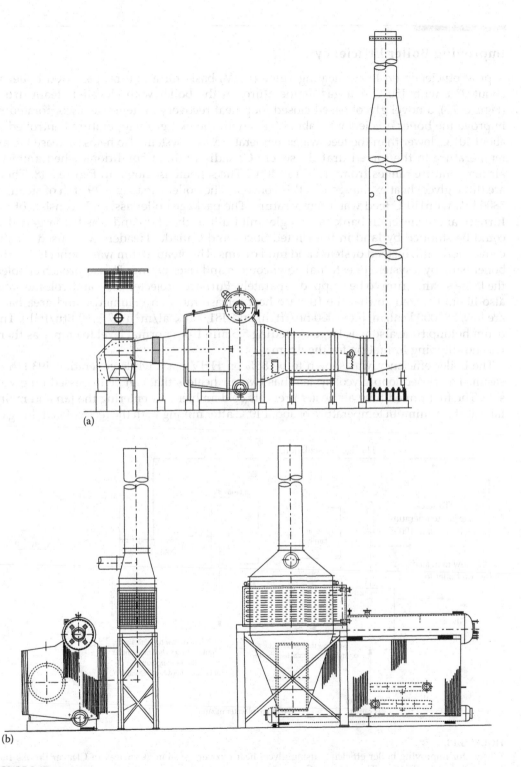

FIGURE 3.17
(a) D-type boiler with horizontal gas flow economizer and (b) D-type boiler with economizer in vertical gas flow direction. (Courtesy of Cleaver Brooks Inc., Engineered Boiler Systems, Thomasville, GA.)

Improving Boiler Efficiency

Typical efficiency on lower heating value (LHV) basis for a natural gas–fired boiler is about 93% with 150°C exit gas temperature. In the boiler with elevated steam drum (Figure 3.3), a novel glycol-based closed-loop heat recovery system was incorporated to improve the boiler efficiency by about 2.5% with the exit gas temperature controlled at about 101°C, lower than the feed water temperature! The system also helps increase the air temperature to the forced draft in severe Canadian winter conditions when ambient air temperature ranges from –10°C to –40°C! The scheme is shown in Figure 3.18. There are three glycol heat exchangers in this concept. The boiler capacity is 193 t/h of steam at 3800 kPa with110°C feed water temperature. The package boiler assembly consists of the furnace and convection bank as a single unit built in the shop and was the largest that could be shipped by land in the United States and Canada. Headers were used for the convection bank instead of steam and mud drums. The steam drum was connected to the boiler bank by a system of external downcomers and riser piping. This approach enabled the large steam drum to be shipped separately. Furnace projected area and volume were also liberally sized so that the furnace heat release rates on volumetric and area basis are low, 587,000 kcal/m³ h (66,000 Btu/ft³ h) and 481,000 kcal/m² h (177,000 Btu/ft² h). The drum holdup time also was large exceeding 5.5 min from normal level to empty as there was no shipping limitation for the drum.

The boiler efficiency was shown to be 86% on HHV basis while generating 193 t/h of steam. The closed-loop glycol system has two air heaters that can be bypassed on glycol side. The first finned tube air heater preheats cold ambient air entering the fan and maintains it at a minimum temperature of about 10°C after mixing with the recirculated flue gas.

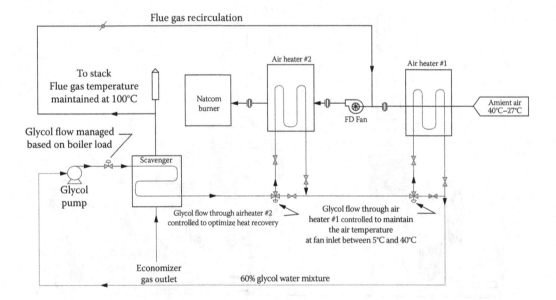

FIGURE 3.18
Scheme for improving boiler efficiency using glycol heat recovery system. (Courtesy of Cleaver Brooks Inc., Engineered Boiler Systems, Thomasville, GA.)

TABLE 3.3

Field Data for the Elevated Drum Boiler

Parameter	100% Load	120% Load
Steam flow, t/h	160	192
Steam pressure, kPa	3751	3751
Feed water in	110	110
Oxygen in flue gas,%	2.3	2.2
NO_x ppmv	46	44
Flue gas recirculation, %	10	11
Ambient temperature, °C	21	22
At air heater exit, °C	66	82
Exit gas temperature, °C	103	108

Note: Typical exit gas temperature would have been 150°C–160°C.

The second air heater, located after the fan, preheats the mixture of flue gas recirculated for NO_x control and air from the fan. This glycol bypass system maintains the stack gas temperature at about 101°C. If the ambient air temperature is high, as it is in summer, then either air heater may be shut off and the glycol flow bypassed on the tube side of either air heater, as very hot combustion air will increase NO_x production. The as-tested performance of the plant is summarized in Table 3.3 [2,3].

Figure 3.19 shows a scheme for improving the efficiency of a gas-fired boiler without using any thermic fluid heating system as discussed earlier. Incoming air from the forced draft fan is heated to about 65°C–75°C by a portion of the feed water. (In case there is no NO_x concern, the air temperature can be even further increased and the exit gas temperature from the economizer can be lowered further.) The cooled feed water is heated in another exchanger called scavenger located behind the regular economizer. The increase in air temperature is approximately equal to the drop in flue gas temperature beyond the economizer. If there are no NO_x or acid dew point concerns, then the flue gas can be cooled to a lower temperature. This scheme does not require a heat sink such as makeup water to lower the flue gas temperature from the system. Efficiency improvement can be in the range of 1.5%–2% over conventional plants, and the additional cost of the two finned coils and air/gas pressure drop will pay for itself in a short period depending on local fuel cost and the cost of electricity. (This scheme is being patented by Cleaver Brooks, Nebraska, boiler division.)

Adding Condensate Heater to Improve Boiler Plant Efficiency

As in HRSGs, one may consider adding a condensate heater in case the makeup water flow is high enough as shown in Figure 3.20. A heat exchanger may also be used to lower the feed water temperature and preheat the makeup in a separate heat exchanger or a condensate heater may be used as shown. One should use appropriate materials when water inlet temperature to condensate heater drops below the water dew point as discussed in Chapters 1 and 6. Another option is shown in Figure 3.20a, where a heat exchanger is used to lower the feed water temperature to the economizer thus preheating the makeup water, which enables reduction in deaeration steam.

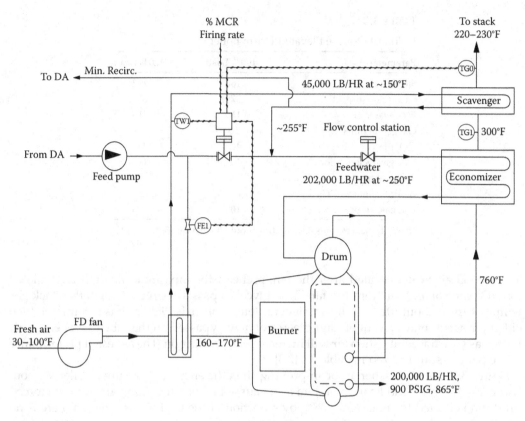

FIGURE 3.19
Scheme to improve boiler efficiency. (Patented by Cleaver Brooks Inc., Engineered Boiler Systems, Thomasville, GA.)

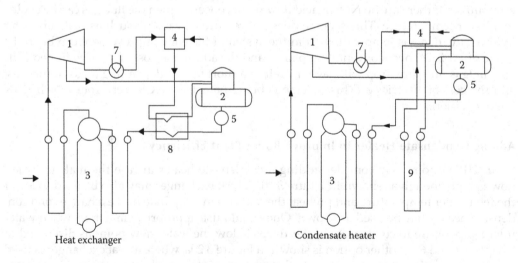

FIGURE 3.20
Scheme to increase boiler plant efficiency using a heat exchanger and lower feed water temperature to economizer. *Note:* (1) turbine, (2) deaerator, (3) HRSG, (4) mixing tank, (5) pump, (6) deaerator coil, (7) condenser, (8) heat exchanger, (9) condensate heater.

TABLE 3.4

Effect of Lower Feed Water Temperature

Item	Case 1	Case 2
Steam flow, kg/h	90,740	90,740
Steam pressure, barg	27.6	27.6
Feed water temp., °C	110	60
Blowdown,%	1	1
Exit gas temp., ±°C	153	115
Efficiency, % LHV	93.2	95.0
Efficiency, % HHV	84.1	85.74
Energy to steam, MM kcal/h	50.78	55.37
Fuel LHV, MM kcal/h	54.48	58.28

Example 3.2

A 90,740 kg/h (200,000 lb/h) natural gas–fired D-type package boiler is generating saturated steam at 27.6 barg (400 psig) with feed water at 110°C (230°F) in a cogeneration plant. Customer is using a lot of steam in the deaerator and wants to know if the feed water can be used to preheat the makeup water–condensate mixture entering the deaerator and thus lower the deaeration steam requirements and whether this scheme has any benefit to him. Customer thinks he can cool the feed water to 60°C and the makeup water mixture can enter the deaerator much hotter and thus save his energy requirements. With 60°C feed water, there is no concern about water dew point condensation as natural gas is fired in the boiler with partial pressure of water vapor about 0.18. Water dew point corresponds to 57°C from steam tables.

The boiler performance was simulated with feed water at 110°C and at 60°C. Results are shown in Table 3.4. Note that the boiler size has not changed in these options—only the feed water temperature.

The scheme may be evaluated in a few different ways. If deaeration steam is taken from the boiler itself, then one can increase the boiler steam capacity based on deaeration steam requirement for each case and see the difference. But it is obvious that even though the fuel-fired is 3.8 MM kcal/h more, the additional energy recovered in the entire system due to the higher condensate temperature entering the deaerator will be 90,740 × 1.01 ×(110–60) = 4.58 MM kcal/h. Hence, the scheme is more efficient; one has to consider the additional cost of the heat exchanger and associated piping valves for cooling the feed water, but it is worth considering if an appropriate heat sink is available.

Understanding Boiler Performance

It is important for plant engineers to have an idea of how the various important parameters such as efficiency, exit gas temperature, and water temperature beyond the economizer vary with load. Many refineries have both steam generators and fired HRSGs. Knowing this trend in performance versus load for both the steam generator and HRSG, one can load each of the equipment such that the maximum steam is generated at the lowest fuel consumption. This is discussed in Chapter 4. Presented in Figure 3.21 is the performance chart of a gas-fired steam generator showing how the various parameters vary with load. It may be seen as follows:

1. The efficiency curve shows a parabolic trend. This is due to the combination of casing loss, that increases as the load decreases and the other heat losses that decrease as the load decreases. This trend is also explained in Chapter 1.

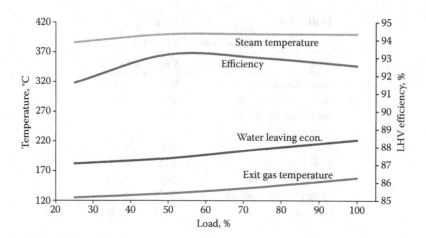

FIGURE 3.21
Steam generator performance versus load.

2. The exit gas temperature and water temperature leaving the economizer increase as the load increases. This is due to the higher mass flow of flue gases at higher gas temperatures as the load increases. Unlike the HRSG, steaming is rare in steam generator economizer as the ratio between air flow and steam flow is maintained by combustion control system. If the economizer is sized so that it does not steam at full load, it will never steam at part loads. The approach point (difference between the saturation temperature and water temperature leaving the economizer) keeps increasing as the load decreases, while it shows the opposite trend in gas turbine HRSGs.

3. The steam temperature increases as the load increases if a convective-type superheater is used. This type of superheater is generally suggested up to about 500°C. For higher steam temperatures, one may consider a combination of radiant and convective superheaters. The features of radiant and convective superheaters are discussed later. The furnace exit gas temperature is lower on oil fuel firing compared to natural gas. Hence, the superheater is sized based on oil firing. With natural gas firing, the gas temperature at the superheater increases, which can be controlled using spray desuperheater.

4. The feed water temperature leaving the economizer is higher at higher loads in a steam generator, while it is lower at higher loads in an HRSG. Hence, we have the *steaming* problem in economizer with HRSGs, while it is absent in steam generators.

Why Economizer Does Not Steam in Steam Generators

In a steam generator, the ratio between the fuel flow and steam flow is maintained at all loads through combustion control; this results in a flue gas flow to steam ratio of about 1.1 at normal loads and about 1.3–1.5 at lower loads. However, in a gas turbine HRSG, the exhaust gas mass flow remains somewhat constant at all loads, while the steam flow varies with exhaust gas inlet temperature. The ratio between exhaust gas and steam can be as high as 7–10 at low loads. At low loads, the HRSG makes lesser steam due to the lower exhaust gas inlet temperature. Since the economizer water flow matches the steam flow, the combination of low water flow through the economizer with large exhaust gas flow results in relatively large energy transfer in the economizer and high enthalpy pickup in the feed water causing it to generate steam or reach temperatures close to saturation. On the other hand, at part loads, the flue gas temperature to the economizer drops off in a steam generator along with flue gas flow. Hence, the water temperature leaving the economizer is lower at lower loads and the *steaming issue* does not arise in steam generators. The approach point (difference between saturation temperature and water temperature leaving the economizer) increases as load decreases, while in an HRSG, it is the other way. This point is also elaborated in Chapter 5 on HRSG simulation.

Understanding Boiler Surface Areas

Many plant engineers and purchase managers make the mistake of evaluating or purchasing boilers solely based on surface areas. The misconception is that more surface area, better the boiler and more *steel* is offered! This is a wrong notion. In HRSGs, for example, we can show that based on fin geometry, one can have 50%–100% more surface area and yet have the same performance!! In steam generators, 15% difference in surface areas is not uncommon based on furnace sizing, location, and type of superheater whether convective or radiant and fin geometry used for the economizer.

Example 3.3

For a 60 t/h, 28 kg/cm²g, 400°C gas-fired steam generator with feed water at 110°C, two boiler designs are being offered, one with a semi-radiant superheater and another with convective superheater. The tube geometry details of both designs are shown in Tables 3.5 and 3.6. (Incidentally, any steam generator or HRSG supplier should provide data for his boiler in this format.) Tube material information may also be included for each section. If there are more heating surfaces, each of them should have the data shown. Streams are the number of tubes that carry the total flow; this has been explained in Appendix B.

TABLE 3.5

Geometric Data with Convective Superheater

Geometry	Screen	Sh1	Evap.	Econ.
Tube OD, mm	50.8	50.8	50.8	50.8
Tube ID, mm	44.7	44	44.7	44.7
Fins/m	0.000	0.000	0.000	197
Fin height, mm	0.000	0.000	0.000	19
Fin thk., mm	0.000	0.000	0.000	1.25
Fin width, mm	0.000	0.000	0.000	4
Fin conductivity	0.000	0.000	0.000	37
Tubes/row	12	12	12	16
Number of rows deep	11	24	60	9
Length, m	2.750	2.7	2.75	4.8
Tr pitch, mm	137	137	137	101
Long pitch, mm	101	125	101	101
Streams	0.000	24.000	0.000	8.000
Parl = 0, countr = 1	0.000	1.000	0.000	1.000

Note: Furnace length = 10.67 m, width = 2.14 m, height = 2.05 m.

TABLE 3.6

Geometric Data with Radiant Superheater

Geometry	Screen	Sh1	Evap.	Econ.
Tube OD, mm	50.8	50.8	50.8	50.8
Tube ID, mm	44.7	44	44.7	44.7
Fins/m	0.000	0.000	0.000	197
Fin height, mm	0.000	0.000	0.000	19
Fin thk., mm	0.000	0.000	0.000	1.25
Fin width, mm	0.000	0.000	0.000	4
Fin conductivity	0.000	0.000	0.000	37
Tubes/row	12	12	12	16
Number of rows deep	2	16	67	12
Length, m	2.750	2.7	2.75	4.8
Tr pitch, mm	137	137	137	101
Long pitch, mm	101	125	101	101
Streams	0.000	24.000	0.000	8.000
Parl = 0, countr = 1	0.000	1.000	0.000	1.000

Note: Furnace length = 9.14 m, width = 1.83 m, height = 2.05 m.

One may see that the overall performance is nearly the same with both the options. The total surface area of the radiant superheater design is about 12% lesser compared to the convective superheater option. The fan power consumption is the same as the back pressure. Efficiency also is close. The heat release rates are reasonable, and though the heat flux with the smaller furnace is more, it is far below the allowable limits for departure from nucleate boiling (DNB) and hence acceptable. The radiant superheater has a higher tube wall temperature, but considering the low steam pressure, the thickness provided is adequate. Life of the radiant superheater may be shorter if we use the Larsen Miller chart for life estimation, but it is likely to exceed the expected 40 years at these temperatures. One has to check the part load performance to see if this is acceptable and make a decision. The convective superheater design, even though it may be more expensive, is preferred for the comfort factor it offers and its longer life. However, the purpose

TABLE 3.7

Summary of Performance for Radiant and Convective Superheaters

Item	Radiant SH	Convective SH
Volumetric heat release rate, kcal/m³ h	928,500	680,560
Area heat release rate, kcal/m² h	482,400	389,352
Exit gas temperature, °C	167	164
Flue gas flow, kg/h	75,618	75,430
Efficiency, % LHV	92.5	92.7
Back pressure, mm wc	236	227
Furnace projected area, m²	98	122
Superheater surface, m²	64	97
Evaporator, m²	399	410
Economizer, m²	1,036	1,221
Economizer fin geometry	118 × 19 × 1.26 × 4	197 × 19 × 1.26 × 4
Furnace width × length × height, m	1.83 × 9.14 × 2.05	2.14 × 10.67 × 2.05
Type of superheater	Radiant	Convective
Superheater tube temperature, °C	560	510

of the exercise is to show that variations in surface areas can be there for the same overall performance, and operating cost and the design with lesser surface area cannot be ignored but properly evaluated (Table 3.7).

Steam Generators for Oil Sands Application

Steam-assisted gravity drainage (SAGD) is considered to be the most viable and environmentally safe recovery technology for extracting heavy oil and bitumen. In the SAGD process, one well is drilled above the bitumen deposit and a second one below the deposit. The upper well is supplied with high-pressure steam, and the lower well collects the heated oil or bitumen that flows out along with any water from the condensation of injected steam. The bitumen and water are pumped out, and they travel to a tank where the two elements are separated. The produced water is then cleaned, and it returns to the boiler where it is converted to steam and reinjected into the well. In the past, warm lime softening process was used to clean the oil field produced water. The produced water cannot be cleaned enough for conventional drum-type boilers so steam is generated in OTSGs making 75–80% quality steam, and hence, 20% of the water is wasted. Environmental regulations are becoming stricter in Alberta causing oil companies to focus on water conservation. As a result, a more viable and environmentally friendly technology called evaporator technology is replacing the warm line softening treatment process.

Evaporator technology produces higher-quality water compared to warm lime softening. The process takes the clarified water and evaporates it out and then condenses it back to produce feed water. It also conserves water and opens up the steam production options to drum-type boilers generating saturated steam. Evaporator water can be run through an OTSG or drum-type boiler (for medium pressure) or through Cleaver Brooks's newly developed forced circulation oil sands steam generator (FC-OSSG). The special design feature of the FC-OSSG is that it operates like an OTSG but can be mechanically cleaned or pigged by conventional means.

The FC-OSSG addresses the key requirements of oil industry professionals:

- Ability to handle the inevitable water quality upsets
- Capability of operating with evaporator-produced water while delivering best-in-class efficiency
- Ability to handle capacities up to 227 t/h of steam (500,000 lb/h)
- Meeting stringent regulatory requirements for water usage and emissions
- Features reliable steam generators that can work with available produced water
- Easy to maintain—all circuits are piggable (a metal piece is sent through each circuit to clean the inside surface, which is called a pig and the process is called pigging)

Figure 3.4 shows the steam generator for oil sands applications. Figure 3.22 shows the scheme of the forced circulation boiler. There are two parallel circuits, one for the furnace section and another for the convective evaporator section. The circulation system is designed for an average CR of about 5. By varying the flow through the radiant or convective section, one can adjust the CR in either path. The tube wall temperatures in the furnace and evaporator section are measured, which gives an indication of any tube-side buildup of solids. The CR can be increased in that circuit if need arises.

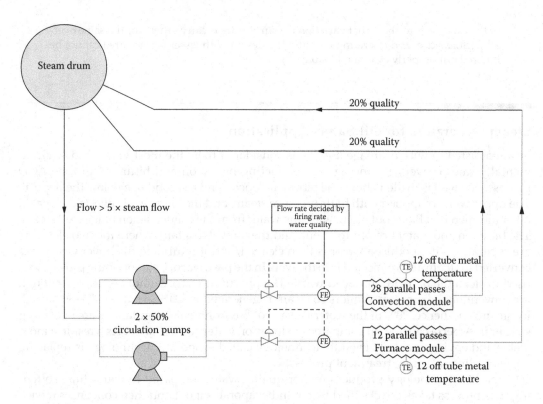

FIGURE 3.22
Scheme of forced circulation boiler (FC-OSSG) for oil field application. (Courtesy of Cleaver Brooks Inc., Engineered Boiler Systems, Thomasville, GA.)

The issue of water quality upsets is one that is important to oil industry professionals. Existing technologies repeatedly fall short in this area. Even though a system may be operating with evaporated water, which is supposed to be pure, it may still contain oil content above the acceptable limit for drum boilers. This, along with the inevitable water quality upsets, can cause fouling of the heat transfer tubes. Cleaning drum boilers requires access to the inside drums, which are not piggable, so tube cleaning can be costly and time-consuming. Alternatively, the OTSG can handle water quality upsets, but the equipment is limited in size. An OTSG can handle only up to 136 t/h (300,000 lb/h) of steam (Figure 3.23).

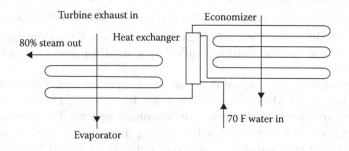

FIGURE 3.23
OTSG generating 80% quality steam in oil fields.

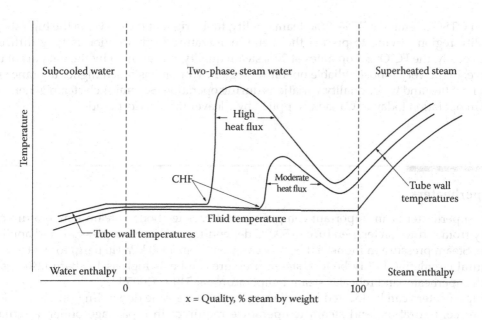

FIGURE 3.24
Typical steam quality versus tube wall temperature in boiling process.

The FC-OSSG is a forced circulation boiler (Figure 3.4). It uses circulation pumps to force the steam water mixture through the furnace and evaporator circuits. A CR of 5–8 is achieved depending on the actual resistance to flow in the circuits. Hence, the steam quality at the exit of the furnace or evaporator is about 12%–20% only compared to 80% for an OTSG. Hence, allowable heat fluxes for DNB are much higher as shown in Figure 3.24 and as also discussed in Chapter 2. Higher fouling compared to OTSG can be handled by OSSG. The tube-side velocities are designed to be very high (on the order of 2.5 m/s). Hence, deposition of solids is ruled out, and frequency of cleaning and potential for failure will be less compared to an OTSG.

In case of upsets or sudden increase in feed water conductivity and turbidity over limits set, the control system reduces the firing rate on the boiler and increases the water flow through the tube to protect it. Increasing the amount of water flowing through the tube reduces the steam quality in the heat transfer circuits, resulting in lesser concentration of impurities and reduced metal temperatures. Heat flux also is reduced due to lowering of heat input. The boiler keeps running, but at a reduced firing rate. Then, if it exceeds another preset value, the operator trips the boiler to protect it.

Unlike in an OTSG, whereby mostly steam travels through the tube, the FC-OSSG is designed so that mostly water flows through the tube. The metal temperature of a water-filled tube closely matches the water temperature rather than the hot gas temperature. For a constant heat flux (rate at which heat is transferred), the tube wall temperature is a function of the steam quality. At lower steam qualities, the tube predominantly is filled with water, and the metal temperature is closer to the water temperature and is stable. As the steam quality increases, the metal temperature also increases, and beyond a certain point, it starts dropping down, due to the very high velocities in the tube caused by the difference in densities of steam and water. Figure 3.24 shows the relationship between steam quality and metal temperature.

An OTSG operates at 75%–80% steam quality, to the right of the curve, in the high steam quality region. Trying to predict the metal temperature in this region is very difficult. Conversely, the FC-OSSG operates at 20% steam quality, represented by the flat part of the curve, in the very robust, reliable operating region. Thus, it has much higher tolerance for any heat flux and water quality variations during operation. Several such steam generators are in operation today in Canada, supplied by Cleaver Brooks Corporation.

Superheaters

The superheater is an important component of a package boiler. Steam temperature can vary from saturation temperature to 500°C depending on the steam pressure and application. Steam pressure in industrial boilers can vary from 1000 kPa (10 barg) to 103 barg. In natural circulation utility boilers, steam pressure can be as high as 165 barg (2400 psig) with superheater and reheater steam temperatures at 540°C (1005°F).

Superheaters can be located in radiant or convection zone depending on application, boiler configuration, and steam temperature required. In a package boiler, a radiant superheater receives direct radiation from the flame in the furnace, while convective superheater is shielded by several rows of screen tubes as shown in Figure 3.25. A convective superheater with horizontal tubes is shown in Figure 3.26a and b, while in Figure 3.27, an inverted-loop superheater is shown with vertical tubes. Note that the tube lengths vary between the headers. The tubes in the middle of the convection bank are the shortest and hence will have more steam flow compared to the longer tubes at the ends. However, one should also note that the maximum heat transfer occurs in the middle of the section and tapers off at the ends. Hence, a self-compensating effect is provided by this design, and this design is used in almost all package boilers. The horizontal tube design has the same length of tubes in all streams, and hence, the nonuniformity in steam flow is minimal. The shorter tube length and more sections may increase the cost of manufacturing, but both designs are used in the industry. Superheaters can also be of semi-radiant type with a portion of the superheater receiving furnace radiation. If the degree of superheat required is very low, say, 10°C–20°C, the superheater may be even located between the evaporator and the economizer of the package boiler.

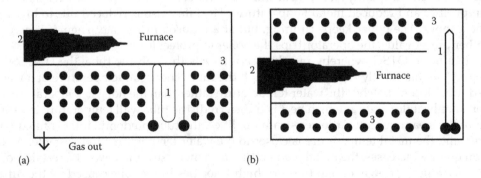

FIGURE 3.25
Location of (a) convective and (b) radiant superheater. *Note:* (1) superheater, (2) burner, and (3) screen, evaporator.

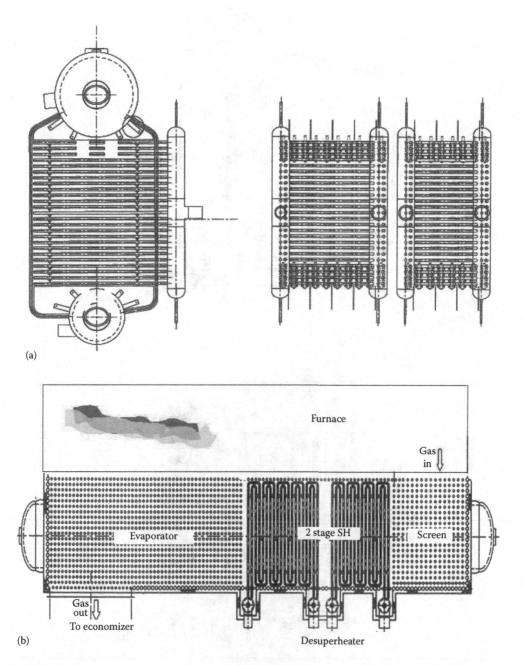

(a)

(b)

FIGURE 3.26
(a) Horizontal tube superheater elevation and (b) arrangement of a convective superheater with horizontal tubes. (Courtesy of Cleaver Brooks Inc., Engineered Boiler Systems, Thomasville, GA.)

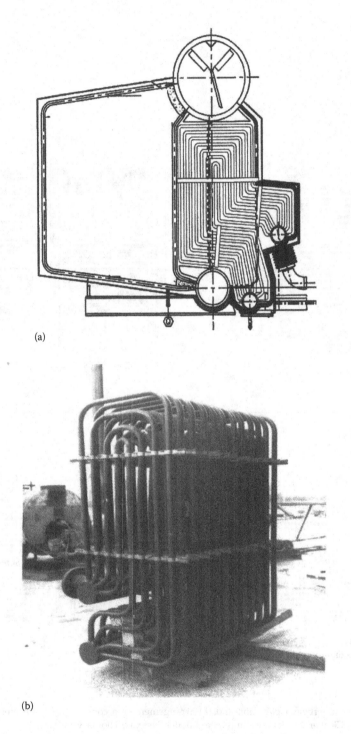

(a)

(b)

FIGURE 3.27
(a) Arrangement of a convective superheater with vertical tubes, (b) inverted-loop superheater.

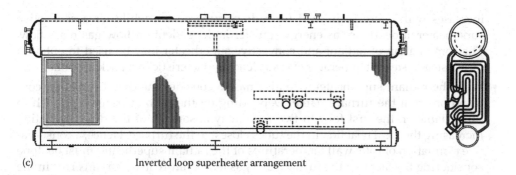

(c)   Inverted loop superheater arrangement

FIGURE 3.27 (*Continued*)
(c) Its location in package steam generator. (Courtesy of Cleaver Brooks Inc., Engineered Boiler Systems, Thomasville, GA.)

Radiant versus Convective Superheaters

Radiant superheaters, which are typically located near the furnace exit region, are often used by several boiler manufacturers. Radiant superheaters have to be designed very carefully because they operate in a much harsher environment compared to a convective superheater. The following issues have to be considered:

- It is difficult to estimate the furnace exit gas temperature accurately. Variations in fuel analysis, excess air, FGR rates, burner location and design, and flame shape can affect this value and consequently the temperature distribution across the furnace exit plane as discussed in Chapter 2. The temperature predicted can be off from the estimated value by even 50°C–110°C as correlations for furnace exit gas temperature are not accurate and established like the heat transfer correlations inside tubes or outside tubes.

- The turning section where the superheater is often located also adds to nonuniformity in gas velocity and temperature profiles, which can affect the superheater performance. Several boiler suppliers locate the superheater at the turning section, and performance prediction is a challenge due to variation in gas velocity and temperature profile across the cross section.

- If the furnace length and cross section are not properly designed or if the burner is not properly tuned, the flame may even lick the radiant superheater tubes resulting in overheating or even failure. I have seen plant engineers or the boiler supplier blame the burner supplier for the superheater failures or overheating, while the superheater itself could have been poorly designed with inadequate streams or steam velocity inside the tubes. This issue has to be understood by the plant engineers and hence discussed later.

- If one looks at the slope of the chart of furnace exit gas temperature versus heat release rate (see Chapter 2), at part loads, the furnace exit gas temperature does not decrease much compared to full load. Hence, the direct radiation from the flame is still significant while the steam flow is reduced. Also at part loads, the nonuniformity in gas-side velocity profile will be more due to the lower gas velocity at the superheater. This fact coupled with the low steam-side pressure drop and flow distribution inside the superheater tubes will impact the tube wall temperatures often leading to overheating of tubes. Hence, one should be careful while using radiant superheaters at low loads. As can be seen from Figure 3.28, the steam temperature increases with decrease in load for radiant superheater, while for a convective superheater, it decreases as energy transfer is dependent on flow gas mass flow. The combination of radiant and convective superheater may be used to obtain a rather steady steam temperature versus load characteristic in a package boiler.

- With the radiant superheater, one also has to consider the direct radiation contribution from the furnace flame. Depending on the ratio of transverse pitch to tube diameter, the first four to five rows may absorb all of the direct radiation increasing the local heat flux in the tubes closest to the furnace. Hence, a conservative estimation of tube wall temperatures of the radiant superheater must be done considering the fact that the furnace exit gas temperature itself has an error in the estimation of over 70°C.

The radiant superheater also absorbs more direct radiation at part loads from the furnace, increasing the steam temperature as well as the tube wall temperature as illustrated with an example in Chapter 2. At low loads, the flow distribution inside the tubes will also not be good. A few tubes can have lower than average steam flow due to lower steam velocity and pressure drop and higher gravity loss. Hence, radiant superheaters are prone to overheating unless carefully designed.

- The convective superheater, on the other hand, is located beyond several screen tubes, and the gas temperature and velocity profile at the superheater inlet region are more uniform and predictable, and hence, the superheater performance can be evaluated more accurately. At part loads, the gas temperature is significantly lower and the effect of flow nonuniformity does not cause overheating of tubes as in the case of radiant superheaters. However, due to the lower log-mean temperature difference (LMTD), the surface area required will be more than that of the radiant design. The materials used can be of lower grade as the tube wall temperatures will be lower than that of the tubes in the radiant section.

- In heavy oil–fired package boilers, use of radiant superheater is not encouraged as slagging of ash components may cause deposits and plugging of gas path and high-temperature corrosion failure of tubes. Wide-spaced screen section followed by a convective superheater is a better design option.

- It will also be shown later that the effect of variations in excess air or FGR does not impact the steam temperature in a convective superheater much, while a radiant superheater is impacted significantly. This can lead to underperformance of the superheater when FGR is introduced at a later data.

- During boiler startup, the gas temperature entering the superheater is a variable that limits the firing rate and hence increases the startup time as steam flow is absent through the superheater, and the tubes can easily attain the gas temperature in a short time. With the convective superheater located beyond several rows of screen tubes, the gas temperature would be much cooler, and hence, the firing rate can be higher reducing the startup time and associated fuel costs. Chapter 6 on miscellaneous calculations gives an example of the time taken by a superheater to attain the maximum tube wall temperature with a given amount of flue gas flow at a given temperature.

Hence, convective superheaters (Figures 3.26 and 3.27) are preferred in package boilers. They may require a little more heating surface due to lower LMTDs, but they are more reliable, and life of these superheaters is much longer. They may also need lower-grade alloy tubes compared to the radiant design.

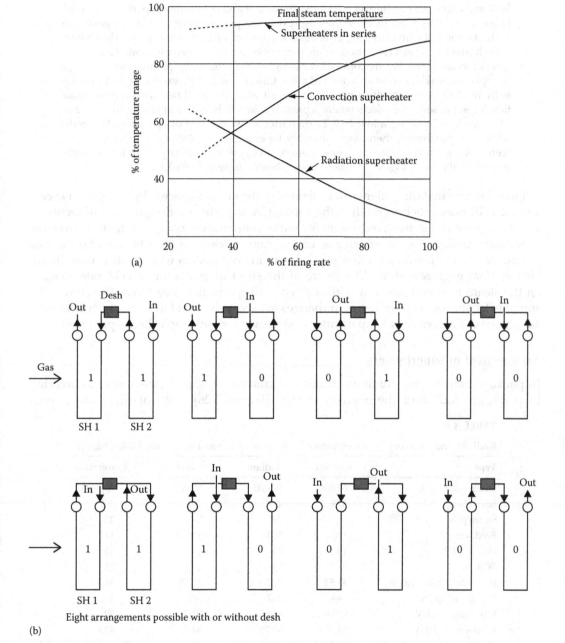

(a)

(b)

Eight arrangements possible with or without desh

FIGURE 3.28
(a) Typical variation in temperature rise versus load for radiant and convective superheaters and (b) two-stage superheater options.

How Emissions Impact Radiant and Convective Superheaters

Example 3.4

FGR and excess air can impact a boiler with radiant and convective superheaters in different ways. Presented in the following text is the performance of a steam generator with radiant and convective superheaters. The boiler capacity is the same. The radiant superheater has no screen section, while the convective superheater is located beyond 18 rows of screen tubes. We are comparing the variation in steam temperature for each type of superheater with excess air and FGR rates. Case 1 is with 10% excess air, and case 2 is with low NO_x requirement using 15% excess air with 25% FGR rate. This type of situation is seen in several projects where a plant is asked to lower its emissions of NO_x and CO emissions by using a low NO_x burner after a few decades of operation. When the boiler was purchased, there were probably no emission regulations. The superheaters were designed without steam temperature control system. The purpose of this example is to show the advantage of convective superheater design (Table 3.8).

It may be seen that the boiler with radiant superheater is impacted by changes in excess air and FGR rates much more than the convective superheater design. The difference in steam temperatures is much more with the radiant superheater compared to the convective superheater design. The boiler exit gas temperature increases as the total mass flow of flue gas increases, and hence, the efficiency is lower in both boilers when FGR is introduced. Hence, plant engineers should be aware of the effect of excess air or FGR rate increase on the steam temperatures and with the type of superheater they have. The convective superheater shows more resilience to changes in excess air and FGR rate, which is another advantage of the convective superheater design over radiant, and is hence preferred.

Arrangement of Superheaters

In package boilers, two common designs of convective superheaters are the inverted loop (Figure 3.27) and the horizontal tube (Figure 3.26). Both of them are proven

TABLE 3.8

Radiant and Convective Superheater Performance When Excess Air, FGR Change

Type	Radiant	Radiant	Convective	Convective
Steam flow, kg/h	60,000	60,000	60,000	60,000
Steam temp., °C	526	495	404	407
Steam press., barg	28	28	28	28
Feed water, °C	110	110	110	110
Excess air, %	10	15	10	15
FGR, %	0	25	0	25
Boiler duty, MM kcal/h	43.52	42.53	39.75	39.67
Exit gas temp., °C	149	167	148	166
Efficiency, % LHV	93.66	92.58	93.7	92.6
Efficiency, % HHV	84.76	83.78	84.8	83.8
Heat input, MM kcal/h	51.57	51.0	47.1	47.54
Back press., mm wc	150	170	150	170
Flue gas flow, kg/h	79,000	101,800	72,100	95,000
Furnace exit gas, °C	1,329	1,165	1,329	1,165

Note: The radiant superheater is located after 1 screen row while the convective is located after 18 rows. Surface area is unchanged when FGR is introduced.

designs. The choice is based on capacity, width of convection section, shipping dimensions to name a few.

Superheaters can be of single stage when steam temperature can float with load or with two stages as shown in Figure 3.28. A desuperheater may be added in between the two stages as required. In this figure, we can see that there are eight possible arrangements depending upon where steam enters the superheaters and flow direction, whether counter-flow or parallel-flow. Using these options, one can arrive at the superheater design meeting the steam temperature needs with a low tube wall temperature and hence higher life. Appendix E shows the Larsen Miller charts for creep life for T11, T22 materials.

Sometimes three-stage superheating with two attemperators in between may be required if there are operating cases with large variations in steam temperatures between them. For example, this situation arises in a process boiler where a large amount of import steam is superheated in the boiler superheater. The amount of imported steam may be much larger than what the boiler makes without the import steam, say, four or five times. Then, if the steam temperature is required to be the same in both cases of with and without import steam and if the superheater is designed for the case with import steam and then when import steam is absent, the steam temperature will go up significantly and a large amount of spray is required to maintain the temperature at the desired value. An example of this is given elsewhere in the book.

Steam Temperature Control

In package boilers, steam temperature is controlled typically from 60% to 100%. If a large load range is desired, then the superheaters may be accordingly designed. In some applications, steam temperature may be required to be maintained at even 10% load. Convective superheaters, due to their low tube wall temperature and heat flux at lower load, are preferred in such cases. There are numerous methods for steam temperature control.

1. Desuperheaters using demineralized water for spray between stages. It may be noted that spraying downstream of the superheater is not recommended as mixing of water and steam may not be good, and if the steam is used in a steam turbine, it can damage the blades. In addition, this practice allows the steam temperature as well as the superheater tube wall temperature to run up hotter and then controls the final temperature. Figure 3.29 shows the arrangement of superheaters with intermediate spray. In case demineralized water is not available, steam can be condensed and used as shown in b. A portion of the feed water from the economizer inlet is used to condense the required steam quantity in a separate heat exchanger as shown. This is called spraying with *sweet water*. The steam-side pressure drop in the first stage of the superheater and the pressure drop in the parallel path, namely, the heat exchanger plus control valve, has to be equal for the sweet water to be injected into the desuperheater. In order to have a good margin for the control valve loss, the exchanger is located several meters above the desuperheater.

2. A portion of the steam between two stages of superheaters may be taken and cooled inside the drum and then mixed with the other portion to maintain the final steam temperature. This is shown in Figure 3.30. A heat exchanger located

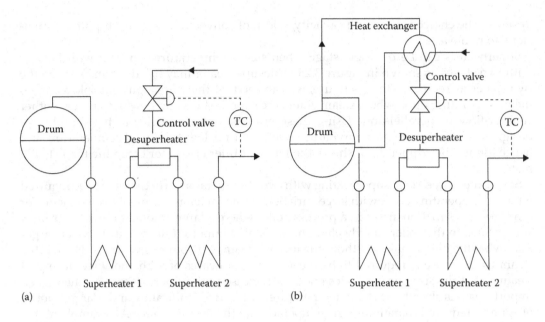

FIGURE 3.29
(a) Interstage attemperator with spray water and (b) condensed sweet water.

inside the drum cools the superheated steam. The amount of steam to be cooled is determined by the final steam temperature, and control valves facilitate that. Chapter 6 illustrates the sizing of this type of exchanger located in the steam drum with an example.

3. One may use the combustion air to cool the steam between stages as shown in Figure 3.31. Drainable superheaters should be used with this concept; else condensed water in the exchanger can cause flow problems. Finned tube bundle is preferred for the exchanger if its duty is large as the exchanger can be compact. Based on the final steam temperature required, the damper air flow is adjusted.

4. The superheaters may be located as shown in the flue gas path in parallel with another heating surface such as economizer (Figure 3.32). Damper may be modulated to obtain the desired steam temperature at a given load. One should ensure that the superheaters are oversized so that bypassing flue gases lowers the steam temperature. This system is not seen in package boilers, but large utility boilers use the concept due to the large space availability in such designs.

5. Three-stage desuperheating may be adopted in some high steam temperature applications to lower the tube wall temperature before spray (Figure 3.33). If a two-stage desuperheater is used, the tube wall temperature ahead of the desuperheater can be much higher compared to the three-stage desuperheater. The three-stage desuperheater system, however, is more expensive and adds to the cost of controls. This system is generally seen in steam generators or waste heat boilers where a large amount of import steam is required to be superheated as discussed earlier or when sudden surges are expected in flue gas flow as in some kiln-based heat recovery boilers discussed in Chapter 4.

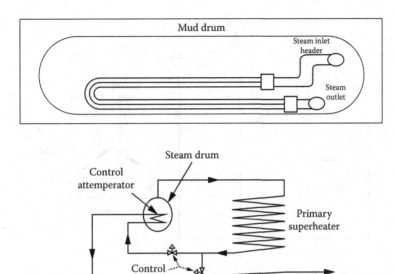

FIGURE 3.30
Steam flow through the drum exchanger is adjusted to maintain the final steam temperature.

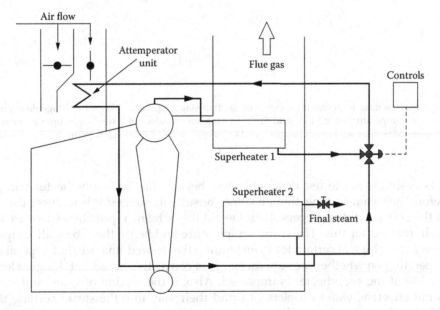

FIGURE 3.31
Combustion air cools steam in a finned exchanger for steam temperature control.

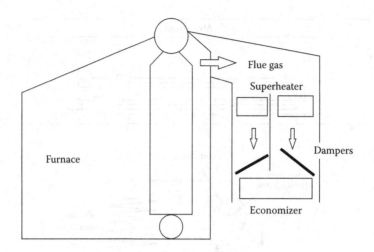

FIGURE 3.32
Gas bypassing using dampers to control final steam temperature.

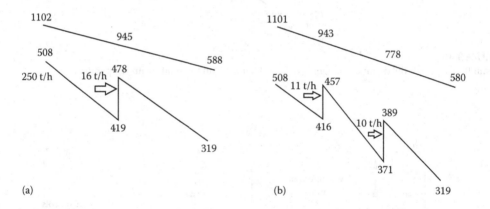

FIGURE 3.33
Three-stage superheating with two attemperators. (a) Two-stage superheater with interstage de superheater. Max. tube wall temperatures are 547°C and 519°C in each superheater. (b) Three-stage superheater with two desuperheaters. Max. tube wall temperatures are 544°C, 498°C, and 419°C in each superheater. Spray water flow is higher.

6. It is desirable not to use desuperheaters beyond the final superheater and just before the steam turbine, though some consultants suggest this to lower the cost of the boiler. Plant engineers should avoid this scheme if possible. There are two main reasons for this. The steam temperature and hence the tube wall temperature will be high at certain loads (exceeding the desired final steam temperature) depending on whether the superheater is of convective or radiant design. Hence, the life of the superheater is impacted. Also, if the mixing of water and steam is not effective, water droplets can find their way into the steam turbine thus affecting its performance. Hence, interstage attemperation is always preferable to downstream attemperation, or one should have the performance evaluated at all operating loads and accept the variations in steam temperatures without desuperheater control.

Steam Generators with Import and Export Steam

Saturated steam is generated in several waste heat boilers in a refinery as there are numerous waste gas sources at low to medium temperatures. However, when a refinery buys a steam generator, it would like to investigate if their new boiler can handle import steam from the waste heat boilers and what the implications are as far as the superheater design is concerned. Detailed calculations are required in such projects as the superheater has to handle different amounts of steam depending on the load and the amount of import steam. The tube wall temperature and spray quantities are the most affected variables in this situation. Figure 3.33 shows two cases for the same flue gas and final steam flow, one with a two-stage superheater and another case with three-stage superheater with two desuperheaters. It is seen that the steam temperature as well as the tube wall temperature in a two-stage design will be higher before the final stage compared to the three-stage option. If tube wall temperature and material selection is a concern, then the three-stage superheating may be looked into but is more expensive and adds to the cost of control system, piping, and attemperators. The spray quantity is also likely to be more. The idea of discussing this here is to inform plant engineers that if they come across a situation like this, both options may be tried. A plant may like to superheat some import steam after the steam generator or waste heat boiler is installed. Then, calculations have to be done to see how much additional spray is required in case of no import steam. Some modifications to the existing superheater may be required. If it is a new boiler, then the plant should obtain all the performance results for cases with and without import steam and obtain an idea of the tube wall temperatures at various locations and spray quantities, before ordering the boiler!

One should also be careful while designing steam generators with export steam. Often plants purchase steam generators, and after a few years, there may arise a need for saturated steam for process heating applications, and one option is to take off saturated steam from the drum and superheat the remaining steam. If the steam generator had been designed originally to handle export steam, then it is fine; else, one has to ensure that the superheaters are not impacted by lower flow through the superheater. Due to the higher duty, more fuel input is required, and the gas inlet temperature to the superheater will be higher. Even if fuel input is maintained the same, the steam flow through the superheater is reduced, which will increase its tube wall temperature. An analysis should be done before proceeding with this option. Chapter 4 shows a similar example of the performance of a waste heat boiler with and without export steam, and a similar analysis may be done for the steam generator.

Flow in Parallel Paths

Superheaters use multiple streams as discussed in Appendix B in order to ensure that the flow in each tube is reasonable from heat transfer and pressure drop considerations. During the design process, an average steam flow per tube is assumed and the sizing calculations are carried out. The tube wall temperature and steam-side pressure drop are estimated. Then part load calculations are done to see if there are concerns.

However, when the tube lengths are unequal or when some path has a tube of different size (this happens in large utility or large industrial boilers where the superheater tube lengths may run into several hundred meters), the flow per tube will not be the same in each tube.

TABLE 3.9

Details of Each Pass of Superheater

Tube Pass	Inner Diameter of Tube, mm	Equivalent Length, m
1	50	122
2	44	107
3	50	112

Or during superheater maintenance or repair, plant engineers may not have a particular tube size for replacement and may use whatever is available. One has to use the electrical analogy to estimate the flow in each parallel pass. Then, one has to perform thermal performance evaluation to find out the maximum tube wall temperature in the most vulnerable tube path.

Example 3.5

There are three tubes connected between two headers of a superheater. The tube details are shown in Table 3.9. Determine the flow in each path. Total flow = 7000 kg/h and steam pressure = 55 bara at 400°C.

In the first step, we assume that the average specific volume of steam is same in all the tubes. The basic equation for tube-side pressure drop (see Appendix B) is

$$\Delta P = 810 \times 10^{-6} \times fL_e vw^2/d_i^5$$

$\Delta P_1 = \Delta P_2 = \Delta P_3$ as tubes are between the same headers (1, 2, and 3 refer to the three different lengths). Also,

$W_1^2 fL_{e1}/d_i^5 = W_2^2 fL_{e2}/d_i^5 = W_3^2 fL_{e3}/d_i^5 = K^2$ a constant and $W_1 + W_2 + W_3$ = total flow = 7000

$W_1^2 \times 122 \times 0.0195/0.05^5 = W_2^2 \times 107 \times 0.020/.044^5 = W_3^2 \times 112 \times 0.0195/0.05^5$ or

$W_1^2 \times 7{,}612{,}800 = W_2^2 \times 12{,}976{,}285 = W_3^2 \times 6{,}988{,}800 = K^2$

$W_1 \times 2759 = W_2 \times 3602 = W_3 \times 2644 = K$

Solving, W_1 = 2492 kg/h, W_2 = 1908 kg/h, and W_3 = 2600 kg/h.

Now the second step is another thermal design check to see the temperature pickup in each tube based on the flow and surface area. The next step is to do another run correcting for the specific volumes in each tube using the earlier equation. A computer program handles this sort of calculation well when the number of parallel paths is high, say, 6 or more.

Steam Inlet and Exit Nozzle Location

The steam nozzle inlet in a superheater header should preferably be located in the middle of the header or if it is long (say, above 4 m) at two quarter points with 50% of flow in each inlet to ensure steam flow is uniform inside the tubes. End inlet or exit connection as shown in Figure 3.34 is not recommended though it may be convenient from piping layout viewpoint. The static head difference between headers impacts the steam flow through each tube, and end connection as shown in arrangement 1 of Figure 3.34 shows the largest difference in steam flow between the first and last tubes, resulting in overheating of tubes close to steam inlet end.

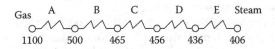

FIGURE 3.34
Temperature distribution across various resistances at superheater exit. *Note:* (A) gas film, (B) gas-fouled layer, (C) tube wall, (D) steam-fouled layer, and (E) steam film.

Off-Design Performance of Superheater

Example 3.6

The following details are available for an inverted-loop superheater shown in Figure 3.27a, and the plant engineer wants to evaluate the performance when the gas flow and inlet conditions to the superheater are known. That is, compute the duty, exit gas temperature, exit steam temperature, and tube wall temperature. The superheater is located beyond several rows of screen tubes in a D-type boiler, and hence, direct radiation from the furnace is not considered.

Solution

Steam pressure at superheater exit = 100 kg/cm^2 a. Gas flow entering the superheater (based on natural combustion calculations) = 250,000 kg/h at 1100°C (measured). % volume of flue gas is CO_2 = 8.29, H_2O = 18.17, N_2 = 71.08, O_2 = 2.46. Saturated steam of 200,000 kg/h is entering the superheater, and flow direction is counterflow. Fouling factor on steam side = 0.0002 m^2 h °C/kcal and that on gas side is 0.0004 m^2 h °C/kcal.

Superheater tube geometry details: 50.8 × 44 mm T11 tubes, 14 tubes/row, 15 rows deep, $S_T = S_L$ = 100 mm. Average (height) length of tube is 8 m, 35 streams. For the meaning of streams, see Appendix B. Surface area = 3.14 × 0.0508 × 14 × 15 × 8 = 268 m^2.

Let us assume a steam-side pressure drop of 4 kg/cm^2. Hence, steam pressure at inlet = 104 kg/cm^2 a. Saturation temperature is 312°C. Assume an average gas temperature of 970°C in the superheater (to be checked later).

The flue gas properties have to be evaluated at the average film temperature. As a first trial, we may assume it as 0.475 (tg$_1$ + ts$_1$) = 0.475 × (1100 + 312) = 670°C. From Appendix F, the flue gas properties are C_p = 0.3075 kcal/kg °C, μ = 0.1373 kg/m h, and k = 0.055 kcal/m h °C.

Using the Grimson's correlation for convective heat transfer coefficient from Appendix C, we have Nu = 0.229 × Re$^{0.632}$ using B and N values of 0.229 and 0.632, respectively.

Gas mass velocity G = 250,000/[14 × (0.1 − 0.0508) × 8] = 45,369 kg/m^2 h (9,258 lb/ft^2 h).
Hence, Re = Gd/μ = 45,369 × 0.0508/0.1373 = 16,786

Nu = h$_c$d/k = 0.229 × 16,786$^{0.632}$ = 107.1 or h$_c$ = 107.1 × 0.055/0.0508 = 116 kcal/m^2 h °C

Calculate the nonluminous heat transfer coefficient as discussed in Appendix D.

Beam length L = 1.08 × [0.1 × 0.1 − 0.785 × 0.0508 × 0.0508]/0.0508 = 0.1695 m.
Factor K = (0.8 + 1.6 × 0.181) × (1 − 0.00038 × 1243) × (0.0829 + 0.1817)/[(0.0829 + 0.1817) × 0.1695]$^{0.5}$ = 0.719. ϵ_g = 0.9 × (1 − e$^{0.719 × 0.1695}$) = 0.103. h$_n$ = 5.67 × 10^{-8} × 0.103 × (1243^4 − 6.6^4)/(1243 − 660) = 18.59 kcal/m^2 h °C.

Hence, $h_o = 18.59 + 116 = 134.59$ kcal/m^2 h °C.

Tube-side coefficient $h_i = 0.0278 \times C \times (200,000/35/3,600)^{0.8}/0.044^{1.8}$.

Assume an average steam temperature of 360°C for estimating h_i. C from Appendix B, Table B.4, is 382. Then, $h_i = 4250$ W/m^2 K = 3655 kcal/m^2 h °C (745 Btu/ft^2 h °F)

$$1/U_o = 50.8/44/3655 + 0.0002 \times (50.8/44) + 0.0508 \ln(50.8/44)/2/35 + 0.0004 + 1/134.59$$

$$= 0.000316 + 0.000231 + 0.0001042 + 0.0004 + 0.00743 = 0.008481 \text{ or}$$

$$U_o = 117.9 \text{ kcal/m}^2 \text{ h °C}$$

Let us use an average gas specific heat at 900°C and an average steam specific heat between 400°C and saturation. Enthalpy at 100 kg/cm^{2}a at 400°C = 740.8 kcal/kg and that of saturated steam at 312°C is 650 kcal/kg. Hence, $C_{ps} = (740.8 - 650)/(400 - 312) = 1.0318$ kcal/kg °C

$$\in = [1 - \exp\{-NTU \times (1 - C)\}]/[1 - C \exp\{-NTU \times (1 - C)\}]$$

$$W_g C_{pg} = 250,000 \times 0.328 = 82,000 \quad W_s C_{ps} = 200,000 \times 1.0318 = 206,360. \quad WC_{min} = 82,000$$

$$C = WC_{min}/WC_{max} = 82,000/206,360 = 0.397. \quad NTU = UA/WC_{min} = 117.9 \times 268/82,000$$
$$= 0.384$$

Hence, $\in = [1 - \exp\{-0.384 \times (1 - 0.397)\}]/[1 - 0.397 \exp\{-0.384 \times (1 - 0.397)\}] = 0.301$

Hence, energy transferred $Q = 0.301 \times 82,000 \times (1100 - 312) = 19.45 \times 10^6$ kcal/h

Hence, $T_{g2} = 1,100 - 19.45 \times 10^6/(250,000 \times 0.328) = 863$°C

$$T_{s2} = 312 + 19.45 \times 10^6/(200,000 \times 1.0318) = 406°C$$

We have to correct for the average specific heats of gas and steam at the revised gas and steam temperatures and make another run, but the duty will be close to the earlier one, and hence, let us proceed with the tube wall temperature estimation.

Tube Wall Temperature Estimation

Let us compute the tube wall temperature at the superheater exit. Being counter-flow, gas temperature = 1100°C, steam temperature = 406°C. Assume tube outer wall is 475°C. C factor at 406°C from Appendix B is 342. Note that the tube-side heat transfer coefficient decreases at higher steam temperatures at high pressures, while the gas-side heat transfer coefficient is higher at higher temperatures, compounding the problem of tube wall temperature (see Figure B.3).

$h_c = 3655 \times 342/382 = 3272$ kcal/m^2 h °C (taking the ratio of C values from previous calculation for h_i).

Assume the film temperature is 750°C (this has to be revised later). From Appendix F, $C_p = 0.3127$, $\mu = 0.1451$, $k = 0.059$. $G = 45,369$. $Re = 45,369 \times .0508/.1451 = 15,883$.

$Nu = 0.229 \times 15,883^{0.632} = 103.5 = h_c \times 0.0508/.059$ or

$h_c = 120$ kcal/m^2 h °C (24.4 Btu/ft^2 h °F)

K = (0.8 + 1.6 × 0.181) × (1 – 0.000381 × 1373) × (0.0829 + 0.1817)/[(0.0829 + 0.1817) × 0.1695]$^{0.5}$ = 0.649. ϵ_g = 0.9 × (1 – e$^{0.649 \times 0.1695}$) = 0.093. h_n = 5.67 × 10^{-8} × 0.093 × (1373^4 – 7.48^4)/(1373 – 748) = 27.34 W/m^2 K = 23.5 kcal/m^{2}h °C (4.8 Btu/ft^{2}h °F)

1/U_o = 50.8/44/3272 + 0.0002 × (50.8/44) + 0.0508 ln(50.8/44)/2/35 + 0.0004 + 1/143.5

= 0.0003529 + 0.000231 + 0.0001042 + 0.0004 + 0.006969 = 0.008057 or

U_o = 124 kcal/m^{2}h °C (25.3 Btu/ft^{2}h °F)

Heat flux based on tube OD = 124 × (1100 – 406) = 86,135 kcal/m^{2}h (31,646 Btu/ft^{2}h).

Temperature drop across inside steam film = 86,135 × 0.0003529 = 30°C (54°F)
Drop across inside fouling layer = 86,135 × 0.000231 = 19.9°C (36°F)
Hence, tube inner wall temperature = 406 + 30 + 20 = 456°C (853°F)
Drop across tube wall = 86,135 × 0.0001042 = 9°C (16°F)

Hence, tube outer wall temperature = 456 + 9 = 465°C (869°F) (used for checking oxidation limits). Table 3.10 shows the allowable temperatures of certain tube materials.

Tube mid-wall temperature = 460°C (860°F) (used for calculating tube wall thickness)
Drop across gas fouling layer = 86,135 × 0.0004 = 34°C (61°F)
Drop across gas film = 86,135 × 0.006969 = 600°C (1080°F)

Considering various nonuniformities due to gas flow, due to steam flow due to header arrangement (Figure 3.35), or variations in tube lengths and margin for safety, one may add a certain value based on experience. Let us use 25°C as margin, and so maximum tube wall temperature is 485°C (905°F) for tube mid-wall temperature and 490°C (914°F) for outer wall for design. The various thermal resistances are shown in Figure 3.34 at superheater exit.

Estimating the Tube Thickness

Table 3.10 shows that the allowable temperature for T11 material is 565°C. Hence, T11 material is adequate for this superheater. Let us check the thickness.

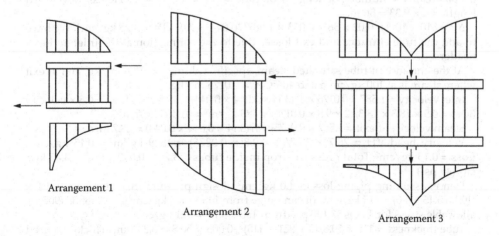

Arrangement 1

Arrangement 2

Arrangement 3

FIGURE 3.35
Flow nonuniformity inside tubes due to header arrangements.

TABLE 3.10

Allowable Material Temperatures

Material	Composition	Temp., °F
SA 178A (erw)	Carbon steel	950
SA 178C (erw)	Carbon steel	950
SA 192 (seamless)	Carbon steel	950
SA 210A1	Carbon steel	950
SA 210C	Carbon steel	950
SA 213-T11	1.25Cr–0.5Mo–Si	1050
SA 213-T22	2.25Cr–1Mo	1125
SA 213-T91	9Cr–1Mo–V	1200
SA 213-TP304H	18Cr–8Ni	1400
SA 213-TP347H	18Cr–10Ni–Cb	1400
SA 213-TP321H	18Cr–10Ni–Ti	1400
SB 407–800H	33Ni–21Cr–42Fe	1500

Per ASME Boiler and Pressure Vessel code sec 1, the following equation may be used to determine the thickness at 485°C (905°F):

$$t = Pd/(2S + P) + 0.005d$$

Design pressure P is arrived at based on pressure drop in the superheater, desuperheater if used, and piping. Let us say the additional pressure drop in the piping and desuperheater is 1 kg/cm². S is the allowable stress.

The superheater pressure drop in kPa is given by $\Delta P = 810 \times 10^{-6} \times fL_e v w^2/d_i^{5}$

w = flow per tube, kg/s = 20,000/3,600/35 = 1.587 kg/s. At average steam temperature of 360°C and at 103 kg/cm²a, specific volume = 0.023 m³/kg.

The effective length of tube = 8 + 2.5 × 1.75/3.28 = 9.33 m (2.5 times inner diameter was taken for the bend loss in feet). (Number of tubes per row × number of rows deep)/streams = number of lengths of tube = 14 × 15/35 = 6. Hence, developed length = 6 × 9.33 = 56 m.

$\Delta P = 810 \times 10^{-6} \times 0.02 \times 56 \times 0.023 \times 1.587^2/0.044^5 = 319$ kPa = 3.25 kg/cm². We have to add the tube entrance and exit losses due to the connection at the inlet and exit headers.

At the entrance of tube saturated steam, specific volume = 0.0176 m³/kg, and at exit where steam is at 100 kg/cm²a and 406°C, v = 0.0275 m³/kg.

Inlet velocity = 1.587 × 0.0176 × 4/(3.14 × 0.044 × 0.044) = 18.3 m/s. Loss = 0.5 × velocity head = 0.5 × 18.3 × 18.3/(2 × 9.8 × 0.0176) = 485 kg/m² = 0.0485 kg/cm².

Steam velocity at exit = 1.587 × 0.0275 × 4/(3.14 × 0.044 × 0.044) = 28.7 m/s.

Velocity head VH = 28.7 × 28.7/(2 × 9.8 × 0.0275) = 1529 kg/m² = 0.15 kg/cm². Loss = 0.15 kg/cm². Total pressure drop inside tubes = 3.25 + 0.049 + 0.15 = 3.45 kg/cm²(49 psi).

Hence, assuming piping loss as 1.0 kg/cm², design pressure may be taken as 1.1 × (100 + 3.45 + 1.0) = 115 kg/cm². (It can range from 108 to 115 kg/cm²g.) At 485°C (905°F), allowable stress for T11 is 13,100 psi (from Table 3.11) = 921 kg/cm².

Tube thickness = 115 × 50.8/(2 × 921 + 115) +0.005 × 50.8 = 3.23 mm. Thickness used is 3.4 mm. One may use a higher thickness after evaluating the flow in unequal tube lengths, or the design pressure may be slightly lowered to give some margin for corrosion allowance if required.

TABLE 3.11

Allowable Stress Values, ASME 2009, S

Temp., °F	SA178a, SA192	SA213 T11	SA213 T22	SA210 A1	SA213 T91	SA213 TP304H	SA213 TP316H	SA213 TP321H	SA213 TP347H
600	13.3	15.7	16.6	17.1	23.7	16.6	16.9	18.3	17.0
650	12.8	15.4	16.6	17.1	23.4	16.2	16.8	17.9	16.6
700	12.4	15.1	16.6	15.6	22.9	15.8	16.8	17.5	16.3
750	10.7	14.8	16.6	13	22.2	15.5	16.8	17.2	16.1
800	9.2	14.4	16.6	10.8	21.3	15.2	16.8	16.9	15.9
850	7.9	14.0	16.6	8.7	20.3	14.9	16.8	16.7	15.7
900	5.9	13.6	13.6	5.9	19.1	14.6	16.7	16.5	15.6
950	4	9.3	10.8	4.0	17.8	14.3	16.6	16.4	15.4
1000	2.5	6.3	8.0	2.5	16.3	14.0	16.4	16.2	15.3
1050		4.2	5.7		12.9	12.4	16.2	12.3	15.1
1100		2.8	3.8		9.6	9.8	14.1	9.1	12.4
1150		1.9	2.4		7	7.7	10.5	6.9	9.8
1200		1.2	1.4		4.3	6.1	7.9	5.4	7.4
1250						4.7	5.9	4.1	5.5
1300						3.7	4.4	3.2	4.1
1350						2.9	3.2	2.5	3.1
1400						2.3	2.5	1.9	2.3
1450						1.8	1.8	1.5	1.7
1500						1.4	1.3	1.1	1.3

Case Study of a Superheater with Tube Failure Problems

An inverted-loop package boiler superheater similar to that shown in Figure 3.27a was having frequent tube failure problems at high loads as well as at low loads. The following information was gathered during the site visit:

1. The superheater was located at the turning section of the furnace without screen tubes. This is a vulnerable location and should have been avoided for reasons discussed earlier regarding radiant superheaters. The furnace exit gas temperature is difficult to predict accurately, and hence, a large variation from estimated value is possible. Hence, the gas temperature entering the superheater could be ±70°C off from the predicted value. The direct radiation from the furnace will also increase the tube wall temperatures of the tubes close to the turning section, which were failing often. The nonuniformity in gas flow at the turn is also not helping the situation. The plant engineers were wondering if the burner flame shape was causing the problem and the boiler supplier was also supporting this view as he did not want to acknowledge that the design of the superheater was vulnerable. This is a common problem in many plants with radiant superheaters.

2. It was noted that superheater did not have multiple passes but just a single pass with steam entering from top header and leaving at the bottom header. With such a design, the desired steam temperature is reached at the outlet header more or less in all the tubes. The external furnace radiation plus a high steam temperature

at the outlet header compounded the problem and resulted in overheating of the tubes close to the furnace exit.

Locating a superheater beyond several rows of screen tubes has its own advantages as discussed earlier. The convective superheater is a proven and safe design. The screen tubes absorb the nonuniformity in gas flow and the external direct radiation from the furnace exit and ensure that a much cooler gas reaches the superheater. The performance is more predictable. The LMTD will be lower, and hence, a little more surface area is required. However, due to the lower tube wall temperatures, such a design will require a lower grade of tube material unlike the radiant superheater. The use of multiple passes and a parallel-flow configuration should have been used for this design as it ensures that the final steam temperature is attained in the tubes further away in a cooler gas zone.

3. In order for an inverted-loop superheater to operate well at low loads continuously, the steam-side pressure drop must be high enough at low loads to overcome gravity loss. However, it was found that the number of streams was too many resulting in low tube-side pressure drop. Hence, the superheater was failing at high loads, and the tubes near the furnace receiving direct radiation from the furnace were the worst hit.

Under the circumstances, the only way to improve the situation is to introduce a baffle in the inlet header and make the superheater a two-pass design with parallel-flow configuration as shown in Figure 3.36. The final steam temperature will be reached in the tubes in the much cooler gas temperature zone, which will ensure a lower tube wall temperature and hence better life. The steam-side pressure drop inside the tubes will be eight times more and was found to be acceptable as a lower steam pressure was required for

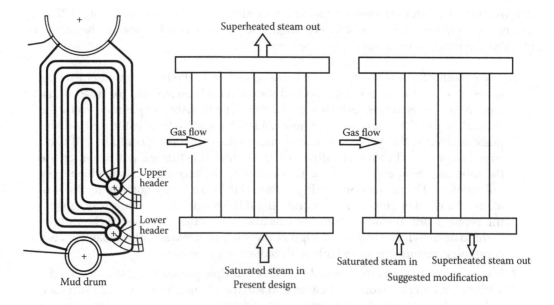

FIGURE 3.36
Suggested modification for the superheater to improve steam-side flow distribution.

the process. The steam velocity also was reasonable considering the original design had a very low steam velocity inside the tubes.

Problem at Low Loads with Inverted-Loop Superheaters

In tall inverted-loop superheater of the type shown earlier, one of the concerns is that the steam-side pressure drop should be larger than the gravity head to ensure flow in the downward direction; this situation arises at low loads in circuits with downward flow when the friction loss is low and is on par with gravity loss.

Figure 3.37 shows the variation of friction and gravity loss as a function of mass flow of steam in downward and upward flow section of a superheater. When steam flows upward, both the gravity loss and friction loss are additive, and hence, the sum is always positive as shown and the curve is monotonic. In the downward flow section, at low flows, the enthalpy pickup in a semi-radiant superheater will be more than in a convective superheater, and the specific volume increases resulting in lower density of steam and lower gravity loss. As the steam flow increases, the temperature and specific volume of steam reach a steady value, and the density of steam will also reach a high value and be leveled out, while the friction loss increases as the square of the flow. The gravity loss versus mass flow curve will have a shape as shown with two likely operating points at low loads. It shows that at higher mass flows, we do not have a problem (as far as stagnation is concerned, but overheating due to low steam velocity or direct furnace radiation is another issue) as we get into the region where the friction loss overtakes the gravity loss and be in the positive slope region. However, at low loads, it is showing an ambiguous trend where for a given pressure drop, there are two possible flows, which is an unstable characteristic; any small nonuniformity in gas-side or steam-side flows can upset the flows in a given tube and place the tube in a vulnerable spot resulting in overheating due to stagnation or even reverse flow.

This trend may be exhibited even with a convective superheater with low steam-side pressure drop located behind a large screen section, but due to the lower gas temperature at low loads, the superheater is in a much safer region. When consultants specify a low steam-side pressure drop, they should also be concerned about the superheater design and the lowest load it can operate without possible stagnation concerns.

It may be noted that the curve of the furnace exit gas temperature versus heat input has a small slope, and hence, even at low loads, the energy absorbed by the superheater in the

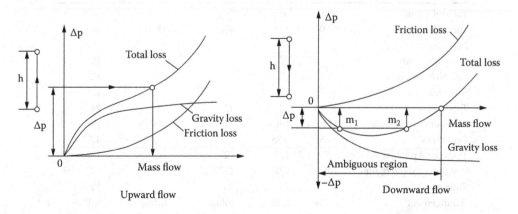

FIGURE 3.37
Variation of pressure drop with flow in downward flow section of superheater.

semi-radiant zone is significant (that is why the author insists on a large screen section ahead of the superheater and always design a convective-type superheater). Hence, the design should have ensured that at the minimum load, the steam pressure drop is high enough to overcome gravity loss; else, a horizontal tube design or some other configuration should have been selected for the superheater by the boiler designer.

To overcome the problem at low loads, a baffle was suggested in the inlet header as shown in Figure 3.36. The introduction of the baffle as shown increased the tube-side velocity and friction loss about eight times more; due to the higher tube-side heat transfer coefficient, the tubes were also cooler at full load. As the flow direction is now parallel-flow, the first-pass exit will see a lower steam temperature as well as low tube wall temperature, and the second-pass exit is located in a much cooler gas temperature region minimizing the external radiation concerns. The final desired steam temperature is reached in a much cooler gas temperature zone, and hence, the tubes will be running cooler than in the option without a baffle in the header. A large turndown is also now possible due to the higher pressure drop.

Steam Generator Design and Performance

Plant engineers who have visited Appendices A through F and reasonably familiar with heat transfer calculations should be able to evaluate the performance of steam generators using the procedure outlined here. By performance, we mean the gas–steam temperature profiles, exit gas temperature, efficiency at any given load given the tube geometry (as shown in Table 3.12), furnace dimensions, fuel analysis, and excess air used for combustion.

A typical flowchart for the process of performance evaluation (also design) of a steam generator is shown in Figure 3.38. The procedure to predict the thermal performance of furnace, screen, superheater, evaporator, economizer, and air heater is outlined in

TABLE 3.12

Steam Generator Tube Geometry Data

	Screen	Final SH	Pry SH	Evaporator	Economizer
Tube OD, mm	50.8	50.8	50.8	50.8	50.8
Tube ID, mm	44.0	43.0	43.0	44.0	44.0
Fins/m					197
Fin height, mm					19
Fin thickness, mm					1.27
Serration, mm					4.37
Tubes/row	12	12	12	12	24
Number of rows deep	15	16	12	46	12
Length, m	3.65	3.35	3.35	3.65	4.87
Transverse pitch, mm	127	122	122	127	101
Longitudinal pitch, mm	101	127	127	101	101
Streams		48	36		12
Arrangement	Inline	Inline	Inline	Inline	Inline
Direction of flow		Counter	Counter		Counter

Note: Furnace length = 9.76 m, height = 3.96 m, width = 2.44 m. Projected area = 142 m².

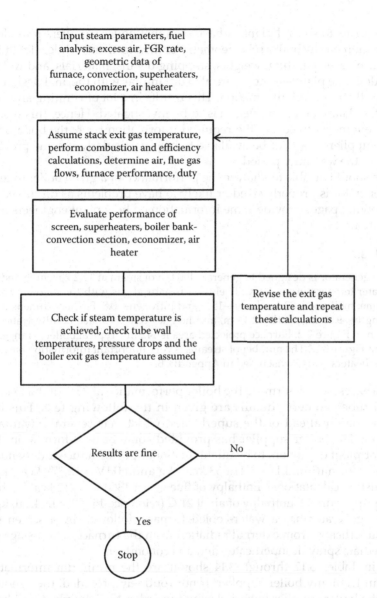

FIGURE 3.38
Flow diagram for evaluating steam generator performance.

Chapter 2 and in Appendices B through F. By integrating the performance of each heating surface, the overall performance is arrived at.

Data Required for Performing Steam Generator Analysis

Plant engineers often face problems with their steam generators and seek the help of the boiler suppliers or consultants to understand and resolve the problem. Often, the basic data of the steam generator required for performing the analysis may not be available with the plant as the boiler supplier often does not provide them at the time of purchasing the steam generator. Hence, the following discussion will help plant engineers obtain the desired data from the boiler supplier (before ordering the boiler) for performing an analysis at a later date.

During my visits to several plants, I have seen that boiler suppliers provide bare minimum information on the boiler tube geometry or arrangement, while a lot of information on nonpressure parts, painting, weights, shipping details, materials, and welding insulation is provided. The plant engineers are also unfamiliar with thermal design aspects and do not ask for the required information. This results in a lot of running around to get the tube geometry data when a problem has to be investigated. Hence, this discussion will enable plant engineers to collect the required information *before* the boiler is ordered as often boiler suppliers may not be available to help particularly if some problem arises a few years after the warranty period.

Hence, one should be able to analyze the performance of a given boiler to understand if the steam generator is properly sized or likely to have problems as discussed in the preface. The following pages provide some information on how plant engineers may perform these calculations.

Example 3.7

A steam generator is designed to generate 100 t/h of steam at 42.2 kg/cm^2 g and 400°C. Feed water temperature is 116°C. Fuel fired is natural gas with the following analysis by volume: methane—83.4, ethane—15.8, and nitrogen—0.8. Furnace dimensions: furnace length—9.75 m, width—2.44 m, and height—3.96. Furnace is fully water cooled as shown in Figure 2.2. Furnace projected area is 142 m^2. Typical boiler arrangement is shown in Figure 3.17. The number of streams is very important information particularly for superheaters. This is discussed in Appendix B.

Using 15% excess air, determine the boiler performance at 25%–100% load. The boiler furnace and tube geometry details are given in the following text. This boiler has a good screen section ahead of the superheater, which has several advantages as discussed earlier. The boiler supplier has provided some basic information. However, it is possible for plant engineers to evaluate the boiler performance independently using some of this information. LHV = 11,653 kcal/kg and HHV = 12,879 kcal/kg (see chapter on combustion calculations). Enthalpy of flue gas at 1360°C = 412 kcal/kg based on the analysis (see Appendix F). Enthalpy of air at 21°C (reference 15 °C) = 1.3 kcal/kg.

This steam generator has a water-cooled furnace, followed by a screen section that shields the superheater from external radiation from the furnace, a two-stage superheater with intermediate spray desuperheater, and an economizer.

The data in Tables 3.12 through 3.14 shown are the minimum information a plant should obtain from any boiler supplier. If not routinely provided, they should demand them from the boiler supplier before placing the order. Else the plant will be facing lots of difficulties later on when a consultant is asked to evaluate the boiler thermal performance. Even if the boiler is operating at a different load, one can apply the methodology discussed here and evaluate the complete boiler performance. Critical field data such as % oxygen in flue gas, exit gas temperature, and any other measured data such as steam, water flows, and temperatures should be available to verify the calculations. The plant engineer can perform the following studies:

1. Predict the boiler performance at any load or with any other fuel firing or excess air or even changes in feed water temperature or steam pressure.
2. Analyze any heat transfer surface such as evaporator, economizer, or superheater for problems such as under- or overperformance.

TABLE 3.13

Detailed Boiler Performance—From Computer Program

Surface	Screen	Final SH	Pry SH	Evaporator	Economizer
Gas temp. in ±5°C	1360	1072	892	778	465
Gas temp. out, ±5°C	1072	892	778	465	154
Gas sp ht, kcal/kg °C	0.3328	0.3244	0.3163	0.3024	0.2815
Duty, MM kcal/h	11.78	7.16	4.44	11.60	10.78
U, kcal/m² h °C	117.81	109.29	104.92	105.23	40.61
Surface area, m²	105	103	77	322	2395
LMTD, C	953	639	550	343	111
Gas press. drop, mm wc	55.22	83.47	54.63	97.00	45.10
Max gas vel., m/s	51	49	43	33	16
Tube wall temp. ±5°C	322	511	386	284	123
Fin tip temp. ±5°C	322	511	386	284	136
Weight, kg	2,774	3,298	2,473	8,506	16,071
Fluid temp. in °C	222	286	256	222	116
Fluid temp. out ±5°C	256	400	313	256	222
Press. drop, kg/cm²	0.00	0.50	0.71	0.00	0.93
Fluid velocity, m/s	0.0	24.4	26.9	0.0	1.7
Fluid ht. tr. Coefft.	9780	1299	1799	9780	9390
Foul factor, gas	0.0002	0.0002	0.0002	0.0002	0.0002
Foul factor, fluid	0.0002	0.0002	0.0002	0.0002	0.0002
Spray, kg/h	3260				

3. Check the tube wall temperature of superheater, steam temperature, and spray quantity.

4. Check boiler efficiency.

Table 3.14 shows the summary performance of the boiler at various loads, as provided by boiler supplier.

For example, during operation, if the exit gas temperature at full load is not close to 154°C based on the fuel, excess air, and water and steam conditions provided earlier, one can review the complete performance and see if fouling on tube side could be an issue. Heat transfer coefficients may be calculated by the methods discussed in Appendices B through E and then checked against the values indicated by the boiler supplier. Large variations may be challenged. Some variations in heat transfer coefficient may be expected as there are several correlations available for flow inside, outside tubes. However, the boiler supplier should have fine-tuned these correlations based on field data from several of his boilers in operation.

If superheater tube wall temperatures from the field data are much higher than indicated, then tubeside fouling is likely; carryover of water particles by steam is possible from the drum. The superheater could also have been oversized. Gas-side flow maldistribution also may be looked into. The screen section is a good insurance against superheater failures as discussed in this chapter earlier, and hence, a design such as that mentioned earlier with a large screen section followed by superheaters is desirable.

TABLE 3.14

Performance at Various Loads

Boiler Load, %	100	75	50	25		100
Boiler duty	64.68	48.51	32.34	15.98	MM kcal/h	34.79
Amb. temp.	21.1	21.1	21.1	21.1	°C	21.1
Relative humidity	60	60	60	60	%	60
Excess air	15	15	15	35	%	15
Flue gas recirculation	0	0	0	0	%	0
Fuel input (HHV)	77.58	57.89	38.52	19.36	MM kcal/h	94.35
Heat rel. rate (HHV)	823,241	614,296	408,750	205,403	kcal/m^3h	1001,179
Heat rel. rate (HHV)	546,978	408,152	271,584	136,476	kcal/m^2h	665,206
Steam flow	99,819	74,864	49,909	24,955	kg/h	100,000
Steam pressure	42.2	42.2	42.2	42.2	kg/cm^2g	42.2
Steam temp.	399	399	399	386	±5°C	399
Feed water temp.	116	116	116	116	±5°C	116
Water temp. lvg eco	222	207	190	183	±5°C	116
Blowdown %	1	1	1	1	%	1
Boiler exit gas temp.	469	417	359	306	±5°C	512
Eco exit gas temp.	158	143	132	125	±5°C	512
Air flow	116,909	87,237	58,047	34,243	kg/h	142,180
Flue gas flow	122,934	91,733	61,039	35,746	kg/h	149,506
Spray flow	3260	1869	492	0	kg/h	8540
Flue gas analysis, losses, efficiency, %						
Dry gas loss	4.62	4.14	3.74	4.18	%	16.51
Air moisture	0.09	0.08	0.07	0.08	%	0.31
Fuel moisture	10.55	10.45	10.36	10.31	%	13.19
Casing loss	0.50	0.67	1.00	2.00	%	0.5
Unacc/margin	0.50	0.50	0.50	0.50	%	0.5
Efficiency, LHV	92.56	93.03	93.21	91.67	%	76.25
Efficiency, HHV	83.75	84.17	84.33	82.94	%	69.00
Furnace back press.	369.96	200.82	85.88	27.92	mm wc	481
% vol. CO_2	8.56	8.56	8.56	7.38		8.56
H_2O	17.30	17.30	17.30	15.12		17.30
N_2	71.65	71.65	71.65	72.51		71.65
O_2	2.48	2.48	2.48	4.99		2.48
SO_2	0	0.00	0.00	0.00		0
Fuel flow	6024	4494	2990	1502		7326

Note: The last column shows the performance without the economizer.

Checking Boiler Performance

Often a plant buys a steam generator without obtaining process data as shown earlier. However, if such data are obtained from the boiler supplier, the plant engineer can check the performance not only at full load but at off-design conditions as well.

The plant engineer can learn a lot on thermal performance of boilers by verifying the thermal performance data given by the boiler supplier, or even if he has not been provided the data at a particular load, the plant engineer, using the methods discussed here, can generate the complete performance data. Let us assume that the boiler described earlier is operating at 100% load with the stated fuel and at stated excess air of 15%; gas temperature measurements at the furnace exit and in high gas temperature regions will not be typically accurate. We can reasonably assume that the boiler exit gas temperature is accurate and also the water and steam temperatures. Measurement errors will be lesser as the gas temperatures become lower. Using this premise, let us analyze the complete boiler to check the aforementioned data. One can independently calculate the heat transfer coefficients at various surfaces using the methods described in Appendices B through E; combustion and efficiency calculations as discussed in Chapter 1 should be performed to arrive at the air and flue gas quantities and the flue gas analysis. The oxygen measurement dry or wet may be used to obtain the operating excess air as discussed in Chapter 1.

Then one may obtain the flue gas properties such as specific heat, viscosity, and thermal conductivity as a function of gas temperature as discussed in Appendix F. These basic data will help one start doing the heat transfer calculations to check the boiler performance at any load. Here, since we have the data from the boiler supplier at 100% load, let us see if these are reasonable. If we do not have the boiler supplier's data, there should be some field measurements of gas temperatures and water–steam temperature in an operating unit. These data may be used for verifying the calculations.

Solution

Economizer Performance

Let us start from the economizer and work backward. Flue gas flow = 123,000 kg/h and analysis is % volume CO_2 = 8.56, H_2O = 17.3, N_2 = 71.65, O_2 = 2.48.

The exit gas temperature is 154°C at 100% load (the summary of performance given in Table 3.14 includes some margin and hence shows a 4°C higher exit gas temperature). Water inlet and exit temperatures are 116 °C and 222 °C. Water flow through the economizer will be steam flow × blowdown-spray water flue = 101,000 − 3,260 = 97,740 kg/h.

The specific heat of water may be taken as 1.04 kcal/kg °C and that of flue gas as 0.2815 kcal/kg °C. One can confirm these from steam tables and flue gas data given in Appendix F.

Duty of economizer = 97,740 × 1.04 × (222 − 116) = 10.81 × 10⁶ kcal/h. We will check this using the number of transfer units (NTU) method.

1. Assume the gas inlet temperature to the economizer (or use the measured field data. Exit gas temperature is an important data in any steam generator or HRSG, and there should be provisions to measure it accurately). Then the duty of the economizer is established as both the exit and inlet fluid temperatures are known. The exit water temperature may be calculated knowing the water flow. Since we have measured the water and steam temperatures, we may also use that information to confirm the estimates.
2. Calculate the U value applying the heat transfer correlations discussed in Appendix E on finned tubes.
3. Using the NTU method, obtain the duty of the economizer. If this matches the assumed duty, then iteration stops; else, inlet gas temperature is varied till convergence occurs. (The procedure provided earlier applies to all heat transfer surfaces.)

Let us, for illustration, use U values shown in Table 3.13 as the purpose is to illustrate how to check the boiler performance and not the calculation for U, which is detailed elsewhere.

$$W_g = 123{,}000 \text{ kg/h}, \quad C_{pg} = 0.2815 \text{ kcal/kg °C}, \quad t_{wo} = 222°C, \quad t_{wi} = 116°C$$

Water flow through the economizer = steam flow × blowdown-spray water flow. Initially, we may assume zero spray and proceed. Here we know the spray water flow and so let us use that.

Duty = $97{,}740 \times (h_{222} - h_{116}) = 97{,}740 \times (227.7 - 117) = 10.81 \times 10^6$ kcal/h (enthalpy of steam from steam tables). The gas inlet temperature = $154 + 10.81 \times 10^6/(123{,}000 \times 0.2815) = 466°C$.

Specific heat of water = 1.044 kcal/kg °C. Using the NTU method,

$$WC_p \text{ for flue gas} = 123{,}000 \times 0.2815 = 34{,}624 \text{ and for water} = 97{,}740 \times 1.044 = 102{,}040$$

$$WC_{min} = 34{,}624 \text{ and } C = WC_{min}/WC_{max} = 34{,}624/102{,}040 = 0.339$$

$$NTU = UA/C_{min} = 40.6 \times 2395/34{,}624 = 2.808$$

$$\in = [1 - \exp\{-NTU \times (1 - C)\}]/[1 - C \exp\{-NTU \times (1 - C)\}] = [1 - \exp\{-2.808 \times 0.661\}]/[1 - 0.339 \times \exp\{-2.808 \times 0.661\}] = [1 - 0.1563]/[1 - .339 \times 0.1563] = 0.89$$

Duty Q = $0.89 \times 34{,}624 \times (466 - 116) = 10.79 \times 10^6$ kcal/h; this closely agrees with the provided data. We may also apply the Q = UAΔT method and verify the results.

Evaporator Performance

See the method discussed in Appendix A for evaporator performance evaluation. The following equation may be used for the performance of the evaporator (see Equation A.13b).

$$\ln[(t_{g1} - t_s)/(t_{g2} - t_s)] = UA/(W_g C_{pg})$$

In our case, U = 105 kcal/m² h °C, A = 322 m², C_{pg} = 0.3024, t_s = 256°C,

Hence, $\ln[(t_{g1} - 256)/(466 - 256)] = 105 \times 322/(123{,}000 \times 0.3024) = 0.909$

$$[(t_{g1} - 256)/(466 - 256)] = \exp(0.909) = 2.482 \text{ or } t_{g1} = 777°C$$

Duty of evaporator = $123{,}000 \times (777 - 466) \times 0.3024 = 11.57 \times 10^6$ kcal/h

Primary Superheater

The steam flow is 100,000 − 3,260 = 96,740 kg/h. U = 105 kcal/m² h °C, A = 77 m².

Steam temperature at exit = 313°C. Hence, the duty of superheater = $96{,}740 \times (714.0 - 668.4) = 4.41 \times 10^6$ kcal/h, where 714.0 and 668.4 kcal/kg are enthalpies of steam at the exit and the inlet of superheater.

The inlet gas temperature is $4{,}410{,}000/(123{,}000 \times 0.3163) + 777 = 891°C$

Specific heat of steam = $(714.0 - 668.4)/(313 - 256) = 0.80$

$$WC_{min} = 123{,}000 \times 0.3163 = 38{,}905 \quad WC_{max} = 96{,}740 \times 0.8 = 77{,}392$$

$$C = 38{,}905/77{,}392 = 0.502$$

$$NTU = UA/WC_{min} = 105 \times 77/38,905 = 0.2078$$

$$\in = [1 - \exp(-0.2078 \times 0.498)]/[1 - 0.502 \times \exp(-0.2078 \times 0.498)] = 0.179$$

$$Q = 0.179 \times (891 - 256) \times 38,905 = 4.43 \times 10^6 \text{ kcal/h}$$
(agrees with value shown in Table 3.13)

Final Superheater

A spray desuperheater cools the steam from 313°C to 286°C in order to attain 400°C at the exit. Let us check the spray quantity.

$96,240 \times 714 + W_s \times 117 = (96,240 + W) \times 695.1$ or W_s, the spray flow = 3146 kg/h, or final steam = 99,886 kg/h or close to 100,000 kg/h.

$$U = 109.3 \text{ kcal/m}^2\text{h °C} \qquad A = 103 \text{ m}^2$$

$$\text{Duty of superheater} = 100,000 \times (766.8 - 695.1) = 7.17 \times 10^6 \text{ kcal/h}$$

$$\text{Gas inlet temperature} = 891 + 7.17 \times 10^6/(123,000 \times 0.3244) = 1071°C$$

$$\text{Specific heat of steam in the range } 286°C - 400°C = (766.8 - 695.1)/(400 - 286) = 0.629$$

$$C_{min} = 123,000 \times 0.3244 = 39,901. \ C_{max} = 100,000 \times 0.629 = 62,900. \ C = 39,901/62,900 = 0.634$$

$$NTU = UA/C_{min} = 109.3 \times 103/39,901 = 0.282$$

$$\in = [1 - \exp(-0.366 \times 0.282)]/[1 - 0.634 \times \exp(-0.366 \times 0.282)] = 0.229$$

$$Q = 0.229 \times (1071 - 286) \times 39,901 = 7.17 \times 10^6 \text{ kcal/h}$$
(agrees with value shown in Table 3.13)

Screen Section

The performance is obtained as for the evaporator.

$$\ln[(t_{g1} - t_s)/(t_{g2} - t_s)] = UA/(W_g C_{pg})$$

$$U = 117.8 \text{ kcal/m}^2\text{h °C}; A = 105 \text{ m}^2 \ C_{pg} = 0.3328 \text{ kcal/kg °C}$$

$$\ln[(t_{g1} - 256)/(1071 - 256)] = 117.8 \times 105/(123,000 \times 0.3328) = 0.302 \text{ or } t_{g1} = 1359°C$$

$$\text{Screen duty } Q = 123,000 \times 0.3328 \times (1359 - 1071) = 11.79 \times 10^6 \text{ kcal/h}$$

The furnace duty may be estimated as shown in Chapter 2, or obtained by difference from total energy absorbed.

The boiler efficiency on LHV basis is 92.56%. Energy absorbed by steam = 64.68 MM kcal/h. The energy absorbed by economizer, evaporator, superheater, screen = 10.79 + 11.57 + 4.43 + 7.17 + 11.79 = 45.75 MM kcal/h. Enthalpy of gas at 1360°C from Table F.10 = 414 kcal/kg.

Furnace duty = $[(64.68/0.9256) - 0.123 \times 414] \times 0.99 = 18.77$ MM kcal/h. Total energy to steam = $45.75 + 18.77 = 64.52$ MM kcal/h, which is close to the total duty of steam. This is a simplified approach and the results give a good estimate of performance. Note that these calculations are iterative in nature. Heat transfer coefficients are functions of gas temperatures, and hence, the U values have to be corrected based on actual average gas or film temperatures. Similarly, the steam-side heat transfer coefficients are a function of the average steam or water temperature, which is again obtained through an iterative process. A few iterations will correct these U values and provide reasonable performance results. A computer program will help speed up these iterations.

One may use this method to evaluate the furnace exit gas temperature and relate it with the correlations or typical charts available for estimating the furnace exit gas temperature. We may then compute the furnace duty using the method discussed in Chapter 2 and then the total duty and steam generation. If there are gas or steam temperatures measurements available at intermediate points, they may also be verified or used as data for cross-checking the calculations. *Note that these calculations may be done at any load even if the boiler supplier has not provided the data for that load.* For example, if a superheater is having high tube wall temperatures, the performance calculations may be done to check the duty as shown earlier using the exit gas temperature measurement alone. Then, the tube wall temperature may be computed as shown in Example 3.6. Comparisons may be made with field data, if any.

The purpose of this exercise is to show that plant engineers can independently perform the complete boiler calculations and verify the proposal data given by the boiler supplier or check the field data and compare the results with similar calculations and see if there are major deviations in duty or gas or steam temperatures at any heating surface, which should be brought to the attention of the boiler supplier.

A procedure to check the boiler performance from exit gas and fluid temperatures is discussed in Appendix A, and the same example is completely worked out.

Evaluating Part Load Performance

Sometimes a plant engineer may like to know what kind of steam temperature can be expected at part loads for which information may not have been provided by the boiler supplier. It will be helpful if the plant engineer can check the performance and compare it with the field data.

Example 3.8

See how the boiler discussed earlier performs at 50% load. The exit gas temperature was measured as 130°C. Ambient temperature is 21°C. Excess air is 15%, and the same fuel is used, and hence, flue gas analysis is % volume $CO_2 = 8.56$, $H_2O = 17.3$, $N_2 = 71.65$, $O_2 = 2.48$.

Solution

Let us use the simplified procedure for estimating the flue gas flow. Efficiency is obtained using the simplified formula as shown in Chapter 1. Boiler duty = 32.45 MM kcal/h. (50,000 kg/h steam at 399°C from 116°C feed water.) Natural gas fuel HHV = 12,879 kcal/kg, and LHV = 11,653 kcal/kg.

Let us start from the furnace end and work toward the economizer so that plant engineers can appreciate both the methods—one working from the economizer end and the other from the furnace end.

Furnace Duty

Since the fuel analysis, excess air, and boiler duty are known, let us obtain basic data such as flue gas flow, furnace duty, and furnace exit gas temperature. From Chapter 1, Equation (1.16a).

Efficiency on HHV basis is $\eta_H = 89.4 - (0.002021 + 0.0351 \times 1.15) \times (130 - 21) - 0.5 = 84.28\%$ and $\eta_L = 99 - (0.00202 + 0.0389 \times 1.15) \times (130 - 21) - 0.5 = 93.4\%$ (note that an additional loss of 0.5% was used to account for higher heat losses at part load). Hence, heat input on HHV basis = 32.45/0.8428 = 38.5 MM kcal/h = 161.2 MM kJ/h. Air flow = $314 \times 1.15 \times 1.013 \times 161.2 = 58,960$ kg/h. Fuel flow = $38.5 \times 10^6/12,879 = 2,989$ kg/h. Flue gas flow = 58,960 + 2,989 = 61,949 kg/h. [A value for combustion on MM kJ/h basis was used for estimating air flow as discussed in Chapter 1. Correction for moisture in air was also used.]

Furnace effective projected area is 142 m^2 (as provided by boiler supplier; 9.76 m long, 3.96 m high, and 2.44 m wide completely water-cooled furnace except for burner openings). Net heat input NHI (it is based on LHV) = $2,989 \times 11,653 \times 0.985 = 34.3$ MM kcal/h. (The value 0.985 is the casing heat loss of 1% plus unaccounted loss of 0.5%, neglecting heat input by air as there is no air heater.) NHI/projected area = $34.3 \times 10^6/142 = 241,549$ kcal/m^2 h = 280.9 kW/m^2 (effective projected area is referred to as EPRS (effective projected radiant surface).

From Figure 2.8 showing furnace exit gas temperature versus NHI/EPRS in Chapter 2, the furnace exit gas temperature is about 1180°C. Enthalpy of flue gas at 1180°C from Appendix F is 351 kcal/kg. Hence, furnace duty = $34.3 \times 10^6 - 61,949 \times 351 = 12.55$ MM kcal/h.

Screen Section

The U value may be calculated using Appendices C and D, or for quick estimates here, let us use the 100% values and correct them. U = $117.8 \times (0.5)^{0.6} = 77.7$ kcal/m^2 h °C (0.5 refers to the 50% load, and a power of 0.6 was used to correct for the load. One can do a detailed calculation to check this value, but for practical purposes, this should be good.) The gas specific heat is assumed as 0.3252 kcal/kg °C. Screen surface area = 105 m^2. Assume drum saturation temperature is 254°C due to the lower pressure drop.

$\text{Ln}[(t_{g1} - t_s)/(t_{g2} - t_s)] = UA/(W_g C_{pg}) = 77.7 \times 105/(61,949 \times 0.3252) = 0.405 = \ln[(1180 - 254)/(t_{g2} - 254)]$. Hence, $t_{g2} = 871$°C. Screen duty = $61,949 \times 0.3252 \times (1180 - 871) = 6.225$ MM kcal/h.

Final Superheater

Let us assume that we can obtain the 399°C final steam temperature. Assume the inlet steam temperature as, say, 300°C. Assumed duty $Q_a = 50,000 \times (766.1 - 705) = 3.055$ MM kcal/h (766.1 and 705 kcal/kg refer to enthalpy of steam at exit and inlet conditions). Assume gas specific heat here as 0.313 kcal/kg °C. The exit gas temperature = $871 - 3.055 \times 10^6/(0.313 \times 61,949) = 713$°C. Let us use a different method here just to show that there are other ways to evaluate the performance of superheaters and economizers other than the NTU method, and the plant engineer can select whatever he is comfortable with. (Note that all calculations are iterative in nature as one has to correct for the gas properties based on the final gas temperature profile.) A = 103 m^2.

For the counter-flow superheater, LMTD = $[(871 - 399) - (713 - 300)]/\ln[(871 - 399)/(713 - 300)] = 442$°C. U = $109.2 \times (0.5)^{0.6} = 72$ kcal/m^2 h °C.

The transferred duty $Q_t = UA\Delta T = 72 \times 103 \times 442 = 3.28$ MM kcal/h. Since the assumed and transferred duties do not match and that the transferred is higher, we can assume, say, a lower steam temperature at the inlet and repeat the exercise. Use 295°C as the steam inlet temperature and try again. $Q_a = 50,000 \times (766.1 - 701.6) = 3.22$ MM kcal/h. Exit gas temperature $= 871 - 3.22 \times 10^6/(0.313 \times 61,949) = 705$°C. LMTD $= [(871 - 399) - (705 - 295)]/\ln[(871 - 399)/(705 - 295)] = 440$°C. $Q_t = 72 \times 10^3 \times 440 = 3.26$ MM kcal/h. This is close enough. Hence, duty $= 3.26$ MM kcal/h.

Primary Superheater

Let us see if we need the desuperheater. $U = 105 \times (0.5)^{0.6} = 69.3$ kcal/m^{2}h °C. $A = 77$ m^2. Assume $C_{pg} = 0.316$ kcal/kg °C. Saturation temperature $= 254$°C, and enthalpy $= 668.6$ kcal/kg.

$Q_a = 50,000 \times (701.6 - 668.6) = 1.65$ MM kcal/h. Exit gas temperature $= 705 - 1.65 \times 10^6/(0.316 \times 61,949) = 620$°C. LMTD $= [(705 - 295) - (620 - 254)]/\ln[(705 - 295)/(620 - 254)] = 388$°C. $Q_t = 77 \times 69.3 \times 388 = 2.07$ MM kcal/h. Hence, we do need the desuperheater as the steam temperature at superheater exit will be much higher than 295°C.

Let us assume we can go to 305°C. The spray required to cool from 305°C to 295°C has to be estimated. Let $W =$ flow in the superheater and so $(50,000 - W)$ is the spray quantity. By heat balance around the desuperheater, $W \times 708.2 + (50,000 - W) \times 117 = 50,000 \times 701.6$ or $W = 49,441$ kg/h. (117 kcal/kg is the enthalpy of spray water, and 708.2 and 701.6 kcal/kg are the enthalpies at 305°C and 295°C, respectively.) Hence, spray flow $= 559$ kg/h.

$Q_a = 49,441 \times (708.2 - 668.6) = 1.957$ MM kcal/h. Exit gas temperature $= 705 - 1.957 \times 10^6/(0.316 \times 61,949) = 605$°C. LMTD $= [(705 - 305) - (605 - 254)]/\ln[(705 - 305)/(605 - 254)] = 375$°C. $Q_t = 375 \times 77 \times 69.3 = 2.00$ MM kcal/h. The assumed and transferred duties are close, and hence, there is no need to repeat the exercise. With a computer program, one can improve the accuracy of these calculations.

Evaporator

$U = 105 \times (0.5)^{0.6} = 69.3$ kcal/m^{2}h °C, $A = 322$ m^2. Let $C_{pg} = 0.2928$, $t_s = 254$°C.

Hence, $\ln[(605 - 254)/(t_{g2} - 254)] = 69.3 \times 322/(61,949 \times 0.2928) = 1.23$ or $t_{g2} = 357$°C.

Duty of evaporator $= 61,949 \times (605 - 357) \times 0.2928 = 4.5$ MM kcal/h.

Economizer

We can try the NTU method here as we know both the gas inlet and water inlet temperatures. Note that the water flow will be steam flow less spray water flow $= 49,441$ kg/h as zero blowdown assumed.

$W_g = 61,949$ kg/h. $C_{pg} = 0.277$ kcal/kg °C. $t_{wi} = 116$°C. Let $t_{w2} = 192$°C.

Duty $= 49,441 \times (h_{192} - h_{116}) = 49,441 \times (195.36 - 117) = 3.874 \times 10^6$ kcal/h (enthalpy of water is taken from steam tables). Average $C_{pw} = (195.36 - 117)/(192 - 116) = 1.031$ kcal/kg °C.

The exit gas temperature $= 357 - 3.874 \times 10^6/(61,949 \times 0.277) = 131$°C.

$U = 40.61 \times (0.5)^{0.6} = 26.8$ kcal/m^2 h °C. Using the NTU method, WC_p for flue gas is $61,949 \times 0.277 = 17,160$ and for water is $49,441 \times 1.031 = 50,974$.

$WC_{min} = 17,160$ and $C = WC_{min}/WC_{max} = 17,160/50,974 = 0.3366$. $(1 - C) = 0.6634$.

$$\text{NTU} = UA/C_{min} = 26.8 \times 2,395/17,160 = 3.74.$$

$$\epsilon = [1 - \exp\{-\text{NTU} \times (1 - C)\}]/[1 - C \exp\{-\text{NTU} \times (1 - C)\}]$$

$$= [1 - \exp\{-3.74 \times 0.6634\}]/[1 - 0.3366 \times \exp\{-3.74 \times 0.6634\}]$$

$$= 0.939 \text{ (counter-flow arrangement)}.$$

Duty $Q = 0.939 \times 17,160 \times (357 - 116) = 3.88 \times 10^6$ kcal/h. Exit gas temperature may be again checked $= 3.88 \times 106/(61,949 \times 0.277) = 131$°C. Exit water temperature $= 116 + 3.88 \times 10^6/49,441/1.031 = 192$°C. These numbers agree with those assumed.

Total boiler duty $= 12.55 + 6.225 + 3.26 + 2.0 + 4.5 + 3.88 = 32.4$ MM kcal/h.

This also matches with the required duty for generating 50,000 kg/h at 399°C from 116°C feed water. Hence, even if performance at a particular load is not provided, plant engineers using the aforementioned approach can estimate the boiler performance. However, it is better to demand and get all the pertinent data from the boiler supplier and then cross-check the results and question the boiler supplier if large differences are seen.

Performance without Economizer

Example 3.9

What happens if the economizer is removed? Can we generate the same amount of steam?

Solution

Applying the same procedure as Example 3.7, one arrives at the results shown in Table 3.14. The following points are noted:

a. The exit gas temperature is much higher, about 512°C versus 154°C at full load with economizer. This is due to the lower efficiency of the boiler, which results in larger fuel flow and hence flue gas flow.
b. The flue gas flow is also much larger than what we had at 100% load with economizer.
c. The energy absorbed by the superheaters is also much higher resulting in a large amount of spray water for steam temperature control, 8500 kg/h versus 3260 kg/h. Also the temperature after spray is getting closer to saturation temperature, which is not a good situation. The superheater can have wet steam in pockets if the mixing of water and steam at the desuperheater is not good and solids can deposit inside the tubes of final superheater, increasing its tube wall temperature leading to its failure.
d. The gas temperature is higher along the entire gas flow path compared to the case with the economizer in operation. Hence, the superheater tube wall temperature will be higher.
e. The flue gas pressure drop will be higher due to the higher mass flow of flue gas and higher flue gas temperatures, 480 mm wc versus 370 mm wc.
f. The boiler efficiency is also much lower due to the higher exit gas temperature of 508°C versus 154°C when economizer is used. A drop of 16% is seen from Figure 3.23.

In conclusion, one should use the economizer or air heater at full load operation; else, the boiler will be operating with larger flue gas flow and higher gas temperature throughout the unit. The other option is to reduce the boiler load to about 70%.

Tube Wall Temperature Estimation at Economizer Inlet

The method of calculating water and acid dew points was presented in Chapter 1. When the flue gas drops below the water dew point, water starts condensing on the heating surfaces. Similarly, when the tube wall temperature drops below the acid dew point in air heater or economizer, acid vapor, if present, starts condensing on the surfaces leading to corrosion and material loss. The tube wall temperature in an economizer is very much dictated by the feed water entering temperature. Similar calculations for minimum tube wall temperature are presented in Chapter 4 for tubular air heaters.

Example 3.10

The performance results of a carbon steel plain tube and finned tube economizer in parallel- and counter-flow arrangement are shown in Table 3.15. Gas flow = 100,000 kg/h at 350°C. At 105°C, 70,000 kg/h water flows inside the tubes. Flue gas is natural gas products of combustion; 38 × 32 mm tubes, 24 tubes/row, 30 deep, 4 m long are used with $S_T = 75$, $S_L = 80$ mm in inline arrangement for plain tubes.

For the finned tube option, same tube size is used with 24 tubes/row, 12 rows deep, 3 m long tubes, $S_T = 100$, and $S_L = 90$ mm in staggered arrangement. Ratio of external to internal surface is 13.44. Inside and outside fouling factors are 0.0002 and 0.0004 m^2 h °C/kcal, respectively. Estimate the tube wall temperature at the cold end. Assume acid dew point temperature is 130°C.

Solution

The U values are computed as explained in Appendices B through E. Then the performance is obtained using either the NTU method or the conventional method of equating assumed duty with transferred duty. Results are shown in Table 3.15. The U_o is the average overall heat transfer coefficient.

Case 1: Finned tubes, parallel-flow
Heat flux inside tubes = 37 × (350 − 105) × 13.44 = 121,833 kcal/m^2 h.
   Tube wall resistance = [0.038/(2 × 35)] ln(38/32) = 0.0000933 m^2 h °C/kcal.
   Resistance due to tube-side heat transfer film = 1/6800 = 0.000147 m^2 h °C/kcal.
   Tube wall temperature = 105 + 121,833 × (0.000147 + 0.0002 + 0.0000933) = 159°C.
   The minimum tube wall temperature is above acid dew point temperature, and hence, this is a viable option provided we find the tube wall temperature above 130°C at low loads also.

Case 2: Finned tube, counter-flow
Heat flux inside tubes = 37 × (170 − 105) × 13.44 = 32,323 kcal/m^2 h.
   Tube wall resistance = [0.038/(2 × 35)] ln(38/32) = 0.0000933 m^2 h °C/kcal.
   Resistance due to tube-side heat transfer film = 1/6800 = 0.000147 m^2 h °C/kcal (due to difference in average water temperature, h_i is different).
   Tube wall temperature = 105 + 32,323 × (0.000147 + 0.0002 + 0.0000933) = 119°C.
   This does not solve the acid vapor condensation problem as the minimum tube wall temperature is below 130°C.

TABLE 3.15

Plain Tube and Finned Economizer Performance Results

Case	Finned-PF	Finned-CF	Plain-PF	Plain-CF	Finned-CF	Plain-CF
Gas temp. in, °C	350	350	350	350	350	350
Gas temp. out, °C	191	170	218	209	188	223
Duty, MM kcal/h	4.23	4.79	3.51	3.75	4.32	3.39
Water temp. in, °C	105	105	105	105	130	130
Water temp. out, °C	165	172	155	158	190	177
U_o, kcal/m^2 h °C	37	37	76.5	76.5	37	77
h_i kcal/m^2 h °C	6800	6800	6600	6600	7240	7110
Surface area, m^2	1167	1167	343	343	1167	343
Min tube wall, °C	159	119	115	109	143	134

Case 3: Plain tubes, parallel-flow

Heat flux inside tubes = $76.5 \times (350 - 105) \times 38/32 = 22{,}257$ kcal/m²h.
  Tube wall resistance = $[0.038/(2 \times 35)] \ln(38/32) = 0.0000933$ m²h °C/kcal.
  Resistance due to tube-side heat transfer film = $1/6600 = 0.0001515$ m²h °C/kcal.
  Tube wall temperature = $105 + 22{,}257 \times (0.0001515 + 0.0002 + 0.0000933) = 115$°C.
  This does not solve the acid vapor condensation problem as the minimum tube wall temperature is below 130°C.

Case 4: Plain tubes, counter-flow

Heat flux inside tubes = $76.5 \times (209 - 105) \times 38/32 = 9448$ kcal/m²h.
  Tube wall resistance = $[0.038/(2 \times 35)] \ln(38/32) = 0.0000933$ m²h °C/kcal.
  Resistance due to tube-side heat transfer film = $1/6600 = 0.0001515$ m²h °C/kcal.
  Tube wall temperature = $105 + 9448 \times (0.0001515 + 0.0002 + 0.0000933) = 109$°C.
  This does not solve the acid vapor condensation problem as the minimum tube wall temperature is below 130°C.

Case 5: Finned tube option, counter-flow with feed water inlet at 130°C using a heat exchanger

Case 6: Plain tube option using a feed water temperature of 130°C and counter-flow.
Calculations for cases 5 and 6 are similar. Therefore, only the results are shown.

Strictly speaking, one should compute the U_o values using the heat transfer coefficients at the inlet/exit gas temperatures (parallel-/counter-flow) and inlet water temperature. However, the objective of this exercise is to show the effect of parallel- versus counter-flow arrangement and the effect of finned and plain tubes. The following points may be noted:

1. With plain tube economizer, there is not much of a difference between parallel- and counter-flow arrangements as the difference in tube wall temperatures is hardly 6°C, which will reduce at lower loads. The tube-side coefficient governs the tube wall temperature. Hence, this method may not avoid condensation of acid vapor. We have to preheat the feed water to a temperature close to the acid dew point as shown in case 6.
2. With finned tubes, due to the higher heat flux with parallel-flow configuration, the tube wall temperature is much higher at the feed water inlet end, 159°C versus 119°C. One should perform these calculations at part load also. It is possible to avoid acid condensation in case the dew point temperature is below the tube wall temperature at the lowest load. The effect of tube-side coefficient is somewhat reduced due to finning. The counter-flow arrangement is not helpful as the tube wall temperature is below the dew point.
3. Raising the feed water temperature to 130°C in finned and plain tube options avoids the condensation concern. The duty with the finned bundle is also slightly more compared to the parallel-flow option of case 2. Hence, one should consider this option rather than going in for parallel-flow configuration, as at low loads too, this option works.

Methods to Minimize Low-Temperature Corrosion Problems

There are two methods to handle the corrosion problem. One is to allow condensation of acid vapor on the economizer but select proper materials to minimize the corrosion rate. This is discussed in Chapter 6. The other option is to avoid condensation itself by using a higher feed water temperature at the economizer inlet.

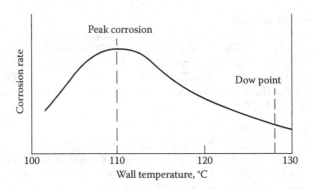

FIGURE 3.39
Tube wall temperature and corrosion rate.

As mentioned earlier, by increasing the lowest tube wall or surface temperature to above the acid dew point, one can prevent deposition of acid vapor on the heating surfaces. This will also ensure that water does not condense on the exchanger surface as the water dew point is much lower than the sulfuric acid dew point. However, this option increases the exit gas temperature from the boiler thus reducing the boiler efficiency.

Figure 3.39 shows the acid corrosion rate versus tube wall temperature based on tests conducted by Land Instruments. This is similar to Figure 1.10. It shows that the corrosion rate is higher not at the acid dew point but at about 10°C–15°C lower temperature. The reason for a higher corrosion rate at lower temperatures is that the acid deposition rate is highest in that area; the rate of condensation of sulfuric acid depends on its concentration and temperature depression (the difference between the metal and acid dew point temperatures). As sulfuric acid condenses out of the flue gas stream, its concentration in the flue gas decreases; this results in maximum acid deposition rate. The general practice in the boiler industry is to design the economizer or air heater prone to low-temperature corrosion so that corrosion is never initiated. Hence, the lowest tube wall temperature is selected to be slightly above the acid dew point. However, this results in a higher boiler exit gas temperature and loss in efficiency. The curve shows that the tube wall temperature may be slightly lower than the dew point, and the increase in efficiency is worth the slight increase in corrosion rate. Note that every 22°C decrease in exit gas temperature results in 1% improvement in efficiency for oil and gaseous fuels.

If the incoming feed water temperature is significantly lower than the acid dew point temperature, the following options are available to the plant engineer.

- A steam-to-water heat exchanger may be used to raise the incoming feed water temperature (Figure 3.40).

- The water leaving the economizer may be used to heat the incoming feed water, or portion of hot water may be recirculated to the economizer inlet as shown in Figure 3.41.

- A coil located inside the steam drum may be used to preheat the feed water to near-acid dew point temperature (Figure 3.42). The sizing procedure for this coil is discussed in Chapter 6.

- In an air heater, a steam coil may be used to preheat the incoming air to increase the minimum cold-end temperature.

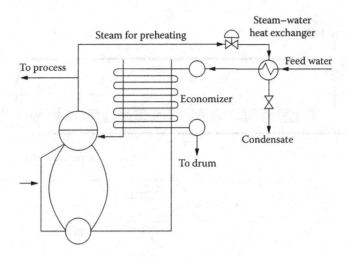

FIGURE 3.40
Using steam for preheating feed water to minimize low end corrosion problem.

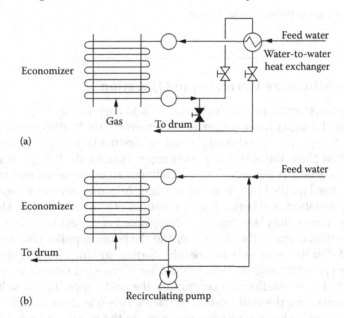

FIGURE 3.41
Using hot water from economizer outlet for preheating incoming feed water or recirculating hot water to increase incoming water temperature. (a) Using a heat exchanger and (b) using a recirculation pump.

- Cold air may be bypassed around the air heater so that the tube wall temperature is higher with the larger flue gas-to-air ratio.
- Some even suggest a parallel-flow economizer where the gas temperature is higher at the cold water end. However, one should perform calculations as shown earlier to ensure that the tube wall temperature is above the acid dew point at all operating loads. A better solution is to raise the inlet water temperature.

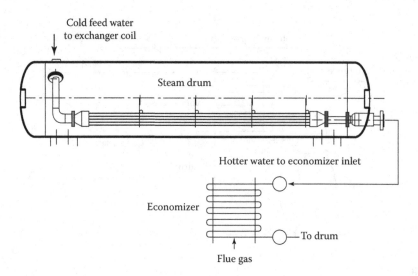

FIGURE 3.42
Using drum coil exchanger to preheat the feed water.

Precautions to Minimize Corrosion in Operation

One method to ensure condensation never occurs is to keep all gas-touched surfaces above the acid dew point through proper design. However, even if the heating surfaces are all above the acid dew point, if proper operational precautions are not taken, corrosion can occur.

In an incineration plant, the acid dew point temperature for the flue gas was estimated as 150°C, and the waste heat boiler was designed without an economizer, and membrane wall construction was used for the boiler with superheater. All surfaces were designed to be above 240°C in normal operation, and hence, it was a thought of as a good design. However, after a few months of operation, the plant engineers complained of deposition of acid and corrosion near the boiler bottom drum at the boiler exit, and the boiler supplier was surprised.

It was learned that the plant was frequently starting up and shutting down the boiler. When the boiler cooled to, say, at 70°C–100°C after being shut down for several hours, it would be restarted with the flue gas containing the acid vapor. This resulted in the corrosive flue gas contacting the cold heating surfaces. This was done frequently, and hence, the surfaces near the boiler exit were more prone to the attack of acid vapor. Note that it takes some time for heating up the boiler from cold to full load conditions, and during this time, acid vapors can deposit on the cold surfaces. The designer then suggested that the boiler should always be started up and shut down on clean gas such as natural gas till the boiler temperature is raised above the dew point and then the sulfur-containing flue gas may be admitted into the boiler. Corrosive flue gas should not contact cold heating surfaces during startup and condense resulting in corrosion. Also, the boiler supplier suggested that a steam sparger be placed inside the drum to keep the boiler warm at, say, close to the dew point temperature. This would require some manual control on the drum level. A damper should also be used at the stack to prevent cold ambient air from contacting the boiler surfaces when the boiler is shut down. These precautions minimized the acid condensation problem.

Water Chemistry, Carryover, Steam Purity

The following discussions on water chemistry and steam purity apply to steam generators as well as waste heat boilers.

Good water chemistry is important for minimizing corrosion and the formation of scale in boilers. Steam-side cleanliness should be maintained in water tube as well as fire tube boilers. Plant engineers should do the following on a regular basis:

1. Maintain proper boiler water chemistry in the drum according to ABMA or ASME guidelines by using proper continuous blowdown rates. Chapter 6 discusses blowdown calculations and also gives the guidelines for feed and boiler water.

2. Ensure that the feed water analysis is fine and that there are no sudden changes in its conductivity or solids content.

3. Check the steam temperature to ensure that there are no sudden changes in its value. A sudden change may indicate carryover of water vapor into steam. The author is aware of a project where the TOC limit in the feed water as recommended by ASME was well exceeded by the plant, and plant engineers were not aware of this. Suddenly, there was lot of foaming in the steam drum and carryover of water into the superheater, which decreased the steam temperature. The steam temperature was fluctuating as the organic compounds were causing varying degrees of foaming and the amount of water carried along with steam was varying. The superheater design and drum sizing were reviewed and found reasonable. The boiler load was also steady, and hence, plant engineers were wondering what the problem was. An investigation was carried out on the feed water chemistry, and organic compounds were ascertained to be the cause. Fortunately, the problem was identified in a short period; the tube-side deposits were found to be minimal, and hence, the superheater tubes did not fail.

4. Check the steam purity from time to time if the steam is used in a steam or gas turbine (for NO_x control). Monitoring the superheater tube wall temperatures is also good practice to ensure scales have not formed inside the tubes. Chapter 3 shows an example of the performance of a fire tube boiler with and without scale formation. Similar calculations on tube wall temperature and duty can be made in water tube steam generators also.

In the process of evaporating water to form steam, scale and sludge deposits form on the heated surfaces of a boiler tube. The chemical substances in the water concentrate in a film at the evaporation surface; the water displacing the bubbles of steam readily dissolves the soluble solids at the point of evaporation. Insoluble substances settle on the tube surfaces, forming a scale and leading to an increase in tube wall temperatures. Calcium bicarbonate, for example, decomposes in boiler water to form calcium carbonate, carbon dioxide, and water. Calcium carbonate has a limited solubility and will agglomerate at the heated surface to form a scale. Blowdown helps remove some of the deposits. Calcium sulfate is more soluble than calcium carbonate and will deposit as a heat-deterrent scale. Most scale-forming substances have a decreasing solubility in water with an increase in temperature. Substances that have a decreasing solubility with decrease in temperature are likely to deposit in steam turbine.

In drum units, some moisture always entrains with the steam as it leaves the water surface in the drum. Thus, drums are equipped with internal steam separators to remove most entrained moisture and return it to the boiler water. However, complete removal of droplets is impossible, and thus trace amounts of solids will carry over to the superheater and main steam. These impurities include chlorides, sulfates, silica, phosphates (in phosphate-treated units), and other compounds. A number of factors can cause carryover to be excessive. These include poor drum design, failed steam separators, improper drainage of separators, high dissolved solids in the boiler water, and excessive ramp rates during startup, among others.

Carryover can also be vaporous; certain compounds will volatilize and transfer the contaminants to steam. Silica and copper (in units with copper–alloy feed water heater tubes) are two notorious agents for vaporous carryover. Vaporous carryover is a function of steam density (pressure) and can be controlled only by limiting the various salts or silica, while mechanical carryover is controlled by using good steam separators and proper velocities in the steam space as discussed in Chapter 6.

Both mechanical and vaporous carryover become more pronounced with increasing pressure; thus, boiler water chemistry guidelines are increasingly stringent at higher pressures. The impurities affect turbine operation. The following list outlines a few of the important issues.

- Silica solubility decreases as steam pressure decreases through the turbine. Thus, silica will deposit on turbine blades, particularly in the HP section.
- Chloride and sulfate salts will deposit, commonly in the LP turbine. These impurities, and chloride in particular, may cause pitting and stress corrosion cracking (SCC) of turbine blades and rotors. The most susceptible locations are the last stages, L-1 and L-0, where early condensate forms. During times of low load operation, early condensate formation may move backward a bit in the LP turbine.
- Sodium hydroxide (NaOH) can also cause SCC of turbine blades.
- In units with copper–alloy feed water heaters that operate at or above 2,400 psig, copper carryover can be quite detrimental. For the most part, copper precipitates in the HP turbine, where even a few pounds of deposition will reduce capacity by perhaps 10 MW or more. This capacity loss can be critical during periods of high power demand.

Steam purity requirements for saturated steam turbines are not stringent. As the steam begins to condense on the first stage of steam turbine water-soluble contaminants carried with the steam do not form deposits. However, there can be erosion concerns due to water droplets moving at high speeds.

If steam is superheated for use in steam turbines, steam purity is critical. Salts soluble in superheated steam may condense or precipitate and adhere to the metal surfaces as the steam is cooled when it expands. Deposition from steam can cause turbine valves to stick. Reduced efficiency and turbine imbalance and lower power output are the other concerns. Deposition and corrosion occur in the *salt zone* just above the saturation line and on the surfaces in wet steam zone. The solubility of all low-volatile impurities such as salts, hydroxides, silicon dioxide, and metal oxides decreases as steam expands in the turbine and is lowest at the saturation line. The moisture formed has the ability to dissolve most of the slats and carry them downstream. The critical region for deposition of impurities in turbines operating on superheated steam is the blade row just upward of the Wilson line.

When a desuperheater is used, the water used for spray should have the same purity as the final steam; else, the solids content of the steam will exceed the desired limits. The other option is to condense the steam and use the condensate for spray as discussed earlier.

Fire Tube Boilers

Oil- and gas-fired fire tube package boilers are widely used in cogeneration plants (Figure 3.43). They generate low-pressure saturated steam, and steam capacity is limited to about 20–35 t/h as larger sizes would involve large-diameter shells and may be uneconomical. The burner is located in the large central pipe called the Morrison pipe, which is a corrugated pipe to handle the differential thermal expansion between the first pass and the other passes that are fixed to the tube sheets. The number of passes can be three or even four to improve the efficiency. An economizer may also be added if required. Figure 3.44 shows a fire tube boiler with a superheater. Note that it is somewhat cumbersome to add a superheater in a fire tube boiler. In a water tube boiler, the superheater can be located after any number of evaporator tubes. However, in a fire tube boiler, it can only be at the end of the second pass or between the boiler and economizer. The flue gas temperature at the exit of the third pass may not be significant to add a superheater, and at part loads, the gas temperature will drop off significantly. In Chapter 4, a fire tube waste heat boiler with a superheater and an economizer beyond the evaporator is shown. However, the steam temperature will be low.

Some boiler suppliers use grooved tubes (Figure 3.45) to improve the tube-side heat transfer coefficient and thus either reduce the number of passes or improve the performance.

FIGURE 3.43
Fire tube boilers in a process plant. (Courtesy of Cleaver Brooks Inc., Thomasville, GA.)

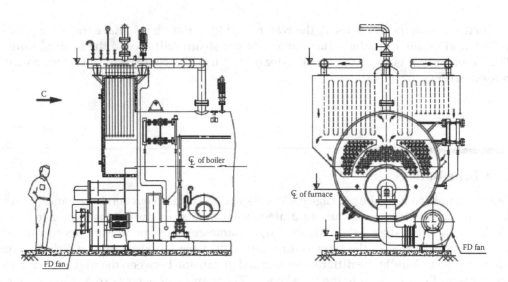

FIGURE 3.44
Fire tube boiler with a superheater. (Courtesy of Thermodyne Technologies, Chennai, India.)

FIGURE 3.45
Grooved tubes used in fire tube boilers to improve energy transfer. (Courtesy of Cleaver Brooks Inc., Thomasville, GA.)

Fire tube boilers are classified as dry back or wet back depending on how the turn-around section of the first pass is built (Figure 3.46). Wet back boilers do not use refractory in the turnaround section thus decreasing the maintenance costs and improving the boiler life; however, they are slightly more expensive. As discussed in Chapter 4 on waste heat boilers, using smaller-diameter tubes in fire tube boilers helps reduce weight and length of the boiler.

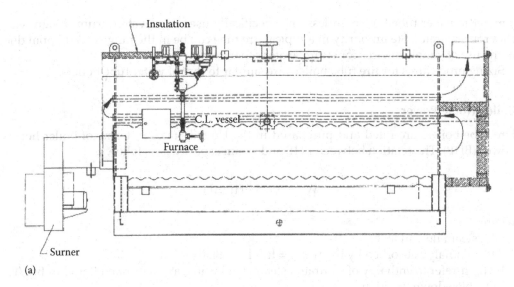

(a)

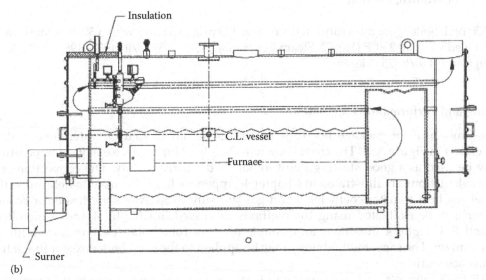

(b)

FIGURE 3.46
(a) Dry back and (b) wet back boilers.

Oil and gaseous fuels are generally fired in package fire tube boilers. Solid fuels such as wood chips have also been fired. The boiler capacity is limited to about 35 t/h of steam as it becomes very expensive to build these boilers beyond this capacity. The heat transfer coefficient when flue gas flows inside the tubes is generally lower than when it flows outside the tubes. Besides, extended surfaces cannot be used to make the boiler compact as in water tube boilers, though grooved tubes help to some extent.

NO_x control methods such as FGR or low NO_x burners have been used with fire tube boilers. Due to the large amount of water inventory compared to an equivalent water tube boiler, fire tube boilers take a little more time to start up. Steam purity is poorer than what you get in water tube boilers unless one specifically uses an elevated drum design with chevron separator. Steam purity in a typical fire tube can be in the range of 3–15 ppm due to the use of simple devices for steam–water separation.

Sizing procedures for fire tube boilers are similar to that of water tube boilers.

Boiler Horse Power

Fire tube boilers are rated and purchased in the United States in terms of boiler horsepower (BHP). The relation between BHP and steam capacity is as follows:

$$W = 33,475 \text{ BHP}/\Delta H$$

where
  W = steam flow in lb/h
  ΔH = enthalpy absorbed by the water = $h_g - h_w + BD(h_f - h_w)$
  h_g, h_w, h_f refer to enthalpy of saturated steam, feed water, and saturated liquid in Btu/lb
  BD = blowdown, fraction

A 500 BHP boiler generates saturated steam at 125 psig (8.6 barg) with a 5% blowdown and with feed water at 230°F (110°C). Steam generation = $500 \times 3,475/[(1,193 - 198) + 0.05(325 - 198)]$ = 16,714 lb/h (7583 kg/h).

Sizing and Performance Calculations

The procedure for package fire tube boilers is similar to that of water tube boilers described in Figure 3.38. The chart shown in Chapter 2 for furnace exit gas temperature may be used as a good starting point to obtain the furnace duty. The convection section calculations are illustrated in Chapter 4. Appendix B may be used to determine the tube-side heat transfer coefficients and gas pressure drop. Any superheater or economizer may be evaluated using the methods discussed in Chapter 4 and Appendices C and E. Chapter 4 illustrates an example of a fire tube boiler performance with an economizer. The same methodology may be applied to the fired boiler except that it has a furnace section.

All the comments and discussions in Chapter 4 on fire tube waste heat boilers will apply to package fire tube boilers also. Using smaller-diameter tubes will reduce the boiler weight and length. Tube thickness required for a given steam pressure is nearly twice that required compared to when pressure acts inside the tubes. The simplified procedures discussed in Chapter 4 may also be used for the performance evaluation of the convection passes.

Computer programs may also be developed for evaluating the performance of the fire tube boiler given the mechanical data. Tables 3.16 and 3.17 show the input data and performance results for a typical fire tube boiler with three passes and an economizer firing natural gas and generating 40,000 lb/h. The efficiency on LHV basis is 94%. The back pressure in the tubes is about 5.3 in wc. The tube geometry may be changed as desired to decrease operating costs or to improve efficiency; performance without the economizer may also be obtained. Part load performance may also be obtained.

TABLE 3.16

Input Steam, Fuel, and Tube Details

Steam Data			
Steam flow, lb/h	40,000		
Steam temp., °F	0		
Fw temp., °F	230		
Blowdown, %	3		
Sh press. drop, psi			
Stm press., psig	250		
Ambient temp., °F	80		
Stack gas temp., °F	400		
Excess air, %	15		
FGR, %	0		
Rel. hum., %	60		
Casing loss, %	0.50		
Unacc. loss, %	0.50		
Fuel: 2—gas, 1—oil, 3—wood	2		
Foul factor, gas	0.001		
Foul factor, steam	0.001		
Fuel temp., °F	75		
Makeup air, °F	75		
Fuel gas analysis, % volume			
Methane	97.00		
Ethane	2.00		
Propane 1	1.00		
Mechanical data–boiler			
Wet back len., in.	24	Dia., in.	80
Number of passes	3		
Pass #	1	2	3
Tube OD, in.	43	4.5	3
Tube ID, in.	46	4.026	2.5
Number	2	95	110
Length, in.	420	420	420
Correction	1	1	1
Economizer: 1—yes, 2—no	1		
Tube OD, in.	2.00	Number of rows deep	8
Tube ID, in.	1.77	Eff. len., ft	6.00
Fins/in.	5.00	St, in.	4.00
Fin ht., in.	0.75	Sl, in.	4.00
Fin thk., in.	0.075	Streams	3
Serr., in.	0.000	Arrgt-in./st	1
Fin K, Btu/ft h °F	25	Confg-cf/pf	1
Tubes/row	12		

Note: Arrgt 1 means staggered and 0 is inline. Configuration 1 is counter-flow, 2 parallel-flow.

TABLE 3.17

Typical Performance Results (Predicted Performance)

Boiler efficiency—per ASME

Excess air	15	%
Amb. Temp.	80	°F
Air temp.	75	°F
Fuel temp.	75	°F
Exit gas temp.	270	±10°F
Dry gas loss	3.55	%
Air moist. loss	0.10	%
Fuel moist. loss	10.56	%
Casing loss	0.50	%
Unacc. and margin	0.50	%
Efficiency, HHV	84.80	%
Efficiency, LHV	94.00	%
Fuel HHV	23,764	Btu/lb
Fuel LHV	21,439	Btu/lb
Boiler duty	40.33	MM Btu/h
Burner duty, HHV	47.56	MM Btu/h
Actual vol. HRR	117,765	Btu/ft³ h
Area HRR, HHV	56,416	Btu/ft² h

Air and flue gas

% Volume	Wet	Dry
CO_2	8.29	10.13
H_2O	18.17	0.00
N_2	71.07	86.86
O_2	2.46	3.01
SO_2	0.00	0.00

Evaporator performance

Pass	1	2	3	
Gas in 1a	3228	2171	635	±10°F
Gas outlet	2171	635	430	±10°F
Gas sp. ht.	0 3558	0.3137	0 2822	Btu/lb °F
Duty	15.79	20.25	2.45	MM Btu/h
Gas press. drop	0.09	0.89	3.57	in wc
Man. ht. flux	24,834	13,110	2124	Btu/ft² h
Max. wall temp.	532	439	411	°F
Overall U	7.64	7.67	10.69	Btu/ft² h °F
Surf. area	916	3505	2520	ft²
Delt	2253	753	91	°F
Avg. gas vel.	37	57	81	ft/s

(Continued)

TABLE 3.17 (*Continued*)

Typical Performance Results

Economizer performance

Gas temp. in	430	±10°F	Gas pr. Drop, in wc	0.72
Out	270	±10°F	Wart pr. drop, psi	4.86
Duty	1.84	MM Btu/h	Tube wall temp, °F	233
T_{w1}	230	±10°F	U_o, Btu/ft²h °F	6.24
T_{w2}	275	±10°F	Tube-side htc, Btu/ft²h °F	1309
Surface area	3497	ft²	Arrgt.	Stagg
Max. gas vel.	29	ft/s	Config.	Counter flow
Air flow	40,022	lb/h		
Flue gas-stack	42,025	lb/h		
Steam flow	40,000	lb/h		
Fluid temp. in	230	°F		
Fluid temp. out	406	°F		
Fluid press.	250	psig		
Foul ftr., gas	0.0010	ft² °F h/Btu		
Foul ftr., stm.	0.0010	ft² °F h/Btu		
Excess air	15	%		
Total duty	40.33	MM Btu/h		

References

1. V. Ganapathy, Specify packaged steam generators properly, *Chemical Engineering Progress*, Sept. 1993, p62.
2. V. Ganapathy et al., Designing large package boilers, *Power*, Feb. 2011, p80.
3. V. Ganapathy et al., Trends in package boiler design, *Power Engineering*, Nov 2011, p100.

4

Waste Heat Boilers

Introduction

Waste heat boilers or heat recovery steam generators form an inevitable part of chemical plants, refineries, power plants, and process systems. They are classified in several ways as seen in Figure 4.1 according to the application, type of boiler used, whether the flue gas is used for process or mainly for energy recovery, cleanliness of the gas, and boiler configuration, to name a few.

The main classification is based on whether the boiler is used for process or energy recovery. Process waste heat boilers are used to cool gas streams from a given inlet gas temperature to a desired exit gas temperature for further processing purposes. The flue gas is not vented to the stack beyond the boiler exit; instead, it is taken to a reactor or convertor for processing. Examples can be seen in sulfuric acid, nitric acid, or hydrogen plants. The exit gas temperature is controlled in such boilers using a gas bypass system at a desired value for downstream processing (Figures 4.2 and 4.3). Energy recovery is not the main objective. In energy recovery applications on the other hand, the objective is to maximize the energy recovery compatible with low-temperature corrosion issues. Examples may be seen in gas turbine heat recovery systems or incinerator, furnace, or kiln exhaust heat recovery systems.

If the gas stream is clean and the gas flow large (above 40,000 kg/h), water tube boilers with finned tubes are used as the boilers are compact (Figure 4.4). In solid or liquid waste incineration systems, the flue gas is generally dirty and may contain corrosive compounds, acid vapors, ash, and particulates. If the ash contains compounds of sodium, potassium, or nonferrous metals, slagging is likely on heating surfaces if these substances become molten. In these cases, bare tube boilers with provision for cleaning with soot blowers or rapping mechanisms are used. A water-cooled furnace that cools the gas stream to a temperature below the ash melting temperature minimizing slagging on the convective surfaces may also be necessary (Figure 4.5).

Special requirements may also affect the design of the boiler. Figure 4.6 shows a waste heat boiler recovering energy from a steel plant kiln. In this boiler handling dusty kiln gas (about 6–10 g/N m^3), the flue gas flow varies with time, and sudden surges occur as a result of which the gas temperature entering the waste heat boiler would go up from 500°C to 900°C in 15–30 s as a result of which the superheated steam temperature would vary significantly during these transients. Locating the superheater as the first heating surface may not be a good idea as we need some heating surfaces to shield the superheater from thermal shock. Hence, the boiler was designed with a large screen section that shields the superheater from these temperature surges and minimizes the steam temperature fluctuations. Three stages of desuperheating were incorporated in the steam system, one between

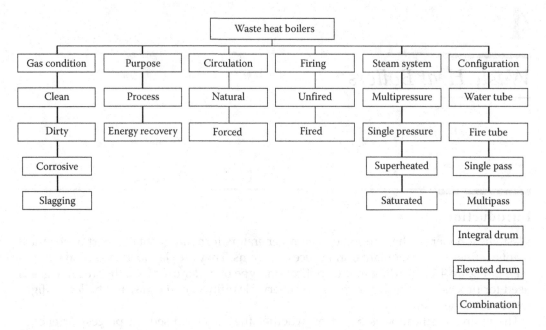

FIGURE 4.1
Classification of waste heat boilers.

FIGURE 4.2
Fire tube boiler for reformed gas application.

the first and second stages of the superheaters and one between the second stage and final superheater, and one emergency desuperheater at the steam exit to handle sudden fluctuations in steam temperature; the drum also was sized with a large holdup time of about 12 min from normal level to empty. As the flue gas is dusty, a ball cleaning system was incorporated in the second pass in which are located the screen section, superheaters, and evaporator. Steel balls will be dropped over the tube bundles at a frequency determined

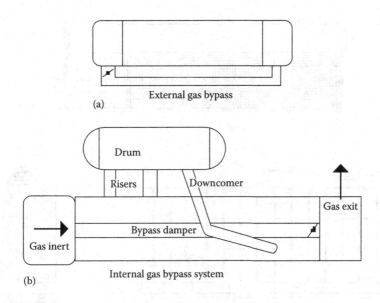

FIGURE 4.3
(a) External and (b) internal gas bypass systems.

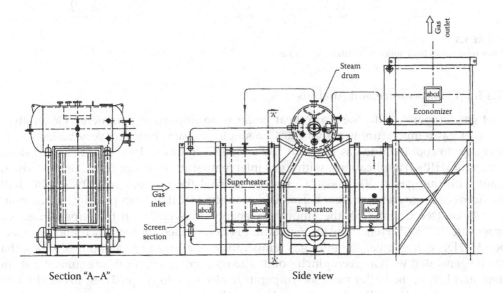

FIGURE 4.4
Water tube waste heat boiler for an incineration plant with screen, superheater, evaporator, and economizer.

by operating conditions, collected at the bottom and sent back to the top of second pass by a pneumatic conveying system (Figure 4.7). Though plain tubes were used for the heating surfaces, the heating surfaces were arranged in a staggered fashion to facilitate better cleaning of tubes. The screen section and evaporator are multistream horizontal tube natural circulation systems. The circulation system was checked carefully considering that horizontal tubes are used in the evaporator. The thermal and performance design of the boiler was done by the author, which was built and erected by a fabricator and is presently in operation in Canada.

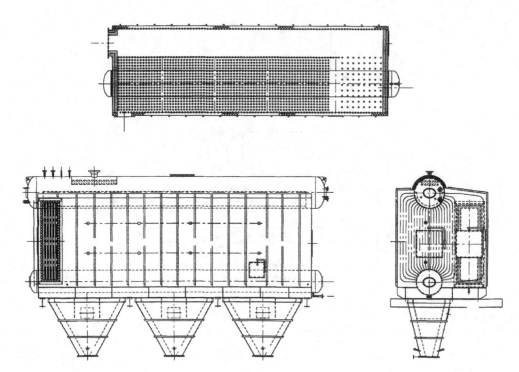

FIGURE 4.5
Water tube waste heat boiler with furnace section.

Gas Inlet Temperature and Analysis

Inlet gas temperature to the waste heat boiler is an important parameter. Generally, if the inlet gas temperature is above 700°C, a single-pressure heat recovery boiler will be adequate to cool the flue gas to about 130°C–150°C. In gas turbine heat recovery steam generator (HRSG) applications, with a gas inlet temperature of, say, 500°C or in a cement plant with flue gas at 300°C–350°C, we cannot cool the flue gas to 130°C–140°C with medium- or high-pressure steam. This is due to the fact that with practical pinch point limitations (see Chapter 5), the amount of energy recoverable in the superheater and evaporator is fixed. Hence, steam generation is determined by the pinch and approach points. Hence, the lower the inlet gas temperature, the lesser the amount of steam that can be generated with a given pinch point. The economizer water flow equals the steam flow, and hence, the boiler exit gas temperature also falls in place. Thus, lower the inlet gas temperature, lower the steam generation and higher the exit gas temperature and vice versa. That is why multiple-pressure steam generation is required to lower the exhaust gas temperature in HRSGs or cement plants. Single-pressure heat recovery system is adequate when the gas inlet temperature is high. Irrespective of the steam pressure, we can cool the flue gas to about 140°C in oil- or gas- or solid fuel–fired steam generators with a single-pressure system (the inlet gas temperature may be taken as the adiabatic combustion temperature due to which a large amount of steam is generated in the furnace and evaporator, and hence, the economizer becomes a very large heat sink cooling the flue gas to 130°C–140°C). Alternatively, one has to consider the Kalina system, discussed later, in which a mixture of ammonia–water is used as the working fluid. In this system, the

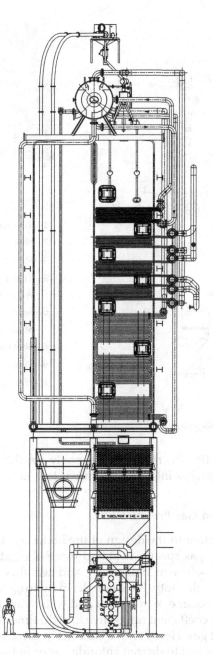

FIGURE 4.6
Water tube waste heat boiler for a steel mill kiln. (Courtesy of Elkam, Alma, Quebec, Canada.)

working fluid has variable boiling points unlike steam, and the exhaust gas can be cooled to a much lower temperature compared to a steam system even if the inlet gas temperature is around 200°C–250°C.

The need for multiple-pressure steam is also discussed in Chapter 5. Depending on the ratio of HP steam to LP steam flow or the ratio of HP to LP steam pressure, one decides if

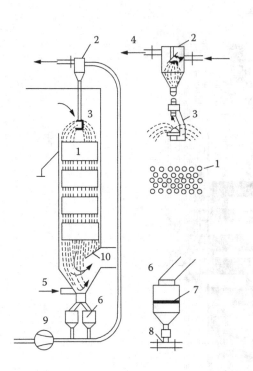

Shot cleaning system:

  1—Tube bank
  2—Shot feeding chamber
  3—Shot spreader
  4—Transport air outlet
  5—Air for dust/shot separation
  6—Separator/shot collector
  7—Strainer
  8—Shot transport nozzle
  9—Transport air fan
  10—Diverting baffle

FIGURE 4.7
Shot cleaning system for waste heat boiler.

single- or multipressure HRSG is required. Generally speaking, multiple-pressure steam generation is required when gas inlet temperature is low, on the order of 500°C or less.

Flue Gas Composition and Gas Pressure

Flue gas analysis is important to the design of the boiler. A large amount of water vapor or hydrogen increases the gas specific heat and thermal conductivity and hence the gas and overall heat transfer coefficient, boiler duty, and heat flux. Presence of SO_2 lowers the gas specific heat and hence the duty. For example, in hydrogen plant reformed gas boilers where we have a large % volume of hydrogen and water vapor in the flue gas at high gas pressure, the heat transfer coefficient can be five to eight times that of typical flue gas from the combustion of natural gas. Heat flux is a concern in these applications.

The presence of chlorine and hydrogen chloride vapor in flue gas is indicative of corrosion potential particularly if a superheater is used and the tube wall temperature is above 475°C. Chlorine attacks carbon steel even at low temperatures as shown in Figure 4.8. Hence, low-pressure saturated steam is generated with flue gas containing chlorine and economizers are avoided. Presence of hydrogen chloride gas (HCl) also indicates high-temperature corrosion. There are instances of high-temperature stainless steel superheaters in several municipal solid waste incineration plants having been destroyed within a few months of operation due to HCl. Hence, modular municipal solid waste incineration plants generate steam at medium pressure and low temperature (about 40 barg, 375°C) enabling longer life of superheater tubes.

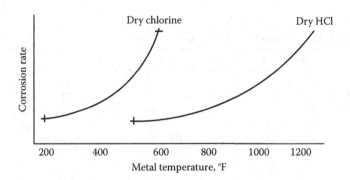

FIGURE 4.8
Chlorine and hydrogen chloride in flue gas cause corrosion problems.

The presence of SO_3 and HCl in the flue gas also suggests low-temperature corrosion problems due to the acid vapor condensation. To minimize low-temperature corrosion concerns, there are several methods that are discussed in Chapter 3. Major decisions such as using membrane wall construction or refractory-lined casing depend on the flue gas analysis and dew point of the flue gas and cost considerations.

More care should be taken when the flue gas contains ash, which has slagging potential. This is common in steel plant effluents or municipal solid waste incineration systems. A furnace to cool the flue gas to below the ash melting temperature as shown in Figure 4.5 or Figure 4.6 is one solution. Some plants recirculate flue gas from the boiler rear and lower the gas inlet temperature to the boiler below the ash melting temperatures. However, this increases the flue gas quantity through the boiler and increases the gas pressure drop, or the boiler cross section may have to be increased. Double-spaced screen section, which prevents molten ash from sticking to the boiler tubes, is recommended, if slagging is envisaged. Tube spacing of 150–200 mm is used at the convective bank inlet as shown for a few rows deep. One can gradually reduce the tube spacing as the gas cools. The superheater, which is more prone to corrosion when deposits stick on to them, is located in a much cooler zone by using adequate number of screen tubes ahead of it.

Flue gas pressure is generally atmospheric in many applications; in pressurized systems (such as gas turbine HRSGs or incineration plants), the flue gas is well contained in the boiler if a membrane wall construction is used. If a suction system with an induced draft fan is used as in some biomass heat recovery projects, then there is possibility of leakage of cold air from the atmosphere into the boiler through man-way doors or hopper openings. This is likely to affect local gas temperatures, and hence, performance prediction may not be accurate. Superheated steam temperature may also be impacted by ingress of cold air into the system.

In reformed gas boilers in hydrogen plants or nitric acid plants, the gas pressure can be very high and hence fire tube boilers are preferred, though special water tube boilers have been built. Table 4.1 shows the flue gas analysis and pressure for different waste gas [1] streams. Alloy steel tubes are recommended when hydrogen is present in flue gas; hence, reformed gas boilers use T11 or T22 tubes for the evaporator to minimize high-temperature concerns. Nelsons chart is often referred to for material selection (Figure 4.9).

In incineration systems involving halogenated compounds, one has to ensure that the combustion temperature of 850°C–900°C is maintained for at least 2 s in the incinerator combustion chamber to destroy the formation of dioxins. Following combustion, the flue

TABLE 4.1

Analysis of Waste Gases from Process Plants, % Volume

Gas	1	2	3	4	5	6	7	8	9	10	11
CO_2	3			9	6	12	7	6		28	0
H_2O	7			18	37	10	8	23	19	7	1
N_2	75	80	81	70	.5	72	76	55	66	59.5	78
O_2	15	10	11	3		6	4		6	5.5	21
SO_2		10	1					6			
HCl							5				
SO_3			7								
CO					8			3			
CH_4					5.5						
H_2S								3			
H_2					43			4			
NO									9		
Press., atm	1	1	1	1	30–50	1	1	1.5	3–10	1	1
Temp., °C	600–100	1000–300	500–250	1100–200	1100–300	1000–200	1350–250	1400–350	850–250	450–150	400–100

Notes: (1) Gas turbine exhaust, (2) raw sulfur gases, (3) SO_3 gases after converter, (4) reformer flue gas, (5) reformed gas, (6) MSW incinerator exhaust, (7) chlorinated plastics incineration, (8) sulfur condenser effluent, (9) nitrous gases, (10) preheater effluent—cement plant, (11) clinker cooler effluent—cement plant.

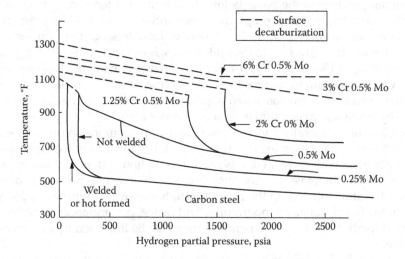

FIGURE 4.9
Nelson's chart for the selection of materials.

gas should be cooled very quickly in the waste heat boiler, typically in less than 0.5 s to ensure dioxins are not formed. Finned water tube boilers have shorter gas path and hence shorter residence time. With fire tube boilers, the residence time may be longer and has to be checked. One option as we shall see later is to use smaller-diameter tubes in fire tube boilers to reduce their length.

Water Tube versus Fire Tube Boilers

A common classification of boilers is whether the gas flows inside the tubes as in fire tube boilers (Figure 4.10) or outside the tubes as in water tube boilers (Figure 4.5). The features of each type are discussed in Table 4.2 [2].

Generally, water tube boilers are suitable for large gas flows in the range of millions of kg/h and can handle high steam pressures and temperatures. Fire tube boilers are suitable for low steam pressures generally below 3500 kPa (35 bar) and low gas flows. Table 4.3 shows the effect of pressure on tube thickness for both types. A given tube thickness can withstand about twice the pressure when it is inside than when outside. Hence, fire tube boilers require large thickness as the steam pressure increases and thus become uneconomical;

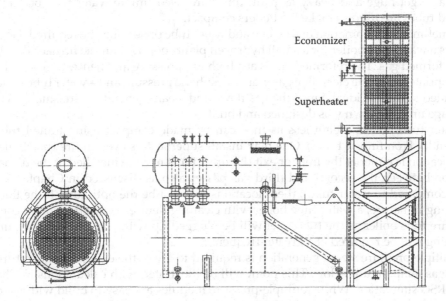

FIGURE 4.10
Fire tube boiler with superheater and economizer. (Courtesy of ABCO Industries, Abilene, TX.)

TABLE 4.2

Comparison of Fire Tube and Water Tube Boilers

Item	Fire Tube	Water Tube
Gas flow	Small—less than 30,000 kg/h	Large—30,000–1,000,000 kg/h
Gas inlet temperature	Low to adiabatic combustion	Low to adiabatic combustion
Gas pressure	As high as 150 barg	500–1000 mm wc
Firing	Possible	Possible
Type of heating surface	Plain tubes	Plain and finned tubes
Superheater location	At inlet or exit	Anywhere in the gas path
Water inventory	High	Low
Heat flux on steam side	Generally low	Can be high with finned tubes
Multiple pressure	No	Yes
Multiple modules	No	Yes
Soot blower location	Inlet or exit	Anywhere in gas path

TABLE 4.3

Tube Thickness versus Steam Pressure

Tube Thickness, mm	External Pressure, barg	Internal Pressure, barg
2.7	40	79
3.05	47.3	92.3
3.5	55	106
3.9	63.5	119
4.6	81	147

Note: 50.8 mm SA 192, SA 178a carbon steel tubes at 371°C.

they also get huge and heavy as plain tubes are used unlike water tube boilers where finned tubes are used to make the boilers compact.

Sometimes, a combination of fire tube and water tube boilers may be required. Figure 4.11 shows a waste heat boiler for a small hydrogen plant cooling both the furnace flue gas and the reformed gas. The reformed gas is at a high gas pressure and hence a fire tube boiler is appropriate, while the clean flue gas is at atmospheric pressure and a water tube boiler with extended surface is ideal. Since the gas flows and steam generation are small, a compact package unit as shown was designed and built.

Water tube boilers weigh less as they can be made compact using finned tubes (see Appendix E on finned tubes). One may install superheaters even in fire tube boilers, but the location is either at the inlet or exit (Figure 4.10). It may be installed at the turnaround section between the second and third gas passes also as discussed in Chapter 3. In any case compared to water tubes, these locations may not be the optimum. If the flue gas is slagging in nature, a water tube boiler with cleaning devices is a better choice compared to a fire tube boiler as the tube inlet will be plugged up with slag in a fire tube unit and cleaning will be required at a greater frequency.

Multiple-pressure steam generation is required to lower the exit gas temperature if the inlet gas temperature is low. This point will be illustrated with calculations in Chapter 5 on HRSG simulation. While multiple-pressure modules are easy to build with water tube boilers, they are rarely seen with fire tubes as it is uneconomical and space requirement will be huge.

Circulation Systems

There can be a variety of circulation systems involving waste heat boilers depending on whether they are fire tube or water tube and whether they have single or multiple evaporators. Figure 4.10 shows a fire tube boiler with external steam drum, downcomers, and risers. Figure 4.11 shows a combination of fire tube and water tube with circulation system. Figure 4.4 shows a water tube boiler with screen and evaporator sections with a superheater in between. The screen and evaporator sections operate in parallel and are connected to the same steam drum by external downcomers and risers. Figure 4.12 shows two fire tube boilers sharing the same steam drum. Figure 4.6 shows a water tube boiler in which the screen section and evaporator sections have multistream horizontal tubes and are connected to the same steam drum by external downcomers and risers. Figure 2.13d shows a large evaporator that required an external steam drum and hence external downcomers, risers, are used.

Circulation calculations are quite involved and preferably done using a computer program as numerous iterations are required before one can arrive at the circulation ratio

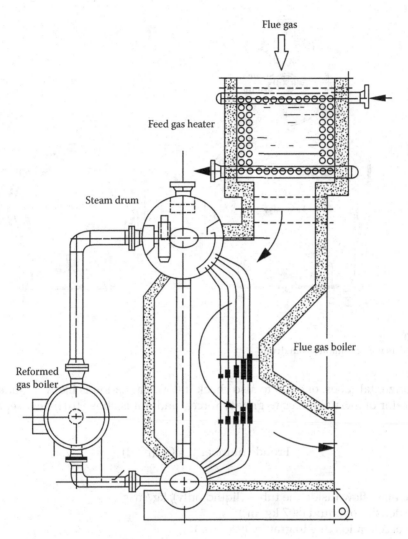

FIGURE 4.11
Combination fire tube and water tube design.

(CR), which is based on the resistance of the downcomers, risers, and evaporator tubes, and the drum location. The procedure is explained in Chapter 2. A forced circulation steam generator using circulation pumps is shown in Chapter 2, while Chapter 3 shows a forced circulation system for an oil field steam generator.

Table 4.4 shows the results from a computer program for the circulation system involving boiler with screen section that has horizontal tubes and multiple streams (Figure 4.6). The screen section generates 13,150 kg/h in the maximum case. It has a screen section with 160 streams (parallel paths); each tube has a developed length of 12.2 m. The distance between the drum centerline and bottom of screen is 5.34 m. There are two downcomers of inner diameter 150 mm and two external risers of inner diameter 190 mm. Their lengths are about 8 m. As discussed in Chapter 2, the CR was assumed, and the various single- and two-phase flow losses were computed, and iterations were carried out till the available head matched the losses as shown. The CR finally arrived at is 12.45.

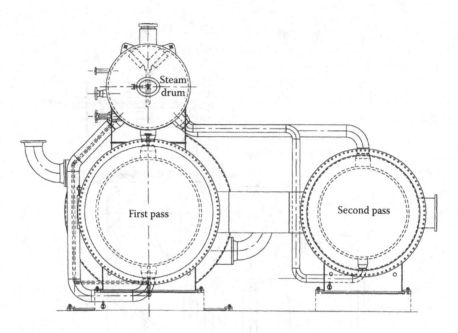

FIGURE 4.12
Fire tube boilers with common steam drum.

With horizontal tubes, one has to also check if flow stagnation is likely. Froude number is an indicator of inertial forces to gravity forces, and if it is above 0.04, flow separation is unlikely.

$$\text{Froude number } F = G^2/(\rho^2 g d)$$

where
G is the mass flow inside the tubes (liquid only), $kg/m^2 s$
ρ is the density of liquid (807 kg/m^3)
g is the acceleration due to gravity (9.8 m/s^2)
d is the tube inner diameter (0.026 m)

In this example, the liquid-only mass flow = $(1 - 1/CR) \times 13{,}150 \times 12.45 = 0.92 \times 13{,}150 \times 12.45 = 150{,}620$ kg/h

$$G = \text{mass velocity} = 150620 \times 4/\left(\pi \times 0.026 \times 0.026 \times 160\right)/3600 = 493\,kg/m^2 s$$

F = 493 × 493/(807 × 807 × 9.8 × 0.026) = 1.46. As this is much higher than the limiting value of 0.04, stagnation is unlikely. The said boiler has been operating well for the past few years. A few boiler suppliers use inclined horizontal tubes to provide a better comfort factor for the flow of two-phase mixture inside tubes at medium and high steam pressures (Figure 4.22c, later in the chapter).

When a circulation pump is used as in forced circulation, boilers the CR is predetermined, say, 4–8 and the circulating pump differential head is selected based on water flow corresponding to CR selected; the pump head is obtained after checking the available head and computing the losses.

TABLE 4.4

Circulation Calculations

Screen Section

Steam pressure	37	kg/cm²g
Sat. temperature	244	°C
Feed water temp.	204	°C
Sp. vol. steam	0.0650	m³/kg
Sp. vol. water	0.0012	m³/kg
Average CR	12.46	
Total static head	5.34	m
Total static head	0.44	kg/cm²
Downcomer losses	0.04	kg/cm²
Riser losses	0.06	kg/cm²
Riser gravity loss	0.12	kg/cm²
Drum internals' loss	0.07	kg/cm²
Mixture temp.	242	°C

Path	Level	Sort	Diameter	Length	Bends	Flow	Velocity	No. Tubes	Press. Drop	Internal dc
1	1	1	150	8	3	163,718	1.56	2	0.04	Downcomer
1	1	1	190	8.5	3	163,718	4.37	2	0.06	Riser

Stream flow	Quality	Accln. loss	Gravity loss	Friction loss	Total flow	Tubes	Tube ID	Boiling ht.	Length	CR
13,150 kg/h	0.079	0.007 kg/cm²	0.061	0.082	164,100 kg/h	160	26.00 mm	1.2 m	12.20 m	12.7

Boiling ht. 0.4 m

Heat Recovery in Sulfur Plants

A sulfur plant forms an important part of a gas processing system in a refinery or gas processing plants. Sulfur is present in natural gas as hydrogen sulfide (H_2S); it is the by-product of processing natural gas and refining high-sulfur crude oils. For process and combustion applications, the sulfur in the natural gas has to be removed. Sulfur recovery refers to the conversion of hydrogen sulfide to elemental sulfur. The most common process for sulfur removal is the Claus process, which recovers about 95%–97% of the H_2S in the feed stream (Figure 4.13). Waste heat boilers are an important part of this process (Figure 4.14a). Due to low gas inlet temperature in sulfur condenser, a single-shell fire tube boiler is often used.

The Claus process used today is a modification of a process first used in 1883, in which H_2S was reacted over a catalyst with air to form elemental sulfur and water. The reaction is expressed as follows:

$$H_2S + 1/2 O_2 \rightarrow S + H_2O$$

Control of this exothermic reaction was difficult, and sulfur recovery efficiency was low. Modifications later included burning a third of the H_2S to produce SO_2, which is reacted with the remaining H_2S to produce elemental sulfur; this process consists of multistage catalytic oxidation of H_2S according to the reactions:

$$2H_2S + 3O_2 \rightarrow 2SO_2 + 2H_2O + heat$$
$$2H_2S + O_2 \rightarrow 2S + 2H_2O$$

Each catalytic stage consists of a gas reheater, a catalyst chamber, and a sulfur condenser as shown in Figure 4.14. In addition to the oxidation of H_2S to SO_2 and the reaction of SO_2 with H_2S in the reaction furnace, many other side reactions occur to yield gas stream containing CO_2, H_2S, SO_2, H_2, CH_4, and H_2O in addition to various species of sulfur. The duty of the boiler behind the sulfur combustor includes both sensible heat from cooling of gas stream from 1426°C (2600°F) to 343°C (650°F) and the duty associated with the transformation of

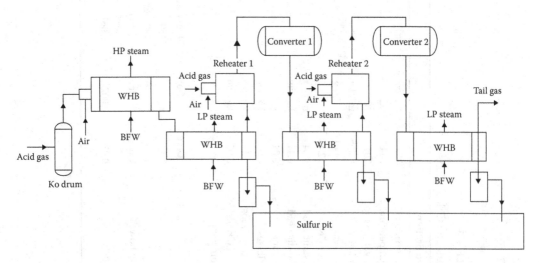

FIGURE 4.13
Claus process for sulfur recovery.

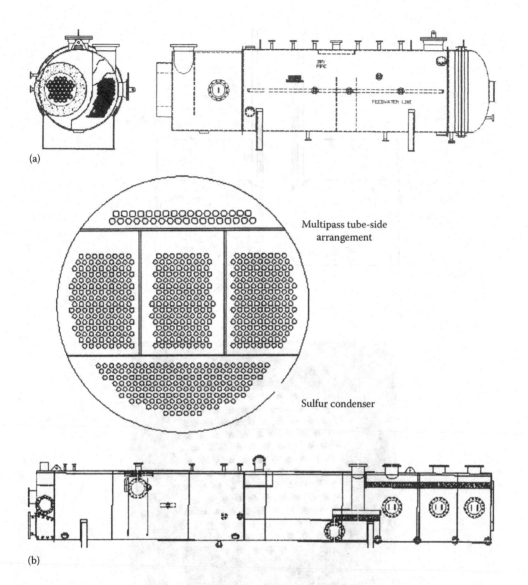

Multipass tube-side arrangement

Sulfur condenser

(a)

(b)

FIGURE 4.14
(a) Two-pass fire tube waste heat boiler for Claus plant. (b) Sulfur condenser with multistreams from reaction generating low-pressure steam.

various species of sulfur. The reaction furnace operates at 982°C–1537°C (1800°F–2800°F), and the flue gas is passed through a waste heat boiler in which saturated steam at about 40 barg is generated. This is typically a two–gas pass design, though single-pass designs are available. The gas is cooled to 650°C (1200°F) in the first pass and finally to 343°C in the second pass.

Figure 4.12 shows the boiler for a large sulfur recovery plant, which consists of two separate shells one for each pass connected to a common steam drum. The external down-comer–riser system ensures adequate circulation and cooling of the tube sheet, which is refractory lined. Ceramic ferrules are used to protect the tube-to-tube sheet joint, as shown in Figure 4.15. The inlet gas chamber is also refractory lined. The casing is kept between 175°C and 205°C (350°F–400°F) through a combination of internal and external insulation

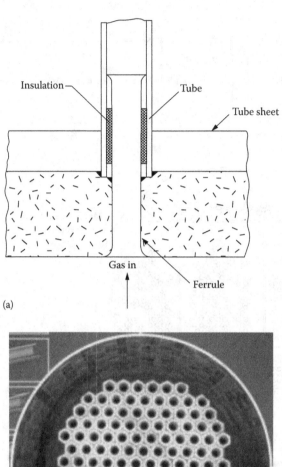

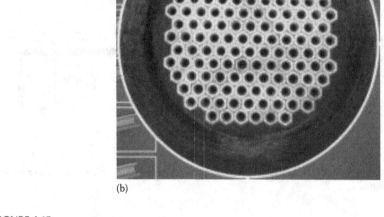

FIGURE 4.15

(a) Tube to tube sheet joint. (b) Picture of tube sheet with refractory and ferrules. (Courtesy of Cleaver Brooks, Thomasville, GA.)

to minimize concerns regarding acid dew point corrosion. The exit chamber is externally insulated. About 65%–70% of the sulfur is removed in the boiler as liquid sulfur by using heated drains. Figure 4.14b shows a single-pass design for the same application. The purpose of using ferrules is explained later. Refractory on tube sheet and ferrules reduce the heat flux through the tube sheet and hence keep the tube sheet cool. Calculations for tube sheet temperature are shown later with and without tube sheet refractory and ferrules.

Though the boiler operates above the sulfur dew point, some sulfur may condense at partial loads and during transient start-up or shutdown mode. The cooled gases exiting the exchanger are reheated to maintain acceptable reaction rates and to ensure that process gases remain above the sulfur dew point and are sent to the catalyst beds for further conversion as shown in Figure 4.13. The catalytic reactors using alumina or bauxite catalysts operate at lower temperatures ranging from 200°C to 315°C. Because this reaction represents an equilibrium chemical reaction, it is not possible for a Claus plant to convert all of the incoming sulfur to elemental sulfur. Therefore, two or more stages are used. Each catalytic stage can recover one half to two-thirds of the incoming sulfur. Acid gas is also introduced at each catalyst stage as shown. The gas stream from each stage is cooled in another low-pressure boiler at 3–5 barg called the sulfur condenser, which condenses some of the sulfur. If the flue gas quantity is small, a single-shell fire tube boiler handles all the streams from the reactors. Each stage has its own gas inlet and exit connections. The outlet gas temperatures of these exchangers are about 175°C. From the condenser of the final catalytic stage, the process stream passes on to tail gas treatment process. Carbon steel tubes are adequate for the boiler for both the low-pressure sulfur condenser and Claus plant heat high-pressure heat recovery boiler.

Heat Recovery in Sulfuric Acid Plant

Sulfuric acid is an important chemical that is manufactured using the contact process (Figure 4.16). Heat recovery plays a significant role in this system, whose main objective is to cool the gas stream to a desired temperature for further processing. Raw sulfur is burned in air in a combustion chamber generating SO_2, O_2, N_2 (see Table 4.2 for gas analysis). The flue gas at about 1040°C and at a pressure of 1200 mm wc passes through a waste heat boiler generating saturated or superheated steam at medium pressure. The boiler could be fire tube or water tube. The flue gas is cooled to about 425°C, which is the optimum temperature for conversion of SO_2 to SO_3. The exit gas temperature from the boiler decreases as the load decreases. In order to maintain 425°C for the reaction to occur, a gas bypass system is incorporated in the boiler. This can be external or internal as shown in Figure 4.3. The flue gas then goes to a converter where SO_2 gets converted to SO_3 in a few stages in the presence of catalysts. The reactions are exothermic, and the gas temperature increases

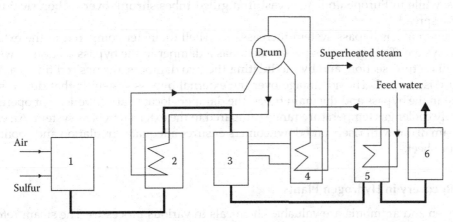

FIGURE 4.16
Scheme for sulfuric acid manufacture. *Note:* (1) Sulfur combustion furnace, (2) waste heat boiler, (3) contact apparatus, (4) superheater, (5) economizer, and (6) absorption tower.

by 20°C–50°C. Air heating or superheating of steam is necessary to cool the gases back to 425°C. After the last stage of conversion, almost all of the SO_2 is converted to SO_3. The gas stream at 480°C containing SO_3 gases is cooled in an economizer before being sent to an absorption tower. The flue gas is absorbed in dilute sulfuric acid to form concentrated sulfuric acid as shown in Figure 4.16. The steam generated in the boiler and superheater is used for power as well as process.

The main boiler behind the sulfur combustor can be of fire tube or water tube design, depending on gas flow. Extended surfaces may also be used if the gas stream has little or no dust. At the hotter end, plain tubes are used followed by tubes with low fin density, and then as the gas cools, tubes with higher fin density are used. Sometimes due to inadequate air filtration and poor combustion, particulates are present in the flue gas, which could preclude the use of finned tubes or cause fouling if finned tubes are used. Casing can be of membrane wall design or refractory lined. The advantage of membrane wall casing is that acid condensation concerns are eliminated as the casing operates at the saturation temperature of steam, which is typically well above the acid dew point. With refractory-lined casing, the heat loss from the casing will be slightly more compared to membrane wall design; however, it is likely to be slightly less expensive. One has also to be concerned about the casing design due to the possibility of low-temperature corrosion if there are cracks in the refractory. The main concern is corrosion due to acid condensation from moisture reacting with SO_3. This is minimized by starting up and shutting down the boiler on clean fuels and avoiding frequent start-ups and shutdowns. If refractory lined, a hot casing design is preferred as discussed in Chapter 6 on miscellaneous calculations. Boiler may be kept in hot standby mode also to minimize acid condensation by injecting steam using a sparger pipe inside the lower drum or placing heating coils inside the drum. Soot blowing is not recommended as it affects the gas analysis and adds moisture to the flue gas. The boiler tubes are generally of carbon steel construction.

The feed water temperature to the economizer must be high above 160°C–170°C to minimize the condensation of acid vapor. If the water vapor content is zero, then the acid condensation concern may not be there, but there is always leakage of air into the system, which can introduce moisture into the system. Carbon steel tubes with carbon steel–welded solid fins have been used in many plants in the United States for the economizers, while in Europe and Asia, cast iron gilled tubes shrunk over carbon steel tubes are widespread.

The internal gas bypass system increases the shell diameter compared to the external bypass system. The internal bypass system has a damper in the bypass section as well as in the main tube section, and by modulating the two dampers, the desired final gas temperature is reached. The advantage over the external bypass system is that due to some cooling in the bypass and the main tubes, the dampers located at the boiler exit operate in a slightly cooler gas temperature range compared to the external bypass system. An external steam drum with risers and downcomers ensures adequate circulation and cooling of the tube sheet.

Heat Recovery in Hydrogen Plants

Hydrogen and ammonia are valuable chemicals in various processes. The steam reforming process is widely used to produce hydrogen from fossil fuels such as natural gas, oil, or even coal as shown in Figure 4.17. There are several variations of the process, but basically, the steam reforming process converts a mixture of hydrocarbons and steam into

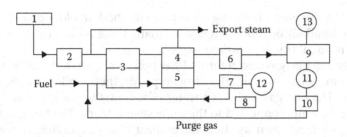

FIGURE 4.17
Steam reforming process in hydrogen plants. 1, Natural gas; 2, sulfur removal; 3, reformer; 4, reformed gas boiler; 5, flue gas boiler; 6, shift converter; 7, air heater; 8, air; 9, CO_2 removal and methanation; 10, pressure swing adsorption; 11, H_2 product; 12, stack; 13, CO_2 by-product.

hydrogen, methane, and carbon dioxide in the presence of nickel catalyst inside tubes. Before entering the reformer, the natural gas has to be desulfurized in order to protect the reformer tubes and catalysts from sulfur poisoning. The desulfurized gas is mixed with process steam, preheated to about 500°C in the flue gas boiler, then sent through the tubes of the reformer. Reactions occur inside the tubes of the reformer at 800°C–950°C.

Reforming pressures range from 20 to 40 atm, depending on the process equipment supplier.

$$C_nH_m + nH_2O \rightarrow nCO + (m/2 + n)H_2$$
$$CH_4 + H_2O \rightleftharpoons CO + 3H_2$$
$$CO + H_2O \rightleftharpoons CO_2 + H_2$$

The overall reaction is highly endothermic, so the reaction heat has to be provided from outside by firing fuel such as natural gas or naphtha outside the tubes. This generates flue gases, typically at 1800°F and atmospheric pressure, that are used to generate high-pressure superheated steam in a water tube waste heat boiler, generally referred to as a flue gas boiler. The flue gases also preheat the steam–fuel mixture and air.

In some processes, the effluents of the primary reformer are led to the secondary reformer, where they are mixed with preheated air. Chemical reactions occur, and the catalysts convert the methane partly to hydrogen. The effluent from the reformer, called reformed gas, is at a high gas pressure, typically 20–40 atm, and contains hydrogen, water vapor, methane, carbon dioxide, and carbon monoxide. This gas stream is then cooled from about 1600°F–600°F in a reformed gas boiler, which is generally an elevated drum fire tube boiler (Figure 4.2) with provision for gas bypass control to maintain the exit gas temperature constant at all loads. The exit gas temperature from the boiler decreases as the duty of the boiler decreases, and the bypass valve adjusts the flow between the incoming hot gases and the cool exit gases to maintain a constant exit gas temperature at all loads. The cooled gases then enter a shift converter, where CO is converted to CO_2 in the presence of catalyst and steam. Additional hydrogen is also produced. The exothermic reaction raises the gas temperature to about 800°F. The CO content is reduced from about 13%–3%. A waste heat boiler referred to as a converted gas boiler cools the gas stream before it enters the next stage of conversion, where CO is reduced to less than 0.3%. The next stage is the methanator, in which catalysts convert traces of CO and CO_2 to methane and water vapor. The H_2, CO, and unreacted methane are then separated. This produces a gas stream that can be recycled to process feed and produce hydrogen of 98%–99% purity that is further purified by the pressure swing adsorption method. In older plants, carbon dioxide is removed in a liquid absorption system and finally the gas goes through a methanation step to remove residual traces of carbon oxides.

In large plants, the flue gas and reformed gas boilers are separate units but have a common steam system, whereas in small hydrogen plants, these boilers can be combined into a single module. The flue gas boiler is a water tube unit; the reformed and converted gas boilers are fire tube units connected to the same steam drum. The flue gas boiler contains various heating surfaces such as the feed preheat coil, evaporator, superheater, economizer, and air heater. The casing is refractory lined, and extended surfaces are used where feasible because the gas stream is generally clean. The steam generated in the reformed gas boiler is often combined with the saturated steam generated in the flue gas boiler and then superheated in the superheater of the flue gas boiler. This is a substantial quantity of steam (often referred to as import steam), so the performance of the superheater must be checked for cases when the import steam quantity diminishes or is reduced to zero for various reasons.

The reformed gas boiler, which handles gases containing a large volume of hydrogen and water at high pressure, operates at high heat flux; the heat transfer coefficient with reformed gases is about six to eight times higher than those of typical flue gases from the combustion of natural gas. Hence, the heat flux at the inlet to the reformed gas boiler is limited to less than 100,000 Btu/ft^{2}h to minimize concerns about vapor formation over the tubes and possible departure from nucleate boiling (DNB) conditions. The gas properties for typical reformed gas and flue gases are listed in Table 4.24. The higher thermal conductivity and specific heat and lower viscosity coupled with higher mass flow per tube lead to higher heat transfer rates and hence higher heat flux in reformed gas boilers. This is discussed in the section on fire tube boilers. Due to the higher heat flux, the circulation system has to be designed with caution. Though refractory on tube sheet and ferrules transfer the high heat flux into the region beyond the tube sheet, the bubble generation near the tube sheet will be intense, and hence, a small riser is generally added near the front tube sheet to ensure that vapor blanketing does not occur near the front tube sheet region.

Combining Solar Energy with Heat Recovery Systems

There are a few plants around the world using solar energy effectively. Solar energy may be used to generate low- to medium-pressure steam. This may be superheated in waste heat boilers or steam generators fired by biomass or flue gas or gaseous fuels. When the solar energy is unavailable, the boiler has to generate additional steam. Boilers can be designed to handle such options. Figure 4.18 shows a scheme of a solar plant operating in conjunction with a waste heat recovery system that can be from biomass or gas turbine HRSG exhaust or any waste heat source.

While traditional solar power facilities need to employ expensive energy storage techniques to ensure continuous operation in all weather conditions, the heat recovery system or steam generator can eliminate that need. Whenever there is not enough direct sunlight to generate solar power, the facility's heat recovery system can be brought on line enabling the continuous production of electricity.

Gas Turbine HRSGs

Gas turbine–based combined cycle and cogeneration plants (Figure 4.19a and b) are found in refineries, chemical plants, power plants, process industry, and cogeneration plants,

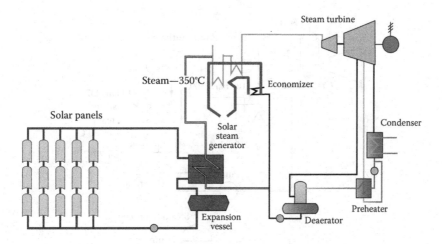

FIGURE 4.18
Solar energy and heat recovery system (typical).

generating steam for power and process or for both. The exhaust energy may also be used for heating industrial heat transfer fluids such as therminol, glycol, and fuel oils. The unfired combined cycle plant with a gas turbine exhausting into a multiple-pressure reheat HRSG supplying steam to the steam turbine is the most efficient electric power generating system today, approaching 60% on lower heating value (LHV) basis. The cost of a combined cycle power plant in the range of 800 MW and above is $550–650/kW, while that of a coal-fired plant is $1200–1400/kW. Several improvements have been made in gas turbine technology such as developments in material technology to handle the high firing temperatures in the range of 1500°C. The HRSGs in combined cycle plants differ from those in cogeneration plants in the following ways:

1. The main steam pressure, reheat steam pressure, medium and low pressure, and temperatures are all optimized in combined cycle plants. In cogeneration plants, it varies from plant to plant depending on process steam demand and plant conditions.

2. Supplementary firing temperatures can vary depending on process steam needs in cogeneration systems, while in combined cycles plants, HRSGs are generally unfired.

3. Steam may be imported into the HRSG superheater or saturated steam may be exported out of the HRSG in cogeneration plants; in cogeneration plants, there can be several heat recovery systems generating medium- to low-pressure saturated steam that has to be superheated, while some application may require saturated steam, which may be taken off the HRSG.

4. Each HRSG design in cogeneration plant is dependent on the needs of the particular plant and steam parameters, and application can vary. The author is aware of a plant where a portion of turbine exhaust gas is used for drying wood chips.

Typical gas turbine combustor operates in the range of 1350°C–1400°C for metallurgical reasons. Hence, a large amount of compressed air is used to cool the flame, which in turn increases the exhaust gas flow. After expansion in the gas turbine, the exhaust gas at about 500°C–550°C and at 150–250 mm wc pressures enters the HRSG. It contains about 6–8% volume of water vapor, but if steam injection is used for gas turbine NO_x control, the water

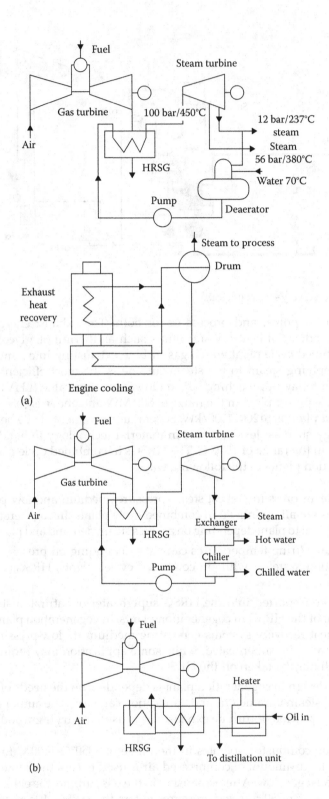

FIGURE 4.19

(a) Gas turbine and engine heat recovery system. (b) Cogeneration systems.

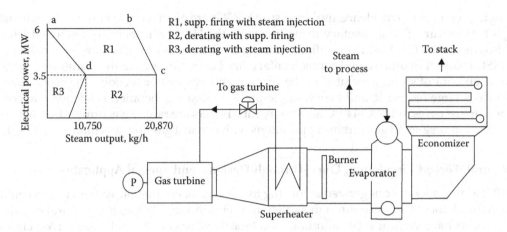

FIGURE 4.20
Cheng cycle system.

vapor can go up to 10%–11%. In systems like the Cheng cycle (Figure 4.20), where significant amount of superheated steam is injected in the gas turbine, the % volume of water vapor can go to about 22%–23%. This naturally impacts the gas specific heat and thermal conductivity and the gas–steam temperature profiles in the HRSG. The oxygen content of typical gas turbine exhaust is 14%–15% volume, and one can use this to add additional fuel in the duct burner or register burner as discussed in Chapter 1. Due to the large ratio of exhaust gas flow to steam in HRSGs, the HRSG cross section will be large though the steam generation is small. The cross section of an HRSG generating the same amount of steam as a steam generator will be about six times larger.

Another aspect of gas turbines is that the exhaust gas flow remains nearly constant with load variations, while the exhaust gas temperature changes. As the exhaust contains 14%–15% volume of oxygen, supplementary firing can be carried to generate large amount of steam without the addition of air. Hence, this is an efficient means of generating additional steam as shown later. A comparison of various types of HRSGs is shown in Table 4.5.

Due to low inlet gas temperature and pinch point limitation as discussed in detail in Chapter 5, the exit gas temperature in an HRSG is a strong function of inlet gas temperature. Lower the inlet gas temperature, higher the exit gas temperature and vice versa in a

TABLE 4.5

Comparison of Unfired and Fired HRSGs

Item	Unfired	Supplementary Fired	Furnace Fired
Gas inlet temp to HRSG, °C	450–550	550–900	900–1400
Gas to steam ratio	5.5–7.0	2.5–5.5	1.2–2.5
Burner type	No burner	Duct burner	Duct register
Fuels fired	None	Gas distillate	Oil–gas–solid
Casing type	Internal insulation	Internal insulation or membrane wall	Membrane wall
Circulation	Natural, forced, once-through	Natural, forced, once-through	Natural
Back pressure, mm wc	150–200	200–275	250–350
Configuration, pressure levels	Single or multiple	Single or multiple	Single
Design	Convective	Convective	Radiant furnace

single-pressure system. Hence, multipressure HRSGs are often required to cool the exhaust gas temperature. Supplementary firing increases the HRSG efficiency as discussed later. Also due to the low LMTD in unfired units, extended surfaces are required to make the HRSG compact in unfired and supplementary-fired units. In furnace-fired units due to the large amount of steam generated in the furnace and evaporator sections, the economizer acts as a large heat sink, and hence, single-pressure steam generation is adequate to cool the exhaust gases to 120°C–140°C levels. Fire tube boilers are rare in gas turbine but may be justified in very small gas turbine applications with exhaust gas flow less than 30,000 kg/h.

Natural, Forced Circulation, Once-Through Designs, and Special Applications

HRSGs are generally categorized according to the type of circulation system, which could be natural, forced, or once-through as shown in Figure 4.21 a through d. Natural circulation units have vertical tubes and horizontal gas flow orientation, whereas forced circulation units have horizontal tubes and vertical gas flow orientation. Once-through can have either horizontal or vertical gas flow orientation. In natural circulation units, the difference in density between the colder, denser water in the downcomer pipes and the hotter, less dense steam–water mixture in the evaporator tubes enables the circulation process. CR can range from 10 to 30 depending on location of drum, system resistance, and steam pressure. Since finned tubes are used in HRSGs, the heat flux inside the tubes will be much higher than in plain tubes. For example, in 50.8×44 tubes with157 fins/in., 15 mm high, 1.5 mm thick solid fins, the ratio of external to internal surface area is 9.64. If $U = 50$ kcal/m^{2}h °C, the heat flux with 500°C gas temperature and 240°C fluid temperature will be $50 \times (500 - 240) \times 9.64 = 125,230$ kcal/m^{2}h (46,000 Btu/ft^{2}h). In fired units, the heat flux could up to nearly double this value, and hence, one has to be concerned about DNB, particularly with horizontal tubes as discussed in Chapter 2 on furnaces and circulation. If the firing temperature exceeds 850°C, typically, a membrane wall furnace is used. Membrane wall furnace is perfect for any firing temperature. The problem of laying insulation, liners, and their maintenance is avoided. Then a few rows of plain tubes are used followed by the superheater if used. Then we have finned tubes with varying fin density to ensure that the heat flux gradually increases along the gas flow direction. A once-through steam generator looks like an economizer coil (Figure 4.21d). As water is converted into steam inside the tubes, the water should have zero solids; else, deposition of salts can occur inside tubes causing overheating of tubes. A once-through HRSG does not have a defined economizer, evaporator, or superheater regions. Slight variations can occur depending on gas inlet conditions. The single-point control for the OTSG is the feed water control valve. Based on inlet gas conditions and exit steam temperature, the water flow (and hence steam flow) is varied. As there is no drum, the water holdup is small compared to drum-type units, and hence, response time to load fluctuations is quick. The OTSG is designed to run dry, and hence, alloy 800 or 825 tubes are used. The high-grade tube material minimizes exfoliation concerns, which are likely with carbon steel and low-grade alloy tubes. (One may fill up the evaporator and economizer of natural circulation units with water and then start the unit. By ensuring a low gas inlet temperature to the superheater, the superheater is also protected well till steam starts generating in the evaporator; hence, carbon steel tubes are adequate for natural circulation unit evaporators and economizer.) When boiler tubes are heated, they form an oxide layer inside the tubes, and when cooler steam flows inside, the oxide particles are dislodged and carried away by steam and deposited in the steam turbine, a process called exfoliation. If this heating takes place for a long time and frequently, alonized or chromized tubes may be used for the superheater.

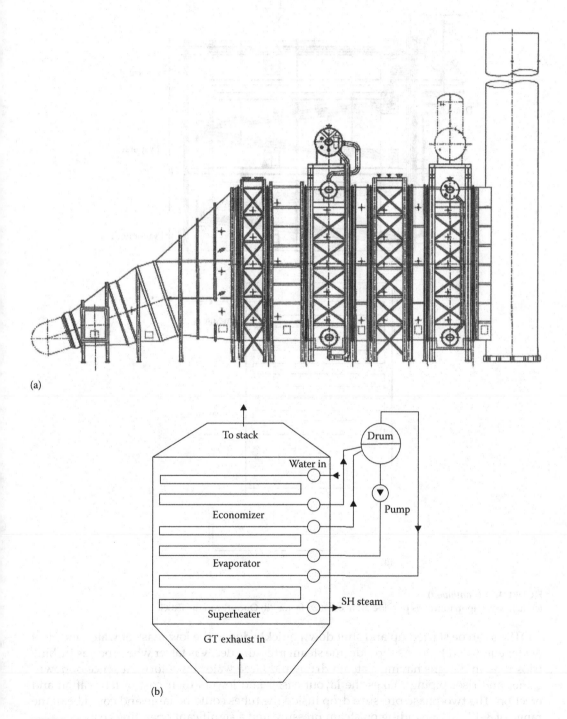

FIGURE 4.21
(a) Natural circulation gas turbine HRSG. (b) Forced circulation gas turbine HRSG. *(Continued)*

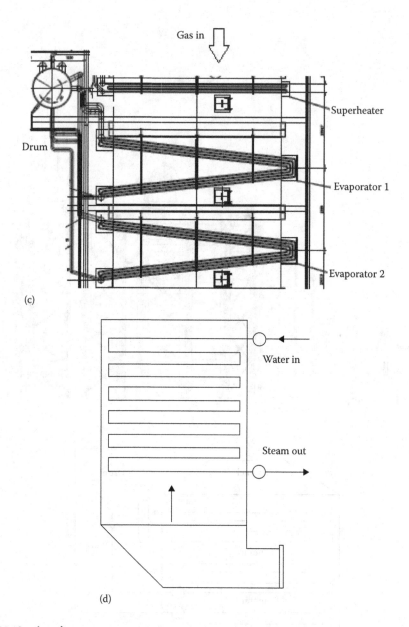

FIGURE 4.21 (Continued)
(c) Inclined evaporator tubes provide more comfort factor. (d) Once-through HRSG.

OTSGs can be started up and shut down quickly due to the low mass of water and steel (water equivalent). On the flip side, the steam pressure decay is faster when the gas turbine trips than in designs having a steam drum and large water inventory. Absence of downcomer and riser piping makes the layout easier and lowers their cost of fabrication and erection. The two-phase pressure drop inside the tubes could be large and could be in the range of 8–12 bar depending on steam pressure and a significant operating cost.

The once-through steam generator (see Chapter 3) used in steam flooding applications generates 80% quality steam at high steam pressures using the energy from gas turbine exhaust as well. These units generate steam at 100–200 barg depending on the depth of oil field. Multiple streams are used in large HRSGs with flow control valve at the water inlet of each stream to ensure that two-phase instability problems with water–steam mixture do not arise. Eighty percent quality steam may also be generated using exhaust gases from gas turbines. The HRSG is also designed to ensure low heat flux inside the tubes by manipulating the fin geometry. Since many HRSGs use horizontal tubes, the allowable heat flux is limited to 270,000 kcal/m² h (100,000 Btu/ft² h or 314 kW/m²). The wet steam (80% quality) ensures that the salts in feed water such as sodium are soluble in the steam–water mixture and are not deposited inside the tubes and carried away with the steam–water mixture. To minimize acid or water dew point concerns, a heat exchanger to preheat the cold water is used as shown.

Figure 4.22a shows a plant with furnace-fired HRSGs, while Figure 4.22b shows the details of its construction. As discussed earlier, the convection sections follow the furnace, and finned tubes of varying density are used in the convection section to keep the tube wall and fin tip temperatures low. Downcomers are located at the rear as shown. The convection bank is followed by an economizer. Figure 4.22c shows a dual-pressure natural circulation HRSG in the field.

Another interesting application in gas turbine plants is the intercooler for compressed air used for combustion (Figure 4.23). Intercooling improves the efficiency of compression process in gas turbine compressors. Hot compressed air from the first stage of the compressor is cooled by feed water (typically high-pressure economizer water) before it is again sent back to the next stage of the gas turbine compressor and combustor. Since the air is at high pressure and the water also is at high pressures, say, 100–150 barg, the heat transfer coil should be placed inside a pressure vessel with high-pressure–high-temperature air, typically at 15–25 barg and at 400°C–450°C outside the tubes and the colder water inside the tubes in counter-flow direction. Finned tubes are used for the coil. Baffles are used to divert the pressurized air over the finned tube bundle and then exit the pressure vessel.

Fluid heaters are another important application of turbine exhaust. Industrial heat transfer fluids such as therminol, glycol, or even hot water are heated by turbine exhaust. A finned coil is used in these applications. The coil looks like an economizer. As many of these fluids have limitation on heat flux and have a low tube-side heat transfer coefficient (compared to water), low fin density not exceeding 98–117 fins/m should be used on the heating surfaces.

Natural versus Forced Circulation HRSGs

In the United States, the trend is to use natural circulation HRSGs, while in Europe, the forced circulation units are the norm. Some thoughts on these designs are outlined in the following text:

1. Natural circulation units do not require a pump for maintaining circulation through the evaporator tubes. Static head available, steam pressure, and system resistance caused by size and length of downcomer, riser pipes determine the CR. The use of circulation pumps in forced circulation units involves an operational and maintenance cost, and their failure for some reason such as power outage or pump failure could shut down the HRSG.

2. The water boils inside vertical tubes in natural circulation units, and the steam bubbles move upward, which is the natural path for them; hence, the tube walls are completely wetted by water (if heat flux is nominal). With horizontal tubes, there is a difference in temperature between the top and bottom portions of the

(a)

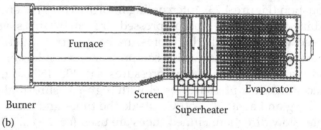

(b)

(c)

FIGURE 4.22
(a) Furnace-fired HRSGs in operation. (b) Furnace-fired HRSG plan view. (c) Dual-pressure natural circulation HRSG. (Courtesy of Cleaver Brooks, Thomasville, GA.)

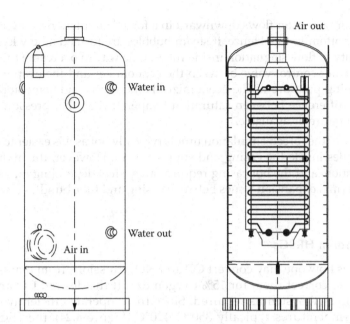

FIGURE 4.23
Pressurized hot air water preheater. (Courtesy of Cleaver Brooks, Thomasville, GA.)

tubes, which could cause thermal fatigue. Also if the selection of streams is not proper and if the velocity of the steam–water mixture is not high enough, the vapor phase can separate from liquid phase leading to steam blanketing and possibly overheating of tubes. This is a possibility when the HRSG is fired and the heat flux inside the tubes is high, particularly when high fin density is used and also if the gas temperature profile at the tube bundle is nonuniform.

3. Natural circulation system with vertical tubes can tolerate higher heat flux generally 50%–80% more than horizontal tube designs. If there is a nonuniformity in gas temperature or heat flux at any cross section, the tube receiving the higher heat flux will have a higher CR due to the greater differential in fluid densities between the more dense fluid in the downcomer circuit and the less dense mixture in the evaporator tubes, which is helpful to even out the imbalances. In forced circulation units, the flow resistance of all tubes is nearly the same as well as the flow of mixture and any severe variation in gas flow and temperature at a cross section can cause some tubes to generate more steam and hence be vulnerable to tube failures.

4. Natural circulation units require more real estate than forced circulation units. The floor space can run into several hundred feet if there are multiple modules and catalysts for NO_x, CO control. In forced circulation units, the floor space may be small, but the height of the HRSG will be large requiring more support steel, ladders, and platforms.

5. The horizontal gas flow configuration of natural circulation units provides an easy way for water washing the highly soluble ammonia compounds formed downstream of the selective catalytic reduction (SCR) when operating with a sulfur-containing fuel. When a vertical gas flow unit is water washed, the slush and corrosive products are likely to deposit over the SCR below the low-temperature surfaces.

6. The economizer water flows downward in a forced circulation unit with gas flowing from bottom to top. Hence, if steam bubbles are formed at low loads, there is a possibility of flow stagnation inside tubes. One way to overcome this is to place the feed water control valve between the economizer and the evaporator so that the operating pressure of the economizer is increased and some approach temperature (difference between saturation temperature at that pressure and water exit temperature) is available.

7. The casing of the forced circulation unit is typically hot as it is easier to contain the tube bundles inside the casing and support them. However, this requires external insulation, and the hot casing requires alloy steel depending on gas temperatures. Thermal expansion issues between casing and tube bundles also have to be reviewed.

Emission Control in HRSGs

Chapter 1 shows how one may convert CO and NO_x emissions from mass to volumetric basis and how to correct them for 15% oxygen dry. If the desired CO and NO_x levels are very low, an SCR system is required. Since these operate efficiently only within a range of gas temperatures typically 350°C–420°C (Figure 4.24), the HRSG should be designed suitably that in unfired as well as in fired modes or at part load operation of the gas turbine, the range of gas temperatures at the SCR can be satisfied. The ammonia grid (Figure 4.25) must be located such that the ammonia injected into the gas stream mixes well with exhaust gas. The use of catalysts affects the HRSG design and performance in a few ways. The gas pressure drop across the CO or NO_x catalyst can be on the order of 25–40 mm wc including the duct work to match the HRSG cross section with that of the catalyst. The evaporator in almost all cases has to be split up into two parts as shown in Figure 4.26 so that the NO_x catalyst can be located in between the two evaporators; the sizing of the evaporators have to be checked at all loads to ensure that the window of gas temperature is obtained at all loads. Figure 4.27 shows how the gas temperature varies at each heating surface as a function of gas inlet temperature. These two evaporators will operate in parallel and are connected to the common steam drum with external downcomers and risers. The lengths of these pipes can be significant and run into, say, 20–40 m depending on HRSG height and location of the two evaporators;

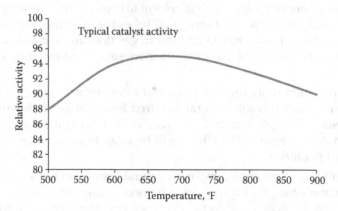

FIGURE 4.24
Operating temperatures of catalysts.

FIGURE 4.25
Ammonia grid. (Courtesy of Nebraska Boiler, Division of Cleaver Brooks. Thomasville, GA.)

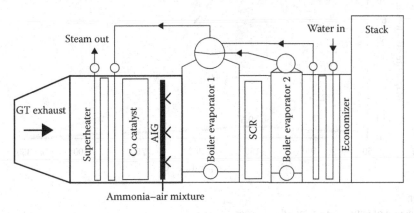

FIGURE 4.26
Arrangement of an HRSG with SCR and CO catalyst.

hence, circulation calculations have to be performed to ensure that the evaporators function well in all modes of operation.

Flow-Accelerated Corrosion

Many utilities and cogeneration plants are finding that flow-accelerated corrosion (FAC) is causing waste age of materials, downtime, and additional maintenance concerns. FAC is defined as the localized rapid metal loss resulting in tube wall thinning of carbon and low

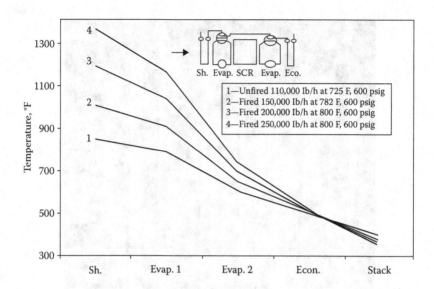

FIGURE 4.27
HRSG gas temperature profiles at various gas inlet conditions.

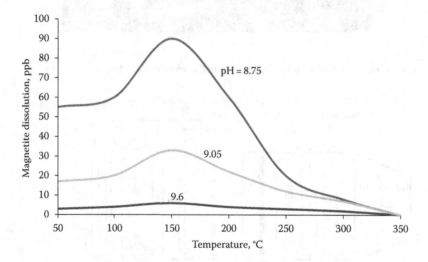

FIGURE 4.28
FAC as a function of temperature and pH of water.

alloy piping. FAC is known to occur in the region of 125°C–220°C as shown in Figure 4.28. In HRSGs, the evaporators and economizers in the LP (low pressure) and IP (Intermediate pressure) sections are affected. FAC is the absence of protective metal oxide layer that when present limits corrosion in boiler systems. Without this protection, the surface is free to react with the passing water (or two-phase flow), and metal loss is rapid. Factors affecting FAC are flow velocity, geometry, metallurgy, dissolved oxygen concentration, and temperature. Feeder connections with sharp bends are common in HRSGs. These multiple feeders connect the downcomer pipes to the evaporator circuits and take the steam–water mixture to the drum.

Past industry practices believed that all of dissolved oxygen must be eliminated from feed water to control corrosion. To deoxygenate the feed water, oxygen was mechanically removed in the condenser–deaerator system with an addition of an oxygen scavenger such as hydrazine to maintain a residual of 40–100 ppb hydrazine. This caused the feed water to become more and more reducing and has produced the opposite effect of producing a protective layer. The normally protective magnetite layer in carbon and low alloy steel dissolves into a stream of flowing water or two-phase flow. Both the pH and temperature and level of dissolved oxygen influence the stability and solubility of the magnetite oxide layer. The difference in wall loss due to FAC is 100 times greater at 1 ppb dissolved oxygen in feed water than at 20 ppb. Hence, maintaining some residual oxygen will reduce the FAC.

In Germany, the maximum oxygen in feed water has been increased from 0.02 to 0.1 (20–100 ppb) mg/kg. Flow velocity is also lowered to reduce the erosion of the magnetite layer. HRSGs have sharp bends in economizers, evaporators that are prone to FAC. FAC also develops at flow disturbances such as elbows, bends, reducers, tees, and steam attemperating lines. Sometimes the bends are made of low chromium alloys to reduce FAC. In oxygenated treatment, oxygen is deliberately introduced in the condensate and feed water system. PH of feed water is raised to 9.2–9.6 for FAC effect to be minimal by an addition of ammonia. It has been found that two-phase FAC is difficult to control chemically and hence 1.25 chromium steels are suggested for such areas. All volatile treatment used in some boiler including once-through units to avoid deposition of solids leads to the dissolution of the protective magnetite.

Water Tube Waste Heat Boilers: Sizing and Performance

Before proceeding with the examples on water tube boiler and fire tube boiler performance, one should review Appendices A through F. Design and off-design calculations, and estimation of heat transfer coefficients inside and outside plain and finned tubes are discussed. Simplified approach to estimating tube-side heat transfer coefficient for several fluids is also explained. With this background, it should be easy to apply the concepts to real-life design issues on both fire tube and water tube boilers.

Optimizing Pinch and Approach Points in HRSGs

Chapter 5 explains the importance of pinch and approach points and how they are selected. Procedure to estimate the steam generation and gas–steam temperature profiles is also discussed. When trying to maximize steam generation, HRSG suppliers should use the lowest pinch and approach points avoiding the possibility of steaming at low loads; cost is also a factor. Plant engineers should also be able to evaluate proposals with different pinch and approach points and choose the best offering. Presented in the following text is an example of HRSG proposal from two different vendors.

Example 4.1

A gas turbine of capacity 7.5 MW has an exhaust gas flow of 100,000 kg/h at 500°C. Gas analysis is % volume $CO_2 = 3$, $H_2O = 7$, $N_2 = 75$, and $O_2 = 15$. In the unfired mode, the HRSG will generate about 15,000 kg/h, while 40,000 kg/h of saturated steam at 40 kg/cm² g is required by the plant, which will be achieved by supplementary firing; the HRSG is likely to operate 50% of the time in unfired mode and 50% of the time in fired mode. The boiler performance in various cases is shown later. Vendor A has

TABLE 4.6

HRSG Designs with Low and High Pinch Points

	Vendor A		Vendor B	
Data Case	Unfired Mode	Fired Mode	Unfired Mode	Fired Mode
Gas flow, kg/h	100,000	100,000	100,000	100,000
Gas temp. in, °C	500	500	500	500
Firing temperature, °C	500	925	500	952
Gas temp. to eco, °C	256	264	271	302
Stack gas temp., °C	168	133	190	168
Gas press. drop, mm wc	159	176	107	121
Feed water in, °C	105	105	105	105
Water to eco, °C	242	189	241	192
Sat temperature, °C	251	251	251	251
Burner duty, MM kcal/h	0	13.0	0	14.0
Steam flow, kg/h	15,565	40,000	14,587	40,000
Evaporator surface, m²	2,951		1,908	
Economizer surface, m²	2,129		1,296	
Boiler duty, MM kcal/h	8.75	22.51	8.21	22.51
Pinch point, °C	5	13	20	51
Approach point, °C	9	62	10	59
No of rows deep, evap.	21		15	
No of rows deep, econ.	14		8	
Efficiency, %	68.8	87.5	64.6	84.2

Note: 50.8 × 44 mm tubes, 24 tubes/row, length of evaporator = 4 m, length of economizer tubes = 3.5 m. staggered arrangement, 101.8 mm pitch, 197 × 19 × 1.5 solid fins. Burner duty is on fuel lower heating value basis.

provided a bigger boiler with a lower pinch point, while vendor B has offered a smaller boiler that requires more fuel than vendor A to generate the same amount of steam. The plant engineer has to evaluate which offering is better (Table 4.6).

Let us use the following cost factors for evaluating the better option; cost of steam: $10/1000 kg; cost of electricity = 11 cents/kWh; cost of fuel = $1/MM kcal/h.

Vendor A has used a lower pinch point and has a bigger evaporator and economizer.

1. Advantage of A over B in steam generation in unfired mode = (15,565 – 14,587) × 4,000 × 10/1,000 = $39,120.
2. Advantage of A over B in fired mode: (14 – 13) × 1 × 4000 = $16,000.
3. Typically, 100 mm wc back pressure increase in the HRSG is equivalent to 1% loss in gas turbine power output. Hence, advantage of B over A is (–107 – 121 + 159 + 176) = 107 or an average of 53.5 mm wc throughout the year = 7,500 × 0.01 × 0.11 × 8,000 × 53.5/100 = $35,310.

Hence, annual savings of design A over design B = 39,120 + 16,000 – 35,310 = $19,810.

One may check the difference in price and compute the payback period for vendor A. It is likely that vendor A is a better option though the initial cost may be slightly more. The payback may not be that long, considering that the cost of instruments, controls, casing, duct, and supporting steel may not be that much different and only the HRSG per se may differ in cost. The purpose of the exercise is to show how to evaluate the variables affecting the operating costs and see if a higher investment in lower pinch point is worth the money.

HRSG Performance and Evaluating Field Data

When a new waste heat boiler is installed, its operating data may differ from that stated in the proposal. There can be several reasons for this such as the following:

- The inlet flue gas flow and temperature may be different from that used in the proposal or guarantee statement. An unfired waste heat boiler simply responds to the inlet gas conditions, and hence, the steam parameters may be different from those expected.

- The plant's operating feed water temperature or pressure may be slightly different from that used for generating the proposal due to plant's limitation, deaerator or water treatment plant's availability. Hence, steam generated or exit gas temperature from boiler may differ from that stated in the proposal.

The question then arises: how can the plant engineer ensure that the boiler has been sized correctly, and will it generate steam at the rated conditions when provided with the rated gas flow and temperature?

Let us consider an unfired HRSG as shown in Figure 4.21a. It was noticed that the operating data were completely different from that predicted by the HRSG supplier. As discussed earlier, one has to investigate the reasons for this. It is likely that the gas turbine is operating at a different load or ambient temperature resulting in variations in exhaust gas flow and temperature from that used for guarantee purposes; the plant may not be able to operate the gas turbine at guarantee load conditions; also, portion of the exhaust gas may have been bypassed as steam demand may not be there during the testing phase and plant does not wish to waste the treated water. While testing a boiler, plant engineers typically shut off the blowdown line and ensure dampers and valves do not leak.

Then the question arises: are the differences between operating data and guaranteed data reconcilable or out of line and whether the HRSG itself was designed properly? If the performance test is not carried out within a few months of installation and if the plant cannot operate the HRSG at the rated conditions, it may not be able to find out if the HRSG was designed correctly or not. The HRSG supplier can blame the plant for not providing proper conditions for testing the HRSG and can walk out without proving if the HRSG design was adequate or not. This is not good from the plant's perspective. The bottom line is how to check if the HRSG was properly sized or not even if the HRSG operates at different parameters from those guaranteed. The following example shows how one can evaluate the performance of an operating HRSG. The procedure may be used for any waste heat boiler. Figure 4.29 shows the logic or flow diagram for off-design performance calculations of waste heat boilers. There are several iterations, and for each change in gas temperature or steam temperatures, the properties have to be evaluated at each component. If there are multiple modules or if the HRSG is fired, the calculations become more tedious.

Example 4.2

An unfired gas turbine HRSG has the data shown in Table 4.7. The design shows over 15.3 t/h of steam at 463°C at 40 kg/cm²g with 105°C feed water. However, in operation, it makes only 11,000 kg/h at 442°C and at 35 kg/cm²g with feed water at 120°C with zero blowdown. The water temperature leaving the economizer is measured as 242°C. The plant is not able to operate at 40 kg/cm²g with feed water at 105°C for various other reasons. Inlet and exit gas temperatures are measured as 500°C and 190°C. Is this performance acceptable? The HRSG supplier says that due to variations in exhaust gas flow and temperature, they are not able to get the desired steam flow of 15.3 t/h at 460°C and that his HRSG design is perfect.

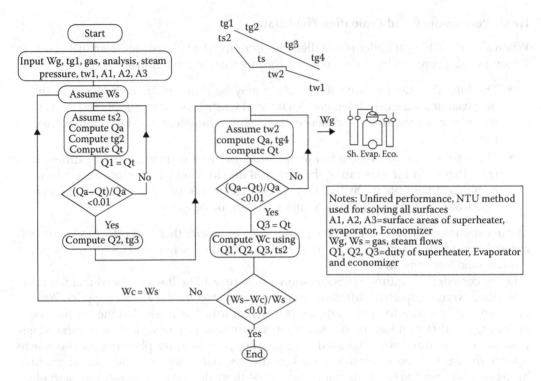

FIGURE 4.29
Logic diagram for HRSG performance evaluation.

TABLE 4.7

Gas Turbine HRSG Tube Geometry Data

Geometry	Sh.	Evap.	Econ.
Tube OD	31.8	50.8	31.8
Tube ID	26	44	26
Fins/in. or fins/m	78	197	197
Fin height	19	19	19
Fin thickness	1.9	1.9	1.9
Fin width	4	4	4
Fin conductivity	35	35	35
Tubes/row	20	20	20
Number of rows deep	6	18	14
Length	4	4	4
Transverse pitch	84	101	84
Longitudinal pitch	101	101	100
Streams	20	0	10

Note: Streams are the number of tubes carrying the steam or water. This is explained in Appendix B.

This type of situation often arises in plants operating HRSGs. The HRSG in all likelihood will operate at parameters slightly different from those stated in the proposal document. Hence, the steam output and the gas–steam temperature profiles will be different from the proposal data. If questioned, the HRSG supplier will cite the gas turbine exhaust conditions for the field performance results, and the gas turbine supplier will suggest that the plant look into the HRSG design. The helpless plant engineer will be better off if he knows how to evaluate the performance himself and get an idea of what is going on.

Tube geometry data is shown in Table 4.7; thermal performance data as shown in Table 4.8 should be obtained from HRSG suppliers for each operating case such as full load and part load of gas turbine and at a few different ambient temperatures, before the HRSG is ordered! If it is a multiple-pressure unit, then such geometric and thermal performance data should be provided for each component in the high- and low-pressure sections. Without a clear performance and geometric data, one cannot evaluate the performance of the HRSG independently. Even if the U values are not given by the HRSG supplier, using the basic equation $Q = UA\Delta T$, one may compute the ΔT (from the gas, steam temperature data), and based on surface area and duty shown, compute the U value. Then one may adjust the U value for gas temperature and gas flow variations as follows (this method of adjusting the U value for off-design conditions is also discussed in Chapter 5 on simulation):

$$U_f = U_d \times \left(\frac{W_f}{W_d}\right)^{0.65} \times \left(\frac{F_{gf}}{F_{gd}}\right) \tag{4.1}$$

where F_g is a factor depending on gas properties $= (C_p^{0.33} k^{0.67}/\mu^{0.32})$.

Subscripts f and d refer to field and design conditions. Gas properties are computed at the average gas temperature of each section as finned tubes are typically used. W refers to gas flow in mass units.

TABLE 4.8

HRSG Performance Data

Process Data Surface	Gas Flow = 110,000 kg/h Design Case		
	Suphtr.	Evap.	Econ.
Gas temp. in, ±5°C	550	482	265
Gas temp. out, ±5°C	482	265	183
Gas spht., kcal/kg °C	0.2756	0.2691	0.2596
U, kcal/m²h °C	52.39	30.96	39.11
LMTD, °F	147	75	42
Duty, MM kcal/h	2.06	6.34	2.33
Surface area, m²	268	2749	1407
Gas press. drop, mm wc	16.53	87.42	41.82
Foul factor, gas	0.0004	0.0004	0.0004
Steam Side			
Steam pres, kg/cm²a	41	42	42
Steam flow, kg/h	15,313	15,313	15,773
Fluid temp. in, °C	253	245	105
Fluid temp. out, ±5°C	463	253	245
Press. drop, kg/cm²	1.5	0.0	0.6
Foul factor, fluid	0.0002	0.0002	0.0002
Spray, kg/h	0		

Gas analysis: % volume $CO_2 = 3$, $H_2O = 7$, $N_2 = 75$, $O_2 = 15$.

If U is not provided, the duty Q and the temperature profiles will be known from which ΔT and U may be estimated for each section such as superheater, evaporator, and economizer. Let us assume that the customer has provided them as shown in Tables 4.7 and 4.8 in our case. One may also use the method discussed in Appendix E for finned tubes and evaluate U for each section and check the values provided by the HRSG supplier. They also should be aware of the fact that fin geometry influences U significantly.

Let us estimate the exhaust gas flow from the inlet and exit gas temperatures and the steam parameters, which are presumed to be more accurately measured than the gas flow. The total energy absorbed by steam = 11,000 × (792.4 − 120.9) = 7.38 × 10⁶ kcal/h (792.4 and 120.9 kcal/kg refer to the enthalpy of superheated steam and feed water, and blowdown was neglected). Gas specific heat at average gas temperature of (500 + 190)/2 = 345°C is 0.266 from Table F.6. Hence, exhaust gas flow = 7,380,000/[0.266 × (500 − 190)× 0.99] = 90,400 kg/h (0.99 refers to casing heat loss of 1%). As discussed in Chapter 6, one may estimate the casing heat loss based on provided insulation thickness; however, for this exercise, 1% heat loss is a good assumption.

Since some variations in exhaust gas temperature profile are present, one may apply correction factor F_g while estimating U from the design data. (The accurate method is to develop a program that estimates U using the detailed procedure discussed in Appendix E on finned tubes after computing the inside and outside heat transfer coefficients and fouling factors.) F_g at average gas temperature of 366°C in design case is 0.4143 and at operating case is 0.4112. Hence, a correction of (0.4112/0.4143) = 0.992 may be applied to U; however, for this exercise, we are not applying temperature correction factor as the main purpose is to show how much steam the HRSG should be making and see if the field data are reasonable. Since the steam flow is significantly lower, the superheater, U, may be reduced by a factor (11,000/15,300)$^{0.15}$ = 0.952 (see Chapter 5 or Ref. [1]).

For superheater, U = 52.39 × 0.952 × (90,400/110,000)$^{0.65}$ = 43.9 kcal/m²h °C

For evaporator, U = 30.96 × (90,400/110,000)$^{0.65}$ = 27.25 kcal/m²h °C

For economizer, U = 39.11 × (90,400/110,000)$^{0.65}$ = 34.43 kcal/m²h °C

Superheater Performance

One may refer to the NTU method discussed in Appendix A. The specific heat of steam between 442°C and saturation of 245°C is (792.4 − 669.3)/(443 − 245) = 0.622 kcal/kg °C (specific heat is enthalpy difference divided by temperature difference). The specific heat of exhaust gas may be taken as 0.273 in the superheater based on the exhaust gas analysis and data in Appendix F on gas properties (and can be checked in the second iteration). C_{max} = 90,400 × 0.99 × 0.273 = 24,432. C_{min} = 11,000 × 0.622 = 6,853.

$$C = \frac{C_{min}}{C_{max}} = \frac{6853}{24432} = 0.28.(1 - C) = 0.72$$

$$NTU = \frac{UA}{C_{min}} = \frac{268 \times 43.9}{6853} = 1.717$$

$$\epsilon = \frac{\left[1 - \exp\{-NTU \times (1 - C)\}\right]}{\left[1 - C\exp\{-NTU \times (1 - C)\}\right]} = \frac{\left\{1 - \exp\left(-1.717 \times 0.72\right)\right\}}{\left(1 - 0.28 \times \exp\left(-1.717 \times 0.72\right)\right)}$$

$$= \frac{0.7095}{0.9187} = 0.772.$$

The Q = 0.772 × 6823 × (500 − 245) = 1.349 × 10⁶ kcal/h

Exit gas temperature of superheater = 1,349,000/90,400/0.99/0.273 = 444°C.

Exit steam temperature = 245 + (1,349,000/6,823) = 442°C

Evaporator Performance

See Appendix A for evaporator performance evaluation.

Using the equation $Ln[(t_{g1} - ts)/(t_{g2} - t_s)] = UA/(W_g C_{pg} \times 0.99)$

$$Ln\left[\frac{(444 - 245)}{(t_{g2} - 245)}\right] = \frac{27.25 \times 2749}{(90400 \times .99 \times 0.268)} = 3.12 \text{ or } t_{g2} = 254°C$$

$$Q = 90,400 \times 0.99 \times (444 - 254) \times 0.268 = 4.56 \times 10^6 \text{ kcal/h}$$

Economizer Performance

C = 11,000 × 1.05655/(90,400 × 0.99 × 0.258) = 0.503 (specific heat of water and gas may be shown to be 1.05655 and 0.258). (1 − C) = 1 − 0.503 = 0.497. C_{min} = 11,000 × 1.05655 = 11,622.

$$NTU = \frac{34.43 \times 1407}{(11000 \times 1.05655)} = 4.168$$

$$\in = \frac{[1 - \exp(-4.168 \times 0.497)]}{[1 - 0.503 \times \exp(-4.168 \times 0.497)]} = \frac{0.874}{0.9366} = 0.933$$

$$Q = 0.933 \times 11622 \times (254 - 120) = 1.453 \times 10^6 \text{ kcal/h}$$

Exit gas temperature = 254 − 1,4540,000/(90,400 × 0.99 × 0.258) = 191°C (agrees with the measured value)

Exit water temperature = 120 + 14,530,000/(11000 × 1.05655) = 245°C (close to steaming)

Total duty transferred = 1.349 + 4.56 + 1.453 = 7.36 MM kcal/h (close to that from field data). Hence, the performance at part load is what can be reasonably expected from the revised gas conditions. Hence, the HRSG may be said to be performing well. One may also in a similar manner review the performance data provided at design conditions and see if 15,300 kg/h of steam at 463°C can be generated under the gas conditions used for performance guarantee. Actual steam generation = 7.36 × 10⁶/(792.4 − 120.9) = 10,960 kg/h agrees with the measured value of 11,000 kg/h.

Hence, the HRSG performance is satisfactory. If the exit gas temperature had been 210°C or the steam generation was, say, 10,300 kg/h, then one may have to investigate further the sizing of the HRSG. Hence, even though the steam parameters such as pressure and feed water temperature have changed slightly, it has been shown by process calculations that the HRSG performance is reasonable. This is a simple and quick estimate. Plant engineers may develop calculation modules correcting for gas and steam properties and generate more accurate numbers.

One can estimate the gas pressure drop also and compare it with the measured value. Calculations with supplementary-fired HRSGs are more involved, as discussed in Chapter 1, but following the same procedure and a few more rounds of iterations, one can evaluate the performance from field data. One may also work from the economizer end and proceed toward the superheater and check the inlet gas temperature.

Efficiency of Waste Heat Boiler, HRSG

The efficiency E of a waste heat boiler is obtained as follows:

E = (energy absorbed by steam and water or other fluids)/(exhaust gas flow
   × enthalpy at entering temperature + fuel input if any on LHV basis)

In Example 4.2, the energy absorbed by steam–water in the superheater, evaporator, and economizer of the HRSG = 10.73 MM kcal/h; exhaust gas flow = 110,000 kg/h at 550°C. Exhaust gas analysis is % volume CO_2 = 3, H_2O = 7, N_2 = 75, and O_2 = 15. The efficiency E = $10.73 \times 10^6/(110,000 \times 140.8)$ = 0.68 or 69%. (Enthalpy of 140.8 kcal/kg was estimated from Table F.12 for the unfired turbine exhaust gas at 550°C.)

In the example shown in Table 4.6, the efficiency in unfired case in column 1: E = 8.75 × $10^6/(100,000 \times 127.1)$ = 68.8%. In the fired case, column 2, E = 22.51 × $10^6/(100,000 \times 127.1 + 13 \times 10^6)$ = 87.5%.

Advantages of Supplementary Firing in HRSGs

As discussed in Chapter 1, there is lot of oxygen in turbine exhaust gases that can be utilized to fire additional fuel to the exhaust gas and thus increase its temperature. Increase in inlet gas temperature generates additional steam in the evaporator, which increases the capacity of the heat sink, which is the economizer. Hence, the exit gas temperature beyond the economizer will be lower in the fired mode compared to the unfired mode. Since additional air is also not added, the lower exit gas temperature helps to increase the fuel utilization significantly. It can be shown through performance calculations that supplementary firing is an excellent way of generating additional steam in HRSGs and fuel utilization is nearly 100% or even more. That is, one can generate additional steam with nearly the same amount or lesser amount of fuel. In conventional steam generators, the LHV efficiency is about 93%, while in HRSGs, it can be close to 100% or even more. A layman's explanation for this is as follows: when we increase excess air in steam generators, the efficiency decreases. In gas turbine HRSGs, we *decrease* the excess air by firing fuel into the oxygen-rich exhaust gas and hence efficiency is higher. In addition, the exit gas temperature from the economizer exit is lower when inlet gas temperature is higher and that contributes to the higher efficiency as well. The economizer acts as a larger heat sink in the fired mode due to the larger flow of water through it and hence is able to cool the exhaust gases to a lower temperature. Chapter 5 on simulation gives examples of improved fuel utilization with supplementary firing.

In Example 4.1, one may check the fuel utilization also. The difference in energy recovered between the unfired and fired cases for vendor A is (22.51 – 8.71) = 13.80 MM kcal/h, while only 13.0 MM kcal/h is fired on LHV basis. That means the fired case is more than 100% efficient. In the case of steam generators, the efficiency on LHV basis was shown to be about 93.5%. That is, by firing in an HRSG, we straightaway get more than 100% fuel utilization. If we look at vendor B, the additional energy from steam = 22.51 – 8.21 = 14.3 MM kcal/h, while 14.0 MM kcal/h is fired! Here again, it is more than 100% efficient. Thus, plant engineers may analyze the HRSG performance in both fired and unfired cases and understand how well the fuel energy is utilized. (The HRSG should have an economizer to be more efficient in the fired mode if it is a single-pressure unit. If we did not have an economizer, the exit gas temperature will be higher without an economizer and hence efficiency in fired mode will be lower.)

Thus, supplementary firing is one way to improve the efficiency of gas turbine HRSGs. That is because we are not adding any combustion air while firing but utilizing the oxygen in the exhaust. In effect, we are reducing the excess air. The heat losses from the boiler are also lowered due to the lower exit gas temperature in the fired case. This is due to the fact that a larger heat sink is available in the form of the economizer in the fired mode with the gas flow remaining the same. Plant engineers while planning future projects with HRSGs in cogeneration plants may consider supplementary- or furnace-fired HRSGs as they are more efficient. Since the cross section of an HRSG is determined by the gas flow, it does not matter whether the steam output is 15 t/h or 60 t/h, as the same exhaust gas flow with different temperatures and analysis flows through the HRSG. The size of drums may be larger for the higher capacity, but the overall cost should not be much more than an unfired unit.

HRSG and Steam Generator Performance

When large amount of steam is required from turbine exhaust, furnace firing is resorted to as discussed earlier. When an HRSG and a steam generator are available in a plant, the plant engineer is faced with the issue of maximizing the steam output with minimum fuel input and how to load each of them in an efficient manner. This is due to the fact that the characteristics of an HRSG and a steam generator versus load are different. Figure 4.30 shows the variations in economizer exit gas temperature, water temperature leaving economizer, and efficiency on LHV basis for both a furnace-fired HRSG and a comparable steam generator. Saturated steam pressure = 30 kg/cm^2 g; feed water temperature = 105°C; fuel is natural gas (methane = 97%, ethane = 2%, and propane = 1%). Excess air of 15% is used in the steam generator. For the gas turbine, the exhaust gas flow = 100,000 kg/h at 485°C. Exhaust gas analysis = % volume CO_2 = 3, H_2O = 7, N_2 = 75, and O_2 = 15. An O-type boiler as shown in Figure 4.22b is used. This has a furnace section followed by convection bank tubes and an economizer. A furnace is required as the firing temperature is about 1250°C in the 100% fired case. A computer program was used in the evaluation.

Optimum Utilization of Boilers

Once the performance characteristics of steam generators and HRSGs are known, one may use the information to generate steam efficiently.

Example 4.3

The performance curves for an HRSG and steam generator of capacity 60 t/h are shown in Figure 4.30. Tables 4.9 through 4.11 give more information.

Let us say that the plant needs 75 t/h of steam. This can be obtained in many ways assuming the boilers are available and capable of operating at any load.

It may be seen that when the HRSG is operated at its maximum capacity, the total fuel consumption is the least. This is a simplistic model. With more HRSGs and more steam generators in a large plant, there will be more options to generate a given amount of steam, and it is a good idea to investigate the optimum loading of each boiler by doing a similar study. Also when planning cogeneration plant HRSGs, it is better to purchase fired HRSGs as the steam generation will be more efficient as seen earlier.

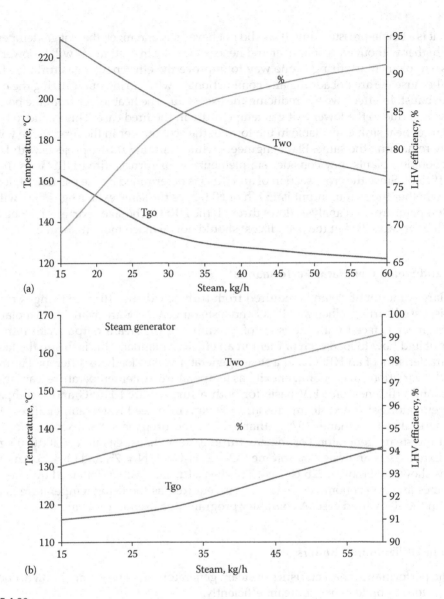

FIGURE 4.30
Load versus performance for an HRSG and a steam generator.

Combined Cycle Plants and Fired HRSGs

It is generally believed that combined cycle plant efficiencies with fired HRSGs are lower than those with unfired HRSGs. The reason is not the poor performance of the HRSG. In fact, the fired HRSG by itself is more efficient than an unfired HRSG. However, the losses associated with the Rankine cycle, particularly when the steam turbine power is a large fraction of the total power output, distorts the results slightly as shown in the following example (it is worked out in British units as all the data were provided in British units, and the purpose is to illustrate the point regarding efficiency).

TABLE 4.9

Steam Generator

Steam, kg/h	15,000	30,000	45,000	60,000
Excess air	35	15	15	15
Duty, MM kcal/h	8.49	16.9	25.4	33.8
water temp. to eco, °C	105	105	105	105
Water temp. lvg. eco, °C	151	151	158	165
LHV efficiency,%	93.36	94.46	94.34	94.07
Exit gas temp., °C	116	119	127	135
Burner fuel, MM kcal/h	8.68	17.78	26.92	35.93

TABLE 4.10

HRSG Performance

Steam, kg/h	15,000	30,000	45,000	60,000
Steam temp., °C	234	234	234	234
Feed water temp., °C	105	105	105	105
Temp leaving eco, °C	229	197	175	162
Duty, MM kcal/h	8.49	16.9	25.4	33.8
Pinch point, °C	5	10	14	17
Approach point, °C	5	37	59	72
Exit gas temp, °C	163	134	126	122
Burner fuel, MM kcal/h	0	7.9	16.3	24.7
Firing temp, °C	485	748	1005	1242

Note: Steam pressure = 30 kg/cm^2 g, fuel-natural gas.

TABLE 4.11

Combined Operation of Steam Generator and HRSG

Total Steam	HRSG Steam	Steam Generator	HRSG Fuel	SG Fuel	Total Fuel
75,000	30,000	45,000	7.9	26.92	34.82
75,000	45,000	30,000	16.3	17.78	34.08
75,000	15,000	60,000	0	35.93	35.93
75,000	60,000	15,000	24.7	8.68	33.38
75,000	37,500	37,500	12.1	22.4	34.50

Example 4.4

A combined cycle plant uses a single-pressure-fired HRSG. The gas turbine used is LM 5000 at 59°F. The following are the basic data:

Exhaust gas flow = 1,030,000 lb/h at 800°F. Exhaust gas analysis: % volume CO_2 = 2.8, H_2O = 8.5, N_2 = 74.4, O_2 = 14.3. Power output = 35 MW. Heat rate = 9649 Btu/kWh.

Steam turbine data: Inlet pressure = 650 psia at 750°F. Exhaust pressure = 1 psia. Efficiency = 80% dropping off by 2%–3% at 40% load.

HRSG data: 230°F feed water, 2% blowdown, 1% heat loss. Steam is generated at 665 psia and 750°F. The HRSG generates 84,400 lb/h in the unfired mode and a maximum of 186,500 when fired to 1200°F. The HRSG performance was simulated using the HRSGS

TABLE 4.12

Cogeneration and Combined Cycle Efficiency with Fired HRSG

Gas Temp.,°F	HRSG Exit, °F	Boiler Duty	Burner Duty	Turbine Power	Cogen Plant η, %	Combined Cycle η, %	Steam, lb/h
800	435	99.8	0	9.2	64.9	44.7	84,400
900	427	129.9	29.6	12.1	67.9	43.8	109.700
1000	423	160	59.1	15.3	70.4	43.2	135.200
1100	420	290.4	90.7	18.2	72.3	42.4	160,960
1200	418	221.0	121.0	21.1	74.2	41.7	186,500

Note: Boiler duty is energy absorbed by steam in MM Btu/h. Burner duty is fuel input to HRSG burner in MM Btu/h on LHV basis.

program (Chapter 5), and results are presented in Table 4.12. The system efficiency in both cogeneration and combined cycle mode is calculated as follows:

Gas turbine fuel input = 35,000 × 9,649 = 337.71 MM Btu/h on LHV basis.

Cogeneration mode efficiency at 900°F = 100 × (35 × 3413 + 129.9)/(337.71 + 29.6) = 67.9%

where 129.9 MM Btu/h is the HRSG output and 29.6 MM Btu/h is the HRSG burner input on LHV basis.

Combined cycle mode efficiency = 100× (35 + 12.1) × 3413/(337.71 + 29.6) = 43.8% where 12.1 is the power from the expansion of steam in the turbine. Table 4.12 shows the results with various firing temperatures. It may be seen that cogeneration plant efficiency improves with firing while that of the combined cycle plant decreases. If you check the additional HRSG fuel versus additional boiler duty, you will again find that the HRSG is more than 100% efficient in generating additional steam.

HRSGs with Split Superheaters

When plants do not have demineralized water or if they prefer to tolerate a modest variation in steam temperature between unfired and fired cases, one interesting option is to split the superheaters so that the final superheater is located between the gas turbine and the duct burner, and the other superheater is located downstream of the burner. The performance of this design in both unfired and fired modes is shown in Tables 4.13 through 4.15. It is seen that the steam temperature is about 350°C in both cases. This example is also evaluated using HRSG simulation method in Chapter 5. For comparison, the geometric data and performance of the HRSG with burner ahead of the superheaters are shown in Table 4.15.

Performance with and without Export Steam

In cogeneration plants, saturated steam is sometimes taken off the drum for process heating purposes. This decision may come after a few years after the boiler is installed, and hence, the boiler supplier may not be around to help the plant. Plant engineers wonder if the lower steam flow through the superheater will result in overheating of tubes. By doing performance calculations, one can check this. Based on the results of the calculation, plant may go ahead with the decision to take the process steam from the drum. Here is an example of such a problem where saturated steam is exported from the HRSG.

TABLE 4.13

Performance of HRSG with Split Superheaters

	Unfired	Fired	Gas Flow, kg/h	400,000
Steam press., kg/cm²g	54.0	54.0	gas temp. in, °C	500
Steam temp. out ±5°C	355	355	CO_2, % volume	2.40
Steam flow, kg/h	48,447	100,000	H_2O	7.00
Feed water temp., °C	109	109	N_2	75.50
			O_2	15.10

Unfired Case

Process Data Surface	Suphtr1	Suphtr2	Evap.	Evap.	Evap.	Econ.
Gas temp. in, ±5°C	500	486	468	444	415	289
Gas temp. out, ±5°C	486	468	444	415	289	213
Gas spht, kcal/kg °C	0.2742	0.2734	0.2753	0.2728	0.2676	0.2611
Duty, NIM kcal/h	1.58	1.87	2.70	3.04	13.41	7.87
Surface area, m²	202	149	199	349	5697	4207
Gas press. drop, mm wc	9.23	2.97	4.71	4.39	36.69	24.14
Foul factor, gas	0.0002	0.0002	0.0002	0.0002	0.0002	0.0002
U, Btu/ft² h °F	49.74	66.89	72.66	54.39	36.79	32.52
LMTD-C	157	188	187	161	64	58
Max gas vel, m/s	26	13	16	17	18	16

Fired Case

Process Data Surface	Suphtr1	Suphtr2	Evap.	Evap.	Evap.	Econ.
Gas temp. in, ±5°C	500	738	697	643	576	307
Gas temp. out, ±5°C	484	697	643	576	307	193
Gas spht, kcal/kg °C	0.2742	0.2911	0.2906	0.2854	0.2770	0.2644
Duty, NIM kcal/h	1.76	4.79	6.32	7.57	29.68	12.05
Surface area, m²	202	149	199	349	5697	4207
Gas press. drop, mm wc	9.10	4.10	6.35	5.69	42.77	24.14
Foul factor, gas	0.0002	0.0002	a 0002	0.0002	0.0002	0.0002
U, Btu/ft² h °F	56.55	76.47	79.52	64.17	40.64	34.48
LMTD-C	154	421	400	339	128	83
Max gas vel, m/s	26	18	21	22	22	17

(Continued)

TABLE 4.13 (Continued)

Performance of HRSG with Split Superheaters

Process Data Surface	Unfired Case						Fired Case					
	Suphtr1	Suphtr2	Evap.	Evap.	Evap.	Econ.	Suphtr1	Suphtr2	Evap.	Evap.	Evap.	Econ.
Tube wall temp., ±5°C	414	331	276	285	288	267	395	362	287	312	314	242
Fin tip temp., ±5°C	455	331	276	312	306	274	446	362	287	377	356	262
Weight, kg	2464	4282	5284	4667	53,590	33,842	2464	4282	5284	4667	53,590	33,842
Fluid temp. in, °C	312	268	261	261	261	109	325	269	224	224	224	109
Fluid temp. out, ±5°C	359	312	269	269	269	261	351	325	270	270	270	224
Pr drop, kg/cm²	0.07	0.18	0.00	0.00	0.00	0.18	0.29	0.78	0.00	0.00	0.00	0.77
Foul factor, fluid	0.0002	0.0002	0.0002	0.0002	0.0002	0.0002	0.0002	0.0002	0.0002	0.0002	0.0002	0.0002
Fluid velocity, m/s	10.8	9.6	0.0	0.0	0.0	0.7	22.3	20.1	0.0	0.0	0.0	1.4
Fluid httr. coefft.	946	1117	9780	9780	9780	4742	1679	1960	9780	9780	9780	8074
Steam flow, kg/h	48,511	48,511	48,511	48,511	48,511	48,511	100,000	100,000	100,000	10,0000	100,000	100,000
Burner duty, MM kcal/h								30.19				

TABLE 4.14

Geometric Data of HRSG with Split Superheaters: Burner
between Superheaters

Geometry	Sh1	Sh2	Evap1	Evap2	Evap3	Econ
Tube OD	38.1	38.1	50.8	50.8	50.8	44.5
Tube ID	30.6	30.6	44.1	44.1	44.1	37.8
Fins/in. or fins/m	0	0	0	78	177	177
Fin height	0	0	0	12.5	12.5	19
Fin thickness	0	0	0	1.9	1.9	1.9
Fin width	0	0	0	0	4.37	4.37
Fin conductivity	0	0	0	35	35	35
Tubes/row	38	38	38	38	38	42
Number of rows deep	4	4	4	2	16	8
Length	8.2	8.2	8.2	8.2	8.2	8.2
Transverse pitch	101	101	101	101	101	92
Longitudinal pitch	101	101	101	101	101	100
Streams	76	76				21
Parl = 0, Countr = 1	1	1	Tubes inline arrgt.			1

Example 4.5

Tables 4.16 and 4.17 show the performance of a waste heat boiler in a chemical plant
as provided by the HRSG supplier. Flue gas flow is 150,000 kg/h at 650°C and has an
analysis % volume $CO_2 = 7$, $H_2O = 12$, $N_2 = 75$, $O_2 = 6$. Boiler generates about 27,684 kg/h
of superheated steam at 40 kg/cm^2 g and at 316°C. Field data show performance close
to these predicted values. Now, the plant engineer wants to withdraw 8,000 kg/h of
saturated steam from the drum for process heating and superheat the balance of steam
when the gas inlet flow is 120,000 kg/h at 600°C with the flue gas analysis remaining the
same as before. What will be the performance and steam temperature?

Solution

This problem is more involved unlike the earlier problem of checking field data. More
iterations are required as we have no idea of the operating conditions at the new gas inlet
conditions. Hence, this is a good exercise for evaluating HRSG off-design performance.

Let us estimate the LMTD and U at each section. Often information on gas–steam
temperature profiles will be available in a boiler plant; if not, one may estimate the gas
temperatures starting from the cold end and move to the hot gas inlet as was done in
Chapter 3 for the steam generator. U may also be estimated for each section by methods
discussed in Appendices C and D. Here since we know the gas and steam temperature
profile, we may estimate the LMTD at each section. Counter-flow configuration is used.
If the duty of each section is not provided, one can use the steam- and water-side data
or field measurements, compute the duty and then estimate the gas temperature drop
in each section. Water or steam flow measurement is typically more accurate than gas
flow measurement, and gas temperature measurements at the cold end are more reliable than those at hot gas inlet. So one may start at the economizer and work backward
to obtain the gas flow and temperature at various sections and arrive at the data given
earlier. As mentioned in several places in this book, plant engineers should demand
such data from the boiler supplier for various load conditions so that they can deal with
problems such as this one, which may arise after several years of operation.

TABLE 4.15

HRSG with Burner ahead of Superheaters

Burner ahead of HRSG

Unfired Case Process Data Surface	Suphtr1	Suphtr2	Evap.	Evap.	Evap.	Econ.	Fired Case Suphtr1	Suphtr2	Evap.	Evap.	Evap.	Econ.
Gas temp. in ±5°C	500	486	468	444	416	290	760	719	680	628	565	308
Gas temp. out-±5°C	486	468	444	416	290	213	719	680	628	565	308	195
Gas spht, kcal/kg °C	0.2742	0.2734	0.2753	0.2729	0.2677	0.2612	0.2925	0.2902	0.2890	0.2855	0.2768	0.2645
Dutr, MM kcal/h	1.58	1.86	2.69	3.00	13.33	7.97	4.75	4.57	6.01	7.12	28.38	11.97
Surface area, m²	149	149	199	344	5618	4146	149	149	199	344	5618	4146
Gas press. drop, mm wc	3.04	2.97	4.71	4.37	36.35	23.79	4.20	4.02	6.24	5.58	42.02	23.84
Foul factor, gas	0.0002	0.0002	0.0002	0.0002	0.0002	0.0002	0.0002	0.0002	0.0002	0.0002	0.0002	0.0002
U, kcal/m²h °C	67.66	66.94	72.70	54.56	37.09	32.83	77.60	76.00	79.19	63.77	40.77	34.75
LMTD-F	157	187	186	160	64	59	411	403	382	325	124	83
Max. gas vel., m/s	13	13	16	16	18	16	18	17	21	21	22	16
Tube wall temp., ±5°C	377	332	277	286	289	268	397	360	288	311	314	244
Fin tip temp., ±5°C	377	332	277	313	308	275	397	360	288	373	353	264
Fluid temp. in, °C	313	270	262	262	262	109	294	271	227	227	227	109
Fluid temp. out, ±5°C	360	312	270	270	270	262	360	325	272	272	272	227
Press. drop, kg/cm²	0.20	0.18	0.0	0.0	0.0	0.2	0.8	0.7	0.0	0.0	0.0	0.7
Foul factor, fluid	0.0002	0.0002	0.0002	0.0002	0.0002	0.0002	0.0002	0.0002	0.0002	0.0002	0.0002	0.0002
Fluid velocity, m/s	10.9	9.5	0.0	0.00	0.0000	0.67	21.9	19.2	0.0	0.0	0.0	1.3
Fluid httr. coefft.	1007	1121	9780	9780	9780	4778	1836	1905	9780	9780	9780	7907
Steam flow, kg/h	48,477						100,000					
Spray, kg/h	0						3.972					
Burner duty, MM kcal/h	0.0						31.2					

	Unfired	Fired	Gas Flow, kg/h	400,000
Steam press., kg/cm²g	54.0	54	Gas temp. in, °C	500
Steam temp. out, ±5°C	360	360	CO_2, % volume	2.40
Steam flow, kg/h	48,447	100,000	H_2O	7.00
Feed water temp., °C	109	109	N_2	75.50
			O_2	15.10

Geometry	Sh1	Sh2	Evap.	Evap.	Evap.	Evap.	Econ.
Tube OD	38.1	38.1	50.8	50.8	50.8	50.8	44.5
Tube ID	30.6	30.6	44.1	44.1	44.1	44.1	37.8
Fins/in. or fins/m	0	0	0	78	177	177	177
Fin height	0	0	0	12.5	12.5	12.5	19
Fin thickness	0	0	0	1.9	1.9	1.9	1.9
Fin width	0	0	0	0	4.37	4.37	4.37
Fin conductivity	0	0	0	35	35	35	35
Tubes/row	38	38	38	38	38	38	42
Number of rows deep	4	4	4	2	16	8	8
Length	8.2	8.2	8.2	8.2	8.2	8.2	8.2
Transverse pitch	101	101	101	101	101	101	92
Longitudinal pitch	101	101	101	101	101	101	100
Streams	76	76	Tubes inline arrgt.				21
parl = o, countr = 1	1	1	1	1	1	1	1

TABLE 4.16

Waste Heat Boiler Performance Data 150,000 kg/h Gas Flow

Process Data Surface	Suphtr.	Evap.	Econ.
Gas temp. in, ±5°C	650	619	330
Gas temp. out, ±5°C	619	330	247
Gas spht., kcal/kg °C	0.2933	0.2850	0.2716
Duty, MM kcal/h	1.36	12.24	3.33
Surface area, m²	43	766	306
Gas pr drop, mm wc	9.61	129.46	70.38
Foul factor, gas	0.0002	0.0002	0.0002
Steam Side			
Steam press., kg/cm²g	41	41	41
Steam flow, kg/h	27,684	27,684	27,684
Fluid temp. in, °C	252	221	105
Fluid temp. out ±5°C	316	252	221
Pr drop, kg/cm²	0.88	0.00	0.21
Foul factor, fluid	0.0002	0.0002	0.0002

TABLE 4.17

Tube Geometry Data

Geometry	Sh.	Evap.	Econ.
Tube OD	31.8	50.8	50.8
Tube ID	25	44	44
Fins/in. or fins/m	0	0	0
Fin height	0	0	0
Fin thickness	0	0	0
Fin width	0	0	0
Fin conductivity	0	0	30
Tubes/row	36	20	20
Number of rows deep	6	60	24
Length	2	4	4
Transverse pitch	100	100	90
Longitudinal pitch	80	80	80
Streams	36	18	10

Superheater $\Delta T = [(650 - 316) - (619 - 252)]/\ln[(650 - 310)/(619 - 252)] = 350°C$. Then $U = Q/A\Delta T = 1.36 \times 10^6/(43 \times 350) = 90.2$ kcal/m²h °C

Evaporator $\Delta T = [(619 - 252) - (330 - 252)]/\ln[(619 - 252)/(330 - 252)] = 187°C$
$U = 12.24 \times 10^6/(766 \times 187) = 85.44$ kcal/m²h °C

Economizer $\Delta T = [(330 - 221) - (247 - 105)]/\ln[(330 - 221)/(247 - 105)] = 125°C$
$U = 3.33 \times 10^6/(306 \times 125) = 87.0$ kcal/m²h °C

Now at 120,000 kg/h gas flow at 600°C, we have to estimate the U values for each section. We may apply the methods discussed in Appendices B through D, or we may use the U values given earlier and correct it for gas flow and temperature. Let us correct the U values for gas flow effect for the first trial. Once we get a preliminary performance and gas–steam temperature profiles, we may use the correlations in Appendices C and D to fine-tune the U values.

$$\text{Superheater } U_p = 90.2 \times (0.8)^{0.6} = 78.9 \text{ kcal/m}^2\text{h °C (0.8 refers to } 120,000/150,000$$
$$\text{of gas flow ratio)}$$

$$\text{Evaporator } U_p = 85.44 \times (0.8)^{0.6} = 74.7 \text{ kcal/m}^2\text{h °C}$$

$$\text{Economizer } U_p = 87 \times (0.8)^{0.6} = 76 \text{ kcal/m}^2\text{h °C}.$$

We have to assume two values—the exit gas temperature from the boiler and the steam temperature—and then fine-tune these later. Let us assume that the exit gas temperature is 240°C and the steam temperature is 370°C. The energy transferred by flue gas to steam = $120,000 \times 0.99 \times 0.28 \times (600 - 240) = 11.97 \times 10^6$ kcal/h, where 0.28 is the average gas specific heat between 600°C and 240°C. The enthalpy of superheated steam at 370°C is 750.48 kcal/kg. Then superheated steam generated is obtained by energy balance. Using zero blowdown, $W_s[(750.48 - 105.9) + 8000 \times (669 - 105.9)] = 11.97 \times 10^6$, where 105.9 is the enthalpy of feed water at 105°C and that of saturated steam is 669 kcal/kg. Hence, $W_s = 11,580$ kg/h.

Superheater may be solved by using either the transferred duty evaluation of $Q_t = UA\Delta T$ or the NTU method. Let us use the transferred duty method.

Superheater

$Q_a = 11,580 \times (750.48 - 669) = 0.9435 \times 10^6$ kcal/h $= 120,000 \times 0.99 \times 0.29 \times (600 - T)$. (Flue gas specific heat of 0.29 kcal/kg °C was used for the superheater region. Hence, exit gas temperature of superheater T = 573°C. Calculate ΔT as we know all the four temperatures at the superheater. $\Delta T = [(600 - 370) - (573 - 251)]/\ln[(600 - 370)/(573 - 251)] = 273°C$. $Q_t = UA\Delta T = 78.9 \times 43 \times 273 = 0.926$ MM kcal/h. It is not close to the assumed duty, but let us proceed.

Evaporator

Evaporator may be solved by using the method shown in Appendix A.
$\ln[(573 - 251)/(T - 251)] = 74.7 \times 766/(120,000 \times 0.99 \times 0.283)$ or T = 310°C. $Q_t = 120,000 \times 0.99 \times 0.283 \times (573 - 310) = 8.842 \times 10^6$ kcal/h (a gas specific heat of 0.283 kcal/kg °C was assumed).

Economizer

Q_a at economizer = $120,000 \times 0.99 \times 0.27 \times (310 - 240) = 2.245 \times 10^6$ kcal/h. Enthalpy pickup of water = $2.245 \times 106/(8,000 + 11,580) = 114.7$ kcal/kg or exit water enthalpy = $105.9 + 114.7 = 220.6$ kcal/kg or water temperature from steam tables is 215°C. (The water flow in economizer is 19,580 kg/h as the 8,000 kg/h of saturated steam has to be added to the superheater steam flow.) (Water-side specific heat = $(220.6 - 105.9)/(215 - 105) = 1.042$ kcak/kg °C. $\Delta T = [(310 - 215) - (240 - 105)]/\ln[(310 - 215)/(240 - 105)] = 114°C$ or $Q_t = 76 \times 306 \times 114 = 2.651 \times 10^6$ kcal/h.) Since the difference is significant, another iteration is required.

The actual duty Q_t of entire HRSG = $(0.9435 + 8.84 + 2.651) = 12.43 \times 10^6$ kcal/h. Revised steam generation = $W_s \times (750.48 - 105.9) + 8,000 \times (669 - 105.9) = 12.43 \times 10^6$ or $W_s = 12,300$ kg/h. Using this value, let us go back to the superheater.

Superheater

Use the NTU method now. Specific heat of steam = $(750.48 - 669)/(370 - 251) = 0.6847$ kcal/kg °C. (WC)steam = $12,300 \times 0.6847 = 8421$ and (WC) flue gas = $120,000 \times 0.99 \times 0.29 = 34,452$.

$$C = \frac{8,421}{34,452} = 0.244. \ (1 - C) = 0.756. \ NTU = \frac{78.9 \times 43}{8421} = 0.4029$$

$\in = [1 - \exp(-0.4029 \times 0.756)]/[1 - 0.244 \times \exp(-0.4029 \times 0.756)] = 0.32$ or $Q_t = 0.32 \times 8421 \times (600 - 251) = 0.94$ MM kcal/h. Steam temperature at exit = $251 + 0.94 \times 10^6/12,300 \times 0.6847 = 362$°C.

Exit gas temperature of superheater = $600 - 0.94 \times 10^6/9,120,000 \times 0.99 \times 0.29) = 573$°C.

Evaporator exit gas temperature and duty will be the same as before as gas entering temperature is unchanged.

Economizer

(WC) on water side = $(12,300 + 8,000) \times 1.042 = 21,152$ and (WC) flue gas = $120,000 \times 0.99 \times 0.27 = 32,076$.

$$C = \frac{21,152}{32,076} = 0.659 \text{ and } (1 - C) = 0.341. \ NTU = \frac{76 \times 306}{21,152} = 1.099$$

$\in = [1 - \exp(-1.099 \times 0.341)]/[1 - 0.659 \times \exp(-1.099 \times 0.341)] = 0.571$ or duty $Q_t = 0.571 \times 21,152 \times (310 - 105) = 2.475 \times 10^6$ kcal/h. Exit gas temperature = $310 - 2.475 \times 106/9,120,000 \times 0.99 \times 0.27 = 233$°C. Exit water temperature is $(105 + 2.475 \times 10^6/21,152) = 222$°C.

Total duty transferred = $0.94 + 8.842 + 2.475 = 12.257 \times 10^6$ kcal/h. Superheated steam generated is obtained from energy balance: $12.257 \times 10^6 = W_s \times (745.8 - 105.9) + 8000 \times (669 - 105.9)$ or $W_s = 12,114$ kg/h. Final steam temperature is 362°C, and its enthalpy is 745.8 kcal/kg. This is a closer estimation of performance. The gas–steam temperature profile is shown in Figure 4.31. The steam temperature increase is only about 46°C and hence will not affect the tube materials used.

One more iteration may be carried out to finalize the complete performance using the appropriate correlations and correct gas and steam properties. However, the purpose of this exercise is to show how plant engineers may evaluate the HRSG performance even if no operating data or performance data are available and they are doing *what if* studies.

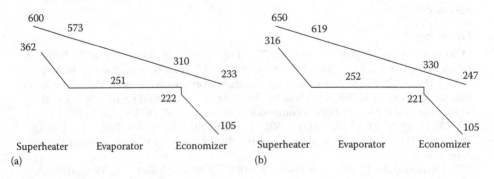

FIGURE 4.31
HRSG gas–steam profiles (a) with and (b) without export steam.

As an exercise, plant engineers may see what happens if saturated steam of 8000 kg/h from another boiler is to be superheated in this boiler at the same inlet gas conditions. What will be the new steam generation and steam temperature when the same amount of steam is imported into the HRSG?

Providing Margins on Heating Surfaces

Some consultants write specifications for waste heat boilers and casually make a demand that 15%–20% additional heating surface should be provided for each component. However, what they fail to do as a follow-up is to ask the boiler supplier to provide performance with and without the margin. One should be careful in multimodule HRSGs as the additional surface upstream can affect the duty and steam temperature of the downstream component if it is a superheater or the duty and water temperature if it is an economizer. Some consultants presume that they can get about 15%–20% more steam generation with about 20% more heating surface, but this is not feasible. Due to the lower LMTD in each section, the duty will not be that much more. Hence, providing some margin is fine, but the expectation of results should be realistic and performance data with and without margins should be obtained before making a decision. Tables 4.18 and 4.19 show an example of an HRSG with about 17% more heating surface but with only 2.4% additional duty. The gas pressure drop also goes by from 75 to 86 mm wc. The steam generation is about 1% more due to the higher steam temperature or, if controlled, will be about 2.4% more. Hence, the margins may be left to the boiler supplier. He has to only prove the guarantee performance results in the field.

Cement Plant Waste Heat Recovery

Cement production is one of the most energy-intensive processes in the world. Nearly 50%–60% of the cost of cement is due to energy cost. Energy is consumed in grinding mills, fans, conveyors, and in the process of calcination. Coal used is in the range of 150–250 kg per ton of cement, and energy used is about 80–125 kWh per ton of cement. A few decades ago, cement plants were venting the waste gases from the preheater and clinker cooler into the atmosphere as they were dusty and were also at low temperatures in the range of 350°C–280°C. Waste heat boilers for such low gas inlet temperatures required large multiple-pressure steam generation to cool the gases to an economical temperature that made the economics unfavorable. However, today, the high cost of energy is forcing all cement plants to recover energy from the preheater and the clinker cooler exhaust gases. The exit gases from rotary kilns, preheater, and calciners are used to heat the incoming feed material, and the gases are cooled to about 300°C–350°C in a four-stage preheater. With more stages, the exhaust temperature can be 200°C–250°C. Part of this gas is used in raw mills and coal mills for drying the coal. The solid clinker coming out of the rotary kiln at 1000°C is cooled to about 120°C using ambient temperature. This generates hot air at about 270°C–310°C. Part of this is used as combustion air in kiln furnaces and rest vented or used in waste heat recovery. Typical flue gas data from a 1500 TPD cement plant is shown later. A scheme of a typical cement plant is shown in Figure 4.32.

A vertical downward gas flow boiler with horizontal tubes is often used for preheater and clinker cooler applications. Vertical tube design with horizontal gas flow natural circulation design is also used. Circulation pumps may not be required if the drum is located properly and adequate streams are used for the evaporator circuit; the tubes may have to be slightly inclined

TABLE 4.18

Performance with and without Additional Surface

With Additional Surface

General Data

Exhaust gas flow	400,000	kg/h	Steam pressure	54.00	kg/cm² g
Exhaust gas temp.	500	±5°C	Steam temp.	378	±5°C
Exhaust gas pressure	1.00	kg/cm² a	Steam flow	48,806	kg/h
Heat loss	1.00	%	Feed water temp.	109	±5°C
Firing temperature	0	±5°C	Process steam	0	kg/h
Burner duty, LHV	0.00	MM kcal/h	Blowdown	1	%
Burner duty, HHV	0.00	MM kcal/h	Eco steaming	0.00	%
Flue gas flow out	0	kg/h	Steam surface	0	m²

Process Data

Surface	Suphtr.	Suphtr.	Evap.	Evap.	Evap.	Econ.
Gas temp. in, ±5°C	500	484	463	439	412	286
Gas temp. out, ±5°C	484	463	439	412	286	206
Gas spht, kcal/kg °C	0.2743	0.2732	0.2754	0.2719	0.2674	0.3
Duty, MM kcal/h	1.79	2.23	2.61	2.91	13.40	8.3
Surface area, m²	187	187	199	344	6321	5183
Gas pr drop, mm wc	3.78	3.68	4.66	4.33	40.56	29
Foul factor, gas	0.0002	0.0002	0.0002	0.0002	0.0002	0

Steam Side

Steam press., kg/cm² g	55	55	55	55	55	55
Steam flow, kg/h	48,806	48,806	48,806	48,806	48,806	49,294
Fluid temp. in, °C	322	270	265	265	265	109
Fluid temp. out, ±5°C	378	322	270	270	270	265
Pr drop, kg/cm²	0.3	0.2	0.0	0.0	0	0
Foul factor, fluid	0.0002	0.0002	0.0002	0.0002	0.0002	0.0002

Without Additional Surface

General Data

Exhaust gas flow	400,000	kg/h	Steam pressure	54.0	kg/cm² g
Exhaust gas temp.	500	±5°C	Steam temp	360	±5°C
Exhaust gas pressure	1.00	kg/cm² a	Steam flow	48,446	kg/h
Heat loss	1	%	Feed water temp.	109	±5°C
Firing temperature	0	°C	Process steam	0.0	kg/h
Burner duty	0.00	MM kcal/h	Blowdown	1.0	%
Furnace duty	0.00	MM kcal/h	Eco steaming	0.0	%
Flue gas flow out	0	kg/h	Steam surface	0	m²

Process Data

Surface	Suphtr.	Suphtr.	Evap.	Evap.	Evap.	Econ.
Gas temp. in, ±5°C	500	486	468	444	416	290
Gas temp. out, ±5°C	486	468	444	416	290	213
Gas spht., kcal/kg °C	0.2743	0.2734	0.2754	0.2729	0.2677	0.2611
Duty, MM kcal/h	1.58	1.87	2.69	3.00	13.32	7.96
Surface area, m²	149	149	199	344	5618	4146
Gas press. drop, mm wc	3.03	2.96	4.70	4.36	36.22	23.71
Foul factor, gas	0.0002	0.0002	0.0002	0.0002	0.0002	0.0002

(Continued)

TABLE 4.18 (*Continued*)

Performance with and without Additional Surface

Steam Side	Suphtr.	Suphtr.	Evap.	Evap.	Evap.	Econ.
Steam press., kg/cm²g	55	55	55	55	55	55
Steam flow, kg/h	48,446	48,446	48,446	48,446	48,446	48,930
Fluid temp. in, C	313	270	262	262	262	109
Fluid temp. out, ±5°C	360	313	270	270	270	262
Press. drop, kg/cm²	0.20	0.18	0.00	0.00	0.00	0.19
Foul factor, fluid	0.0002	0.0002	0.0002	0.0002	0.0002	0.0002

TABLE 4.19

Tube Geometry Data with and without Additional Surface

Geometry	Sh	Sh	Evap	Evap	Evap	Econ
With Additional Surface						
Tube OD	38.1	38.1	50.8	50.8	50.8	44.5
Tube ID	30.6	30.6	44.1	44.1	44.1	37.8
Fins/in. or fins/m	0	0	0	78	177	177
Fin height	0	0	0	12.5	12.5	19
Fin thickness	0	0	0	1.9	1.9	1.9
Fin width	0	0	0	0	4.37	4.37
Fin conductivity	0	0	0	35	35	35
Tubes/row	38	38	38	38	38	42
Number of rows deep	5	5	4	2	18	10
Length	8.2	8.2	8.2	8.2	8.2	8.2
Transverse pitch	101	101	101	101	101	92
Longitudinal pitch	101	101	101	101	101	100
Streams	76	76				21
parl = 0, countr = 1	1	1				1
Without Additional Surface						
Tube OD	38.1	38.1	50.8	50.8	50.8	44.5
Tube ID	30.6	30.6	44.1	44.1	44.1	37.8
Fins/in. or fins/m	0	0	0	78	177	177
Fin height	0	0	0	12.5	12.5	19
Fin thickness	0	0	0	1.9	1.9	1.9
Fin width	0	0	0	0	4.37	4.37
Fin conductivity	0	0	0	35	35	35
Tubes/row	38	38	38	38	38	42
No deep	4	4	4	2	16	8
Length	8.2	8.2	8.2	8.2	8.2	8.2
Transverse pitch	101	101	101	101	101	92
Longitudinal pitch	101	101	101	101	101	100
Streams	76	76				21
parl = 0, countr = 1	1	1				1

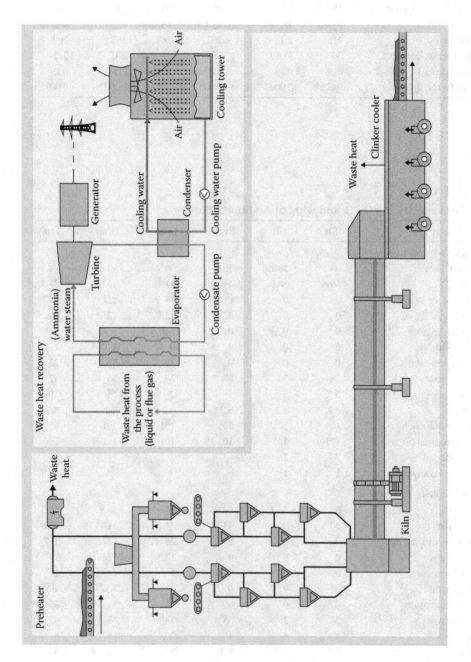

FIGURE 4.32
Typical cement plant heat recovery system.

to the horizontal. Since the gas temperature is quite low to start with, medium-pressure steam in the range of 15–20 barg is generated with steam temperature about 20°C–30°C lower than the gas inlet temperature. It is shown in Chapter 5 on HRSG simulation why with low inlet gas temperature, exit gas temperature from boiler will be high and vice versa. Since a single-pressure unit cannot cool the flue gas to an economical low temperature, low-pressure steam for deaeration is also generated in the clinker cooler boiler. Condensate preheating is also carried out to minimize the steam for deaeration. Many of these streams do not contain acid vapors, and hence, water vapor dew point alone limits the heat recovery.

Since the preheater exhaust gas contains high dust content, rapping mechanisms are used for cleaning the boiler tubes to prevent the dust from settling and fouling the tubes. Fouling factor in the range of 0.002–0.02 m^2 h °C/kcal is used in the design (0.01–0.1 ft^2 h °F/Btu). The nature of fouling is shown in Figure 4.33a. When the cleaning mechanism is used, fouling temporarily reduces and the performance improves for a short duration till the dust settles on the heating surfaces again. The frequency of cleaning is determined by operating experience. A higher gas velocity helps to lower fouling as the dust can get carried away with the flue gas, while too high a gas velocity can also lead to erosion of tubes due to the presence of ash particulates (Table 4.20).

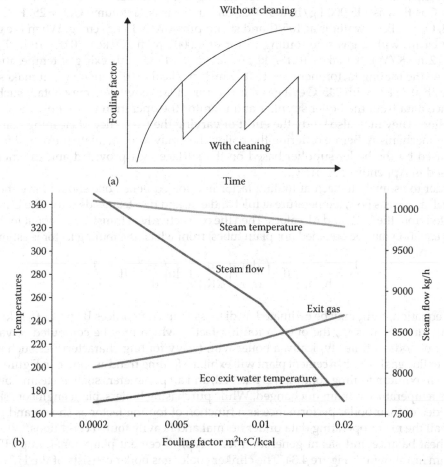

FIGURE 4.33
(a) Fouling characteristics in a boiler.- (b) Effect of fouling factor on performance of a waste heat boiler.

TABLE 4.20

Flue Gas from Preheater and Clinker Cooler

Source	Preheater Flue Gas	Clinker Cooler
Mass flow, kg/h	85,000	100,000
Gas temperature, °C	320–400	280–370
% volume CO_2	28	0
H_2O	11	1
N_2	58	78
O_2	3	21
Dust content, mg/N m^3	75–90	5–10

Clinker cooler boilers often use finned tubes with fin density varying from 2 to 3 fins/in. as the exhaust gas is mostly air with low dust content. The fouling factor in this boiler often ranges from 0.002 to 0.004 m^2 h °C/kcal (0.01–0.02 ft^2 h °F/Btu).

Since fouling factor is a complex factor in the design of any boiler with dusty flue gas, it is often based on experience from the operation of similar units with similar raw materials. Presented in Figure 4.33b is the performance of a preheater boiler at various fouling factors. Gas flow is 145,000 kg/h at 370°C. Gas analysis is % volume $CO_2 = 28$, $H_2O = 11$, $N_2 = 58$, $O_2 = 3$. Feed water is at 105°C and steam pressure is 13 kg/cm^2 g. When the boiler is very clean, with a gas-side fouling factor of 0.0002 m^2 h °C/kcal (0.001 ft^2 h °F/Btu) (0.000172 m^2 K/W), it makes 10,187 kg/h steam at 343°C with exit gas temperature of 200°C. As the fouling factor increases to 0.02 m^2 h °C/kcal (0.0172 m^2 K/W), it makes only 7660 kg/h at 321°C with 245°C exit gas temperature. Plant engineers may obtain such performance data from the boiler supplier and monitor the operation to see how the fouling is trending. They may also study the effect of varying the frequency of cleaning using the rapping mechanism. Before ordering the boiler, they may also calculate the actual fouling factor used by the boiler supplier based on the surface area provided and the methods discussed in Appendices A and C.

In order to estimate the actual fouling factor in a fouled boiler, one should have the field data such as the gas flow, temperature at inlet and exit, and the steam-side duty. The U may be computed from the Q, A, and ΔT values. Then the convective heat transfer coefficient in the following equation may be obtained (for plain tubes) from which the fouling factor is estimated.

$$\frac{1}{U_o} = \frac{d}{h_i/d_i} + ff_i \times \left(\frac{d}{d_i}\right) + \left(\frac{d}{2K_m}\right) \times \ln\left(\frac{d}{d_i}\right) + ff_o + \frac{1}{h_o}$$

In this equation, h_i, h_o may be estimated as discussed in Appendices B and C. ff_i is known. The only unknown is ff_o, the outside fouling factor, which may be computed at various loads or periods of time. Typically, a boiler with heavy fouling characteristics such as the preheater flue gas boiler in a cement plant will exhibit a fouling trend as shown in Figure 4.33a. Often, a maximum fouling condition is reached, and all parameters such as steam flow and exit gas temperature remain unchanged. While purchasing boilers, plant engineers should get an idea of their boiler performance as a function of fouling factor as shown and get an idea of all the major operating data under normal and heavily fouled conditions.

The heat balance and steam generation for a typical cement plant generating HP and LP steam are shown in Figure 4.34. The clinker cooler gas boiler consists of the HP superheater, evaporator, and economizer and a condensate heater. The preheater gas boiler consists of HP evaporator and LP superheater and evaporator. The economizer of cooler flue

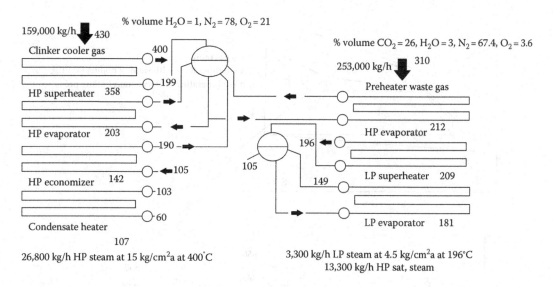

FIGURE 4.34
Gas–steam temperature profiles in a cement plant from cooler gas and preheater gas streams.

gas boiler handles the feed water heating of both the boilers and preheats the feed water from 105°C to 190°C and sends it to the HP drum operating at 15 kg/cm²g. The preheater HP evaporator generates about 13,500 kg/h of HP steam and the cooler boiler about the same amount, and the total flow of 26,800 kg/h is superheated in the cooler boiler to 400°C. As mentioned earlier, one has to generate steam at dual pressures to recover additional energy from low gas temperature streams. Hence, in order to improve the overall energy recovery, such combining of water and steam flows is necessary, which is why the Kalina system is more attractive.

Kalina Cycle

The Kalina cycle uses a mixture of 70% ammonia and 30% water as the working fluid instead of water, which has a constant boiling point. The problem with constant boiling temperature of water–steam is that with a minimum pinch point, the amount of energy recovered from a low-temperature heat sources such as diesel engine or cement plant will be very small, and hence, the heat recovery system will be uneconomical. Figure 4.35 shows the gas–fluid temperatures for a 500 psia steam and Kalina system with an inlet gas temperature of 550°F. With a, say, 30°F pinch and 10°F approach, the exit gas temperature will be about 474°F. This is a thermodynamic restriction, and more surface areas can probably lower the pinch point slightly but cannot bring down the exit gas temperature to much below 450°F. However, with the Kalina cycle, the same 500°F inlet temperature permits the gases to be cooled to 200°F. This is feasible as the ammonia – water mixture boils over a range of temperatures, and the boiling curve follows the gas temperature line while the pinch point restricts the energy that can be recovered with steam. Since the boiling occurs over a range of temperatures, it is possible to recover more energy from the same low inlet gas temperature. The amount of energy recovered is (550 – 200)/(550 – 474) = 4.6 times

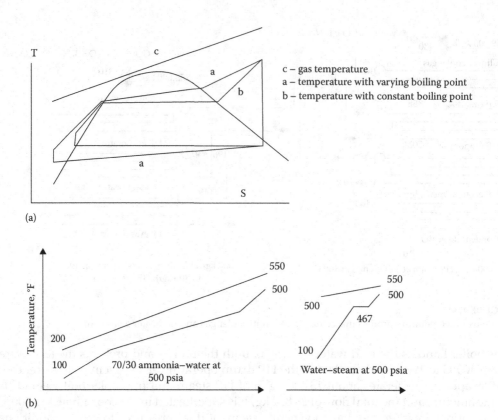

FIGURE 4.35
Gas–fluid temperature profile with steam and ammonia–water mixture.

more. A key feature of Kalina cycle is the ability to vary the ammonia–water concentration throughout the power plant to optimize energy conversion and to add heat recuperative stages for increased efficiency providing a richer concentration throughout the heat acquisition stage and leaner composition in the low-pressure condenser system.

In a typical steam-based Rankine cycle, the energy loss associated with the condenser is large. Also the energy recoverable is much lower when the gas inlet temperatures are low. Hence, Kalina cycle is a good bet when the flue gas inlet temperature to the waste heat boiler is low on the order of 250°C–350°C. As can be seen from Figure 4.35a, the heat is added and rejected at varying temperatures, which reduces these losses. The distillation condensation subsystem (DCSS) changes the concentration of the working fluid enabling the condensation of the vapor from the turbine to occur at lower pressures. The DCSS without any other source of energy brings back the concentration of the mixture back to the 70%–30% level for recovering energy from the waste heat boiler. The heat recovery vapor generator (HRVG) is similar in design to the HRSG. Carbon steel tubes are adequate for the superheater, evaporator, and economizer.

Advantages of Kalina Cycle

1. This is ideal for low inlet gas temperature heat sources in the range of 200°C–300°C. The ammonia–water mixture has a varying boiling and condensing temperature that enables the fluid to extract more energy from the hot gas stream than with a

steam system that has a constant boiling and condensing temperature; pinch and approach points set the temperature profile and energy recovery in steam system while the 70–30 ammonia–water mixture can follow the gas temperature profile closely as shown in Figure 4.35b. By changing the working fluid to 45%–55% ammonia–water in the condenser system, condensation of the vapor is enabled at a lower pressure that allows more extraction energy in the vapor turbine. The DCSS accomplishes this.

2. The thermophysical properties of the ammonia–water mixture can be changed by changing the ammonia concentration. Thus, even at high ambient temperatures, the cooling system can be effective, unlike in a steam Rankine system where the condenser efficiency drops off as the cooling water temperature or ambient temperature increases.

3. The ammonia–water mixture has thermophysical properties that cause mixed fluid temperatures to change without a change in heat content. The temperature of water or ammonia does not change without a change in energy.

4. Water freezes at 32°F while pure ammonia freezes at −108°F. Ammonia–water solutions have very low freezing temperatures and hence at low ambient temperatures can generate more power without raising concerns about freezing.

5. The condensing pressure of the ammonia–water mixture is high, on the order of 1.5–2 bara compared to 0.1 bara in a steam Rankine system, resulting in lower specific volumes of the mixture at the steam turbine exhaust and consequently smaller turbine blades. The expansion ratio is about 10 times smaller. This reduces the cost of the turbine–condenser system. With steam systems, the condenser pressure is already at a low value on the order of 1 psia, and hence, further lowering would be expensive and uneconomical.

6. The losses associated with the cooling system are smaller due to the lower condensing duty; hence, the cooling system components can be smaller and the environmental impact lesser.

7. Standard carbon steel material can be used for the HRVG, piping and ammonia-handling equipment. Only copper and copper alloys are prohibited.

A 3 MW demonstration plant based on Kalina cycle was operating in California for more than 5 years two decades ago (Figure 4.36). In this plant, 14,270 kg/h of ammonia vapor enters the vapor turbine at 110 bara at 515°C and exhausts at 1.45 bara. The main working fluid for the HRVG is the 70–30 ammonia–water mixture while at the condenser it is 42–58 ammonia–water mixture. The leaner fluid has a lower vapor pressure, which allows for additional turbine expansion and greater work output. The ability to vary this concentration enables the plant performance to be varied and improved irrespective of the cooling water temperature.

Following the expansion in the turbine, the vapor is at too low a pressure to be completely condensed at the available coolant temperature. Increasing the pressure would increase the temperature and hence reduce the power output. Here is where the DCSS comes in. DCSS system consists of demister separator, recuperative heat exchangers, high- and low-pressure condensers, and control system. DCSS enables condensing to be achieved in two stages, first forming an intermediate leaner mixture leaner than 70% and condensing it, then pumping the intermediate mixture to higher pressure, reforming the working mixture, and condensing it as shown for admission into the HRVG. In the process of reforming the mixture (back to 70%), additional energy is recovered from the exhaust stream that

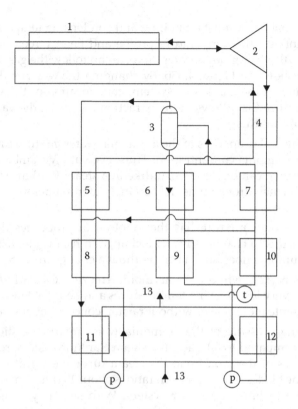

FIGURE 4.36
Canoga Park HRVG. 1, HRVG; 2, turbine; 3, flash tank; 4, final preheater; 5, HP preheater; 6, second recuperator; 7, vaporizer; 8, HP preheater; 9, first recuperator; 10, LP preheater; 11, HP condenser; 12, LP condenser; 13, cooling water; t, throttling device; p, pump.

increases the power output. Calculations show that the power output can be increased by 10–15% in the DCSS compared to the steam-based Rankine cycle.

The HRVG for the Kalina system is simply a once-through steam generator with an inlet for the 70% ammonia liquid mixture, which is converted to vapor at the superheater. The tube-side pressure drop in an HRVG is typically large. About 60 barg is the exit pressure, and about 10% of this is the loss in the superheater, evaporator, and economizer. The two-phase pressure drop in the evaporator particularly in the preheater will be high as several bare tube rows are required for recovering energy from the dusty gas compared to the clinker cooler gas stream with its finned tubes. Proper tube-side velocities have to be chosen to ensure that separation of vapor from liquid does not arise. Plain tubes are used if fouling is a concern as in the preheater flue gas boiler. Finned tubes with a low fin density of 2 to 3 fins/in. is used in the clinker cooler gas stream; the tube-side heat transfer coefficients are lower than steam water exchangers, and hence, using more fin density than 2 to 3 fins/in. will be counterproductive as shown in Appendix E.

Figure 4.37 compares the energy recovery potential of Kalina cycle with that of a steam system for a clinker cooler air stream. It is seen that due to the constant boiling temperature of steam and with 8°C pinch point and 7°C approach, the exit gas temperature is 236°C and energy recovered is only 11 MW. The Kalina cycle on the other hand with its 70–30 ammonia–water mixture with variable boiling temperature follows the gas temperature profile closely as a result of which the exit gas temperature could be dropped to 137° and the energy recovered is 23.5 MW. It is true that the tube-side coefficients with ammonia

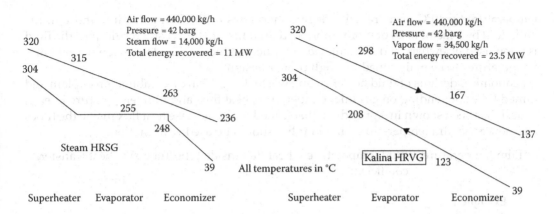

FIGURE 4.37
Comparison of energy recovery from steam and ammonia–water mixture.

water mixture will be lower than that of steam–water and as a result of which the HRVG will have more rows deep as large fin density cannot be used for the superheater, evaporator, and economizer in the clinker cooler heat recovery system; also due to the presence of dust, large fin density is not recommended. Considering the potential for large heat recovery, the Kalina cycle is attractive when gas inlet temperature gets smaller. For the steam system to recover more energy, one should think of multiple-pressure system and combining of flows in the economizer as shown earlier in another system.

Fluid Heaters and Film Temperature

Industrial heat transfer fluids are often heated by exhaust gases from incinerators, turbine exhaust, or waste flue gas from various sources. It is basically a coil or tube bundle like an economizer consisting of tubes in parallel- or counter-flow direction with plain tubes or finned tubes if the duty is large. The fin geometry is selected with great care as one should be concerned about the heat flux inside the tubes, which can increase the film temperature as well as the tube wall temperature! Hence, one of the important considerations in the design of these hot oil or fluid heaters is the film temperature limitation. Many fluids such as glycol–water solutions, hydrocarbon oils, silicone oils, molten salts, and liquid metals have limitations on both bulk fluid temperature and film temperature. If the film temperature is exceeded by using improper fin geometry, then thermal degradation of the fluid can set it. Two fluid heaters may operate at the same bulk temperature or have the same duty, but one can have a much higher film temperature due to the use of poor fin geometry or high gas velocity, which results in high heat flux inside the tubes. Appendix E discusses the effect of fin geometry on superheater tube wall temperatures.

The fluid film temperature is not usually measured directly, and it must be estimated by the designer of the exchanger. Two different heaters may operate at the same heat duty, fluid flow rate, and bulk fluid outlet temperature, and yet they may have significantly different film temperature profiles. The film temperature is an important value as fuel degradation begins in the fluid film, as this is where the fluid is hottest. For typical thermal oils, the fluid decomposition rate doubles for roughly every 10°C increase in fluid temperature. It is not unusual for the maximum fluid film temperature to exceed the heater outlet temperature by 40°C or more, with degradation rates in the film exceeding those in the bulk fluid by a factor of 20 or more. Shortened fluid life can be a costly result since the fluid is often used elsewhere and recirculated. Suppliers of thermic fluids provide the safe film temperatures to be maintained through proper design.

Economical designs of fluid heaters using waste flue gas have a combination of plain and finned tubes depending on gas inlet temperature, heat flux, and film temperature in each zone. It has been shown in Appendix E that finned tubes increase heat flux inside the tubes and hence the film temperature. Finned tubes should be used with caution.

Film temperature = fluid temperature + heat flux inside tube/tube-side heat transfer coefficient.

Example 4.6

A thermal fluid of 450,000 kg/h has to be heated from 313°C to 340°C. Duty is 9040 kW. The heat source is 45,000 kg/h of clean incinerator exhaust at 1000°C. Flue gas analysis is % volume $CO_2 = 4.5$, $H_2O = 10$, $N_2 = 75$, $O_2 = 10.5$. Flue gas is clean. Fouling factors used = 0.00017 m^2 K/W on gas and fluid sides. The following design was provided by a vendor. Check the film temperature.

The information from the vendor is shown in Figure 4.38. To check the results, one must first compute the gas and fluid properties. The following fluid properties were

Coil performance							
Surface	Coil3	Coil2	Coil1	Geometry	Coil3	Coil2	Coil1
Gas temp in ±10°F	1,832	1,409	958	Tube OD, in.	3.500	3.500	3.500
Gas temp out ±10°F	1,409	958	754	Tube ID, in.	3.046	3.046	3.046
Gas sp ht, Btu/lb°F	0.3022	0.2887	0.2780	Fins/in	0.000	1.500	3.000
Duty, MM Btu/h	12.55	12.79	5.57	Fin height, in.	0.000	0.500	0.750
U, Btu/ft²h°F	11.01	9.83	5.97	Fin thk, in.	0.000	0.050	0.050
Surface area, ft²	1173	2,406	3,829	Fin width, in.	0.000	0.000	0.000
LMTD, °F	972	541	244•	Fin conductivity	24	24	24
Gas pr drop-in wc	0.23	0.22	0.24	Tubes/row	16	16	16
Max gas vel-ft/s	51	43	35	No deep	8	6	4
Fluid temp in, F	624.8	604	595	Length, ft	10.000	10.000	10.000
Fluid temp out±10°F	645	625	604	Tr pitch, in.	6.000	6.000	6.000
FLuid pr drop, psi	3.6	2.6	1.7	Long pitch	6.000	6.000	6.000
Fluid velocity, ft/s	6.7	6.6	6.5	Streams	16	16	16
Fluid ht tr coefft	394	382	374	Parl = 0, countr = 1	1	1	1
Max film temp, °F	684	690	648	Arrangement	ST	ST	ST
Max heat flux	15,275	24,864	16,268	Fluid flow, lb/h	992000		
Max tube temp, °F	721	751	684	% volume-fluegas $H_2O = 10$, $N_2 = 75$, $O_2 = 10.5$			
Max fin tip temp	721	891	772				
Gas foul factor	0.001	0.001	0.001				
Gas flow, lb/h	99.200						

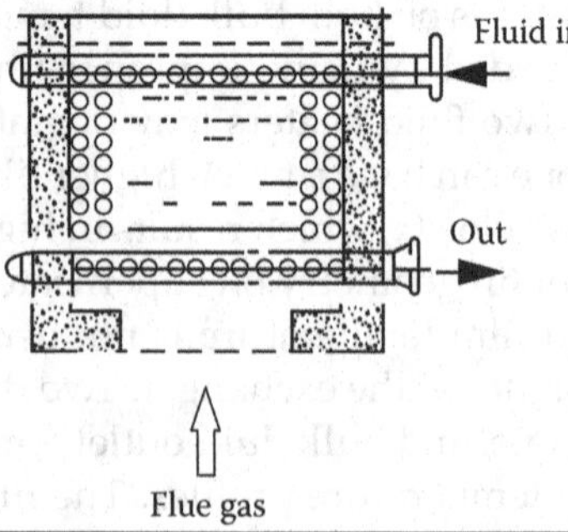

FIGURE 4.38
Thermal performance information for fluid heater.

TABLE 4.21

Properties of Thermic Fluid

	Thermic Fluid		Flue Gas	
Temperature	313	340	1000	400
Specific heat, J/kg K	2571	2665	1291	1150
Viscosity, kg/m h	1.01	0.8296	0.1714	0.1127
Conductivity, W/m K	0.092	0.089	0.0806	0.0482
Density, kg/m³	820	800	0.269	0.509

obtained from the fluid manufacturer, and the flue gas properties were estimated using method discussed in Appendix F and are shown in Table 4.21.

The coil is divided into three sections. The hot front end has plain tubes, while the next section has tubes with 1.5 fins/in. and the last section has 3 fins/in.

One may use the methods discussed in Appendices C and E to evaluate the gas-side heat transfer coefficients. The tube-side coefficient at the hot end exit as shown in Appendix B is estimated using $h_i = 0.0278 \, w^{0.8} \, C/d_i^{1.8}$

Flow per tube (with 16 streams) w = 455,000/16/3,600 = 7.9 kg/s

$$C = \left(\frac{C_p}{\mu}\right)^{0.4} k^{0.6} = \left(\frac{2665 \times 3600}{0.8296}\right)^{0.4} \times 0.089^{0.6} = 156.6 \text{ at } 340°C \text{ and } 145.6 \text{ at } 313°C$$

$h_i = .0278 \times 7.90^{0.8} \times 151/.0774^{1.8} = 2194 \text{ W/m}^2 \text{ K } (385 \text{ Btu/ft}^2 \text{ h } °F)$. This is an average value for the entire coil, but Figure 4.39 shows the breakup for each of the three sections. Let us use these values.

The gas-side heat transfer coefficient at the highest gas temperature may be estimated by the methods described in Appendices C and D. Heat flux inside the tubes for the bare tube section = 11.01 × (1,832 − 645) × 3.5/3.046 = 15,016 Btu/ft²h. Hence, film temperature = 15,016/394 + 645 = 683°F. Similarly, one may compute the film temperatures in each section. One may also check the gas-side heat transfer coefficients using Appendices C and E for plain and finned tubes and Appendix D for nonluminous heat transfer coefficient.

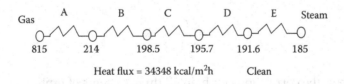

Heat flux = 34348 kcal/m²h   Clean

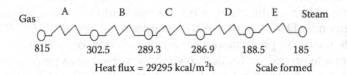

Heat flux = 29295 kcal/m²h   Scale formed

FIGURE 4.39

Temperature profile in clean and fouled conditions. *Note:* (A) Gas film, (B) gas-fouled layer, (C) tube wall, (D) steam-fouled layer, (E) steam wall.

The maximum film temperature must be checked against the allowable film temperature provided by the supplier of the oil or fluid. A margin of 10°C–15°C below the allowable value may be prudent.

Design of Fire Tube Boilers

Fire tube boilers are widely used in heat recovery applications as discussed earlier in chemical plants, refineries, and incineration plants. The procedure for design and performance evaluation is illustrated as follows.

In order to size a fire tube boiler, the overall heat transfer coefficient U must be determined. The overall heat transfer coefficient U_o (based on tube OD) is given by the basic equation as discussed in Appendix A.

$$\frac{1}{U_o} = \frac{d}{h_i d_i} + ff_i \left(\frac{d}{d_i} \right) + ff_o + \left(\frac{d}{2K_m} \right) \ln\left(\frac{d}{d_i} \right) + \frac{1}{h_o} \tag{4.2}$$

The inside coefficient h_i consists of convective coefficient h_c and nonluminous coefficient h_n.

$$h_i = h_c + h_n \tag{4.3}$$

h_c is obtained as shown in Appendix B.

$$h_c = 0.0278 \, w^{0.8} \left(\frac{C_p}{\mu} \right)^{0.4} \frac{k^{0.6}}{d_i^{1.8}} = \frac{0.0278 \, C w^{0.8}}{d_i^{1.8}} \tag{4.4}$$

where

$$C = \left(\frac{C_p}{\mu} \right)^{0.4} k^{0.6} \tag{4.5}$$

Appendix B shows the correlations for h_c in all three systems of units.

The gas properties are evaluated at the average gas temperature. w is the mass flow of flue gas per tube in kg/s. Estimation of h_n, the nonluminous heat transfer coefficient, is discussed in Appendix D. As discussed in Appendix A, a U value of 90%–95% of h_i may be used to evaluate the boiler performance for estimation purposes. However, here we shall discuss the complete procedure for design so those versed with programming skills can develop their own computer code. Once h_i is evaluated, the boiling heat transfer coefficient h_o has to be determined to obtain U_o.

Given the gas flow, inlet temperature, desired exit gas temperature, flue gas analysis, and steam parameters, one may assume a tube count and obtain the tube-side heat transfer coefficients as discussed in Appendices B and D. The outside boiling coefficient is obtained as shown later, though for practical purposes, its effect of U may be ignored.

Boiling Heat Transfer Coefficient h_o

Numerous correlations are available in the literature for the estimation of h_o. It must be understood that the variation among correlations for h_o can be wide, but this will not affect U as discussed in Appendix A, as h_i is much smaller than h_o and hence it is the limiting resistance to energy transfer.

The temperature difference between the tube wall and saturation temperatures is given by the well-known Roshenow correlation:

$$\frac{C_{pf}(t_w - t_s)}{\Delta H_{fg}} = 0.013 \left[\frac{q_o \{g_o \sigma / g(\rho_f - \rho_g)\}^{0.5}}{\mu_f \Delta H_{fg}} \right]^{0.33} \left(\frac{\mu_f C_{pf}}{k_f} \right) \tag{4.6}$$

$$\text{Heat flux } q_o = h_o(t_w - t_s) \tag{4.7}$$

where
C_{pf} is the specific heat of saturated liquid, kcal/kg °C
ρ_f, ρ_g is the density of saturated liquid, vapor kg/m³
σ is the surface tension, kg/m
μ_f is the viscosity of saturated liquid, kg/m h
k_f is the thermal conductivity of saturated liquid, kcal/m h °C
ΔH_{fg} is the latent heat, kcal/kg
t_w, t_s are temperatures of tube wall and steam,°C
q_o is the heat flux, kcal/m² h

Surface tension, thermal and transport properties, and properties of steam are given in Tables F.22 and F.13.

An iterative procedure is required to determine h_o accurately as h_o requires q_o for its estimation, and to estimate U_o (from which q_o is calculated), we need to know h_o! Also the equation may appear tedious as steam properties are required. However, an accurate estimate of h_o is not that important as U is governed by h_i and not by h_o.

Let us look at some other simplified correlations for h_o. For steam pressure P_s in the range of 1–30 atm (a) and heat flux q_o from 50,000 kcal/m² h (0.058 MW/m²) (18,370 Btu/ft² h) to 1000,000 kcal/m² h (1.16 MW/m²) (36,740 Btu/ft² h),

$$h_o = 3 q_o^{0.7} Ps^{0.17 \log Ps} \tag{4.8}$$

For P_s above 30 atma (atmospheres absolute) and q_o from 100,000 to 1×10^6 kcal/m² h,

$$h_o = 4.5 q_o^{0.7} e^{0.01 Ps} \tag{4.9}$$

For pressures in the range of 1–50 atma and q_o from 16,000 to 730,000 kcal/m² h,

$$h_o = 3 q_o^{0.7} P_s^{0.2} \tag{4.10}$$

Example 4.7

Determine the size of fire tube boiler required to cool 45,370 kg/h of flue gases from 815°C to 260°C. Flue gas analysis is % volume $CO_2 = 12$, $H_2O = 12$, $N_2 = 70$, $O_2 = 6$. Flue gas pressure is atmospheric. Steam pressure is 10.5 kg/cm² g. Tubes used: 600 number of 50.8 × 45 mm carbon steel SA 178a. (Flow per tube is based on experience. For this tube ID, about 75 kg/h is reasonable. Based on actual boiler size and gas pressure drop, one may revise this value. A computer program will be helpful to study the options.) Fouling factor on gas side is 0.0004 m²h °C/kcal (0.002 ft²h °F/Btu) and that on steam side is 0.0002 m²h °C/kcal (0.001ft²h °F/Btu). Tube thermal conductivity = 37 kcal/m h °C (24.8 Btu/ft h °F). Use a heat loss of 1%, a blowdown of 3%, and a feed water temperature of 104°C. Saturation temperature is 185°C at the operating pressure of steam.

Solution

First determine the tube-side convective heat transfer coefficient h_c. Gas properties at average gas temperature of 538°C may be estimated using the procedure discussed in Appendix F based on the flue gas analysis. The results are as follows:

$C_p = 0.2876$ kcal/kg °C = 1204 J/kg K, $\mu = 0.1258$ kg/m h = 3.484×10^{-5} kg/m s, k = 0.0476 kcal/m h °C = 0.0552 W/m K. From Appendix B,

$$h_c = \frac{0.0278 \, C \, W^{0.8}}{d_i^{1.8}}$$

w = 45,370/3,600/600 = 0.021 kg/s, d_i = 0.045 m. C = $(1{,}204 \times 10^5/3.484)^{0.4} \times 0.0552^{0.6}$ = 182. $h_c = 0.0278 \times 182 \times 0.021^{0.8}/0.045^{1.8}$ = 61.1 W/m² K (52.55 kcal/m² °C) (10.75 Btu/ft²h °F).

In addition to convective heat transfer coefficient, the nonluminous heat transfer coefficient has to be estimated. However, it will be small as the beam length is the tube inner diameter. The nonluminous heat transfer coefficient is worked out in Example 4.5, Appendix D. $h_n = 3.93$ W/m² K. Hence, $h_i = h_c + h_n = 65$ W/m² K = 55.9 kcal/m²h (11.4 Btu/ft²h °F).

Let us use $U_i = 0.95 \times 55.9 = 53.1$ kcal/m²h °C or $U_o = 53.1 \times 45/50.8 = 47$ kcal/m²h °C in order to estimate the average heat flux and then h_o.

$$q_o = 47 \times (538 - 185) = 16{,}591 \text{ kcal/m}^2 \text{ h.}$$

Using (4.10), $h_o = 3 \times 16{,}121^{0.7} \times 11.1^{0.2} = 4279$ kcal/m²h °C.

Let us estimate q_o using (4.6). Obtain steam properties from Appendix F.

$\sigma = 0.002813$ lb/ft = 0.004186 kg/m from Table F.22. From steam tables and Table F.13, $\mu_f = 0.365$ lb/ft h = 0.543 kg/m h. $\Delta H_{fg} = 858$ Btu/lb = 476.6 kcal/kg. ρ_g, ρ_f = 5.78 and 882 kg/m³. $C_{pf} = 1.05$ kcal/kg °C, $k_f = 0.389$ Btu/ft h °F = 0.5788 kcal/kg °C.
Using (4.6),

$$1.05(t_w - t_s)/476.6 = 0.013[(16{,}591/0.543 \times 476.6) \times (0.004186/876.2)^{0.5}]^{0.33}$$
$$\times (0.543 \times 1.05 \,/0.5788) = 0.00663 \text{ or } (t_w - t_s) = 3.0°C$$
$$\text{or } h_o = 16{,}591/3.0 = 5530 \text{ kcal/m}^2\text{h °C}$$

$$1/U_o = (50.8/45)/55.9 + 0.0004 \times 50.8/45 + (0.0508/2/37) \times \ln(50.8/45) + 0.0002 + 1/5530$$
$$= 0.02 + 0.000452 + 0.000083 + 0.0002 + 0.0001808 = 0.0209 \text{ or } U_o = 47.8 \text{ kcal/m}^2\text{h °C}$$

$q_o = 47.8 \times (538 - 185) = 16{,}882$ kcal/m² h, close to that obtained earlier. The various resistances are in m²h °C/kcal:

Gas-side film: 0.02
Gas-side fouling: 0.000452
Tube wall resistance: 0.000083

Steam-side fouling: 0.0002
Steam-side heat transfer = 0.0001808 (resistances in $m^2 h\ °C/kcal$)

Duty = 45,370 × 0.99 × 0.2876 × (815 − 260) = 7.17 MM kcal/h = 8337.2 kW. From steam tables, saturated steam enthalpy = 664.4 kcal/kg, saturated liquid enthalpy = 187.8 kcal/kg, and feed water enthalpy = 104.3 kcal/kg. Steam generation = 7.17 × 10^6/[(664.4 − 104.3) + 0.03 × (187.8 − 104.3)] = 12,744 kg/h

$$LMTD = \frac{\left[(815-185)-(260-185)\right]}{\ln\left[(815-185)/(260-185)\right]} = 261°C.$$

Surface area required (based on tube OD) = A_o = 7,170,000/261/47.8 = 575 m^2 (A_i = 575 × 45/50.8 = 509 m^2). With 600 tubes, the length required = 575/(3.14 × 0.0508 × 600) = 6.0 m. One may add some margin if required.

Let us now compute the gas-side pressure drop inside the boiler tubes. This consists of three parts: the inlet loss at the tube entrance, the exit loss at tube exit, and the friction loss inside the tubes.

Gas Pressure Drop inside Tubes

To estimate the pressure drop, density of flue gas is required, which in turn requires the molecular weight (MW).

MW = 0.12 × 44 +0.12 × 18 +0.70 × 28 +0.06 × 32 = 28.96. Density at gas inlet at atmospheric pressure = ρ_{g1} = (28.96/22.4) × 273 /(273 + 815) = 0.324 kg/m^3. At exit, density ρ_{g2} = 0.324 × (273 + 815)/(273 + 260) = 0.662 kg/m^3.

Gas velocity at inlet = 4 × (45,270/600/3,600)/0.324/(π × 0.045 × 0.045) = 40.7 m/s; at exit, gas velocity = 19.87, say, 20 m/s (taking ratio of gas densities).

Inlet loss = 0.5 × velocity head = 0.5 × ρ_{g1} × V^2/2/9.8 = 0.5 × 0.324 × 40.6^2/2/9.8 = 13.6 kg/m^2 = 13.6 mm wc (9.8 m/s^2 is acceleration due to gravity). Exit loss = 1 × VH = 0.661 × 20^2/2/9.8 = 13.5 mm wc. Friction loss in tubes: ΔP_g = 810 × 10^{-6} f L v w^2/d_i^5 (from Table B.6). Flow w is in kg/s, L and d_i in m, v the specific volume in m^3/kg, and ΔP_g in kPa. Gas density at average gas temperature of 537.5°C = (28.96/22.4) × 273/(273 + 537.5) = 0.435 kg/m^3 or v = 2.296 m^3/kg.

$$\Delta P_g = \frac{810 \times 10^{-6} \times 0.02 \times 6.1 \times 2.296 \times \left(45370/600/3600\right)^2}{0.045^5} = 0.54\ kPa = 55\ mm\ wc$$

(0.02 if the friction factor [see Appendix B]. 6.1 m is the approximate tube length in the earlier equation including tube sheet thickness.)

Total gas pressure drop = 13.6 + 55 + 13.3 = 82 mm wc. One has to add the inlet and exit vestibule losses and any duct losses. The pressure drop value 82 mm wc is only for the losses inside the tubes. If ferrules are used, the gas inlet velocity will be much higher, and the inlet loss will be much higher depending on the ferrule inner diameter.

Evaluating Tube Wall Temperatures

The maximum heat flux outside the tubes q_o is required to determine the maximum tube wall temperatures. q_o is also important as it gives an idea if DNB is a concern. (In fire tube boilers, heat flux outside the tubes is important as steam is outside the tubes, while for water tube boiler, q_i is important as steam is inside the tubes). DNB is more likely in reformed gas boilers in hydrogen plants where the tube-side heat transfer coefficients can be much higher than the heat transfer coefficients with typical flue gases, about six to seven times. This we shall see later.

The maximum heat flux is generally used to determine the maximum tube wall temperature. U_o has to be computed at the gas inlet temperature. From Appendix F, at 815°C, C_p = 0.3050 kcal/kg °C = 1277 J/kg K, μ = 0.1534 kg/m h = 4.261 × 10^{-5} kg/m s

$$k = 0.0613 \text{ kcal/mh°C} = 0.07128 \text{ W/m K}$$

$$h_c = .0278 \times 0.021^{0.8} \times (1277 \times 10^5/4.261)^{0.4} \times 0.07128^{0.6}/0.045^{1.8} = 67.3 \text{ W/m}^2 \text{ K}$$
$$= 57.8 \text{ kcal/m}^2\text{h °C}$$

h_n using charts and equations from Appendix D = 7.7 W/m² K; hence, h_i = 75 W/m² K = 64.5 kcal/m²h °C. Let us assume U_i = 0.95 × 64.5 = 61.3 kcal/m²h °C or U_o = 54.27 kcal/m²h °C. (U_i × ratio of tube inner to outer diameter = U_o).

$$q_o = 54.27 \times (815 - 185) = 34190 \text{ kcal} / \text{m}^2\text{h}$$

σ = 0.002813 lb/ft = 0.004186 kg/m from Appendix F. From steam tables in Appendix F, μ_f = 0.365 lb/ft h = 0.543 kg/m h. ΔH_{fg} = 858 Btu/lb = 476.6 kcal/kg
ρ_g, ρ_f = 5.78 and 882 kg/m³. C_{pf} = 1.05 kcal/kg °C
k_f = 0.389 Btu/ft h °F = 0.5788 kcal/kg °C. Using (4.6),

$1.05(t_w - t_s)/476.6 = 0.013[(34,190/0.543 \times 476.6) \times (0.004186/876.2)^{0.5}]^{0.33}$
$\qquad \times (0.543 \times 1.05 /0.5788) = 0.008416$ or $(t_w - t_s) = 3.8$°C or

h_o = 34,190/3.8 = 9,000 kcal/m²h °C

$1/U_o$ = (50.8/45)/64.5 + 0.0004 × 50.8/45 +(0.0508/2/37) × ln(50.8/45) + 0.0002 + 1/8997
$\qquad$ = 0.0175 + 0.000452 + 0.000083 + 0.0002 + 0.0001111 = 0.01834 or
U_o = 54.5 kcal/m²h °C

This is close to U_o estimated earlier. Hence, let us proceed with this.

$$q_o = 54.5 \times (815 - 185) = 34,348 \text{ kcal/m}^2\text{h.}$$

Temperature drop across gas film = 34,348 × 0.0175 = 601°C
Temperature drop across gas fouling layer = 34,348 × 0.000452 = 15.5°C
Temperature drop across tube wall = 34,348 × 0.000083 = 2.8°C
Temperature drop across steam-side fouling layer = 34,348 × 0.0002 = 6.9°C
Temperature drop across steam film = 34,348 × 0.0001111 = 3.8°C

Hence, tube inner wall temperature = 815 − 601 − 15.5 = 198.5°C and outer wall temperature = 195.7°C. For design purposes, one may use, say, 10–15°C margin.

Effect of Scale Formation

Due to low pressure of steam in fire tube waste heat boilers, water treatment processes are not elaborate as in water tube boilers and hence scale formation is often likely. Nonsoluble salts such as calcium or magnesium salts or silica present in feed water can deposit as a thin layer on tube surfaces during evaporation thereby resulting in higher tube wall temperatures. Table 4.22 lists the thermal conductivities of a few scale substances.

TABLE 4.22

Thermal Conductivity of Scale

Material	k, W/m K (Btu in./ft² h °F)
Analcite	1.27 (8.8)
Calcium phosphate	3.605 (25)
Calcium sulfate	2.307 (16)
Magnesium phosphate	2.163 (15)
Magnetic iron oxide	2.884 (20)
Silicate scale	0.0865 (0.6)
Boiler steel	44.7 (310)
Firebrick	1.0 (7)
Insulating brick	0.1 (0.7)

ff_o is the thickness of scale/thermal conductivity. If 0.25 mm thick of silicate scale forms outside the boiler tubes, then the outside fouling factor ff_o = 0.00025/0.0865 = 0.00289 m² K/W = 0.00336 m² h °C/kcal. This steam side fouling is used in the example that follows.

Example 4.8

Evaluate the boiler performance in fouled condition when 0.25 mm silicate forms outside the tubes on steam side.

Solution

Assume in the previous example that gas and steam inlet parameters have not changed except for the scale formation in the boiler, then the thermal resistances will be as follows (the slight change in h_o due to change in heat flux may be neglected for now).

$$1/U_o = (50.8/45)/64.5 + 0.0004 \times 50.8/45 + (0.0508/2/37) \times \ln(50.8/45) + 0.00336$$
$$+ 1/8997 = 0.0175 + 0.000452 + 0.000083 + 0.00336 + 0.0001111 = 0.0215 \text{ or}$$
$$U_o = 46.5 \text{ kcal/m}^2 \text{ h °C.}$$

Revised maximum heat flux = $46.5 \times (815 - 185) = 29,295$ kcal/m² h.

$$1.05(t_w - t_s)/476.6 = 0.013[(29,295/0.543 \times 476.6) \times (0.004186/876.2)^{0.5}]^{0.33}$$
$$\times (0.543 \times 1.05 /0.5788) = 0.0080457 \text{ or } (t_w - t_s) = 3.64°C \text{ or}$$

$$h_o = 29,295/3.64 = 8048 \text{ kcal/m}^2 \text{ h °C.}$$

$$1/U_o = (50.8/45)/64.5 + 0.0004 \times 50.8/45 + (0.0508/2/37) \times \ln(50.8/45) + 0.00336$$
$$+ 1/8048$$

$$= 0.0175 + 0.000452 + 0.000083 + 0.00336 + 0.0001242 = 0.0215 \text{ or}$$
$$U_o = 46.47 \text{ kcal/m}^2 \text{ h °C.}$$

Hence, not much different from the value estimated earlier. Summary of resistances:

Gas-side film: 0.0175
Gas-side fouling: 0.000452
Tube wall resistance: 0.000083
Steam-side fouling: 0.00336
Steam-side heat transfer = 0.0001242
Temperature drop across steam film = $29,295 \times 0.0001242 = 3.6°C$
Temperature drop across steam-side fouling layer = $29,295 \times 0.00336 = 98.4°C$

Temperature drop across tube wall = 29,295 × 0.000083 = 2.4°C
Temperature drop across gas fouling layer = 29,295 × 0.000452 = 13.2°C
Temperature drop across gas film =29,295 × 0.0175 = 512.6°C
Tube outer wall temperature =185 + 98.4 + 3.5 = 287°C
Tube inner wall temperature = 815 − 512.6 − 13.2 = 289.2°C

Working from gas side, we have inner tube wall temperature as follows:

$$815 - 29{,}295 \times (0.0175 + 0.000452) = 289°C.$$

Hence, the presence of scale on steam side results in tube overheating, raises thermal stress concerns, and also leads to tube failures. Figure 4.39 shows the temperatures at the gas inlet end in clean and fouled cases. The tube wall runs about 100°C higher than normal.

Estimation of Loss in Performance Due to Fouling

Here is a quick way to estimate the loss in duty due to fouling. Assume that the entire boiler is fouled with a fouling factor of 0.00336 m² h °C/kcal. Plant engineers may use this method to check the fouling factor based on exit gas temperature measurements or the exit gas temperature from a given fouling factor:

From Appendix A, for the clean design case, $\ln[(815 - 185)/(260 - 185)] = UA/W_gC_{pg} = 2.128 = K_1$.

U_o= 47.8 kcal/m² h °C in clean conditions. In the fouled case, U_o = 1/(0.0209 + 0.00336 − 0.0002) = 41.56 kcal/m² h (we adjusted the U_o using the revised fouling factor).

K_2 = 2.128 × 41.56/47.8 = 1.85 (as gas flow, inlet temperature, and surface area are the same and specific heat variation is neglected). Hence, ln [(815 − 1 we can take 85)/(T_2 − 185)] = 1.85 or T_2 = (815 − 185)/6.36 = 185 + 99 = 284°C.

Hence, exit gas temperature in fouled case is 284°C versus 260°C, and the fraction of new duty versus the old duty is (815 − 284)/(815 − 185) = 0.96 or 4% loss in duty. One may compute the actual U values based on gas properties and fine-tune the results, but it appears that about 4%–5% loss in duty is envisaged due to fouling.

Tube Wall Thickness Calculations

Thickness calculation of tubes subject to external pressure is a little more involved. Per ASME Boiler and Pressure vessel code sec 1, the following procedure is used to obtain the thickness or the allowable pressure.

$$P_a = \frac{4B}{3(d/t)} \tag{4.11}$$

where P_a is the maximum allowable external pressure, psi

A, B are factors obtained from ASME code, sec 1, depending on values of d/t and L/d, where L, d, and t refer to length, diameter, and thickness (Figure 4.40).

When d/t <10, A and B are determined from tables or charts. For d/t < 4, A = 1.1/(d/t)²

Two values of allowable pressure are computed. P_{a1} = {(2.167/d/t) −0.0833}B and P_{a2} = 2S_b[(1 − (t/d)]/(d/t), where S_b is the lesser of 2 times maximum allowable stress at design tube temperature or 1.8 times the yield strength of the material at design tube temperature. The smaller of P_{a1} or P_{a2} is then used.

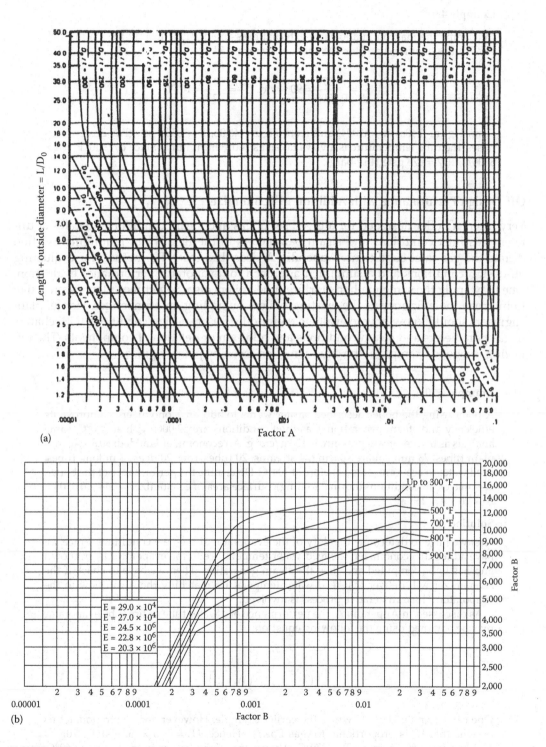

FIGURE 4.40
Tube thickness calculation—ASME. (a) Factor A. (b) Factor B.

Example 4.9

Determine the maximum allowable external pressure at 315°C (600°F) for 3 mm thick sa192 tubes of outer diameter 50.8 mm and length 4.57 m.

$$\frac{L}{d} = \frac{4.57}{0.0508} = 90 \quad \text{and} \quad \frac{d}{t} = \frac{50.8}{3} = 16.7$$

From Figure 4.40a, A = 0.004, and from Figure 4.40b, B = 8500 psi. (For design purposes, the tube temperature is always taken as 700°F minimum.) Since d/t >10, $P_a = 4 \times$ 8500/3/16.7 = 678 psig (46.7 barg).

Off-Design Performance with Addition of Economizer

Very often, a boiler designed for a given set of gas and steam conditions operates at different gas inlet conditions and steam parameters in the field. Plant requirements often dictate the gas inlet conditions. Appendix A describes the NTU method for evaluating off-design conditions and a simplified approach for evaporator performance evaluation. One may apply the same methods for fire tube boilers also. Sometimes plants would like to improve the efficiency of the existing boilers by adding an economizer. Hence, plant engineers should know how to evaluate their boiler for off-design conditions with change in gas parameters and steam conditions and addition or deletion of components. The following is a good example of the situation often faced by plant engineers.

Example 4.10

A plant using the boiler described earlier wants to add an economizer to improve its efficiency and steam generation. Gas inlet conditions are 52,000 kg/h at 850°C, same analysis as before. Steam pressure is 12 kg/cm² g. An economizer is added: 50.8 × 44 mm plain tubes, 76 mm square spacing, 4 streams, 20 tubes/row, 20 deep, 4 m long tubes. Feed water to economizer is as before at 104°C. Determine the performance of the boiler. Assume clean conditions (normal fouling). Surface area of economizer = 3.14 × 0.0508 × 20 × 20 × 4 = 255 m².

Solution

First we have to determine the duty of the evaporator at the new conditions. Since we have added an economizer, the feed water temperature to the evaporator is unknown and has to be computed and then the steam generation. Then we have to check if the economizer will deliver that feed water temperature matching the steam flow to the evaporator. Thus, an iterative process is involved.

As discussed in Appendix A, the performance of a water tube or fire tube evaporator may be obtained using the following equation:

$$\ln\left[\frac{(T_1 - T_s)}{(T_2 - T_s)}\right] = \frac{UA}{(W_g C_{pg})} \tag{4.12}$$

One can compute U as shown in the earlier example. However, for illustration, let us assume that U_o is proportional to (gas flow)$^{0.8}$. Hence, $U_o = 47.8 \times (52,000/45,370)^{0.8} =$ 53.3 kcal/m² h °C. Assume $T_2 = 270$°C. Average gas temperature in the boiler = 560°C. $C_p = 0.29$ kcal/kg °C from Appendix F. Saturation temperature at 12 kg/cm² g = 191°C.

Use 3% blowdown. Hence, $\ln[(850 - 191)/(T_2 - 191)] = 53.3 \times 575/(52,000 \times 0.99 \times 0.29) =$ 2.05 or exit gas temperature from evaporator $T_2 = 276°C$. Hence, evaporator duty = 52,000 $\times 0.99 \times 0.29 \times (850 - 276) = 8.57$ MM kcal/h = 9964 kW.

The revised steam generation is obtained using an iterative process.

Assume that feed water temperature leaving the economizer is 170°C. Enthalpy pickup in evaporator = $(665.4 - 171.8) + 0.03 \times (193.6 - 171.8) = 494.3$ kcal/kg. (Enthalpies of saturated steam, saturated water, and feed water entering the evaporator are, respectively, 665.4, 193.6, and 171.8 kcal/kg from steam tables. Hence, steam generation = 8,570,000/494.3 = 17,337 kg/h. Flow through economizer = $1.03 \times 17,337 = 17,857$ kg/h.

For the plain tube bundle, one may use the methods discussed in Appendix C to estimate the value of U. It may be shown that U = 56 kcal/m² h C. Surface area of the economizer = 255 m². Let us use the NTU method to determine the performance of the economizer as explained in Appendix A.

Let us assume that C_p in the economizer = 0.267 kcal/kg °C (flue gas).

$W_g C_{pg} = 52,000 \times 0.267 \times 0.99 = 13,745$ (0.99 refers to the 1% heat loss). $W_i C_{pi} = 17,857 \times 1.025 = 18,303$ (where 1.025 is the average specific heat of water in the economizer). C = 13,745/18,303 = 0.75. NTU = $56 \times 255/13,745 = 1.0389$

$$\epsilon = \frac{\{1 - \exp[-NTU(1 - C)]\}}{\{1 - C\exp[-NTU(1 - C)]\}} = 0.543$$

Q = $0.543 \times 13,745 \times (276 - 104) = 1.283$ MM kcal/h. Enthalpy pickup of water = $1.283 \times 10^6/17,857 = 71.8$ kcal/kg or exit water enthalpy = 104.3 + 71.8 = 176.1 kcal/kg or temperature of water at economizer exit = 174°C from steam tables. Close to that assumed. Hence, steam generation will be slightly more.

The exit gas temperature from the economizer = $276 - 1.283 \times 10^6/(52,000 \times .99 \times 0.267) = 182°C$. One may check the gas pressure drop in the evaporator as discussed earlier and in the economizer as discussed in Appendix C. Thus, plant engineers may obtain the performance of their boiler with the addition of economizer at any inlet gas condition. One can also develop a computer code or an Excel program using the logic discussed here. As mentioned earlier, due to variations in steam and gas properties with temperature, a few iterations are required. As an exercise, plant engineers may add superheater upstream of the evaporator and see how to evaluate this option!

The computer output from a vendor is shown in Table 4.23. Slight variations may be seen due to approximations made.

Simulation of Fire Tube Boiler Performance

One may simulate the fire tube boiler performance at any gas inlet condition using the concept that U is proportional to $W_g^{0.8}$. Effect of gas properties are neglected in the first iteration. The effect of nonluminous heat transfer coefficient is also presumed to be small. The advantage of this method is that U need not be evaluated in detail.

Using the equation $\ln[(T_1 - T_s)/(T_2 - T_s)] = UA/(W_g C_{pg})$ and substituting $KW_g^{0.8}$ for U and simplifying, we have $\ln[(T_1 - T_s)/(T_2 - T_s)] = K/(W_g^{0.2})$.

One can estimate the constant K for the boiler using known conditions and then predict the exit gas temperature and duty at any other condition of gas flow or inlet gas temperature. The effect of flue gas properties is neglected, and hence, this is an approximate method, but for quick estimates, it is reasonable.

TABLE 4.23

Performance of Fire Tube Boiler with Economizer

Gas Data		
Gas flow	52,000	kg/h
Gas inlet	850	°C
Gas press.	0	kg/cm²g
Fouling ftr.	0.0002	m² h °C/kcal
CO_2	12	%
H_2O	12	%
N_2	70	%
O_2	6	%
Heat loss	1	%
Steam data		
Fw. temp.	104	°C
Stm press.	12	kg/cm²g
Foul ftr.	0.0002	m² h °C/kcal
Blowdown	3	%
Mechanical Data—Boiler		
Sort	1	
Pass	1	
Tube OD	50.8	mm
Tube ID	45	mm
Number	600	
Length	6	m
Ferrule ID	45	mm
Correc. factor	1	

Economizer Mech. Data					
Tube OD	50.80	mm	Number of rows deep	20	
Tube ID	44.00	mm	Eff. length	4.00	m
Fin density	0.00	fins/m	Tr pitch	76.00	mm
Fin height	0	mm	Long pitch	76.00	mm
Fin thickness	0	mm	Streams	5	
Serration	0.000	mm	Arrangement	inline	
Fin ther. con.	0	kcal/m h °C	Configuration	counter flow	
Tubes/row	20				

Evaporator Performance		
Pass	1	2
Gas inlet	850	°C
Gas outlet	273	±5°C
Gas sp ht.	0.2883	kcal/kg °C
Duty	8.52	MM kcal/h
Gas press. drop	114.00	mm wc
Max. ht. flux	39,589	kcal/m² h
Max. wall temp.	208	°C
overall U	61.19	kcal/m² h °C
Surf. area	509	m²
Delt.	278	°C
Max. gas vel.	48	m/s
Gas vel.-ferrule	45	m/s

(*Continued*)

TABLE 4.23 (*Continued*)

Performance of Fire Tube Boiler with Economizer

Economizer Performance

Gas temp. in	274	±5°C	Gas press. drop	24.00	mm wc
Gas temp. out	179	±5°C	Water press. drop	0.53	kg/cm²
Duty	1.28	MM kcal/h	Min. Wal. Temp.	107	°C
Wat. temp. in	104	°C	Overall U	58.92	kcal/m² h °C
Wat. temp. out	176	±5°C	Tube-side htc	4557	kcal/m² h °C
Surf. area	255	m²	Arrangement	inline	
Max. gas vel.	11	m/s	Flow direction	Counter-flow	
Fw. temp. in	104	°C			
Stm. Temp.	191	°C			
Steam flow	17,472	kg/h			
Gas flow	52,000	kg/h			
Tot gas press. drop	137.00	mm wc			

Example 4.11

Let us solve the earlier example for evaporator performance using the simulation approach.

First, the constant K is evaluated using known conditions. W_g= 45,370 kg/h, T_1 = 815°C, T_2 = 260°C, t_s = 185°C

$$K = \ln\left[\frac{(815 - 185)}{(260 - 185)}\right] = 2.13 = K/W_g^{0.2} = K/45370^{0.2}. \quad \text{Hence, } K = 18.17$$

New conditions: W_g = 52,000 kg/h. T_1 = 850°C, t_s = 191°C.

ln [(850 − 191)/(T_2 − 191)] = 18.17/52,000$^{0.2}$ = 2.07 or (850 − 191)/7.93 = (T_2 − 191) or T_2 = 274°C. Very close to that obtained in the previous example. Thus, this is yet another useful method for plant engineers for obtaining quick estimates of fire tube boiler performance.

Effect of Gas Analysis on Performance

With common flue gases in general, at atmospheric pressure, U_o will be in the range of 50–70 W/m² K. However, the U_o value will be much higher (about five to seven times) and so also the heat flux with reformed gas (Table 4.1). Table 4.24 compares the performance of a typical flue gas and reformed gas boiler.

One can see from the earlier text that the heat transfer coefficient with reformed gas is much higher than that of the typical flue gas due to higher specific heat and thermal conductivity and lower viscosity. Also, due to high gas pressure, more mass flow per tube can be handled for similar gas pressure drop. Table 4.25 compares the heat transfer coefficients and heat fluxes. Hence, one should check the critical heat flux (CHF) to ensure that DNB condition is not reached with reformed gas boilers and that the circulation system is well designed.

TABLE 4.24

Comparison of Reformed Gas and Flue Gas Boiler Gas Properties

Gas Type	Reformed Gas		Flue Gas	
Gas press, atma	26	26	1	1
Temperature, °C	900	400	900	400
% volume CO_2	6	6	12	12
H_2O	37	37	12	12
N_2	0.5	0.5	73.5	73.5
O_2			2.5	2.5
H_2	43	43		
CO	8	8		
CH_4	5.5	5.5		
C_p, kcal/kg °C	0.710	0.624	0.31	0.279
μ, kg/m h	0.136	0.0865	0.160	0.110
k, kcal/m h °C	0.1755	0.1052	0.065	0.041
$C = (C_p/\mu)^{0.4}k^{0.6}$	0.682	0.57	0.253	0.213

TABLE 4.25

Comparison of Reformed Gas and Flue Gas Boiler Performance

Gas	Flue Gas	Reformed Gas
Gas flow, kg/h	50,000	50,000
Gas pressure, kg/cm²g		
Gas temperature in, °C	900	900
Gas temperature out, °C	384	386
Gas specific heat	0.2948	0.6657
Duty, MW	8.67	19.57
Gas pressure drop, mm wc	85	109
Steam flow, kg/h	14,400	32,500
Steam pressure, kg/cm² g	40	40
Tubes OD × ID	50.8 × 44	50.8 × 44
Tube count	700	200
Length, m	4.2	6
Surface area, m²(ID)	406	166
U_o, W/m²K	57	313
Max heat flux, q_o W/m²	41,970	229,500

Note: Steam pressure = 40 kg/cm² g, feed water 104°C. Allowable heat flux in reformed gas boilers is in the range of 300–350 kW/m².

Critical Heat Flux q_c

With flue gas boilers in general, the heat transfer coefficients as well the heat flux outside the tubes are low, typically less than 30,000 kcal/m² h (35 kW/m²), and hence, DNB is not an issue. In the case of reformed gas, the heat transfer coefficient and the heat flux will be much higher, about five to eight times more, and hence, the allowable heat flux has to be checked.

There are a few widely used correlations for CHF in pool boiling. These are mainly for clean tubes and clean water. In practice, oxides and deposits on tubes will limit the allowable heat flux to about 15%–20% of that calculated. Many boiler suppliers use their own margins and correction factors for CHF based on experience.

Kutateladze recommends the following correlation for q_c:

$$\frac{q_c}{(\rho_g H_{fg})} = 0.16 \left[\frac{\sigma(\rho_f - \rho_g)g}{\rho_g^2} \right]^{0.25} \tag{4.13}$$

Zuber's correlation is also widely used:

$$\frac{q_c}{(\rho_g H_{fg})} = 0.13 \left[\frac{\sigma(\rho_f - \rho_g)g}{\rho_g^2} \right]^{0.25} \left[\frac{\rho_f}{(\rho_f + \rho_g)} \right]^{0.5} \tag{4.14}$$

(H_{fg} is the latent heat, kcal/kg, q_c in kcal/m^2 s, g = 9.81 m/s^2, ρ_f, ρ_g the density of saturated water and steam in kg/m^3).

Motsinki's correlation takes the following form:

$$q_c = 30937 \, P_c \left(\frac{P_s}{P_c} \right)^{0.35} \left(1 - \frac{P_s}{P_c} \right)^{0.9} \tag{4.15}$$

(here, q_c is in kcal/m^2 h).

Example 4.12

Determine the CHF for the boiler discussed in Example 4.7.

Solution

Using Zuber's correlation, at 10.5 kg/cm^2 g, σ = 0.04077 N/m, ρ_f = 881.8 kg/m^3, ρ_g = 5.78 kg/m^3, H_{fg} = 476.6 kcal/kg for steam from Appendix F.

Then, q_c = 0.13 × 5.78 × 476.6 × 3600[.0477 × 9.81 × (881.8 − 5.78)/5.78^2]$^{0.25}$ × (881.8/887.6)$^{0.5}$ = 2,405,000 kcal/m^2 h = 893,600 Btu/ft^2 h (the 3600 factor converts the kcal/m^2 s to kcal/ m^2 h).

Using Motsinki's correlation, q_c = 30,836 × 226(11.5/226)$^{0.35}$ × (1 − 11.5/226)$^{0.9}$ = 2,344,560 kcal/m^2 h = 871,000 Btu/ft^2 h. As mentioned earlier, these values are based on research and laboratory tests. In actual practice, a value of about 15%–20% of this is used based on field experience.

A correction factor is applied to CHF in pool boiling as discussed later in case of tube bundles.

Example 4.13

A fire tube boiler operates at 10.5 kg/cm^2 g with the following data.

Tube OD = 50.8 mm, 600 number, 6.1 m long. Shell inner diameter = 2133 mm. Surface area of tubes based on tube OD = 583 m^2.

Compute a factor ψ = DbL/A (Db = bundle diameter, m; L = length, m; A = surface area on OD basis, m^2) = 2.133 × 6.1/583 = 0.022. (ψ corrects for bundle effect.)

Correction factor F is obtained from the following equation:

$$\log F = 0.8452 + 0.994 \log y = 0.8452 - 0.994 \times 1.6575 = -0.802 \text{ or } F = 0.157$$

Hence, the corrected heat flux = $0.157 \times 2{,}405{,}000 = 377{,}585$ kcal/m² h (138,800 Btu/ft² h) (439 kW/m²). It may be noted that the actual heat flux was lower than 20,000 kcal/m² h, significantly lower than the allowable limit. In case of reformed gas boilers, however, the actual heat flux may be in the range of 200,000–300,000 kcal/m² h, and one may check the actual and allowable heat fluxes more carefully and ensure that the actual heat flux is much lower than the allowable CHF by a reasonable margin, say, 20%. One can lower the actual heat flux by using a lower gas velocity, which reduces U_o and hence q_o. One may also use a larger-diameter tube that lowers U_o and hence the heat flux. However, this may add to the boiler weight and length as shown.

Effect of Tube Size on Boiler Design

In general, boilers with smaller tubes have higher heat transfer coefficients and weigh less compared to those with larger tubes and vice versa. However, one has to check the labor costs as more tubes may be required.

Example 4.14

In a fire tube waste heat boiler, gas flow = 50,000 kg/h at 850°C in and 273°C out. Gas analysis is as given in Example 4.7. Table 4.26 shows two designs one using 50.8 mm tubes and another with 31.8 mm tubes for the same duty and gas pressure drop. Two cases are considered for each tube size, namely, different gas velocities. Operating steam pressure = 12 kg/cm² g (saturation temperature = 191°C), feed water temperature = 104°C, gas-side fouling = 0.0004, and steam side = 0.0002 m² h °C/kcal. Calculations were performed as discussed earlier, and the results are shown in Table 4.26.

It is seen that for the same duty and gas pressure drop, the smaller tube design is much shorter. It also weighs less. So if there are length or weight considerations in locating or handling fire tube boilers, a smaller tube design may be used. It is also easier to ship and handle due to the lower weight. However, the number of tubes is

TABLE 4.26

Effect of Tube Size and Gas Velocity on Boiler Design

Tube Size OD × ID	50.8 × 45	50.8 × 45	31.8 × 27	31.8 × 27
Gas flow, kg/h	50,000	50,000	50,000	50,000
Gas inlet temperature, °C	850	850	850	850
Gas exit temperature, °C	273	273	273	273
Number	600	425	1,690	1,200
Length, m	6	6.6	3.3	3.6
U, kcal/m² h °C (tube ID)	58.8	75.4	63.2	81.2
Max gas inlet velocity, m/s	46	65	46	64
ΔP, mm wc	105	225	105	225
Duty, MW	9.5	9.5	9.5	9.5
Shell diameter, m	1,900	1,676	2,000	1,720
Surface area, m² (tube ID)	509	397	473	366
Tube weight, kg	13,400	10,400	10,800	8,400
Shell weight, kg	3,600	3,500	2,100	1,650
Additional fan power, k_w	—	20	—	20

much more, and hence, labor cost may be higher depending on the location where the boiler is manufactured. Surface area with smaller tubes is lower due to the higher heat transfer coefficient. Hence, one should not compare boiler designs based on surface areas alone. (This dictum is true whether it is a fire tube or water tube plain or finned tube boiler!)

With higher gas velocity, one can make the boiler smaller. But the cost of moving the gas through the boiler has to be computed and the cost of fan power consumption must be evaluated to see if this is a good idea. If the gas velocity is increased from 46 to 64 m/s, the number of tubes is lesser and length is more as also the gas pressure drop. The additional cost of $225 - 105 = 120$ mm wc gas pressure drop will be significant over a period of time, but the initial cost will be lower. Hence, it is important that plant engineers while buying a boiler ensure that the duty and gas pressure drop are the same in all the designs; else, they will be comparing apples and oranges.

Let us compute the additional fan power consumption.

$$P = 2.7 \times 10^{-6} W H_w / \rho \, \eta \tag{4.16}$$

where

P is the fan power consumption
H_w is the fan head in mm wc
ρ is the density of flue gas at fan, kg/m^3
η is the efficiency of the fan and motor combination
W is the flue gas flow in kg/h.

$$W = 50,000 \, kg/h, \quad H_w = 120 \, mm \, wc, \quad \rho \, at \, 30°C = 1.16 kg/m^3 \quad and \quad \eta = 0.7.$$

Hence, $P = 2.7 \times 10^{-6} \times 50,000 \times 120/(01.16 \times 0.7) = 20$ kW. If electricity costs 15 cents/kWh, then for 8000 h operation in a year, the additional cost of power = $20 \times 0.15 \times 8,000 = \$24,000$, which is very significant. Hence, plant engineers should be aware of both the operating and the initial cost of their boilers.

Estimating Tube Bundle Diameter

Tubes of fire tube boilers are typically arranged in inline or triangular fashion. The ratio of tube pitch to diameter can vary from 1.25 to 2 depending on tube size and the manufacturer's practice.

Looking at the triangular pitch arrangement, we see that half the tube area is located within the triangle, whose area is given by $0.5 \times 0.86p^2 = 0.433p^2$. Total area occupied by all tubes = $0.866p^2N$.

If tube bundle diameter is D, then $3.14D^2/4$ = area of bundle = $0.866p^2N$ or $D = 1.05pN^{0.5}$.

Similarly, for the square pitch, the area occupied by the tubes is p^2. Hence, bundle area = $3.14D^2/4 = Np^2$ or $D = 1.128pN^{0.5}$. In practice, some additional space is added for manufacturing reasons and for refractory provision.

Simplified Approach to Evaluating Performance of Fire Tube Boilers

Plant engineers may use the following procedure discussed to evaluate quickly the design or performance of fire tube boilers. Plants, for example, re-tube their old boilers or modify the existing ones and would like to know how the performance will be affected when tube geometry data or steam parameters change. It is a fact that a computer program gives more accurate results. However, plant engineers may not have access to such programs, and hence, shortcuts such as these shown later help them in evaluating new designs or performance of existing ones. Based on these shortcuts, they may also develop some computer code or Excel worksheet. Note that this method neglects the contribution of nonluminous heat transfer coefficient, which is anyway not significant in fire tube boilers due to the small inner diameter of tubes, which is the beam length. The overall heat transfer coefficient U_i may be written as

$$U_i = h_c = \frac{0.0278C\ w^{0.8}}{d_i^{1.8}} \tag{4.17}$$

w = flow per tube in kg/s = W/N, where N is the number of tubes and W the total flow in kg/s. $Q/\Delta T = U_i A_i = 0.0278 A_i\ C\ w^{0.8}/d_i^{1.8} = 0.0873C\ d_i\ N\ LW^{0.8}/(N^{0.8}d_i^{1.8})$ (substituting $A_i = \pi d_i LN$). Rewriting,

$$\frac{Q}{(\Delta TCW^{0.8})} = \frac{0.0873N^{0.2}L}{d_i^{0.8}} \tag{4.18}$$

Q = duty in W, L is the length in m, d_i the tube ID in m, and $C = (C_p/\mu)^{0.4}k^{0.6}$, where C_p is in J/kg K, μ in kg/m s, and k in W/m K.

For the evaporator, $\Delta T = (T_1 - T_2)/\ln[(T_1 - T_s)/(T_2 - T_s)]$ and $Q = WC_p(T_1 - T_2)$ neglecting heat loss. Substituting these in (4.17) and simplifying, we have

$$\ln\left[\frac{(T_1 - T_s)}{(T_2 - T_s)}\right] = \frac{0.0873CN^{0.2}L}{\left(C_p d_i^{0.8}W^{0.2}\right)} \tag{4.19}$$

Example 4.15

Let us check the results of Example 4.7 using this shortcut. W = 45,370 kg/h = 12.6 kg/s. N = 600, L = 6.1 m, C_p = 1,204 J/kg K, C = 182, d_i = 0.045 m, T_1 = 815°C, T_s = 185°C. Solve for T_2.

$$\ln\left[\frac{(815 - 185)}{(T_2 - 185)}\right] = \frac{0.0873 \times 182 \times 600^{0.2} \times 6.1}{\left(1204 \times 0.045^{0.8} \times 12.6^{0.2}\right)} = 2.083$$

or T_2 = (815 − 185)/8.03 or T_2 = 263°C. Close to 260°C using the detailed calculations.

If the gas pressure drop equation is considered, we have

$$\Delta P = \frac{0.08262\ f\ L_e v\ w^2}{d_i^5} \tag{4.20}$$

where
    w is the flow/tube in kg/s
    L_e is the equivalent tube length in m
    d_i is the tube inner diameter in m
    f is the friction factor
    v is the specific volume in m³/kg
    ΔP in mm wc as discussed in Appendix B

If inlet and exit losses are included, $L_e = L + 60d_i$. Also substituting for w = W/N,

$$\Delta P = \frac{0.08262\ f\ (L + 60\ d_i)\ v\ W^2}{(N^2 d_i^5)} \tag{4.21}$$

Substituting for N from this in (4.19) and simplifying, we have

$$\ln\left[\frac{(T_1 - T_s)}{(T_2 - T_s)}\right] = \frac{0.068\ C\ f^{0.1}\ (L + 60di)^{0.1}\ v^{0.1}L}{(\Delta P^{0.1} di^{1.3} C_p)} \tag{4.22}$$

This formula relates the duty with gas pressure drop. Equations 4.19 and 4.22 help one arrive at quick solutions to the major boiler configuration or performance.

In the conventional method, the following are the steps (though the use of a computer program makes things easier).

1. Given the gas flow, and inlet and exit gas temperatures, one first assumes w, flow per tube, or N from which w is obtained. w, the flow per tube, is based on experience.
2. h_c and h_n and then U, ΔT are computed.
3. The required surface area is obtained, and the tube length L is computed.
4. The gas pressure drop is then evaluated. If one does not like these values, steps 1–4 are repeated.

The simplified approach gives a reasonable geometry for a desired performance quickly or the performance for a given tube geometry in one step.

Example 4.16

In a fire tube boiler with 600 tubes with tube size 38 × 33 mm and length 5 m, 30,000 kg/h of flue gas with the same analysis shown in Example 4.7 has to be cooled from 700°C. Steam pressure is 12 kg/cm² g. Determine the exit gas temperature and gas pressure drop.

$T_1 = 700°C$. $T_s = 191°C$. L = 5 m, $d_i = 0.033$ m, W = 8.33 kg/s. Let $C_p = 1195$ J/kg K, C = 181 as before. (The gas properties may be corrected after obtaining the average gas temperature.) Average gas specific volume is 2.19 m³/kg.

Using (4.19),

$$\ln\left[\frac{(T_1 - T_s)}{(T_2 - T_s)}\right] = \frac{0.0873 C N^{0.2} L}{(C_p d_i^{0.8} W^{0.2})}$$

$$\ln\left[\frac{(700 - 191)}{(T_2 - 191)}\right] = \frac{0.0873 \times 181 \times 600^{0.2} \times 5}{(1195 \times 0.033^{0.8} \times 8.33^{0.2})} = 2.382$$

Or $T_2 = 238°C$. The duty and steam generation may be obtained from this as shown earlier.

Using (4.22),

$$\ln\left[\frac{(T_1 - T_s)}{(T_2 - T_s)}\right] = \frac{0.068 C f^{0.1} (L + 60di)^{0.1} v^{0.1}L}{(\Delta P^{0.1} di^{1.3} C_p)}$$

$$2.382 = \frac{0.068 \times 181 \times 0.022^{0.1} \times (5 + 60 \times 0.033)^{0.1} \times 5 \times 2.19^{0.1}}{(\Delta P^{0.1} \times 0.033^{1.3} \times 1195)}$$

or

$$\Delta P = 136 \text{ mm wc.}$$

These values may be checked out using detailed calculations.

Example 4.17

If L = 4 m and gas pressure drop = 100 mm wc due to plant space limitations, what is T_2 and N in the aforementioned situation? Other data such as gas flow and analysis may be assumed to be the same. These types of *what if* studies may be quickly performed with these equations.

Using (4.22),

$$\ln\left[\frac{(700-191)}{(T_2-191)}\right] = \frac{0.068 \times 0.022^{0.1} \times (4+1.98)^{0.1} \times 2.19^{0.1} \times 4 \times 181}{(100^{0.1} \times 0.033^{1.3} \times 1195)} = 1.935.$$

Hence,

$$T_2 = 265°C. \text{ Using } (4.19), 1.935 = \frac{0.0873 \times 181 \times N^{0.2} \times 4}{(1195 \times 0.033^{0.8} \times 8.33^{0.2})} \text{ or } N = 648$$

Thus, the entire geometry is arrived at. One may obtain the duty and steam generation as shown earlier. These methods are suggested for quick evaluation of performance or geometry or to study the impact of varying any tube geometry data.

Estimating Tube Sheet Thickness

The shell diameter for the boiler in Example 4.7 is obtained as follows: N = 600 tubes, 50.8 × 45 mm. Tube pitch p = 70 mm triangular. D = minimum diameter of shell = 1.05 × 70 × $\sqrt{600}$ = 1800 mm. Use, say, 150 mm between the outer most tubes and the shell inner diameter. Hence, shell ID = 2100 mm. Let the design pressure including static head = 14 kg/cm^2 g = 199 psig.

ASME code provides a simple formula for the tube sheet thickness; t = p$\sqrt{(P/2.2/S)}$, where P is the design pressure, psig, and S the allowable stress. Up to 650°F, for carbon steel plates, S = 17,500 psi may be used. p = 70 mm.

Hence, t = 70 × (199/2.2/17,500) = 5.03 mm. Considering corrosion allowances on gas and steam side and rigidity, let us use 25 mm plate.

Estimating Tube Sheet Temperature with Refractory and Ferrules

When flue gas temperature entering the boiler exceeds 750°C, refractory on tube sheet and ferrules are recommended (Figure 4.16). Ferrules, typically made of ceramic material, direct the heat flux from the gas inlet chamber to the water region and prevent the tube sheet from getting overheated as discussed earlier. Gas inlet temperature is 815°C or 1499°F. Let the tube sheet refractory surface be at 1202°F or 650°C (this will be verified later).

Let the inlet vestibule have an inner diameter of 2.2 m and length = 1.5 m. The % volume of CO_2 = 8, H_2O = 18. Beam length = 3.4 × volume/surface area = (3.4 × π × 2.2^2 × 1.5/4)/(π × 2.2 × 1.5 + π × 2.2^2/4) = 1.37 m = 4.49 ft.

P_c = 0.08 × 4.49 = 0.359 atm ft P_w = 0.18 × 4.49 = 0.808 atm ft. From Hottel's charts (Appendix D), $\in_c$ = 0.12. $\in_w$ = 0.23. Hence, $\in_g$ = 0.12 + 0.23 − 0.035 = 0.315.

$$h_n = \sigma \in_g \frac{\left[T_g^4 - T_o^4\right]}{(T_g - T_o)} \tag{4.23}$$

σ is Stefan Boltzman constant = 5.67×10^{-8} W/m^2 K^4

$h_n = 5.67 \times 10^{-8} \times 0.315 \times [10.88^4 - 9.23^4]/(1088 - 923) = 73$ W/m^2 K (12.81 Btu/ft^{2}h °F)

The thermal conductivity of the 100 mm refractory layer on tube sheet from data provided by refractory supplier is 11.6 Btu in./ft^2 h °F = 1.672 W/m K. Its thermal resistance = 0.1/1.672 = 0.06 m^2 K/W.

Let boiling heat transfer coefficient = 5721 kcal/m^{2}h °C = 6652 W/m^2 K.

Tube sheet thermal resistance = 0.1/37 = 0.0027 m^2 K/W (thermal conductivity of tube sheet, which is of carbon steel material SA 515-60, is taken as 37 W/m K and thickness = 0.1 m).

The sum of all thermal resistances = 1/73 + 0.06 + 0.0027 + 1/6652 = 0.0137 + 0.06 = 0.0137 + 0.06 + 0.0027 + 0.00015 = 0.07655 m^2 K/W. Heat flux across tube sheet = (815 – 185)/0.07655 = 8230 W/m^2 (neglecting heat flux from tubes. Tubes are protected by ceramic ferrule with ceramic paper wrap so that no energy is transferred to the tube sheet from gas flowing through the tube sheet. However, if there is no refractory and ferrule, then there will be a contribution due to this heat flux as discussed later).

Temperature drop across steam film = 8230 × 0.00015 =1.3°C
Temperature drop across tube sheet = 8230 × 0.0027 = 22.3°C
Temperature drop across refractory = 8230 × 0.06 = 494°C
Temperature drop across gas film = 8230/73 = 113°C

Hence, tube sheet maximum temperature = 185 + 1.3 + 22.3 = 209°C and refractory is at 815 – 113 = 702°C. With a margin of, say, 20°C, tube sheet temperature will be 229°C. Design temperature is typically 343°C. One may revise the calculations correcting for the thermal conductivity as a function of refractory temperature. However, due to the significant difference between the allowable and actual tube sheet temperature, we leave the exercise to the readers. A more accurate method is to use finite element analysis, but for quick estimates, this gives a reasonable value.

Calculation of Tube Sheet Temperature without Refractory on Tube Sheet

Let us assume that the tube sheet is at 350°C without the refractory. Then, the nonluminous heat transfer coefficient between the flue gas and tube sheet may be estimated as

$$h_n = \frac{5.67 \times 10^{-8} \times 0.315 \times \left[10.88^4 - 6.23^4\right]}{(1088 - 623)} = 48 \text{ W/m}^2 \text{ K} \left(8.43 \text{Btu/ft}^2 \text{ h }^\circ\text{F}\right)$$

The flue gas transfers some energy to the tube sheet through frontal radiation and also through convection in the gas entry region. Appendix B gives an idea of the heat transfer coefficient at the gas entry region, which can be typically 2.5 times the convective heat transfer coefficient for fully developed flow. At 815°C, the convective heat transfer coefficient = 67.3 W/m^2 K and the nonluminous heat transfer coefficient = 7.7 W/m^2 K. The effective coefficient at the tube sheet = 2.5 × 67.3 + 7.7 = 176 W/m^2 K.

$$\text{Area occupied by tubes} = \frac{\pi \times 600 \times 0.05 \times 0.05}{4} = 1.178 \text{ m}^2$$

Area occupied by tube sheet = π × 2.1^2/4 = 3.46 m^2. Hence, an area of (3.46 – 1.178) = 2.282 m^2 receives energy with h_n = 48 W/m^2 K while 1.178 m^2 receives energy with an h_e = 176 W/m^2 K. Effective heat transfer coefficient = (2.282 × 48 + 1.178 × 176)/3.46 = 91.55 W/m^2 K (resistance = 0.0109 m^2 K/W). The sum of all thermal resistances = 0.0109 + 0.0027 + 0.00015 = 0.01375 m^2 K/W (we are considering only the gas-side resistance, tube sheet resistance, and steam film resistance and neglecting fouling effects that are small).

$$\text{Heat flux through tube sheet} = \frac{(815-185)}{0.01376} = 45{,}818 \text{ W/m}^2$$

$$\text{Temperature drop across gas film} = 45818 \times 0.0109 = 499^\circ\text{C}.$$

Hence, tube sheet temperature = 815 − 499 = 316°C. Compare this to 209°C with refractory. Hence, using refractory and ferrules, help the tube sheet operate at a low temperature. With a margin of 20°C, the tube sheet could be at 336°C–340°C. One has to also consider expansion and stresses at these higher temperatures. Hence, as a precaution, refractory and ferrules are used whenever gas inlet temperature exceeds 750°C. Based on steam pressure and saturation temperature, one may estimate the tube sheet temperature and ensure that it is less than 350°C (carbon steel limit); else, alloy steel tube sheet may be required adding to the cost of manufacturing and adding to expansion and additional stresses. A more accurate method is to use finite element analysis, but for quick estimates, the earlier procedure is adequate.

Air Heaters

Air heaters are used in steam generators when low BTU fuels are fired or in a few waste heat boilers for preheating combustion air. Incineration plants and reformer furnaces also use preheated air. Decades ago, they were used in boilers firing oil, gas, and solid fuels. However, with NO_x emission limitations throughout the world, as discussed in Chapter 1, they are used only if the fuel combustion process warrants it. If the gaseous fuel has a low heating value or if the solid fuel has significant amount of moisture, then hot air is required for drying the fuel and also to ensure combustion with a stable flame. A gas to gas heater, which is similar to the tubular air heater (Figure 4.41), is also used in incineration plants.

Air heaters can be of tubular type or regenerative or heat pipes (Figure 4.41). In the tubular air heater, air or flue gas could flow inside the tubes. If the flue gas contains dust or ash particles, it is preferable to make the flue gas flow inside the tubes so that cleaning is easier. The air then takes a multipass route outside the tubes.

One of the concerns with air heaters is low-temperature corrosion at the cold end. The tube wall temperature at the cold end falls below the acid dew point temperature of the flue gas if the inlet air temperature is low or during part loads when both the air and flue gas flows are reduced. Steam is often used to preheat air to mitigate this concern and raise the air inlet temperature, which in turn raises the flue gas exit temperature and thus the average tube wall temperature. Some boiler suppliers use corten steel tubes to minimize corrosion concerns.

There are two types of regenerative air heaters, one in which the heater matrix rotates (Ljungstrom) and the other in which the connecting air and flue gas duct rotate (Rothemuelle). The energy from the hot flue gas is transferred to the slowly rotating matrix (1–5 rpm) made of enamel or alloy steel material, which absorbs the heat and transfers the energy to the cold air as it rotates. The elements are contained in baskets, which makes

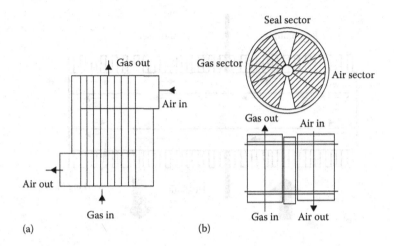

FIGURE 4.41
(a) Tubular and (b) regenerative air heaters.

cleaning or replacement easier. The air- and gas-side pressure drops are high in both types of air heaters, adding to the fan power consumption. Due to the low air- and gas-side heat transfer coefficients and low overall U and the low LMTD, the surface area required is large. However, in the case of Ljungstrom air heater, a lot of surface area can be packed into each basket and hence can be made compact while the tubular air heater (where finned tubes are not effective as discussed in Appendix E) will be huge often as large as the steam generator itself.

One of the problems with regenerative-type air heaters is the leakage (from 5% to 10%) from air to flue gas side (air is often at higher pressure), and this affects the performance even with good design of seals.

In case of low-temperature corrosion, the cold-end baskets can be made of alloy or corten material, while in the case of tubular type, the entire tube has to be replaced.

Heat Pipes

Heat pipes were introduced in the market about 50 years ago. It consists of a bundle of tubes filled with a working fluid such as toluene, naphthalene, or water and sealed (Figure 4.42). Heat from the flue gas evaporates the working fluid collected in the lower end of the slightly inclined pipes (6°–10° from the horizontal), and the vapor flows to the condensing section, where it gives up the heat to the incoming combustion air. The condensed fluid returns by gravity to the evaporator section assisted by an internal capillary wick that is essentially a porous surface or circumferentially spiraled groove of proprietary design. The process of evaporation and condensation continues as long as there is a temperature difference between the air and the flue gas.

In a typical design, there is a divider plate at the middle of the tubes that supports the tube and also maintains a seal between the hot flue gas and cold air. Pipes are finned to make the heat transfer surface compact. Finned surfaces are used as the tube-side coefficient is very high due to the evaporation and condensation processes. Fin density is based on cleanliness of the gas stream.

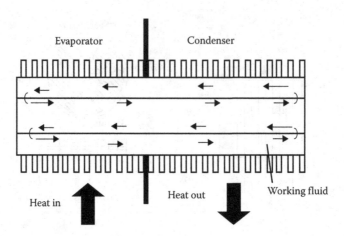

FIGURE 4.42
Arrangement of heat pipe.

Heat pipes have several advantages over conventional air heaters:

1. They are compact and weigh less than other air heaters due to the use of extended surfaces.

2. They have zero leakage because pipes are stationary and the divider plate is welded to the tubes.

3. No auxiliary power is needed.

4. No rotating parts and hence no maintenance.

5. They have low corrosion potential. Owing to the isothermal behavior of the pipes, minimum tube metal temperature is higher than in other types of air heaters. By selecting proper working fluids, it is possible to maintain the cold end above the acid dew point. The pipes also operate at constant temperature along the entire length due to evaporation and condensation process.

6. Gas- and air-side pressure drops are generally lower than in tubular or regenerative air heater due to the compactness of the design.

Design of Tubular Air Heaters

The design of tubular air heaters is discussed later. This is an iterative process. One has to assume the cross section or the number of tubes and length and then arrive at the required length of the tubes based on duty. If they do not match, another iteration is called for. Then air and gas pressure drops are checked, and if they are high and unacceptable, yet another iteration is required. A well-written computer program can do both design and off-design calculations.

Let W_a and W_g be the air and flue gas quantities. (This procedure may be applied to any two gas streams.) Often, flue gas flows inside the tubes while air flows across the tubes in cross-flow fashion as shown in Figure 4.42. Carbon steel tubes of OD 1.5–3 in. are used. Thickness is less than 0.08 in. due to the low pressure of air and flue gas. Tubes are arranged in inline fashion as staggered arrangement is not economical as discussed in Appendix C since plain tubes are used. Finned tubes are not recommended as both the gas- and air-side heat transfer coefficients are of the same order and fins become ineffective as the tube-side heat transfer coefficient is reduced as shown in Appendix E. Air and flue gas velocities range from 10 to 20 m/s at full load. Number of passes outside the tubes can be from one to three depending on the duty and allowable air- and gas-side pressure drops. It does not matter whether flue gas or air flows inside the tubes if the gas stream is clean. When there are ash particles or dust, it is better to have the flue gas inside the tubes as cleaning becomes easier.

Design involves manipulating the number of tubes wide, number of tubes deep, and tube length to meet the desired duty. In off-design calculations, one checks the duty and exit air and gas temperatures for the given geometry with different gas and air flows.

One has to go through both the Appendices B and C to see how the heat transfer coefficients for flow inside and outside tubes are computed.

Heat Transfer Inside and Outside Tubes

First the number of tubes is determined based on the following equation. Assumption for a reasonable tube-side velocity is made (see Equation B.5).

$$N = \frac{1.246 W_g}{(\rho d_i^2 V)} \tag{4.24}$$

where
W_g is the flue gas flow, kg/s
ρ is the average density of flue gas, kg/m^3
d is the tube inner diameter, m
V is the gas velocity, m/s

Then based on flow/tube, the tube-side heat transfer coefficient is determined (see Appendix B), neglecting nonluminous heat transfer coefficient,

$$h_c = h_i = 0.0278 w^{0.8} \frac{(C_p/\mu)^{0.4} k^{0.6}}{d_i^{1.8}} = \frac{0.0278 C(W_g/N)^{0.8}}{d_i^{1.8}} \tag{4.25}$$

$C = (C_p/\mu)^{0.4} k^{0.6}$ and N is the total number of tubes through which the flue gas (or air) is flowing. N = number of tubes wide × number of tubes deep and W_g the gas flow in kg/h. properties C_p—kcal/kg °C, μ—kg/m h, k—kcal/m h °C, h_c—kcal/m^{2}h °C.

The heat transfer coefficient outside the tubes may be obtained using correlations discussed in Appendix C. Here too, since the temperatures are not high, nonluminous heat transfer coefficient may be neglected. (A computer program may be written to evaluate the nonluminous coefficient also, but here for illustration, we are limiting to convective heat transfer coefficients.)

$$Nu = 0.35 Re^{0.6} Pr^{0.3} \tag{4.26}$$

where $Re = Gd/\mu$, $Pr = \mu Cp/k$, $Nu = h_c d/k$.

Simplifying, $h_o = 0.35\, FG^{0.6}/d^{0.4}$ where $F = (C_p/\mu)^{0.3} k^{0.7}$

The gas properties are evaluated at the film temperature, which is approximated as follows: $t_f = (3t_a + t_g)/4$ when air flows outside the tubes or as $(3t_g + t_a)/4$ when flue gas flows outside tubes. Some use the average of gas and air temperatures for preliminary estimates.

Example 4.18

At 375°C, 225,000 kg/h of flue gas flows inside the tubes of an air heater, while 200,000 kg/h of air at 25°C flows outside the tubes. Using tubes of size 50.8 × 46.4 mm size a suitable tubular air heater if air is to be heated to 180°C. The flue gas analysis is typical natural gas products of combustion with % volume

$$CO_2 = 8.29, \quad H_2O = 18.17, \quad N_2 = 71.08, \quad O_2 = 2.46.$$

$$MW = 8.29 \times 44 + 18.17 \times 18 + 71.08 \times 28 + 2.46 \times 32 = 27.6.$$

Let tubes be arranged in inline fashion with $S_T = 82$ mm and $S_L = 70$ mm. Fouling factor on air and gas sides is 0.0002 m²h °C/kcal.

Solution

The duty of the air heater = 200,000 × 0.2455 × (180 − 25) = 7.61 MM kcal/h = 225,000 × 0.99 × (375 − t) × 0.2831 or exit flue gas temperature = 254°C. (The specific heats of air and flue gas were obtained from Appendix F at the average air and flue gas temperatures. One percent casing heat loss was assumed and hence the 0.99 factor.)

Computing h_i

The number of tubes is estimated assuming some flue gas velocity. Let us use 22 m/s. Density ρ of flue gas at average gas temperature of (375 + 254)/2 or 314°C = 27.6 × 101,325/ (273 + 314)/8314 = 0.573 kg/m³. Then, N = 1.246W_g/(ρd_i^2V) = 1.246 × 62.5/(0.573 × 0.0464 × 0.0464 × 22) = 2860. Gas properties at 314°C: C_p = 0.2831 kcal/kg °C, μ = 0.0988 kg/m h, k = 0.0371 kcal/m h °C by interpolation from Tables F.6 through F.8.

$$h_c = h_i = 0.0278 w^{0.8} \left(C_p/\mu\right)^{0.4} k^{0.6}/d_i^{1.8} = 0.0278\, C\left(W_g/N\right)^{0.8}/d_i^{1.8}$$

$$= 0.0278 \times \left(225,000/2,860\right)^{0.8} \times \left(0.2831/0.0988\right)^{0.4} \times 0.0371^{0.6}/0.0464^{1.8} = 48.5\ \text{kcal/m}^2\,\text{h°C}$$

Computing h_o

Approximate film temperature = (3 × 102 + 314)/4 = 155°C using average air temperature of 102°C and flue gas temperature of 314°C. (One may compute the actual film temperature after estimating h_i and h_o and revise the calculations if necessary.)

Air properties at 155°C: C_p = 0.2477 kcal/kg°C, μ = 0.087 kg/m h, k = 0.0305 kcal/m h °C by interpolation from Tables F.1 through F.3.

Let number of tubes wide N_w = 50, number deep N_d = 2860/50 = 57. This is just the first trial. We have to check the tube length and gas-, air-side pressure drops and then review this design. These calculations are for illustrative purposes only. We have to assume a height of tubes H; compute h_o and then U_o and then see if the air heater will transfer the required duty. Let tube length L = 3 m.

$$G = \frac{200,000}{\left[50 \times \left(0.082 - 0.050.8\right) \times 3\right]} = 42,735\ \text{kg/m}^2\,\text{h}$$

$$Re = \frac{42,735 \times 0.0508}{0.087} = 24953. \quad Pr = \frac{0.087 \times 0.2477}{0.0305} = 0.7065$$

$$Nu = 0.35 \times 24,953^{0.6} \times 0.7065^{0.3} = 137.1 = h_o d/k.$$

$$h_o = 137.1 \times 0.0305/0.0508 = 82.3\ \text{kcal/m}^2\,\text{h°C}$$

Tube wall resistance is computed as shown in Appendix A.

$1/U_o = 1/82.3 +0.0002 + 0.000066 + 0.0002 \times 50.8/46.4 + (1/48.5) \times 0.0508/0.0464$
$= 0.01215 + 0.0002 + 0.000066 + 0.000219 + 0.02257$
$= 0.0352$ or $U_o = 28.4$ kcal/m² h °C (0.000066 is the resistance of tube wall)

To calculate the corrected LMTD, use the correction factor from Figure A.1. R = (25 − 180)/(254 − 375) = 1.28. P = (254 − 375)/(25 − 375) = 0.345. F = 0.93 for single-pass cross-flow exchanger.

$$\text{LMTD} = \left[\frac{(229-195)}{\ln(229/195)} \right] = 211.5°C. \quad \text{Corrected LMTD} = 0.93 \times 211.5 = 197°C.$$

Required surface area A = 7.61 × 10⁶/197/28.4 = 1360 m² = 3.14 × 0.0508 × 2860 × L or L = 2.98 m. Provided height is 3 m. Hence, the exchanger can transfer the duty of 7.61 MM kcal/h.

Pressure Drop outside Tubes

For cross-flow outside tubes, $\Delta P = 0.204 f G^2 N_d / \rho_a$ from Appendix C.

G = 42,375/3,600 = 11.77 kg/m² s. Density ρ_a at 103°C is 28.74 × 101,325/8,314 /(273 + 102) = 0.934 kg/m³. N_d = 57 rows deep. S_T/d = 82/50.8 = 1.614 S_L/d = 70/50.8 = 1.378

Reynolds number computed at average air temperature of 103°C: Re = 42,735 × 0.0508 /0.0806 = 26,934.

Friction factor f (see Appendix C) for inline arrangement = $Re^{-0.15}$ [0.044 + 0.08 S_L/d/ $\{(S_T/d - 1)^{(0.43+1.13d/S_L)}\}$ = $26,934^{-0.15}$ [0.044 + 0.08 × 1.378/{(1.614 − 1)$^{(0.43+1.13//1.378)}$} =0.053.

$$\Delta P = 0.204 \times 11.77^2 \times 57 \times 0.053/0.934 = 91 \text{ mm wc}$$

Pressure Drop inside Tubes

Pressure drop inside tubes involves estimation of inlet and exit losses due to gas velocity and the friction loss inside the tubes (Appendix B).

$\Delta P = 0.08262 f L_e v w^2 / d_i^5$, where w is the flow per tube in kg/s, L the tube length in m, di the tube inner diameter in m, and ΔP is in mm wc.

The average density of flue gas = 0.573 kg/m³ from above at 314°C. Specific volume v = 1.745 m³/kg. f may be taken as 0.02 from Appendix B. w = 225,000/2,860/3,600 = 0.0218 kg/s.

$$\Delta P = 0.02862 \times 0.02 \times 3 \times 1.745 \times (0.0218)^2 / 0.0464^5 = 7 \text{ mm wc}$$

Density at 375°C = 27.6 × 101,325/(273 + 375)/8,314 = 0.519 kg/m³, and at an exit temperature of 254°C, density = 0.638 kg/m³ by taking ratio of temperatures.

Gas inlet velocity = V_1 = 1.246 $W_g/\rho/N/d_i^2$ = 1.246 × (225,000/3,600)/0.519/2,860/.0464² = 24.4 m/s. Velocity head VH_1 = 24.4 × 24.4 × 0.519/2/9.8 = 15.8 mm wc. Inlet loss = 0.5 VH_1 = 8 mm wc. Exit gas velocity at 254°C = 24.4 × 0.519/0.638 = 19.8 m/s. VH_2 = 19.8 × 19.8 × 0.638/2/9.8 = 12.8 mm wc. Exit loss = 1 velocity head = 12.8 mm wc. Total pressure drop = 8 + 7 + 13 = 28 mm wc.

Checking the Design Using the NTU Method

Appendix A describes the elegant NTU method, which has been used in several performance evaluation calculations throughout the book.

In our case, the flue gas is unmixed while air can be considered mixed. From Appendix A, the effectiveness factor from Table A.4 is

$$\varepsilon = 1 - \exp\{-1 / C[1 - \exp(-NTC \times C)]\}$$

$$WC_{min} = 200{,}000 \times 0.2455 = 49{,}100$$
$$WC_{max} = 225{,}000 \times 0.99 \times 0.2831 = 63{,}060$$
$$C = 49{,}100 / 63{,}060 = 0.7786$$
$$NTU = UA / WC_{min} = 28.4 \times 1360 / 49{,}100 = 0.7866$$
$$\varepsilon = 1 - \exp\left\{-\left[1 - \exp(-0.7866 \times 0.7786)\right] / 0.7786\right\} = 0.445$$

$Q = 0.445 \times 49{,}100 \times (375 - 25) = 7.647$ MM kcal/h. Agrees closely with the design duty.

Tube Wall Temperature

One has to check if the lowest tube wall temperature is below or close to the acid or water dew point temperature to ensure that corrosion does not occur due to condensation of water or acid vapor in which case suitable materials as discussed in Chapter 3 should be used. At the cold end, Figure 4.43 shows the temperature distribution.

One should estimate the actual h_i and h_o values corresponding to 254°C flue gas and 25°C air temperatures and use that U_o to compute the heat flux and lowest tube wall temperatures. However, for illustrating the procedure, let us use the same h_i and h_o as earlier.

$$\text{Heat flux } q_o = 28.4 \times (254 - 25) = 6503 \text{ kcal/m}^2\text{h}.$$

Temperature drop across the gas film = 6503 × 0.0227 = 147.6°C.
Temperature drop across the inside fouling layer = 6503 × 0.000219 = 1.4°C
Temperature drop across tube wall = 6503 × 0.000066 = 0.4°C.
Temperature drop across air fouling layer = 6503 × 0.0002 = 1.3°C
Temperature drop across air film = 6503 × 0.01215 = 79°C
Hence, tube outer wall temperature = 25 + 79 + 1.3 = 105.3°C.

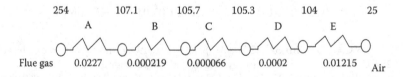

FIGURE 4.43
Temperature distribution at the cold end of air heater. *Note:* (A) Gas film, (B) gas-fouled layer, (C) tube wall, (D) air-fouled layer, and (E) air film.

Air film temperature is $25 + 79 = 104°C$. One may revise the runs and compute h_i and h_o based on corrected air film and average gas temperatures. However, tube wall will be at about 105°C, which is above the water dew point which for 18.17% volume water vapor corresponds to 58°C or 136°F from steam tables. If sulfuric acid vapor had been present, then we would have to increase the air inlet temperature; steam–air exchangers are commonly used for increasing the cold-end temperature as at part load conditions, the tube wall temperature will be even lower.

Part Load Performance

Example 4.19

Determine the performance of the aforementioned air heater when 130,000 kg/h of flue gas with same analysis flows inside the tubes at 330°C while air flow is 100,000 kg/h at 25°C.

Solution

The NTU method may be applied as it is direct and gives the duty from which the exit gas and air temperatures may be computed.

U_o value may be obtained as a first approximation as follows: With a computer program, one may obtain accurate U values accounting for variations in gas properties with temperature. Let us use the values of h_i and h_o calculated earlier and apply multiplication factors for an initial estimate.

$h_i = 48.5 \times (130,000/225,000)^{0.8} = 31.3$ kcal/m²h °C. Flue gas–side resistance corrected for tube OD = 50.8/46.4/31.3 = 0.035 m²h °C/kcal.

$h_o = 82.3 \times (100,000/200,000)^{0.6} = 54.3$ kcal/m² h °C. $1/h_o = 0.0184$ m² h °C/kcal.

$1/U_o = 0.0184 + 0.0002 + 0.000066 + 0.000219 + 0.035 = 0.05388$ or $U_o = 18.56$ kcal/m² h °C

$WC_{min} = 100,000 \times .24 = 24,000$. $WC_{max} = 130,000 \times 0.275 = 35,750$. C = 24,000/35,750 = 0.671 (Specific heat values were assumed but have to be corrected based on actual gas and film temperatures.)

$$NTU = 18.56 \times 1360 / 240,000 = 1.0517$$
$$\in = 1 - \exp\left[-\{1 - \exp(-1.0517 \times 0.671)\} / 0.671\right] = 0.53$$
$$Q = 0.53 \times 24,000 \times (330 - 25) = 3.88 \text{ MM kcal/h.}$$

Exit air temperature = $25 + 3.88 \times 10^6/24,000 = 187°C$ and exit gas temperature = $330 - 3.88 \times 10^6/35,750 = 221°C$. With a computer program, one can compute the air and gas properties at the corrected average film and gas temperatures. This exercise is left as an exercise to the readers. The tube wall temperature will also be lower now.

Heat flux at cold end (using the same U_o though the actual U_o will be lower at the cold end) = $18.56 \times (221 - 25) = 3628$ kcal/m²h.

Low-end tube wall temperature = $25 + 3628 \times (0.0002 + 0.0184) = 92.5°C$. Hence, we see that the tube wall temperature is much lower at lower loads.

Appendix E illustrates with an example why fins are not used in air heaters or gas to gas exchangers where the tube-side and gas-side heat transfer coefficients are in the same range of values.

Specifying Waste Heat Boilers

While developing specifications for waste heat boilers, consultants and plant engineers should have in mind the following points regarding thermal design and performance aspects.

1. The type of boiler is based on several aspects discussed earlier. While these are general guidelines, experience is the key. Though steam and gas parameters often dictate the type, whether fire tube or water tube, it is likely that even with small gas flow, a water tube may be economical, while even with large gas flows, a fire tube may be a better option. In special cases, a discussion with various boiler suppliers will help.

2. The process that generates the hot flue gas should be described and the resulting nature of the waste gas stream. With clean gas, finned tubes can be used, but if it is a dusty gas with high fouling tendency, plain tubes have to be used with a high fouling factor. A fire tube boiler with dusty (but not slagging) flue gas may be easier to clean if flue gas is inside the tubes. Some process boilers such as those in hydrogen plants require exit gas temperature control, and hence, this information should be made available to the boiler designer.

3. If the flue gas has particulates that can cause slagging of salts, then the melting temperatures of the ash should be determined in a laboratory and results conveyed to the boiler company. If the flue gas temperature entering the boiler is higher than the ash melting temperature, then the front end of the water tube boiler should be carefully designed with wide-spaced plain tubes followed by single-spaced tubes. Fire tube boilers are better avoided as the tubes will be difficult to clean if slag forms on the tube sheet refractory or over tubes. Superheater if required is preferably located after a furnace section and screen tubes to avoid fouling and corrosion, and provision should be made for cleaning the tube bundles. If dust content is high, then erosion of tubes is likely, and gas velocities in the boiler must be low on the order of 8–12 m/s. With clean flue gas such as gas turbine exhaust or flue gas from incineration of gases or fumes, gas velocity is limited by gas pressure drop considerations and can be even in the range of 20–40 m/s.

4. Desired steam purity should be mentioned, particularly if steam generated is used in a gas or steam turbine. Boiler water must be maintained within limits as discussed in Chapter 6. If steam at low pressure (1000–3000 kPa) is used for process and if the gas flow is not large, a fire tube boiler with integral steam space may be adequate. However, even with fire tube boilers, if steam is used in a turbine, an elevated drum boiler is a better option.

5. The extent of optimization required and the cost of fuel, electricity, and steam should be indicated, particularly with large gas flow units. Simply stating that energy recovery should be maximized is not adequate. If Supplier A cools the gas stream to, say, 250°C and supplier B to 200°C, the plant must know how to evaluate the options. Also the difference in gas pressure drop should be evaluated and annualized.

6. If steam for deaeration is taken from the boiler, then it should be so mentioned as it will affect the superheater size. Similarly, if import steam is likely to be superheated in the waste heat boiler, it should be specified as it will affect the superheater size and cost.

7. Flue gas analysis including any sulfur compounds must be specified as it affects the economizer or air heater if used; the minimum feed water temperature to be used is also determined by corrosion considerations. The exit gas temperature or efficiency of energy recovery is affected by the feed water inlet temperature, and hence, all these issues are interlinked. The presence of hydrogen chloride can cause high-temperature corrosion, and hence, superheater steam temperature selected should preferably be below 400°C. The duty of the boiler is also impacted by the flue gas analysis as the specific heat is a function of gas constituents such as water vapor, sulfur dioxide, and hydrogen.

8. Flue gas flow should be clearly stated in mass units such as kg/s or lb/h and not in volumetric units as this can lead to some differences in the estimation of gas density and mass flow. The energy balance or duty recovered is dependent on mass flow of flue gas, and any misunderstanding in its value can lead to differences in boiler duty and size.

9. Emission regulations and limits on CO, NO_x, SO_x, if any, should be stated in the specifications itself so the boiler can be designed appropriately and cost estimation can be reasonably accurate.

10. Any cycling requirements should be stated upfront as this has an impact on tube failures and method of welding tubes to headers, drums, and use of refractory in the boiler. Also if the gas flow and inlet gas temperature are likely to fluctuate, then this should be discussed with the boiler supplier. A large screen section and a furnace will help dampen the effect of these fluctuations.

11. Feed water used for steam temperature control should be demineralized and preferably have zero solids. If not, a sweet water condenser system as described in Chapter 3 may be required adding to the cost of the boiler.

12. The type of casing for the boiler is important whether refractory lined or membrane wall. Start-up time and casing corrosion issues are dependent on the type of casing used.

Waste Heat Boiler Data (Thermal Design)

The information that should be provided to the boiler supplier by the end user as well as the information to be provided by the boiler supplier in his proposal is discussed here. This may not be complete but a good starting point.

1. Waste flue gas flow (kg/s), gas inlet temperature (°C), and gas analysis at various loads or modes of operation.

2. Flue gas pressure, kPa, and maximum allowable gas pressure drop in the boiler, mm wc.

3. Flue gas analysis in % volume CO_2, H_2O, N_2, O_2, SO_2, CO, HCl, CH_4, SO_2, SO_3, H_2S.

4. Ash or dust content, mg/Nm^3.

5. Ash analysis and melting temperatures.

6. Describe the process of generating flue gas and its nature:
   a. Dirty and slagging in nature.
   b. Whether dusty, sticky in nature or not, or clean.
   c. Corrosive, erosive.
   d. Is the exhaust gas flow and temperature fluctuating or steady. If so, provide a time versus gas flow, temperature curve.

7. Steam pressure, kPa, steam temperature desired, °C, feed water temperature, °C.

8. Is steam temperature control required? If so, range of load and acceptable range of steam temperatures.

9. Any export or import steam requirements? Where is steam for deaeration taken from?

10. Feed water chemistry and analysis. Desired steam purity. Where is steam used in a turbine or for process?

11. Space limitations.

12. NO_x, CO, UHC, SO_x levels at inlet to boiler and desired values at boiler exit.

13. If exhaust gas is from gas turbine, is duct firing required or desired steam generation. Is fresh air firing required? Maximum acceptable steam in fresh air–fired case.

14. Burner fuel type and analysis, heating values.

TABLE 4.27

Tube Geometry Data

Geometry	Sh1	Sh2	Evap1	Evap2	Evap3	Econ.
Tube OD	38.1	38.1	50.8	50.8	50.8	44.5
Tube ID	30.6	30.6	44.1	44.1	44.1	37.8
Fins/in. or fins/m	0	0	0	78	177	177
Fin height	0	0	0	12.5	12.5	19
Fin thickness	0	0	0	1.9	1.9	1.9
Fin width	0	0	0	0	4.37	4.37
Fin conductivity	0	0	0	35	35	35
Tubes/row	38	38	38	38	38	42
Number of rows deep	4	4	4	2	16	8
Length	8.2	8.2	8.2	8.2	8.2	8.2
Transverse pitch	101	101	101	101	101	92
Longitudinal pitch	101	101	101	101	101	100
Surface area	*	*	*	*	*	*
Streams	76	76				21
parl = 0, countr = 1	1	1				1
Material	T22	T11	Sa192		Sa192	Sa192
Inline-staggered	Inline	Inline	Inline		Inline	Inline
Furnace dimensions (if used)	Length =	Width =	Height =			

Note: * to be filled by boiler supplier.

15. Cost (or credit) of additional 25 mm wc pressure drop, 10 kW fan power consumption, 100 kW duct burner fuel consumption, credit for 5% additional steam generation.

16. While obtaining quotes for a water tube boiler, information should be obtained for each heating surface in the format shown in Table 4.27.

17. If a liquid/fluid is heated by flue gas, the liquid/fluid flow, inlet and exit temperatures desired, pressure and thermal and transport properties such as specific heat, viscosity, and thermal conductivity at inlet and exit temperatures as well its molecular weight/density should be provided by the client or end user.

References

1. V. Ganapathy, *Waste Heat Boiler Deskbook*, Fairmont Press, Atlanta, GA, 1991, p2.
2. V. Ganapathy, *Industrial Boilers and Heat Recovery Steam Generators*, CRC Press, Boca Raton, FL, 2003, p54.

5

HRSG Simulation

Introduction

Heat recovery steam generator (HRSG) simulation is an interesting concept used to arrive at the thermal design as well as off-design performance of unfired, fired, simple or complex, multimodule, multipressure gas turbine HRSGs without physically sizing them. That is, one need not know about tube sizes, tube lengths, fin geometry, surface area, or tube spacing and yet obtain the valuable information such as duty of each component of the HRSG, steam generation, burner fuel input in case of fired HRSGs, performance with fresh air fan firing and performance at various gas turbine loads, and complete gas–steam temperature profiles. Since energy transfer in almost all HRSGs is in convective mode, simulation techniques can be applied with good accuracy. One need not be an HRSG designer to determine the performance of any HRSG. Anyone who understands heat balance calculations can perform HRSG simulation. Hence, any plant or process engineer can obtain a lot of performance-related information about his HRSG even before it is purchased or designed. Simulation may also be used to find out more information about an operating HRSG; for example, how will it perform if any gas- or steam-side parameter is changed or whether the HRSG is meeting performance guarantees as promised based on operating data. In the case of multimodule or multipressure HRSGs, simulation helps one to arrive at an optimum HRSG configuration and how the various heating surfaces should be arranged to maximize energy recovery. Examples in the following pages will illustrate the earlier points. Simulation will be helpful not only to plant or process engineers but also to HRSG suppliers who can optimize the HRSG configuration before even physically designing it.

What Is HRSG Simulation?

Knowing exhaust gas flow, temperature, gas analysis, and steam parameters, one can establish gas–steam temperature profiles and duty of each section such as superheater, evaporator, and economizer in the design mode. Then, in the off-design mode, determine how the HRSG performs when exhaust gas conditions or any of the steam parameters change. In other words, one can thermally design an HRSG and evaluate its performance without knowing about its geometry per se [1]. HRSG suppliers do what is called physical design. With physical design, one should know the HRSG configuration and the tube and fin geometry details; then, the overall heat transfer coefficient U is estimated for each component based on gas velocity, tube geometry, fin configuration, tube spacing, tube length, and then, the required surface area for each component is estimated using the equation $A = Q/(U\Delta T)$. Once the design mode is established, the number of transfer units (NTU) method is used to arrive at the off-design performance of the complete HRSG or any of its components as discussed in Chapter 4 and in Appendix A. With simulation process, we do not compute U but compute the term $UA = Q/\Delta T$ for each heating surface, which is possible from gas–steam temperature profiles alone. Then, in the off-design mode, we correct this term (UA) for each surface such as superheater, evaporator, or economizer for variations in gas flow, temperature, and analysis and use the NTU method to predict its off-design performance. Thus, the geometry of the HRSG need not be known to obtain performance information. The results obtained from simulation will be close to that obtained using physical design as will be shown later.

Note that simulation will not give results such as surface area per se. The value (UA) is a proxy for surface area. The product (UA) is obtained in design and off-design cases. Since U varies with gas flow, analysis, and temperatures, (UA) will vary depending on these parameters. Also gas pressure drop or tube wall temperature is not evaluated as physical design data are not used in simulation calculations. One can get a good idea of the overall thermal performance as shown in the following examples.

Since simulation calculations are tedious particularly if multimodule HRSGs are involved, a program has been developed by the author. Figure 5.1 shows how one may arrive at complex HRSG configurations by combining basic modules such as a superheater, an evaporator, and an economizer. These modules may be arranged in any order to optimize the HRSG performance. The concept of common economizer and superheater is used, which allows an economizer to feed an evaporator located anywhere in the gas path or a superheater to be fed by an evaporator located anywhere in the gas path. This enables one to simulate complex multimodule HRSGs such as those shown in the figure.

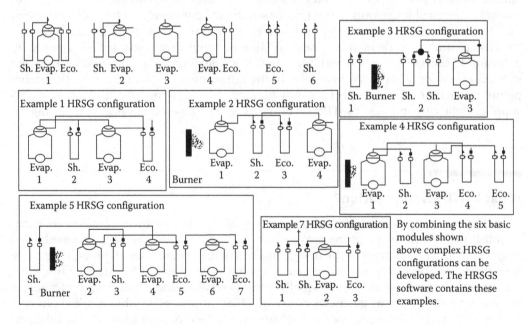

FIGURE 5.1
By combining various modules shown earlier, complex HRSG configurations can be simulated.

Applications of HRSG Simulation

- Given the exhaust gas flow and temperature, the first question on any process engineer's mind is how much steam can be generated at a given pressure and temperature and what the exit gas temperature is. In the design mode, using the simulation concept, one can determine the gas–steam temperature profiles, duty of each component, and steam generation. Design condition is typically the unfired mode of HRSG operation at the design or guarantee point. This is done by simply selecting pinch and approach points at each evaporator as explained later. Hence, by simply assuming pinch and approach points, we get a lot of information about the HRSG.

- We can obtain the off-design performance and see what happens to the HRSG gas–steam temperature profiles when exhaust gas flow or temperature conditions or any of the steam parameters change. This can in part load or fired mode of operation. This situation arises very often in any plant. There is only one design case, but there can be numerous off-design cases, both unfired and fired.

- One can estimate fuel required for generating a given quantity of steam in the fired mode without approaching the HRSG suppliers! Thus this is also a great planning tool.

- Check if the economizer is likely to generate steam at low loads. One can change approach temperature and redesign the HRSG if necessary (increase approach point in design mode or decrease the pinch point and ensure that steaming does not occur at low loads).

- Plant engineers can evaluate different gas turbines to see which one matches the steam needs of the plant better. There can be many gas turbines that can deliver a certain amount of power that the plant needs, but their exhaust gas conditions may be different. The exhaust gas flow can be higher or lower and so also the exhaust gas temperature. By doing a simulation study, plant engineers can get an idea of the steam generation, fuel required in fired mode, and efficiency of HRSG in various modes of operation for each gas turbine in consideration. Using this information, they can zero in on any particular gas turbine that meets their power as well as steam needs better.

- Plant or process engineers can also manipulate the HRSG configuration or arrangement to maximize the energy recovery. They can relocate the heating surfaces such as superheater, evaporator, or economizer and see if the energy recovery can be improved; this manipulation is necessary when there are multi-modules in the HRSG. The HRSG arrangement can be optimized, and then, consultants can develop specifications based on this configuration. One should not expect HRSG suppliers to spend time on this type of analysis as they may not have the time and also they know less about the plant than the plant engineers concerned. For example, process steam may be taken out of the low-pressure (LP) evaporator for deaeration or heating purposes, or import steam from other boilers can be superheated in the HRSG. Plant engineers can quickly find out how this affects the HRSG performance without waiting for a reply from the HRSG supplier.

- Field performance of an existing HRSG may also be evaluated and correlated with guarantee values to see if the performance is acceptable. This is illustrated later. One can also get an idea which surface is underperforming or overperforming.

All these points are illustrated with the following examples.

Understanding Pinch and Approach Points

Two important terms, namely, pinch and approach points should be understood by everyone involved in simulation analysis (see Figure 5.2). When gas inlet temperature to a waste heat boiler is low, on the order of 400°C–600°C, pinch and approach points determine the gas–steam temperature profiles and steam generation. Irrespective of surface areas provided, the exit gas temperature (in a single-pressure HRSG) cannot be predetermined. It falls in place depending on steam pressure and pinch and approach points selected in the design mode. In conventional steam generators, the combustion temperature in the furnace may be taken as the gas inlet temperature and hence is very high (1800°C–1900°C), and hence, pinch point or the steam pressure does not govern the temperature profiles; the heat sink in the form of economizer is very huge as a lot of energy is transferred in the furnace, evaporator, and superheater, and hence, the gas is cooled to any level with appropriate sizing of the economizer. One can obtain any desired exit gas temperature in steam generators irrespective of steam pressure. However, in low gas temperature heat recovery systems, the steam pressure and pinch point play a crucial role in determining the gas–steam temperature profiles, and the exit gas temperature from the economizer cannot be arbitrarily assumed. A temperature profile analysis must be performed to evaluate the steam generation and exit gas temperature. As a thumb rule, higher the inlet gas temperature, lower the exit gas temperature from the economizer and vice versa in a single-pressure HRSG.

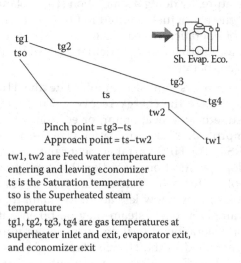

Pinch point = tg3−ts
Approach point = ts−tw2

tw1, tw2 are Feed water temperature entering and leaving economizer
ts is the Saturation temperature
tso is the Superheated steam temperature
tg1, tg2, tg3, tg4 are gas temperatures at superheater inlet and exit, evaporator exit, and economizer exit

FIGURE 5.2
Explanation of pinch and approach points.

TABLE 5.1

Suggestions on Pinch and Approach Points

Evaporator Type	Plain Tubes	Finned Tubes	For Both
Gas inlet temp., °C (°F)	Pinch point, °C (°F)	Pinch point, °C (°F)	Approach,°C (°F)
650–900 (1200–1650)	60–85 (108–153)	20–35 (36–63)	20–40 (36–72)
375–650 (705–1200)	40–60 (72–108)	5–20 (9–36)	5–20 (9–36)

Note: Pinch and approach points are differences in temperatures and hence 1.8 × pinch point in °C = pinch point in °F.

As seen in Figure 5.2, pinch point is the difference between the gas temperature leaving the evaporator and saturation temperature, and approach point is the difference between the saturation temperature and water temperature leaving the economizer. Once the pinch and approach points are selected, the entire HRSG gas–steam temperature profiles, duty of each section, and UA of each section are obtained. Note that we select pinch and approach points only once, in the design mode. The design mode should be the unfired mode at the design ambient temperature or the guarantee point even if the HRSG operates in the fired mode all the time. In off-design cases, the pinch and approach points fall in place. Table 5.1 gives some suggestions on selecting pinch and approach points.

The earlier suggestions are based on author's experience. When the gas inlet temperature is very high, say, 900°C (incineration plants), a low pinch point is not feasible as it will lead to temperature cross situation (explained later). Also, when plain tubes are used, it is difficult to get a low pinch point as the HRSG will be huge and impractical to build. When we are required to maximize steam generation, we select low pinch and approach points such as 5°C–7°C pinch point in finned tube gas turbine HRSGs. Approach point selection is based on whether the HRSG is likely to operate at low loads and whether economizer steaming is an issue. This is discussed later. If there are multiple evaporator modules and steam generation in the second pressure level is more important and is to be guaranteed, then a high pinch point can be selected for the first evaporator and a low pinch point for the second evaporator.

Many process engineers and consultants are of the view that any desired exit gas temperature can be obtained in single-pressure HRSGs. This is wrong. Gas–steam temperature profile is a function of steam pressure and the exit gas temperature t_{g4} cannot be preselected. This is explained later.

Estimating Steam Generation and Gas–Steam Temperature Profiles

Assumption of pinch and approach points will enable one to arrive at the complete gas–steam temperature profiles, steam generation, and duty in each section such as superheater or evaporator as mentioned earlier. This is illustrated by an example.

Example 5.1

Exhaust gas flow from a gas turbine is 100,000 kg/h at 500°C. Gas analysis is % volume $CO_2 = 3$, $H_2O = 7$, $N_2 = 75$, $O_2 = 15$. Steam at 41 kg/cm²a and at 375°C is required to be generated using 105°C feed water. Use a blowdown of 1%. Determine the gas–steam temperature profiles, duty of each section, and steam generation.

Solution

Let us assume 10°C (18°F) as the pinch point and 7°C (13.6°F) as the approach point. Assume that the superheater pressure drop is 1 kg/cm²; the drum pressure will be 42 kg/cm² a or saturation temperature is 252°C. Hence, gas temperature leaving the evaporator is 262°C and water temperature entering the evaporator is 252 − 7 = 245°C.

From Appendix F, the exhaust gas specific heat at the average gas temperature of (500 + 262)/2 or 381°C is 0.268 kcal/kg °C. Hence, energy absorbed by the superheater and evaporator = 100,000 × 0.99 × 0.268 × (500 − 262) = 6.32 MM kcal/h (0.99 refers to the 1% heat loss assumed). From steam tables, the enthalpy of superheated steam at 375°C and 41 kg/cm²a is 753.4 kcal/kg and that of feed water at 245°C is 253.5 kcal/kg. Saturated liquid and vapor enthalpy at 42 kg/cm²a are 261.8 and 668.8 kcal/kg, respectively. One can obtain the steam generation W_s using the energy absorbed by the superheater and evaporator as follows:

$$6.32 \times 10^6 = W_s[(753.4 − 253.5) + 0.01 \times (261.8 − 253.5)] = 500 \, W_s \text{ or } W_s = 12,640 \text{ kg/h}$$

(the blowdown effect is taken care by the second term).

The energy absorbed by the superheater = 12,640 × (753.4 − 668.8) = 1.069 MM kcal/h (1.24 MW). Using a gas specific heat of 0.274 kcal/kg °C at the superheater, we obtain the gas temperature drop in the superheater as 1.069 × 10⁶/(100,000 × 0.99 × 0.274) = 39.4°C or gas temperature leaving the superheater is = 460.4°C. The evaporator duty by difference is (6.32 − 1.069) = 5.251 MM kcal/h = 6.11 MW.

$$\text{The economizer duty} = 1.01 \times 12,640 \times (253.5 − 105.8) = 1.886 \text{ MM kcal/h} \, (2.19 \text{ MW}).$$

Gas temperature drop across the economizer = 1.88 × 10⁶/(100,000 × .99 × 0.26) = 73.3°C or exit gas temperature from the economizer = 262 − 73.3 = 188.7°C. Thus, the gas and steam temperatures are obtained simply by assuming pinch and approach points. Note that specific heat of gas at the average gas temperature in a given surface is required for the calculation of the gas temperature drop, and hence, the temperature profile evaluation involves a few iterations and is best done using a computer program. The results from HRSG simulation program developed by the author is shown in Figure 5.3. Slight differences are noticed in the results as the program evaluates the gas specific heats more accurately; a few iterations are required to estimate the inlet–exit gas temperatures at each section and the accurate gas specific heat at each section. But the idea is to show that by simply assuming the pinch and approach points, one can establish a design. Note that the program also computes the efficiency of the HRSG as explained in Chapter 3. We can now evaluate the UA value for each section as the log-mean temperature difference (LMTD) for each section is known (counter-flow arrangement is typical in HRSGs for superheater and economizer). Using the results from the program, let us compute the various terms required to perform an off-design evaluation.

Q_1 = 1.24 MW = 1,069,000 kcal/h, Q_2 = 6.11 MW = 5,251,000 kcal/h, Q_3 = 2.19 MW = 1,892,000 kcal/h (sections 1, 2, and 3 are the superheater, evaporator, and economizer, respectively).

Superheater

LMTDs: ΔT_1 = [(500 − 375) − (460 − 252)]/ln[(500 − 375)/(460 − 252)] = 163°C. $(UA)_1 = Q_1/\Delta T_1$ = 1,069,000/163 = 6,558. Program shows a value of 6,607.

Evaporator

LMTD ΔT_2 = [(460 − 252) − (262 − 252)]/ln[(460 − 252)/(262 − 252)] = 65.2°C. $(UA)_2 = Q_2/\Delta T_2$ = 5,251,000/65.2 = 80,536. Program shows a value of 80,645.

HRSG performance—Design case

Sh. Evap. Eco.
Project—eg1   Units—Metric case—eg1 Remarks-

Amb. temp., °C = 25 Heat loss, % = 1 Gas temp. to HRSG C = 500 Gas flow, kg/h = 100,000
% vol CO_2 = 3. H_2O = 7. N_2 = 75. O_2 = 15. SO_2 =. ASME eff., % = 64.85 tot duty, MW = 9.6

Surf.	Gas temp. in/out °C		Wat./Stm. in/out °C		Duty MW	Pres. kg/cm²a	Flow kg/h	Pstm. %	Pinch °C	Apprch. °C	US kcal/h °C	Module no.
Sh.	500	460	252	375	1.25	41.	12,702	100			6,607	1
Evap.	460	262	245	252	6.13	42.	12,702	100	10	7	80,645	1
Eco.	262	188	105	245	2.2	42.7	12,829	0			45,310	1

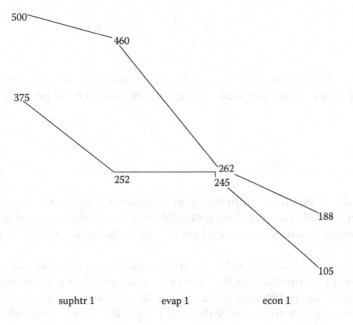

FIGURE 5.3
Example 5.1: gas–steam temperature profile in design mode.

Economizer

LMTD ΔT_3 = [(262 – 245) – (189 – 105)]/ln[(262 – 245)/(189 – 105)] = 42°C. $(UA)_3 = Q_3/\Delta T_3$ = 1,892,000/42 = 45,047. Program shows 45,310. The variations are mainly due to rounding off the LMTD values and slight variations in gas specific heats. However, the purpose is to illustrate the methodology. These data will come in handy later when we discuss off-design calculations.

One should understand that the HRSG heating surfaces have been determined *indirectly* when we selected the pinch and approach points. (UA) is in fact a proxy for the surface area.

Why Cannot We Arbitrarily Select the Pinch and Approach Points?

Let us see the basic equations involved in arriving at the gas–steam temperature profiles. From Figure 5.2, considering the superheater and evaporator,

$$W_g \times C_{pg} \times (t_{g1} - t_{g3}) = W_s (h_{so} - h_{w2})$$ (5.1)

Considering the entire HRSG and neglecting blowdown and heat losses,

$$W_g \times C_{pg} \times (t_{g1} - t_{g4}) = W_s (h_{so} - h_{w1})$$ (5.2)

Dividing (5.1) by (5.2) and neglecting the variations in C_{pg}, we have

$$\frac{(t_{g1} - t_{g3})}{(t_{g1} - t_{g4})} = \frac{(h_{so} - h_{w2})}{(h_{so} - h_{w1})} = K$$ (5.3)

Note that factor K is a function of steam–water properties alone. For steam generation to occur and for a feasible temperature profile, two conditions must be met:

$$t_{g3} > t_s$$ (5.4)

$$t_{g4} > t_{w1}$$ (5.5)

It is possible that if pinch and approach points are arbitrarily selected, one of these conditions will not be met. This is called temperature cross situation. K value has been computed for various steam pressures and presented in the following text for a specific case of an HRSG.

It is seen from Table 5.2 that as the steam pressure increases, the exit gas temperature increases. If steam is superheated, then the exit gas temperature is even higher for the same pressure. This is due to the simple fact that with a superheater for a given pinch point, the energy absorbed by the superheater and evaporator is the same, while the enthalpy absorbed by steam is more and hence less steam is generated. Lesser water flow through economizer in turn means smaller heat sink and hence higher exit gas temperature.

TABLE 5.2

K Values (482°C Gas Inlet, 110°C Feed Water, 11°C Pinch, and 8.3°C Approach)

Pressure kg/cm² g (kPa) (psig)	Steam Temp., °C (°F)	Sat. Temp., °C (°F)	K	Exit Gas, °C (°F)
7.03 (689) (100)	Saturated	170 (338)	0.904	149 (300)
10.54 (1033) (150)	Saturated	185.5 (366)	0.870	156 (312)
17.58 (1723) (250)	Saturated	208 (406)	0.8387	167 (332)
28.1(2755) (400)	Saturated	231 (448)	0.7895	178 (352)
28.1 (2755) (400)	315 (600)	232 (450)	0.8063	186 (366)
42.2 (4137) (600)	Saturated	254 (489)	0.740	189 (372)
42.2 (4137) (600)	399 (750)	255 (491)	0.7728	203 (397)

Hence, one cannot arbitrarily select an exit gas temperature in a single-pressure HRSG. A temperature profile analysis is required. One may develop such a table for any exhaust gas conditions. The author has seen a few specifications where consultants call out for an exit gas temperature, which cannot be obtained at a given pressure due to the limitation of pinch point (say, 140°C exit gas temperature from economizer at 28 kg/cm²g saturated steam conditions with about 480°C–490°C gas inlet temperature and 110°C feed water inlet).

Example 5.2

Check the exit gas temperature for 689 kPa (100 psig) saturated steam using 11°C pinch and 8.3°C approach points.
t_s = 170°C. t_{g3} = 181°C. Using (5.3), $(482 - 181)/(482 - t_{g4}) = 0.904$ or t_{g4} = 149°C as shown in Table 5.2. We cannot get much lower than this.

Example 5.3

Why cannot we obtain an exit gas temperature of 149°C when the steam pressure is 4137 kPa (600 psig) and steam temperature is 399°C? K is 0.7728.

Using (5.3), let us see what should be t_{g3} to obtain this desired t_{g4}. $(482 - t_{g3})/(482 - 149) = 0.7728$. t_{g3} = 225°C. This is an example of temperature cross situation. The pinch point is negative. Saturation temperature is 255°C, while the gas temperature leaving the evaporator is 225°C, which is not feasible. Hence, exit gas temperature should be higher than 203°C.

Example 5.4

What should be done to obtain an exit gas temperature of 149°C in the earlier case? Let us raise the inlet gas temperature to say 870°C. Then,
$(870 - t_{g3})/(870 - 149) = 0.7728$ or t_{g3} = 312°C. This is feasible as t_{g3} is above the saturation temperature of 254°C. So supplementary firing helps increase the steam generation and thus provides a larger heat sink at the economizer that enables cooling of the exhaust gas to a lower temperature. Higher the gas inlet temperature, lower the economizer exit gas temperature in single-pressure HRSGs.

Example 5.5

Is it possible to have a pinch point of 11°C with 870°C inlet gas temperature?
t_{g3} is then 255 + 11 = 266°C.
$(870 - 266)/(870 - t_{g4}) = 0.7728$ or t_{g4} = 88°C, which is impossible as it is below the feed water temperature of 110°C, another case of temperature cross! So low pinch points cannot be achieved with high inlet gas temperatures. This was also suggested in Table 5.1. Hence, these are basic principles to be kept in mind by process engineers while trying to develop gas–steam temperature profiles for their HRSGs. The important point to be kept in mind is that pinch and approach points should be selected in the unfired mode of HRSG. Then, the values suggested in Table 5.1 can be achieved.

Why Should Pinch and Approach Points Be Selected in Unfired Mode?

In order to arrive at a *design*, one should start with the unfired exhaust gas conditions as shown in Example 5.1, even though the HRSG may operate in fired mode all the time. This is because experience has shown that the unfired HRSG, with an exhaust gas temperature ranging from 450°C to 600°C, can be simulated with pinch and approach points in the range shown in Table 5.1 and the surface areas of the HRSG will be *reasonable*, practical, and feasible; starting the analysis in the fired mode can cause temperature cross issues as shown earlier as it is difficult to assume a practical pinch point for design purposes.

1. A realistic gas–steam temperature profile will result if we start with unfired HRSG exhaust conditions. We can ensure that the steam temperature can met at the design condition and without economizer steaming. In the fired mode, the steam temperature can only be higher, which can be easily controlled. If one starts the analysis in the fired mode, the steam temperature has to be checked in the unfired cases, involving more analysis and waste of time.

2. Based on experience, boiler size has been found to be reasonable and feasible with the use of pinch and approach points of 5°C–10°C in the unfired mode. If one selects these values with a higher gas inlet temperature (>600°C), then even if there is no temperature cross, the size may be uneconomical.

3. Steaming in the economizer is likely in the normal unfired operating mode if the pinch and approach points had been selected in the fired mode. One has to then evaluate the off-design performance in the unfired mode to ensure that there is no steaming in the economizer and then again check the fired performance.

4. The amount of spray cannot be predetermined in the *design* mode with ease. It is difficult to estimate the extent of over-surfacing required without checking the performance in the unfired mode. Hence, this is an unnecessary and extra exercise that can be avoided if we straightaway start with unfired mode. Spray desuperheating is required in the fired case if the steam temperature is met in the unfired case due to the higher gas temperature and LMTDs. Also one may not know if the steam temperature can be met in the guaranteed unfired case. Hence, several iterations are required to arrive at the final design. Hence, always start with unfired mode to fix the HRSG heating surfaces, even if the HRSG is fired.

Simulation of HRSG Evaporator

In order to explain what simulation is, a simple example is given.

Example 5.6

The HRSG evaporator in Example 5.1 operates with a gas flow of 90,000 kg/h, and the gas inlet temperature to the evaporator is 470°C (versus 460°C). Steam pressure is 38 kg/cm²a. What is likely to be its duty and the exit gas temperature?

Solution

Let us use the concept that U is proportional to $W_g^{0.65}$. Effects of gas properties are neglected in the first iteration. The nonluminous heat transfer coefficient is also presumed to be small. The advantage of this method is that U need not be evaluated in detail. The following equation has been derived in Appendix A for an evaporator:

$$\ln\left[\frac{(T_1 - T_s)}{(T_2 - T_s)}\right] = \frac{UA}{(W_g C_{pg})} \tag{5.6}$$

U is proportional to $W_g^{0.65}$. Since the surface area is the same, and neglecting variations in gas properties, we may write

$$\ln\left[\frac{(T_1 - T_s)}{(T_2 - T_s)}\right] = \frac{K_e}{W_g^{0.35}} \tag{5.7}$$

where

$$K_e = f(A/C_{pg}) \tag{5.8}$$

where K_e is a function of the surface area and gas specific heat of each surface.

For the design conditions, W_g = 100,000 kg/h, T_1 = 460°C, t_s = 252°C. T_2 = 262°C. Calculate constant K_e for the evaporator from (5.7):

$$K_e = 100,000^{0.35} \times \ln\left[\frac{(460 - 252)}{(262 - 252)}\right] = 170.7$$

For the new conditions, W_g = 90,000 kg/h, T_1 = 470°C. t_s = 246°C.

Hence, $\ln[(470 - 246)/(T_2 - 246)] = 170.7/90,000^{0.35} = 3.149$ or $T_2 = 256$°C. Hence, Duty = 90,000 × 0.267 × 0.99 × (470 − 246) = 5.09 MM kcal/h = 5.92 MW.

This is a good estimate only as variations in gas properties have not been considered, but the exercise shows what simulation is and how it may be applied to real-life problems to obtain speedy results.

Supplementary Firing

Supplementary firing is an excellent way to generate additional steam in cogeneration plants. Steam generators firing oil or gas have a lower heating value (LHV) efficiency of about 93%, while it is nearly 100% when the HRSG operates in the fired mode. That is, if 8 MW of additional energy is to be added to steam, a maximum of 8 MW of fuel on LHV basis alone need be fired in an HRSG. With a steam generator, one would fire about 6% more fuel or about 8.5 MW fuel unless it has a condensing economizer, which is a special case. To illustrate this point, an example is provided.

Example 5.7

Saturated steam is generated at 40 kg/cm² g in a boiler. In the unfired mode with an exhaust gas of 100,000 kg/h at 500°C, the HRSG makes about 15.4 t/h. Determine the fuel required to generate 25 t/h and 40 t/h of steam.

Solution

The HRSG was simulated using a pinch point of 8°C and an approach point of 7°C. Results are summarized in Tables 5.3 and 5.4, and Figures 5.4 and 5.5 show the off-design performance runs.

We see that the additional fuel requirement is less than the boiler duty in the fired modes. To generate 25,000 kg/h, we are using only 5.84 MW of fuel energy while the additional duty is 6.3 MW! Similarly, to generate 40 t/h, we are using only 15.53 MW

TABLE 5.3

HRSG Performance—Design Case

Surf.	Gas Temp. in/out, °C		Wat./Stm. In/Out, °C		Duty, MW	Press., kg/cm² a	Flow, kg/h	Pstm., %	Pinch, °C	Appr. ch., °C	US, kcal/h °C	Module No.
Evap.	500	259	244	251	7.48	41.0	15,454	100	8	7	91,709	1
Econ	259	171	105	244	2.63	41.7	15,454				65,692	1

Note: Project—Example 5.7 Units—METRIC Case—Example 5.7 Remarks—Amb. temp., C = 20 heat loss, % = 1 gas temp. to HRSG C = 500 gas flow, kg/h = 100,000
% vol CO_2 = 3. H_2O = 7. N_2 = 75. O_2 = 15. SO_2 =. ASME eff, % = 68.41, tot. duty, MW = 10.1

TABLE 5.4

Off-Design Fired Performance

Case	Design	Off-Design 1	Off-Design 2
Steam	15,400	25,000	40,000
Boiler duty, MW	10.1	16.4	26.3
Burner duty, MW	0	5.84	15.53
Additional boiler duty, MW	0	6.3	16.2
Exit gas temperature, °C	171	150	139
Pinch point, °C	8	12	18
Approach point, °C	7	33	62

additional fuel while the additional duty is 16.2 MW! That means we are generating the additional steam at more than 100% efficiency in these cases! Hence, supplementary firing is more than 100% efficient. This was also seen in Chapter 4. The exit gas temperature is lower in the fired mode for reasons given in Chapter 4. Again, the economizer is acting as a bigger heat sink when we increase the steam flow. Hence, the HRSG is more efficient in the fired mode. In a steam generator, the exit gas temperature increases with increase in steam generation as discussed in Chapter 4, while in an HRSG, it is the other way around. By firing fuel in an HRSG, we are in effect reducing excess air as we do not add air for this fuel but only utilize the excess oxygen in the exhaust gas.

Note that we assume the pinch and approach points only in the design mode. In fired cases, the pinch and approach points fall in place as the surface area has been selected by assuming pinch and approach points. We see that the pinch is 18°C and approach is 62°C in the 40 t/h case. How will one know what pinch and approach points to use if we had started the design in the fired mode? Yet another reason that we should always start the design in the unfired mode!

Off-Design Performance Evaluation

Figure 4.29 shows the procedure for evaluating the performance of an HRSG in off-design [2] mode. In physical design, U is computed for each heating surface at the off-design case, and the NTU method is applied to arrive at the performance. In simulation, we correct the design UA value using appropriate factors for gas properties and gas flow and use the NTU

HRSG performance—Off—Design case

Evap. Eco.

Project—eg2   Units—Metric case—eg2 Remarks-

Amb. temp., °C = 20 Heat loss, % = 1 Gas temp. to HRSG C = 500 Gas flow, kg/h = 100,000
% vol CO_2 = 3. H_2O = 7. N_2 = 75. O_2 = 15. SO_2 =. ASME eff., % = 79.42 tot duty, MW = 16.4

Surf.	Gas temp. in/out °C		Wat./Stm. in/out °C		Duty MW	Pres. kg/cm² a	Flow kg/h	Pstm. %	Pinch °C	Apprch. °C	US kcal/h °C	Module no.
Burn	500	671	0	0	5.84	0	422	0				
Evap.	671	262	216	250	13.	41.	25,033	100	12	33	96,598	1
Eco.	262	150	105	216	3.38	41.7	25,033				62,895	1

Stack gas flow = 100,422 % CO_2 = 3.72 H_2O = 8.41 N_2 = 74.45 O_2 = 13.41 SO_2=.
Fuel gas: vol%
Methane = 97 Ethane = 2   Propane = 1
LHV - kcal/cu m = 105 LHV - kcal/kg = 11,910 avg air - kg/h = 0

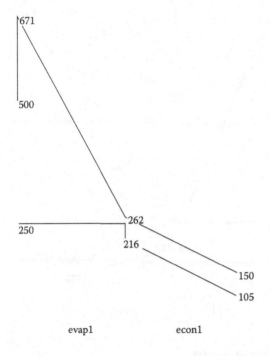

FIGURE 5.4
Off-design mode generating 25,000 kg/h steam.

method or heat balance to arrive at the duty of each section. First, we compute the factor F_g, which reflects gas properties for each surface as shown later. Then, we compute the transferred duty Q_p in the off-design mode either using the expression $Q_p = (UA)\Delta T$ or the NTU method described in Appendix A. For the superheater, a small correction for actual steam flow to design steam flow is used; for the evaporator and the economizer, this can be neglected. W_{sp}, the steam flow in off-design case, may be known or assumed; if assumed, then a few iterations are required to arrive at the final results as shown in Figure 4.29.

HRSG performance—Off—Design case

Evap. Eco.

Project—eg2 Units—Metric case—eg2 Remarks-

Amb. temp., °C = 20 Heat loss, % = 1 Gas temp. to HRSG C = 500 Gas flow, kg/h = 100,000
% vol CO_2 = 3. H_2O = 7. N_2 = 75. O_2 = 15. SO_2 =. ASME eff., % = 86.7 tot duty, MW = 26.3

Surf.	Gas temp. in/out °C		Wat./Stm. in/out °C	Duty MW	Pres. kg/cm²a	Flow kg/h	Pstm. %	Pinch °C	Apprch. °C	US kcal/h °C	Module no.
Burn	500	935	0	0	15.53	0	1,121	0			
Evap.	935	268	188	250	22.29	41.	40,057	100	18	62	104,441	1
Eco.	268	139	105	188	3.98	41.7	40,057				63,339	1

Stack gas flow = 10,1122 % CO_2 = 4.89 H_2O = 10.7 N_2 = 73.55 O_2 = 10.83 SO_2=.
Fuel gas: vol%
Methane = 97 Ethane = 2   Propane = 1
LHV - kcal/cu m = 105 LHV - kcal/kg = 11,910 avg air - kg/h = 0

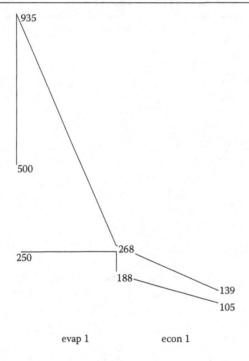

evap 1   econ 1

FIGURE 5.5
Off-design mode generating 40,000 kg/h steam.

From the relation between $(UA)_d$, the design (UA) value, and the $(UA)_p$, (UA) value in the off-design case may be obtained as follows:

$$(UA)_d = \left(\frac{Q}{\Delta T}\right)_d \tag{5.9}$$

where Q and ΔT are the duty and LMTD in the design case as obtained from Example 5.1; this term (UA) is obtained for each heating surface such as superheater, evaporator, and economizer. Then, we obtain the $(UA)_p$, the (UA) value in off-design or performance mode by using correction factors for gas flow, gas analysis, and steam flow. The steam flow correction is not required for evaporator and economizer.

$$(UA)_p = (UA)_d \times \left(\frac{W_{gp}}{W_{gd}}\right)^{0.65} \times \left(\frac{F_{gp}}{F_{gd}}\right) \times \left(\frac{W_{sp}}{W_{sd}}\right)^{0.15} \tag{5.10}$$

(Subscript d stands for design mode and p for off-design or performance mode.)

Then, we may either use the equation $Q_p = (UA)_p \Delta T_p$ or the NTU method to obtain the duty in the off-design mode. The gas property factor F_g is obtained as follows:

$$F_g = \frac{C_p^{0.33} k^{0.67}}{\mu^{0.32}} \tag{5.11}$$

Example 5.8

Let us check the off-design of the HRSG discussed in Example 5.1 when the firing temperature is 750°C. Results from the computer program are shown in Figure 5.6. Exhaust gas flow is 100,634 kg/h. Gas analysis after the burner is % volume $CO_2 = 4.08$, $H_2O = 9.11$, $N_2 = 74.17$, $O_2 = 12.62$. Steam pressure is the same. Determine the gas–steam profiles and steam flow. It is seen that the steam temperature is 439°C and steam flow is 23,471 kg/h.

Solution

Several iterations are required to solve the aforementioned problem. However, to explain the off-design calculation procedure, the final results are used.

The NTU method will be applied. For the superheater, the terms $(WC)_{min}$ and $(WC)_{max}$ have to be first obtained. The enthalpy of superheated steam at 439°C is 789.4 kcal/kg and that of saturated steam is 668.8 kcal/kg. Hence, specific heat of steam is $(789.4 - 668.8)/(439 - 255) = 0.655$ kcal/kg °C. Gas-side specific heat based on the analysis may be shown to be 0.29 kcal/kg°C at the average gas temperature in the superheater.

Gas side $(WC) = 100,640 \times 0.29 \times 0.99 = 28,893$. Steam side $(WC) = 23,471 \times 0.655 = 15,373$.
$(WC)_{min} = 15,373$. $C = 15,373/28,893 = 0.532$. $(1 - C) = 0.468$

$NTU = (UA)_p/C_{min}$. $(UA)_p$ is the corrected (UA) for the fired case. To obtain this, we need to know the $(UA)_d$ in the unfired or design case; from Figure 5.3, it is 6607 for the superheater. Using the gas properties from Appendix F, it may be shown that at 480°C (average gas temperature in the design mode of superheater), $F_g = 0.160$ and in the off-design mode (at 700°C), $F_g = 0.1777$. Then, using (5.10),

$$(UA)_p = 6,607 \times (100,634/100,000)^{0.65} \times (0.1777/0.16) \times (23,471/12,700)^{0.15} = 8079.$$

$NTU = 8,079/15,373 = 0.5255$ (see Appendix A)

$$\in = \left[1 - \exp\{-NTU(1-C)\}\right] / \left[1 - C\exp\{NTU(1-C)\}\right]$$

$$= \left[1 - \exp(-0.5255 \times 0.468)\right] / \left[1 - 0.532\exp(-0.5255 \times 0.468)\right]$$

$$= 0.2167/0.5833 = 0.3715$$

$Q_1 = $ duty of superheater $= 0.3715 \times 15,373 \times (751 - 255) = 2.833$ MM kcal/h $= 3.294$ MW

Exit gas temperature $= 751 - 2.833 \times 10^6/100,634/0.99/0.29 = 652°C$.

Exit steam temperature $= 255 + 2.833 \times 10^6/23,471/0.655 = 439°C$

HRSG performance—Off—Design case

Sh. Evap. Eco.

Project—eg3 Units—Metric case—eg3 Remarks -

Amb. temp., °C = 25 Heat loss, % = 1 Gas temp. to HRSG C = 500 Gas flow, kg/h = 100,000
% vol CO_2 = 3. H_2O = 7. N_2 = 75. O_2 = 15. SO_2 =. ASME eff., % = 79.42 tot duty, MW = 18.7

Surf.	Gas temp. in/out °C		Wat./Stm. in/out °C		Duty MW	Pres. kg/cm²a	Flow kg/h	Pstm. %	Pinch °C	Apprch. °C	US kcal/h °C	Module no.
Burn	500	751	0	0	8.77	0	633	0				
Sh.	751	652	255	439	3.32	41.	23,471	100			8,082	1
Evap.	652	273	216	255	12.16	44.6	23,471	100	17	39	85,906	1
Eco.	273	168	105	216	3.19	45.3	23,706				45,577	1

Stack gas flow = 100,634 % CO_2 = 4.08 H_2O = 9.11 N_2 = 74.17 O_2 = 12.62 SO_2=.
Fuel gas: vol%
Methane = 97 Ethane = 2   Propane = 1
LHV - kcal/cu m = 105 LHV - kcal/kg = 11,910 avg air - kg/h = 0

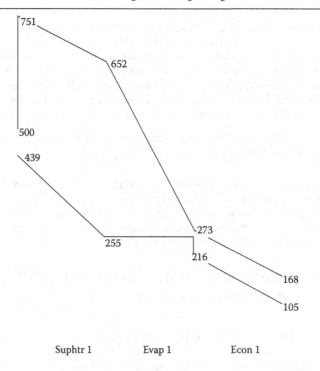

Suphtr 1   Evap 1   Econ 1

FIGURE 5.6
Fired performance of Example 5.1 HRSG.

Evaporator

Assume that F_g values have been computed in the design and off-design modes.

$$(UA)_p = 80,645 \times (0.16/0.1509) \times (100,634/100,000)^{0.65} = 85,860.$$

$$\ln\left[(652-255)/(t_{g2}-255)\right] = 85,860/100,634/0.2763/0.99 = 3.119 \text{ or } t_{g2} = 273\degree C$$

$$Q_2 = Duty = 100,634 \times 0.99 \times (652-273) \times 0.2763 = 10.432\,MM\,kcal/h = 12.13\,MW$$

Economizer

$$(UA)_p = 45,310 \times (0.1412/0.14) \times (100,634/100,000)^{0.65} = 45,886.$$

Gas side (WC) = 100,640 × 0.2625 × 0.99 = 26,153. Enthalpy of water at 216°C and 105°C are 221.2 and 105.9 kcal/kg, respectively. Hence, C_{pw} = (221.2 – 105.9)/(216 – 105) = 1.0387 kcal/kg °C. Gas side specific heat is 0.2625 kcal/kg °C.

$$(WC)_{gas} = 10,0634 \times 0.99 \times 0.2625 = 26,152 \, (WC)_{water} = 23,706 \times 1.0387 = 24,623$$

$$C_{min} = 24,623. \; C = 24,623/26,152 = 0.9287. \; NTU = 45,886/24,623 = 1.8635$$

$$\epsilon = \left[1 - \exp\{-NTU(1-C)\}\right]/\left[1 - C\exp\{NTU(1-C)\}\right]$$

$$= \left[1 - \exp(-1.8635 \times 0.0713)\right]/\left[1 - 0.9287\exp(-1.8635 \times 0.0713)\right] = 0.666$$

$$Q_3 = \text{duty of economizer} = 0.666 \times 24,623 \times (273-105) = 2.755\,MM\,kcal/h = 3.2\,MW$$

$$\text{Exit gas temperature} = 273 - 2.755 \times 10^6/100,634/0.2625/0.99 = 168\degree C$$

$$\text{Exit water temperature} = 105 + 2.755 \times 10^6/24,623 = 217\degree C$$

Total duty = 2.833 + 10.432 + 2.755 = 16.02 MM kcal/h. Steam generation based on enthalpy pickup = 16.02 × 10⁶/(789.4 – 105.9) + 0.01 × (265.9 – 221.2)] = 23,422 kg/h

The results appear to be fine, as seen in Figure 5.6. Note that the manual method is tedious as several iterations are required to arrive at the final results, particularly if one had to compute the firing temperature and burner duty for a desired amount of steam. In such a case, we have to assume some firing temperature and compute the burner duty and exhaust gas analysis (as shown in Chapter 1); then, check the duty of each heating surface and steam generation, and if the required and estimated steam flows do not match, another revision of firing temperature is required, and the earlier process has to be repeated. Hence, a computer program is required to handle such problems. The purpose of the exercise is to explain the procedure for off-design calculations, knowing the design data and thermal performance.

How Accurate Is Simulation?

In order to check how the simulation process compares with the physical design and performance calculations, an HRSG was physically designed to generate the steam parameters shown in Example 5.1, and its performance was checked in the off-design mode. The results are shown in Tables 5.5 through 5.7. In the unfired design mode based

TABLE 5.5

Geometric Data of HRSG

Geometry	Sh.	Sh.	Evap.	Evap.	Econ.
Tube OD	38.1	50.8	50.8	50.8	44.5
Tube ID	30.6	44.1	44.1	44.1	37.8
Fins/in. or fins/m	0	78	78	177	177
Fin height	0	12.5	12.5	15	19
Fin thickness	0	1.9	1.9	1.9	1.9
Fin width	0	0	0	4.37	4.37
Fin conductivity	0	30	35	35	35
Tubes/row	24	24	24	24	24
Number of rows deep	4	2	2	18	14
Length	3.5	3.5	3.7	3.7	3.5
Transverse pitch	101	101	101	101	101
Longitudinal pitch	101	101	101	101	101
Streams	12	12			4
Parl = 0, countr = 1	1	1			1
Arrangt.	Inline	Inline	Inline	Inline	Inline

TABLE 5.6

Unfired Performance

Surface	Suphtr.	Suphtr.	Evap.	Evap.	Econ.
Process data					
Gas temp. in, ±5°C	500	486	460	425	267
Gas temp. out, ±5°C	486	460	425	267	189
Gas spht., kcal/kg °C	0.2743	0.2732	0.2751	0.2670	0.2598
Duty, MM kcal/h	0.38	0.72	0.95	4.19	2.01
Surface Area, m²	40	93	98	2152	1770
Gas press. drop, mm wc	2.64	4.07	3.45	37.87	22.93
Foul factor, gas	0.0002	0.0002	0.0002	0.0002	0.0002
Steam side					
Steam press., kg/cm² g	42	44	43	43	44
Steam flow, kg/h	12,691	12,691	12,691	12,691	12,818
Fluid temp. in,°C	327	254	253	253	105.00
Fluid temp. out, ±5°C	377	327	255	255	253
Press. drop, kg/cm²	1.51	0.11	0.00	0.00	0.96
Foul factor, fluid	0.0002	0.0002	0.0002	0.0002	0.0002

TABLE 5.7

Fired Performance

Surface	Suphtr.	Suphtr.	Evap.	Evap.	Econ.
Process data					
Gas temp. in, ±5°C	746	711	648	576	280
Gas temp. out, ±5°C	711	648	576	280	164
Gas spht., kcal/kg °C	0.2918	0.2890	0.2854	0.2759	0.2626
Duty, MM kcal/h	1.01	1.83	2.07	8.12	3.03
Surface area, m²	40	93	98	2152	1770
Gas press. drop, mm wc	3.60	5.40	4.39	43.94	22.72
Foul factor, gas	0.0002	0.0002	0.0002	0.0002	0.0002
Steam side					
Steam press., kg/cm² g	42	48	47	47	51
Steam flow, kg/h	23,500	23,500	23,500	23,500	23,735
Fluid temp. in, C	361	260	227	227	105
Fluid temp. out, ±5°C	438	361	261	261	227
Press. drop, kg/cm²	5.7	0.4	0.0	0.0	3
Foul factor, fluid	0.0002	0.0002	0.0002	0.0002	0.0002
Spray, kg/h	0				

on the tube geometry, the steam generation is 12,715 kg/h at 377°C. In the fired mode, the firing temperature is 746°C to generate 23,500 kg/h of steam at 437°C. The burner duty is 7.38 MM kcal/h on LHV basis. (See Chapter 1 on how to compute duct burner duty.) Hence, the simulation results agree closely with the physical design. Since nonluminous heat transfer coefficients are neglected in the simulation calculations, it is slightly conservative. The physical design done using a computer program evaluates both the convective and nonluminous heat transfer coefficients and hence more accurate. The physical design also enables one to estimate the gas- and steam-side pressure drops and the tube wall temperatures, while the simulation process stops with the thermal design and gas–steam temperature profiles. However, it is a useful tool to obtain a basic idea of various parameters involved.

Steaming in Economizer

Economizer steaming is a common problem in gas turbine HRSGs. This arises at low or part load operation of gas turbine when the exhaust gas temperature is low while the gas mass flow is nearly the same. As an example, Figure 5.7 shows the results from the simulation program when the exhaust gas temperature is 385°C instead of 500°C, other parameters remaining the same as in Example 5.1. It is seen that due to the lower amount of steam generated, the economizer reaches saturation temperature and some steam is also formed.

HRSG performance—Off—Design case

Sh. Evap. Eco.

Project—eg3   Units—Metric case—eg2 Remarks -

Amb. temp., °C = 25 Heat loss, % = 1 Gas temp. to HRSG C = 385 Gas flow, kg/h = 100,000
% vol CO_2 = 3. H_2O = 7. N_2 = 75. O_2 = 15. SO_2 =. ASME eff., % = 47.4 tot duty, MW = 5.3

Surf.	Gas temp. in/out °C		Wat./Stm. in/out °C		Duty MW	Pres. kg/cm²a	Flow kg/h	Pstm. %	Pinch °C	Apprch. °C	US kcal/h °C	Module no.
Sh.	385	368	251	336	.52	41.	7,253	100			5,776	1
Evap.	368	256	250	251	3.38	41.3	7,253	100	5	0	78,576	1
Eco.	256	211	105	250	1.37	42.	7,325	1.58			45,511	1

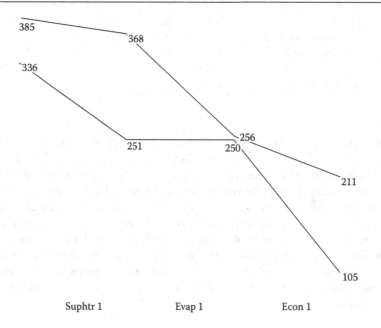

FIGURE 5.7
Steaming economizer.

When the gas inlet temperature is low, the steam generated in the superheater and evaporator is significantly reduced, and hence, a small amount of steam is generated compared to the design case. Here, we see only 7,253 kg/h of steam versus 12,700 kg/h in the unfired normal case. As the gas mass flow is unchanged, the heat transfer coefficient in the economizer is unchanged as gas-side flow governs U (see Appendix A). Hence, the enthalpy pickup on water side is rather high, and the exit water temperature reaches the saturation temperature and causes steaming. One can estimate the surface area involved in steaming (see Figure 5.8) and ensure that the flow of steam in those tubes is not from top to bottom but from bottom to top so that the bubbles can flow smoothly. There are a few methods to handle steaming economizer that can cause tube vibration or deposition of salts inside tubes or flow blockage inside tubes.

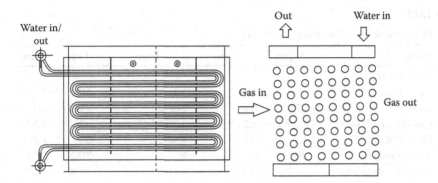

FIGURE 5.8
Steaming portion of economizer has upward water flow direction.

Methods to Minimize Steaming in Economizer

1. Increase the blowdown rate and hence the mass flow through the tubes to minimize the amount of steaming. This can be done if the steaming occurs for a short period and not a prolonged affair.

2. Increase steam flow through the HRSG by supplementary firing. Once the exhaust gas temperature is increased, more steam is generated in the evaporator and more flow through the economizer. This reduces the enthalpy pickup of water, and hence, steaming is avoided.

3. Place the feed water flow control valve between the economizer and the evaporator so that the economizer operates close to the feed pump discharge pressure, which has a higher saturation temperature, and hence, steaming is avoided.

4. Calculate the surface area in which steaming occurs. The HRSG supplier should do this exercise. Then, ensure that the water flow direction in the steaming surface is upward and not downward as shown in Figure 5.8.

5. Provide bypass on gas side at the economizer so that the duty of the economizer is reduced. This is a loss of energy, and the system is easy to install in small HRSGs. Bypassing the entire HRSG is another method, but the energy loss is much more.

In practice, one has to have an idea if the gas turbine is likely to operate at part loads for a long time and check the performance at this condition before the HRSG is designed and fabricated. If steaming is likely, then the economizer size can be reduced (resulting in loss of efficiency). If part load operation of the gas turbine is for short duration only, then we may use methods 1–3 suggested earlier to overcome this problem. Steaming in economizer is a nuisance and may cause vibration of tubes or flow blockage and deposition of salts present in feed water.

In the economizer mentioned earlier, the gas flows horizontally. The final section of the economizer is likely to steam. Hence, the water flow direction is from bottom to top in that portion. However, such an arrangement of economizer may not be always feasible particularly in large HRSGs. The simulation program results as well as results from physical design are shown in Figure 5.7 and Table 5.8, respectively. The amount of steaming is slightly different due to the consideration of nonluminous heat transfer.

TABLE 5.8

Performance of a Steaming Economizer

Surface	Suphtr	Suphtr	Evap	Evap	Econ
Process data					
Gas temp. in, ±5°C	385	379	368	349	259
Gas temp. out, ±5°C	379	368	349	259	211
Gas spht., kcal/kg °C	0.2681	0.2676	0.2689	0.2649	0.2601
Duty, MM kcal/h	0.15	0.31	0.50	2.36	1.24
Surface area, m²	40	93	98	2152	1770
Gas press. drop, mm wc	2.22	3.49	3.02	35.20	23.66
Foul factor, gas	0.0002	0.0002	0.0002	0.0002	0.0002
Steam side					
Steam press., kg/cm²g	41	41	41	41	41
Steam flow, kg/h	7270	7270	7270	7270	7343
Fluid temp. in, °C	306	251	251	251	105
Fluid temp. out, ±5°C	338	306	252	252	251
Press. drop, kg/cm²	0.5	0.0	0.0	0.0	0
Foul factor, fluid	0.0002	0.0002	0.0002	0.0002	0.0002
General data					
Exhaust gas flow	100,000	kg/h	Steam pressure	40.00	kg/cm²g
Exhaust gas temp.	385	±5°C	Steam temp.	338	±5°C
Exhaust gas pressure	1.00	kg/cm² a	Steam flow	7270	kg/h
Heat loss	1.00	%	Feed water temp.	105	±5°C
Firing temperature	0	±5°C	Process steam	0	kg/h
Burner duty, LHV	0.00	MM kcal/h	Blowdown	1	%
Burner duty, HHV	0.00	MM kcal/h	Eco steaming	2.94	%
Flue gas flow out	100,000	kg/h	Steam surface	521	m²

Field Data Evaluation

In Chapter 4, we saw the example of an HRSG operating at conditions different from those stated in the proposal, and we made an evaluation of its performance and found that the off-design performance was acceptable and that the HRSG would perform as stated in the proposal guarantee within margins of error. We shall evaluate the same problem using HRSG simulation methods so that plant engineers can appreciate how useful this tool is.

Example 5.9

An HRSG is designed to generate 15,313 kg/h of steam at 40 kg/cm²g and at 463°C with 105°C feed water using 110,000 kg/h of gas turbine exhaust at 550°C. The exit gas temperature was stated as 183°C.

The plant is unable to operate at the earlier parameters for several reasons. The gas turbine is not operating at full load as the plant does not need power. The feed water temperature is 120°C not 105°C, and steam pressure required is only 35 kg/cm²g. The plant is generating 11,000 kg/h of steam at 442°C. Plant engineers want to know if this performance can be accepted or whether the gas turbine or HRSG supplier should be questioned further regarding the performance.

Solution

Here is a simple way to check if the HRSG performance is reasonable. Let us do a heat balance and find out the exhaust gas flow. From Chapter 4, the gas flow required to generate 11,000 kg/h of steam at 442°C from 120°C feed water was shown to be 90,400 kg/h. Gas analysis used % volume $CO_2 = 3$, $H_2O = 7$, $N_2 = 75$, $O_2 = 15$.

The HRSG was simulated in the design case. Then, in the off-design case, the lower gas flow and temperature conditions were inputted. Results are shown in Figure 5.9a and b.

HRSG performance—Design case

Sh. Evap. Eco.
Project—eg4 Units—Metric case—eg4 Remarks -

Amb. temp., °C = 25 Heat loss, % = 1 Gas temp. to HRSG C = 550 Gas flow, kg/h = 110,000
% vol $CO_2 = 3$. $H_2O = 7$. $N_2 = 75$. $O_2 = 15$. $SO_2 =$. ASME eff., % = 68.89 tot duty, MW = 12.4

Surf.	Gas temp. in/out °C		Wat./Stm. in/out °C		Duty MW	Pres. kg/cm²a	Flow kg/h	Pstm. %	Pinch °C	Apprch. °C	US kcal/h °C	Module no.
Sh.	550	481	252	463	2.39	41.	15,337	100			13,965	1
Evap.	481	264	244	252	7.42	42	15,337	100	12	7	86,475	1
Eco.	264	184	105	244	2.62	42.7	15,337	0			52,013	1

Gas–steam temperature profiles

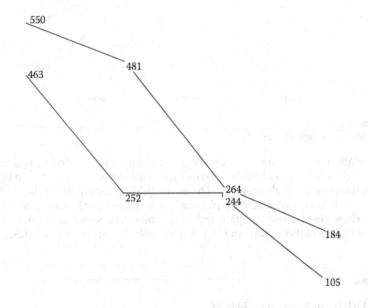

(a)   Suphtr 1   Evap 1   Econ 1

FIGURE 5.9
(a) Design information for an installed HRSG.   (*Continued*)

HRSG performance—Off—Design case

Sh. Evap. Eco.

Project—eg4 Units—Metric case—eg4 Remarks -

Amb. temp., °C = 25 Heat loss, % = 1 Gas temp. to HRSG C = 500 Gas flow, kg/h = 94,000
% vol CO_2 = 3. H_2O = 7. N_2 = 75. O_2 = 15. SO_2 =. ASME eff., % = 64.03 tot duty, MW = 8.6

Surf.	Gas temp. in/out °C		Wat./Stm. in/out °C		Duty MW	Pres. kg/cm²a	Flow kg/h	Pstm. %	Pinch °C	Apprch. °C	US kcal/h °C	Module no.
Sh.	500	445	243	439	1.56	36.	10,967	100			11,458	1
Evap.	445	252	242	243	5.34	36.6	10,967	100	8	1	75,171	1
Eco.	252	191	120	242	1.65	37.3	10,967				45,708	1

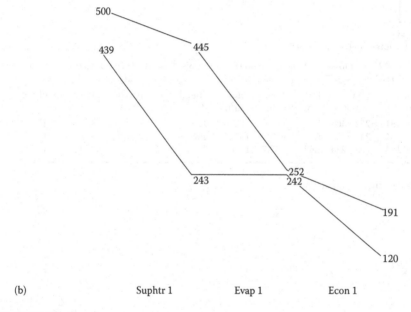

(b)   Suphtr 1   Evap 1   Econ 1

FIGURE 5.9 (*Continued*)
(b) Field performance of the installed HRSG.

The simulation results show that 11,000 kg/h of steam at 439°C can be expected from 90,400 kg/h of exhaust gas at 500°C. The exit gas temperature is expected to be 191°C, and steam temperature is about 439°C. The actual field data show 11,000 kg/h of steam at 442°C, which is also obtained by physical performance evaluation (see Tables 5.9 and 5.10). These results also compare well confirming that simulation can be used to obtain a lot of information about an *yet-to-be-built* HRSG or an operating HRSG.

Single- or Multiple-Pressure HRSG

When steam at dual pressure is required in cogeneration projects, often consultants assume that a dual-pressure HRSG is the way to go without doing any analysis; they even develop specifications around this multipressure design, which is more expensive than

TABLE 5.9

Physical Design of HRSG at 550°C

General data					
Exhaust gas flow	110,000	kg/h	Steam pressure	40.0	kg/cm² g
Exhaust gas temp.	550	±5°C	Steam temp.	463	±5°C
Exhaust gas pressure	1.00	kg/cm² a	Steam flow	15,367	kg/h
Heat loss	1	%	Feed water temp.	105	±5°C
Surface	**Suphtr.**	**Evap.**	**Econ.**		
Process data					
Gas temp. in, ±5°C	550	481	265		
Gas temp. out, ±5°C	481	265	184		
Gas spht., kcal/kg °C	0.2756	0.2691	0.2596		
Duty, MM kcal/h	2.06	6.34	2.30		
Surface area, m²	268	2749	1407		
Gas press. drop, mm wc	16.53	87.41	41.90		
Foul factor, gas	0.0004	0.0004	0.0004		
Steam side					
Steam press., kg/cm² g	41	42	42		
Steam flow, kg/h	15,367	15,367	15,367		
Fluid temp. in, °C	253	246	105		
Fluid temp. out, ±5°C	463	253	246		
Press. drop, kg/cm²	1.53	0.00	0.52		
Foul factor, fluid	0.0002	0.0002	0.0002		

a single-pressure HRSG [3]. Sometimes it is possible to meet the specific steam needs using a single-pressure HRSG. The controls and instrumentation are also simpler than for a multiple-pressure unit. Here is an example of how simulation process helps one arrive at the HRSG configuration when there is a demand for dual-pressure steam.

Example 5.10

A cogeneration project requires 100,000 kg/h of steam at 350°C and at 40 kg/cm² g. Gas turbine exhaust = 453,000 kg/h at 540°C with an analysis of % volume $CO_2 = 3$, $H_2O = 7$, $N_2 = 75$, $O_2 = 15$. It also needs 7000 kg/h of saturated steam at 10 kg/cm²g for process. The consultant is suggesting a multiple-pressure HRSG with a high-pressure module consisting of a superheater, evaporator, and economizer followed by a low-pressure evaporator and a common economizer that feeds both the modules to optimize energy recovery.

The HRSG is first simulated in the dual-pressure unfired mode to generate about 8000 kg/h LP steam and in the fired mode generates close to 7000 kg/h LP steam. In the unfired mode, the exit gas temperature is 161°C. In the fired mode, the burner fuel input is 25.9 MW and exit gas temperature is 155°C. The plant is going to operate most of the time in the fired mode only. Results are shown in Figure 5.10a and b.

An HRSG supplier who is more resourceful comes up with the suggestion of a single-pressure HRSG as shown later. He wants the plant to take off steam from the drum and pressure-reduce it. The degree of superheat is about 5°C, and hence, this is a less expensive option. His simulation results are shown in Figure 5.11a and b.

The exit gas temperature is about 149°C in the fired mode. The high pressure (HP) steam generated is 105,072 kg/h, and of this, 98,000 kg/h is saturated steam is taken to the superheater, and the desuperheater spray used compensates for the total steam

TABLE 5.10

Off-Design Performance Matching Field Data

General Data			
Exhaust gas flow	90,500	kg/h	Steam pressure	35.00	kg/cm² g
Exhaust gas temp.	500	±5°C	Steam temp.	442	±5°C
Exhaust gas pressure	1.00	kg/cm² a	Steam flow	11,038	kg/h
Heat loss	1.00	%	Feed water temp.	120	±5°C

Surface	Suphtr.	Evap.	Econ.
Process Data			
Gas temp. in, ±5°C	500	444	253
Gas temp. out, ±5°C	444	253	190
Gas spht., kcal/kg °C	0.2732	0.2674	0.2595
Duty, MM kcal/h	1.38	4 59	1.45
Surface area, m²	268	2749	1407
Gas press. drop, mm wc	10.75	57.61	28.00
Foul factor, gas	0.0004	0.0004	0.0004
Steam side			
Steam press., kg/cm² g	36	36	37
Steam flow, kg/h	11,038	11,038	11,038
Fluid temp. in, °C	244	244	120
Fluid temp. out, ±5°C	442	245	244
Press. drop, kg/cm²	0.9	0.0	0.3
Foul factor, fluid	0.0002	0.0002	0.0002
Spray, kg/h	0		
Geometry	Sh.	Evap.	Econ.
Tube OD	31.8	50.0	31.8
Tube ID	26	44	26
Fins/in. or fins/m	78	197	197
Fin height	19	19	19
Fin thickness	1.9	1.9	1.9
Fin width	4	4	4
Fin conductivity	35	35	35
Tubes/row	20	20	20
Number of rows deep	6	18	14
Length	4	4	4
Transverse pitch	04	101	84
Longitudinal pitch	101	101	100
Streams	20	0	10
Parl = 0. countr = 1	1	0	1
Arrangement	Inline	Inline	Inline

of 100,000 kg/h. (105,072 – 98,000) = 7,072 kg/h of steam is taken from the drum and pressure-reduced for the LP steam process.

The fuel input is even lower, 25 MW. In the fired mode, the exit gas temperature is lower. Hence, the plant may review this option seriously. If we add up the total surface areas (US values), the single-pressure HRSG has a total US value of 775,855 in the unfired mode (for comparison, unfired mode US is used), while the US value in multiple-pressure design is 604,573. With finned heating surfaces, this difference will not result in a more expensive design. A few more rows of finned tubes may be required in the economizer and the evaporator. However, ductwork, piping, controls, valves, and instruments will be less expensive; the LP steam drum is also eliminated.

The purpose of this exercise is to show that plants should not jump to the conclusion that if steam at dual pressure is required, then dual-pressure HRSG is the only option. Now it is likely that there are some situations where multiple-pressure steam generation is the right thing to do. For example, if the plant wanted 100,000 kg/h HP steam and maximum amount of LP steam, then dual-pressure HRSG is the way to go as it is more efficient than a single-pressure HRSG.

HRSG performance —Design case

Sh. Evap. Eco. Evap.   Eco.
Project—MM1   Units—Metric case—B Remarks -

Amb. temp., °C = 30 Heat loss, % = 1 Gas temp. to HRSG C = 540 Gas flow, kg/h = 453,000
% vol CO_2 = 3. H_2O = 7. N_2 = 75. O_2 = 15. SO_2 =. ASME eff., % = 72.54 tot duty, MW = 52.8

Surf.	Gas temp. in/out °C		Wat./Stm. in/out °C		Duty MW	Pres. kg/cm²a	Flow kg/h	Pstm. %	Pinch °C	Apprch. °C	US kcal/h °C	Module no.
Sh.	540	503	252	350	5.29	41.	64,742	100			20,796	1
Evap.	503	280	243	252	31.32	42.5	64,742	100	28	9	265,367	1
Eco.	280	240	175	243	5.63	43.2	64,742	0			97,361	1
Evap.	240	207	175	183	4.49	11.	7,944	100	24	7	101,418	2
Eco.	207	162	105	175	6.05	50.	72,686	0			119,631	3

Gas–steam temperature profiles

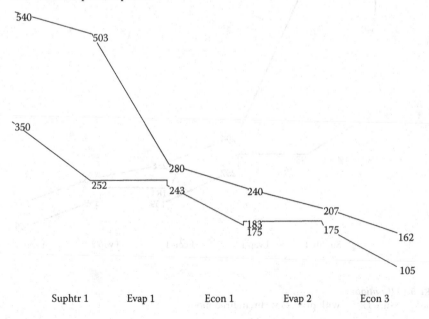

(a)

FIGURE 5.10
(a) Dual-pressure HRSG with process steam. *(Continued)*

HRSG performance—Off—Design case

Sh. Evap. Evap. Eco.   Eco.

Project—MM1 Units—Metric case—B Remarks -

Amb. temp., °C = 20 Heat loss, % = 1 Gas temp. to HRSG C = 540 Gas flow, kg/h = 453,000

% vol CO_2 = 3. H_2O = 7. N_2 = 75. O_2 = 15. SO_2 =. ASME eff., % = 79.77 tot duty, MW = 78.7

Surf.	Gas temp. in/out °C		Wat./Stm. in/out °C		Duty MW	Pres. kg/cm²a	Flow kg/h	Pstm. %	Pinch °C	Apprch. °C	US kcal/h °C	Module no.
Burn	540	705	0	0	25.93	0	1,872	0				
Sh.	705	638	255	350	10.19	41.	100,159	100			23,664	1
Desh	678	678	322	298	0	42.6	2,607	0				
Evap.	638	296	227	255	49.31	44.2	97,551	100	40	27	277,557	1
Eco.	296	236	159	227	8.27	44.9	97,551	-			98,269	1
Evap.	236	205	159	183	4.2	11.	7,180	100	22	23	101,809	2
Eco.	205	155	105	159	6.75	50.	104,731	-			119,936	3

Stack gas flow = 454,873 % CO_2 = 3.7 H_2O = 8.38 N_2 = 74.46 O_2 = 13.44 SO_2=.

Fuel gas: vol %

Methane = 97 Ethane = 2 Propane = 1

LHV - kcal/cv m = 105 LHV - kcal/kg = 11,910 aug air - kg/h = 0

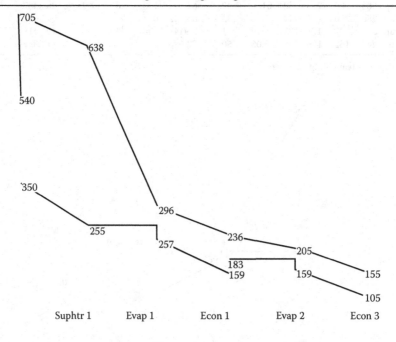

(b)

FIGURE 5.10 (*Continued*)

(b) Dual-pressure HRSG with process steam in fired mode.

Example 5.11

If LP steam required is 15,000 kg/h and HP steam is still 100,000 kg/h at 350°C, which option is better, the dual pressure or single pressure?

Analysis was performed using the simulation program, and results for single-pressure option are presented in Figure 5.12a and b. The single-pressure option is not efficient now. If we take off 15,000 kg/h of steam from the drum as before, 57,535 kg/h of steam at 350°C is generated, and the exit gas temperature is 166°C. In the fired mode, when 15,000 kg/h is taken off the drum, (112,564 – 15,000) of saturated steam is taken to the superheater, and if we add the 2,334 kg/h of spray, we obtain the desired final steam

---

HRSG performance—Design case

Sh. Evap.Eco.

Project—MMG Units—Metric case—B Remarks -

Amb. temp., °C = 30 Heat loss, % = 1 Gas temp. to HRSG C = 540 Gas flow, kg/h = 453,000
% vol CO_2 = 3. H_2O = 7. N_2 = 75. O_2 = 15. SO_2 =. ASME eff., % = 71.31 tot duty, MW = 51.9

Surf.	Gas temp. in/out °C		Wat./Stm. in/out °C		Duty MW	Pres. kg/cm²a	Flow kg/h	Pstm. %	Pinch °C	Apprch. °C	US kcal/h °C	Module no.
Sh.	540	503	252	350	5.18	41.	63,412	100			20,336	1
Evap.	503	260	247	252	34.17	42.5	71,412	100	8	5	416,536	1
Eco.	260	168	105	247	12.53	43.2	71,412	0			338,983	1

Gas steam temperature profiles

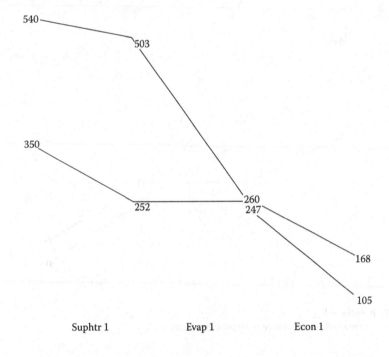

Suphtr 1　　　　Evap 1　　　　Econ 1

(a)

FIGURE 5.11

(a) Unfired single-pressure HRSG with process steam.　　　　　　　　(*Continued*)

HRSG performance—Off—Design case

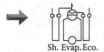

Sh. Evap. Eco.

Project—MMG   Units—Metric case—B Remarks -
Amb. temp, °C = 30 Heat loss, % = 1 Gas temp. to HRSG C = 540 Gas flow, kg/h = 453,000
% vol CO_2 = 3. H_2O = 7. N_2 = 75. O_2 = 15. SO_2 =. ASME eff., % = 80.23 tot duty, MW = 78.7

Surf.	Gas temp. in/out °C		Wat./Stm. in/out °C		Duty MW	Pres. kg/cm²a	Flow kg/h	Pstm. %	Pinch °C	Apprch. °C	US kcal/h °C	Module no.
Burn	540	700	0	0	25.31	0	1,826	0				
Sh.	700	634	255	350	9.87	41.	100,035	100			23,181	1
Desh.	673	673	319	301	0	42.7	2,012	0				
Evap.	634	266	230	255	52.81	44.4	105,072	100	11	25	434,294	1
Eco.	266	149	105	230	15.99	45.1	105,072				339,488	1

Stack gas flow = 454,826 % CO_2 = 3.68 H_2O = 8.35 N_2 = 74.47 O_2 = 13.48 SO_2 =.
Fuel gas: vol %
Methane = 97 Ethane = 3
LHV - kcal/cv m = 105 LHV - kcal/kg = 11,922 aug air - kg/h = 0

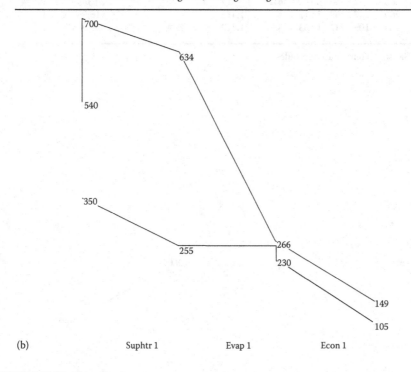

(b)   Suphtr 1   Evap 1   Econ 1

FIGURE 5.11 (*Continued*)
(b) Single-pressure fired performance with process steam.

of 100,000 kg/h. The fuel required for this is 29.67 MW. The multiple-pressure option as seen in Figure 5.13b requires 28.7 MW and exit gas temperature is 138°C. In the unfired mode also, we generate 61,800 kg/h compared to 57,535 kg/h steam at 350°C. Hence, as the ratio of LP steam flow to HP steam flow increases, dual-pressure HRSG will be more attractive. Simulation helps to see which option is better suited for the particular case. The surface area equivalent of the single-pressure HRSG is 785,988, while that of the dual-pressure HRSG is 741,258. One can convert the fuel cost differential to dollars and see the payback period. The exercise shows that sometimes a single-pressure HRSG can meet the steam needs of a plant and sometimes cannot, and one has to evaluate these options before suggesting a dual-pressure design.

HRSG performance—Design case

Sh. Evap. Eco.

Project—GG Units—Metric case—B Remarks -

Amb. temp., °C = 30 Heat loss, % = 1 Gas temp. to HRSG C = 540 Gas flow, kg/h = 453,000
% vol CO_2 = 3. H_2O = 7. N_2 = 75. O_2 = 15. SO_2 =. ASME eff., % = 71.67 tot duty, MW= 52.1

Surf.	Gas temp. in/out °C		Wat./Stm. in/out °C		Duty MW	Pres. kg/cm²a	Flow kg/h	Pstm. %	Pinch °C	Apprch. °C	US kcal/h °C	Module no.
Sh.	540	506	251	350	4.69	41.	57,535	100			18,288	1
Evap.	506	259	246	251	34.85	41.5	72,535	100	8	5	420,532	1
Eco.	259	166	105	246	12.59	42.2	72,535	0			347,168	1

Gas–steam temperature profiles

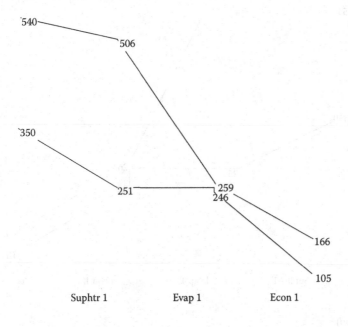

(a)

FIGURE 5.12
(a) Single-pressure HRSG in unfired mode with 15k kg/hLP steam. *(Continued)*

HRSG performance—Off—Design case

Sh. Evap. Eco.

Project—GF   Units—Metric case—B Remarks -
Amb. temp., °C = 30 Heat loss, % = 1 Gas temp. to HRSG C = 540 Gas flow, kg/h = 453,000
% vol CO_2 = 3. H_2O = 7. N_2 = 75. O_2 = 15. SO_2 =. ASME eff., % = 81.63 tot duty, MW= 83.6

Surf.	Gas temp. in/out °C		Wat./Stm. in/out °C		Duty MW	Pres. kg/cm²a	Flow kg/h	Pstm. %	Pinch °C	Apprch. °C	US kcal/h °C	Module no.
Burn	540	727	0	0	29.67	0	2,140	0				
Sh.	727	662	252	350	9.84	41.	99,898	100			21,407	1
Desh	701	701	320	298	0	41.7	2,334	0				
Evap.	662	264	224	252	57.51	42.4	112,564	100	11	28	441,804	1
Eco.	264	145	105	224	16.26	43.1	112,564	-			347,627	1

Stack gas flow = 455,141 % CO_2 = 3.8 H_2O = 8.58 N_2 = 74.38 O_2 = 13.22 SO_2 =.
Fuel gas: vol %
Methane = 97 Ethane = 3
LHV - kcal/cv m = 105 LHV - kcal/kg = 11,922 aug air - kg/h = 0

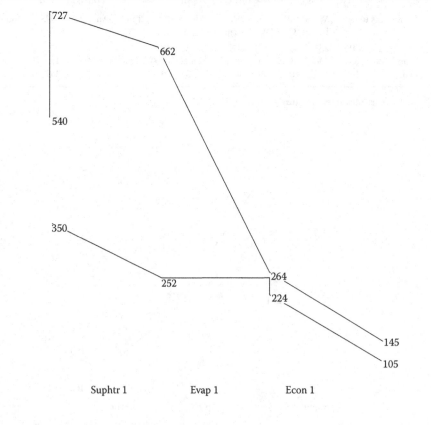

(b)

             Suphtr 1   Evap 1   Econ 1

FIGURE 5.12 (*Continued*)
(b) Single-pressure fired with 15k kg/h process steam.

Split Superheater Design

HRSG simulation may be used to evaluate the best option for a particular steam plant, and then based on the studies, one can suggest to the HRSG supplier or to the consultant what the HRSG configuration should look like. If simulation analysis is not done, it is doubtful if HRSG suppliers or consultant will come up with such solutions.

A plant had a peculiar problem. It did not have demineralized water for desuperheating steam, and it did not want to spend money on a more expensive sweet water condensing system (see Chapter 3). It wanted an HRSG that could meet its steam temperature

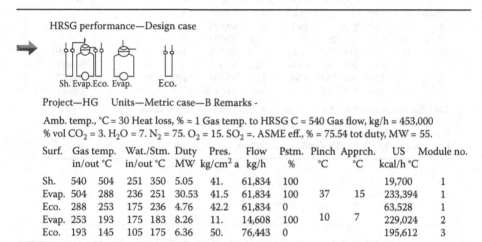

HRSG performance—Design case

Sh. Evap.Eco. Evap.   Eco.

Project—HG   Units—Metric case—B Remarks -

Amb. temp., °C = 30 Heat loss, % = 1 Gas temp. to HRSG C = 540 Gas flow, kg/h = 453,000
% vol CO_2 = 3. H_2O = 7. N_2 = 75. O_2 = 15. SO_2 =. ASME eff., % = 75.54 tot duty, MW = 55.

Surf.	Gas temp. in/out °C		Wat./Stm. in/out °C		Duty MW	Pres. kg/cm² a	Flow kg/h	Pstm. %	Pinch °C	Apprch. °C	US kcal/h °C	Module no.
Sh.	540	504	251	350	5.05	41.	61,834	100			19,700	1
Evap.	504	288	236	251	30.53	41.5	61,834	100	37	15	233,394	1
Eco.	288	253	175	236	4.76	42.2	61,834	0			63,528	1
Evap.	253	193	175	183	8.26	11.	14,608	100	10	7	229,024	2
Eco.	193	145	105	175	6.36	50.	76,443	0			195,612	3

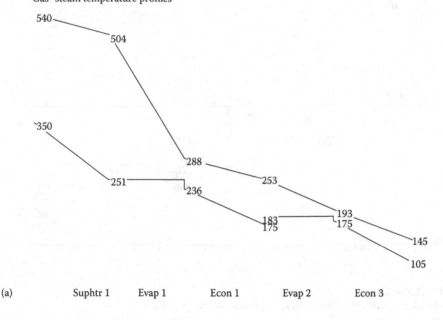

Gas–steam temperature profiles

(a)   Suphtr 1   Evap 1   Econ 1   Evap 2   Econ 3

FIGURE 5.13
(a) Dual-pressure HRSG in unfired mode with 15k kg/h process steam.   (*Continued*)

HRSG performance—Off—Design case

Sh. Evap. Eco. Evap.   Eco.

Project—HG   Units—Metric case—B   Remarks -

Amb. temp., °C = 30 Heat loss, % = 1 Gas temp. to HRSG C = 540 Gas flow, kg/h = 453,000
% vol CO_2 = 3. H_2O = 7. N_2 = 75. O_2 = 15. SO_2 =. ASME eff., % = 82.35 tot duty, MW = 83.5

Surf.	Gas temp. in/out °C		Wat./Stm. in/out °C		Duty MW	Pres. kg/cm²a	Flow kg/h	Pstm. %	Pinch °C	Apprch. °C	US kcal/h °C	Module no.
Burn	540	721	0	0	28.67	0	2,068	0				
Sh.	721	653	252	350	10.22	41.	99,925	100			22,710	1
Desh	694	694	323	296	0	41.6	2,975	0				
Evap.	653	308	219	252	50.01	42.2	96,949	100	55	32	245,380	1
Eco.	308	258	161	219	6.94	42.9	96,949				64,402	1
Evap.	258	193	161	183	8.84	11.	15,187	100	10	21	230,776	2
Eco.	193	138	105	161	7.51	50.	112,136				196,247	3

Stack gas flow = 455,069 % CO_2 = 3.78 H_2O = 8.53 N_2 = 74.4 O_2 = 13.28 SO_2=.
Fuel gas: vol %
Methane = 97 Ethane = 3
LHV - kcal/cv m = 105 LHV - kcal/kg = 11,922 aug air - kg/h = 0

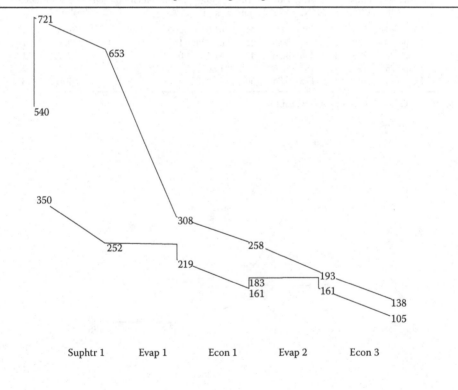

(b)

FIGURE 5.13 (Continued)
(b) Fired dual-pressure HRSG with 15k kg/h process steam.

requirements in both unfired and fired modes without much deviation so that desuperheater spray could be avoided. When we have an HRSG that fires into a two-stage superheater with a desuperheater in between, the steam temperature will go up in the fired mode, and hence, desuperheater spray would be required to control the steam temperature. If demineralized water is unavailable, a sweet water condenser system, which is more expensive, will be required. The tube wall temperature will also be higher in this system as direct radiation from the flame will be significant in the first few rows of the superheater tubes as discussed in Appendix D.

By locating part of the superheater upstream of the burner, one can minimize or even avoid the need for desuperheating. The cross section of the final superheater and the one downstream of the burner may be different, and more ductwork and piping will be required. If we look at the surface areas between the two options (Figures 5.14 and 5.15),

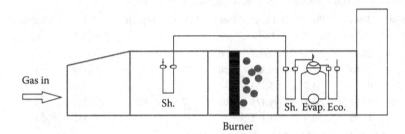

Project—ef Units—Metric case—B Remarks - Unfired case

Amb. temp., °C = 30 Heat loss, % = 1 Gas temp. to HRSG C = 500 Gas flow, kg/h = 400,000
% vol CO_2 = 3. H_2O = 7. N_2 = 75.5 O_2 = 15.1 SO_2 =. ASME eff., % = 60.2 tot duty, MW = 35.6

Surf.	Gas temp. in/out °C		Wat./Stm. in/out °C		Duty MW	Pres. kg/cm² a	Flow kg/h	Pstm. %	Pinch °C	Apprch. °C	US kcal/h °C	Module no.
Sh.	500	485	312	359	1.82	54.	48,525	100			9,984	1
Sh.	485	467	268	312	2.16	54.1	48,525	100			10,018	2
Evap.	467	288	260	268	22.29	54.2	48,525	100	20	7	245,015	2
Eco.	288	211	105	260	9.33	54.9	48,525	0			136,873	2

Project—ef Units—Metric case—B Remarks - Fired case

Amb. temp., °C = 30 Heat loss, % = 1 Gas temp. to HRSG C = 500 Gas flow, kg/h = 400,000
% vol CO_2 = 2.4 H_2O = 7. N_2 = 75.5 O_2 = 15.1 SO_2 =. ASME eff., % = 76.26 tot duty, MW = 73.1

Surf.	Gas temp. in/out °C		Wat./stm. in/out °C		Duty MW	Pres. kg/cm² a	Flow kg/h	Pstm. %	Pinch °C	Apprch. °C	US kcal/h °C	Module no.
Sh.	500	485	333	358	1.91	54.	99,826	100			11,221	1
Burn.	485	748	0	0	36.75	0	2,652	0				
Sh.	748	702	269	333	6.21	55.	99,826	100			125,99	2
Evap.	702	309	222	269	50.82	55.4	99,826	100	39	47	265,329	2
Eco.	309	194	105	222	14.18	56.1	99,826				138,345	2

Stack gas flow = 402,652 % CO_2 = 3.53 H_2O = 9.2 N_2 = 74.63 O_2 = 12.62 SO_2=.
Fuel gas: vol %
Methane = 97 Ethane = 3
LHV - kcal/cv m = 105 LHV - kcal/kg = 11,922 aug air - kg/h = 0

FIGURE 5.14
HRSG with split superheaters.

HRSG performance—Off—Design case
Project–v_{g1} Units–METRIC Case–B Remarks
Amb. temp., °C = 20 heat loss, % = 1 gas temp. to HRSG C = 500 gas flow, kg/h = 400,000
% vol CO_2 =3. H_2O = 7. N_2 = 75. O_2 = 15. SO_2 =. ASME eff, % = 60.53 tot. duty-MW = 35.8

Surf.	Gas Temp. In/Out, °C		Wat./Stm. In/Out, °C		Duty MW	Press. kg/cm²a	Flow kg/h	Pstm %	Pinch °C	Apprch °C	US kcal/h °C	Module No.
Sh.	500	467	268	359	4.0	54.0	48,770	100			20,460	1
Evap.	467	288	261	268	22.32	54.3	48,770	100	20	7	246,650	1
Eco	288	210	105	261	9.46	55.0	48,770	0			141,415	1

HRSG performance—Off—Design case
PROJECT–v_{g1} Units–METRIC Case–B Remarks
Amb temp., °C = 20 heat loss, % = 1 gas temp. to HRSG C = 500 gas flow, kg/h = 400,000
% vol CO_2 = 3. H_2O = 7. N_2 = 75. O_2 = 15. SO_2 =. ASME eff, % = 76.51 Tot. duty, MW = 73.4

Surf.	Gas Temp. In/Out, °C		Wat./Stm. In/Out, °C		Duty MW	Press. kg/cm²a	Flow kg/h	Pstm %	Pinch C	Apprch C	US kcal/h °C	Module No.
Burn	500	763	0	0	36.8	0	2,658	0				
Sh.	763	675	269	359	11.85	54.	99,830	100			25,178	1
Desh	727	727	346	300	0	54.5	5,358	—				
Evap.	675	307	226	269	47.57	55.1	94,471	100	37	42	264,213	1
Econ	307	193	105	226	13.96	55.8	94,471				142,817	1

Stack gas flow = 402,653% CO_2 = 4.13, H_2O = 9.21, N_2 = 74.13, O_2 = 12.51 SO_2 = 0.
Fuel gas: vol%
Methane = 97, ethane = 2, propane = 1
LHV. kcal/cu m = 105 LHV. Kcal/kg = 11,910 aug air, kg/h = 0

FIGURE 5.15
Two-stage superheater downstream of burner.

the difference is not much. The (US) values of both options are close to each other. In order to see how much the steam temperature varies between the fired and unfired modes, a simulation was performed.

Example 5.12

Exhaust gas flow from a gas turbine is 400,000 kg/h at 500°C. % Volume CO_2 = 2.4, H_2O = 7, N_2 = 75.5, O_2 = 15.1. About 48,500 kg/h of steam at 54 kg/cm²a and at 355°C ± 5°C is required in the unfired mode and 100,000 kg/h of steam at the same temperature in the fired mode. Natural gas is the burner fuel, and feed water is at 109°C. See if the steam temperatures can be met in both modes of operation without control.

Solution

Using the simulation program, the arrangement shown in Figure 5.14 was simulated. The final superheater stands aloof and is located ahead of the burner. A module consisting of superheater, evaporator, and economizer generates superheated steam and feeds the final superheater. In the unfired mode, 312°C steam temperature was selected as the primary superheater exit temperature and the final superheater was simulated to give 359°C. In the fired mode, which is the off-design case, the steam temperature is 333°C and the final steam temperature is 358°C. Hence, this scheme is feasible. Chapter 4 on waste heat boilers shows the physical design of this HRSG. One can see that there is a good agreement between the simulation results and the actual results from the physical design. Hence, simulation offers an easy method to conceive complex HRSG arrangements without physically designing them. Once a concept or arrangement is arrived at through simulation, one can design the HRSG with the help of an HRSG supplier. It is doubtful if HRSG suppliers will be able to arrive at these options straightaway without results from such simulation studies unless they have had experience with similar units before.

When the burner is located ahead of a two-stage superheater with desuperheater in-between, the performance is shown in Figure 5.15. It may be seen that the firing temperature is higher as the entire boiler duty is handled downstream of the burner. This results in higher steam temperature, which is controlled by the desuperheater. Due to the higher firing temperature and direct radiation from the flame, it is likely that the tube wall temperature of the superheater is higher than in the previous case of split superheaters. The overall performance is nearly the same. The burner duty has also not changed. If desuperheater is not used in the case of downstream superheaters, the steam temperature will increase by about 50°C and so will the tube wall temperature.

Fresh Air Firing

Often fresh air firing is resorted to generate steam when the gas turbine is not in operation. However, it should be noted that using ambient air for generating steam in an HRSG is not an efficient process and must be used sparingly or for emergency reasons only and not for continuous operation. It must be clearly understood that in HRSGs, the ratio of gas flow to steam is much larger than in conventional steam generators. Hence, a large amount of energy is wasted by raising the temperature of air from ambient conditions and exhausting the gas at a higher temperature at the stack. One may wonder that we do the same thing in a steam generator. However, keep in mind that the ratio of flue gas to steam is about 1.1 in a steam generator, while in an HRSG, it will be in the range of 6–8, and so a large amount of exhaust gas is heated from ambient conditions relative to steam generation and then vented to the atmosphere. The sizing of the HRSG should be such that the exhaust gas flow in fresh air mode also should be in the same range as the exhaust gas flow with gas turbine operation or else the gas velocities can be much lower causing nonuniformities in gas flow at the heat transfer surfaces and poor heat transfer. It can be slightly lower than the turbine exhaust gas flow say by 10%–20% but not much lower. The following examples give an idea of the performance in fresh air mode with two different air flows, and one can see the issues here.

Example 5.13

The HRSG in Example 5.1 is to be operated in fresh air mode. Check the performance when 100,000 kg/h of air and 80,000 kg/h of air are used to generate 12,000 kg/h.

The results for both the cases are presented in Figure 5.16a and b. It can be seen that the efficiency is higher when the air flow is lower and vice versa. However, the firing temperature increases with lower air flow. If the HRSG had been designed to accept higher firing temperatures, then we can further lower the air flow and generate the same amount of steam more efficiently. However, we cannot reduce the air flow too much due to gas maldistribution concerns. When sizing the fan, about 80% of the turbine exhaust flow should be adequate. While locating the fan and duct, one should ensure that the flow nonuniformity is minimized as the fan is generally located at right angles to the main duct or at an angle. Grid plates with variable openings may be inserted in the duct,

Project—hh Units—Metric case—B Remarks -

Amb. temp., °C = 25 Heat loss, % = 1 Gas temp. to HRSG C = 25 Gas flow, kg/h = 80,000
% vol CO_2 = . H_2O = 1. N_2 = 78. O_2 = 21. SO_2 =. ASME eff., % = 69.85 tot duty, MW = 9.2

Surf.	Gas temp. in/out °C		Wat./Stm. in/out °C		Duty MW	Pres. kg/cm²a	Flow kg/h	Pstm. %	Pinch °C	Apprch. °C	US kcal/h °C	Module no.
Burn	25	551	0	0	12.92	0	932	0				
Sh.	551	498	251	394	1.34	41.	11,979	100			5,779	1
Evap.	498	260	238	251	5.87	41.9	11,979	100	8	13	70,521	1
Eco.	260	178	105	238	1.97	42.6	12,099				39,211	1

Stack gas flow = 80,932 % CO_2 = 2.05 H_2O = 5.02 N_2 = 76.41 O_2 = 16.49 SO_2=.
Fuel gas: vol %
Methane = 97 ethane = 3
LHV - kcal/cv m = 11 LHV - kcal/kg = 6,623 aug air - kg/h = 0

Sh. Evap. Eco.

Project—hh Units—Metric case—B Remarks -

Amb. temp., °C = 25 Heat loss, % = 1 Gas temp. to HRSG C = 25 Gas flow, kg/h = 100,000
% vol CO_2 = . H_2O = 1. N_2 = 78. O_2 = 21. SO_2 =. ASME eff., % = 62.9 tot duty, MW = 9.

Surf.	Gas temp. in/out °C		Wat./Stm. in/out °C		Duty MW	Pres. kg/cm²a	Flow kg/h	Pstm. %	Pinch °C	Apprch. °C	US kcal/h °C	Module no.
Burn	25	486	0	0	14.08	0	1,015	0				
Sh.	486	449	251	371	1.16	41.	12,018	100			6,497	1
Evap.	449	261	246	251	5.76	41.9	12,018	100	9	5	80,327	1
Eco.	261	190	105	246	2.11	42.6	12,138				45,398	1

Stack gas flow = 101,016 % CO_2 = 1.79 H_2O = 4.52 N_2 = 76.61 O_2 = 17.06 SO_2=.
Fuel gas: vol %
Methane = 97 Ethane = 3
LHV - kcal/cv m = 105 LHV - kcal/kg = 11,922 aug air - kg/h = 0

FIGURE 5.16
Fresh air firing cases with 80k and 100k kg/h air flow.

and by measurement of static pressure, one can get an idea of the velocity profile across the cross section at the burner. The profile should also be checked for turbine exhaust gas. The velocity profile cannot be optimized for both turbine exhaust and fresh air as the flow directions are at nearly 90° to each other. Some compromise must be accepted depending on how often fresh air firing is used. More on fresh air firing is discussed in Chapter 4.

Efficiency of HRSG

The efficiency of the HRSG, E, in gas turbine mode is (Figure 5.3) as follows [4,5]:
E = steam generated/[exhaust gas flow enthalpy + burner fuel input on LHV basis] = 9.6 × 0.86 × 10^6/[100,000 × 127.1] = 65%. The 0.86 factor converts watts to kcal/h. From Table F.12, 127.1 kcal/kg is the enthalpy of the exhaust gas at 500°C.

With fresh air firing using 100,000 kg/h of air at 25°C, E = 9 × 10^6 × 0.86/[100,000 × 2.3 + 14.08 × 0.86 × 106] = 62.7% (2.3 kcal/kg is the enthalpy of air at 25°C).

While with 80,000 kg/h air, it is E = 9.2 × 10⁶ × 0.86/[80,000 × 127.1 + 12.92 × 106 × .86] = 70%.

Thus, fuel utilization is better with lower air flow, but keep in mind that we cannot use much lower air flow for reasons cited earlier.

Cogeneration Plant Application

The steam parameters of combined cycle and cogeneration plants differ significantly:

Steam parameters of combined cycle plant HRSGs have been standardized. HP, MP, and LP parameters are more or less universal for a given plant power output or gas turbine capacity. These HRSGs are generally unfired as discussed in Chapter 4. Fired HRSGs are more common in cogeneration plants. Steam demand is quite high in refineries, chemical plants, or fertilizer plants. Due to the high firing temperatures, single-pressure HRSGs are adequate to cool the exhaust gases to a reasonably low temperature. Condensate heater or a deaerator may be added to maximize energy recovery.

In a chemical or fertilizer plant, saturated steam is generated is generated in numerous boilers and, to superheat the steam, the HRSG is often used. Similarly, saturated steam is taken from the drum for process heating purposes. In such plants, one also has to evaluate the HRSG performance when the export/ import steam is absent. This affects the steam temperature, and the steam temperature may be reduced significantly or large amount of desuperheating may be required as the steam flow is now reduced. HRSG simulation may be used to determine the HRSG performance upfront.

Example 5.14

Exhaust gas flow from a gas turbine is 250,000 lb/h at 1000°F. Gas analysis % volume is $CO_2 = 3$, $H_2O = 7$, $N_2 = 75$, $O_2 = 15$. Superheated steam at 600 psia at 875°F and about 20,000 lb/h of saturated steam are required for process, which is taken off the drum. Predict the performance using 20°F pinch and approach points, 230°F feed water, and 1% blowdown and heat loss. Check what happens when the process steam is not used but sent through the superheater.

It is seen that for the normal performance with export steam (Figure 5.17a), the steam temperature is 875°F. However, when the export steam is not used (Figure 5.17b), the entire steam flow goes through the superheater. Hence, the steam temperature drops to 749°F. It is possible to oversize the superheater and ensure that the desired steam temperature is obtained when all the steam is passing through the superheater and then control the steam temperature when process steam is taken off the drum. This may require even a three-stage desuperheater depending on steam temperature requirements.

Example 5.15

Here is another simple example on simulation. A water tube boiler has inlet and exit gas temperatures of 700°C and 250°C with saturation temperature at 220°C and a gas flow of 100,000 kg/h. If gas flow increases by 20% to 120,000 kg/h at the same temperature of 700°C, how much will the duty increase? Assume gas specific heat is 0.27 kcal/kg °C.

Solution

Use the following equation:

$$\ln\left[(700-220)/(250-220)\right] = UA/100{,}000/0.27/0.99 \text{ or } UA = 74{,}111$$

With 20% increase in gas flow, UA is $1.20^{.65} \times 74{,}111 = 83{,}435$

$$\ln\left[(700-220)/(T-220)\right] = 83435/120000/.27/.99 = 2.6$$

or T = 256°C. Hence, ratio of new duty to earlier duty = $120{,}000 \times 0.27 \times 0.99 \times (700-256)/$ $[100{,}000 \times 0.27 \times 0.99 \times (700-250)] = 1.184$ or 18.4%, slightly less than the 20% increase in mass flow.

HRSG performance—Design case

Sh. Evap. Eco.

Project—study1 Units—British case—B Remarks

Amb. temp., °F = 60 Heat loss, % = 1 Gas temp. to HRSG F = 1,000 Gas flow, Lb/h = 250,000

% vol CO_2 = 3. H_2O = 7. N_2 = 75. O_2 = 15. SO_2 =. ASME eff., % = 68.69 tot duty, MW Btu/h = 42.5

Surf.	Gas temp. in/out °F	Wat./Stm. in/out °F	Duty MMB/h	Pres. Psia	Flow Lb/h	Pstm. %	Pinch °F	Apprch. °F	US Btu/h °F	Module no.
Sh.	1000 935	488 875	4.4	600.	17,883	100			17,407	1
Evap.	935 506	458 488	28.45	615.	37,883	100	20	20	207,256	1
Eco.	506 359	230 458	8.55	625.	38,262	0			126,636	1

Gas–steam temperature profiles

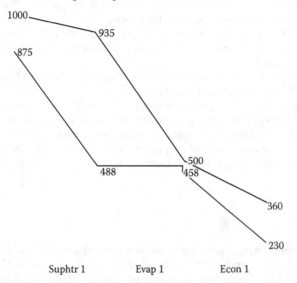

(a)   Suphtr 1   Evap 1   Econ 1

FIGURE 5.17
(a) Design with process steam.

HRSG performance—Off—Design case

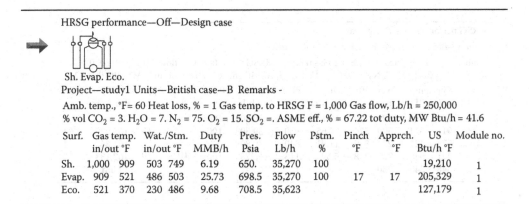

Sh. Evap. Eco.

Project—study1 Units—British case—B Remarks -

Amb. temp., °F= 60 Heat loss, % = 1 Gas temp. to HRSG F = 1,000 Gas flow, Lb/h = 250,000

% vol CO_2 = 3. H_2O = 7. N_2 = 75. O_2 = 15. SO_2 =. ASME eff., % = 67.22 tot duty, MW Btu/h = 41.6

Surf.	Gas temp. in/out °F		Wat./Stm. in/out °F		Duty MMB/h	Pres. Psia	Flow Lb/h	Pstm. %	Pinch °F	Apprch. °F	US Btu/h °F	Module no.
Sh.	1,000	909	503	749	6.19	650.	35,270	100			19,210	1
Evap.	909	521	486	503	25.73	698.5	35,270	100	17	17	205,329	1
Eco.	521	370	230	486	9.68	708.5	35,623				127,179	1

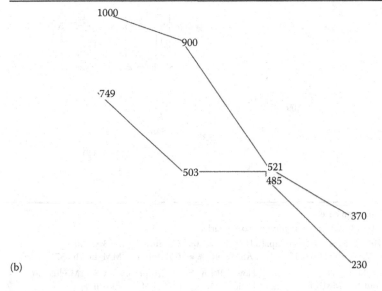

(b)

FIGURE 5.17 (Continued)
(b) Performance without process steam.

Optimizing HRSG Arrangement

Often, dual-pressure steam is required, and the question arises, how should the HP and LP sections be arranged to maximize energy recovery? One option is HP section followed by the LP section as shown in Figure 5.18a. In this example (in British units), about 39,500 lb/h of HP steam at 800 psig is required, and the plant wants the maximum amount of LP steam at 100 psig. Exhaust gas flow is 300,000 lb/h at 1000°F. The pinch and approach points for the HP section were manipulated to give the required 39,500 lb/h of HP steam and low pinch and approach points were used to maximize the LP steam as required by the plant. Figure 5.18a shows that this arrangement gives only about 6200 lb/h of LP steam. The HP and LP sections have their own economizers in this option.

However, another option to maximize the LP steam is to use what is called the *common economizer* concept (Figure 5.18b). That is, by increasing the water flow of the *common*

HRSG performance—Design case

Project—opti   Units—British case—B Remarks -

Amb. temp., °F= 60 Heat loss, % = 1 Gas temp. to HRSG C = 1,000 Gas flow, Lb/h = 300,000

% vol CO_2 = 3. H_2O = 7. N_2 = 75. O_2 = 15. SO_2 =. ASME eff., % = 70.69 tot duty, MM Btu/h = 52.5

Surf.	Gas temp. in/out °F	Wat./Stm. in/out °F	Duty MMB/h	Pres. Psia	Flow Lb/h	Pstm. %	Pinch °F	Apprch. °F	US Btu/h °F	Module no.
Sh.	1000 918	524 750	6.71	815.	39,503	100			21,208	1
Evap.	918 574	509 524	27.65	830.	39,503	100	50	15	165,929	1
Eco.	574 421	230 509	11.91	840.	39,898				101,848	1
Evap.	421 348	328 338	5.59	115.	6,278	100	10	10	162,133	2
Eco.	348 340	230 328	.64	125.	6,341				12,051	2

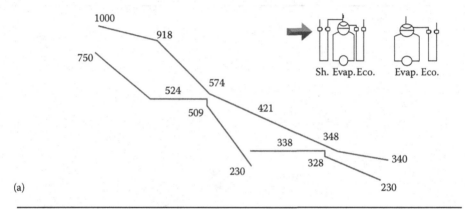

(a)

HRSG performance—Design case

Project—opti   Units—British case—common eco Remarks -

Amb. temp., °F= 60 Heat loss, % = 1 Gas temp. to HRSG F = 1,000 Gas flow, kg/h = 300,000

% vol CO_2 = 3. H_2O = 7. N_2 = 75. O_2 = 15. SO_2 =. ASME eff., % = 76.77 tot duty, MM Btu/h=57.

Surf.	Gas temp. in/out °F	Wat./Stm. in/out °F	Duty MMB/h	Pres. psia	Flow Lb/h	Pstm. %	Pinch °F	Apprch. °F	US Btu/h °F	Module no.
Sh.	1,000 918	524 750	6.71	815.	39,503	100			21,208	1
Evap.	918 574	509 524	27.65	830.	39,503	100	50	15	165,929	1
Eco.	574 473	238 509	7.93	840.	39,898				79,420	1
Evap.	473 348	328 338	9.65	115.	10,835	100	10	10	200,914	2
Eco.	348 281	230 328	5.08	900.	50,842				151,890	3

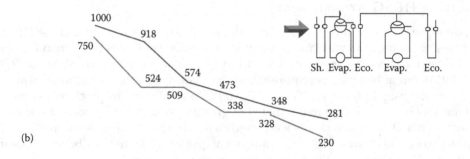

(b)

FIGURE 5.18

(a) HP section followed by LP section. (b) Use of common economizer concept can improve energy recovery.

economizer, which feeds both the HP and LP sections, one increases the heat sink capacity at the exit of LP evaporator. This arrangement increases the surface area of the LP section and common economizer. However, considering the additional amount of LP steam or the 60°F lower stack gas temperature or the 4.5 MM Btu/h additional energy recovery, one can make a quick economic evaluation and determine if this option is worth it. Typically considering the life of the HRSG (30–40 years), this option will pay off in a short period as the cost of the HRSG will be only slightly more. The purpose of this example is to let process and plant engineers know that one can recover additional energy from turbine exhaust gases by rearranging heating surfaces (particularly in a multimodule configuration HRSG), and simulation is a good tool for this exercise. Note that HRSG suppliers may not have time to do such studies to optimize the HRSG configuration.

Application of Simulation to Understand the Effect of Ambient Conditions

Exhaust gas flow and temperature of a gas turbine vary with ambient temperature or inlet air density and load as shown in Tables 5.11 and 5.12. Also depending on whether there is steam or water injection, the exhaust gas analysis may vary. What are important to the HRSG performance are the exhaust gas flow, temperature, and analysis. Using these data, plant engineers may simulate the HRSG performance even before the HRSG is supplied or details of its physical design are known as shown. Figure 5.19 (gas flow has a multiplication factor of 0.1) shows such a study. This information may be used to see if the particular

TABLE 5.11

Gas Turbine Performance versus Ambient Temperature

Ambient, °F	20	40	60	80	100	120
Power, kW	38,150	38,600	35,020	30,820	27,360	24,040
Heat rate, Btu/kWh	9,384	9,442	9,649	9,960	10,257	10,598
Exhaust temp., °F	734	780	797	820	843	870
Exhaust flow, lb/h	1,123,200	1,094,400	1,029,600	950,400	878,400	810,000
% volume CO_2	2.7	2.9	2.8	2.8	2.7	2.7
H_2O	7.6	8.2	8.5	9.2	10.5	12.8
O_2	14.6	14.3	14.3	14.2	14	13.7
N_2	75.1	74.7	74.4	73.8	72.8	70.8

TABLE 5.12

Gas Turbine Performance at Various Loads

Load, %	10	20	30	40	50
Generator, kW	415	830	1,244	1,659	4,147
Exhaust gas flow, lb/h	147,960	148,068	148,170	148,320	148,768
Exhaust temp., °F	562	612	662	712	1,019
% Volume CO_2	1.18	1.38	1.59	1.79	3.04
H_2O	3.76	4.14	4.53	4.93	7.33
O_2	18.18	17.78	17.28	16.88	14.13
N_2	76.9	76.7	76.6	76.4	75.5

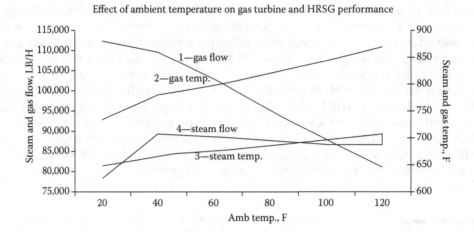

FIGURE 5.19
HRSG performance at different ambient temperatures.

gas turbine is suitable for the plant. It may be difficult to get such information from HRSG suppliers as it is unlikely that they will spend time on such studies before a contract is awarded, and then it may be too late to find out that a different gas turbine would have suited their need better!

Similarly, if we evaluate the HRSG performance at different loads, we may note that at 20% load, the exhaust gas temperature is so low that the economizer will generate steam, and the exit gas temperature will be very high. This has been discussed earlier.

Applying Margins on Exhaust Gas Flow and Temperature

Instrumentation errors or margins should be considered while evaluating the operating data of an HRSG. The end user and the HRSG supplier may arrive at some understanding on how to evaluate the performance test data. For example, the uncertainty in gas flow

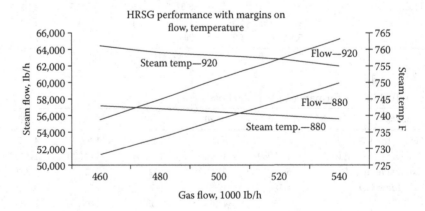

FIGURE 5.20
Typical gas turbine performance with margins.

measurement per power test code ASME PTC 4.4 varies from 3 to 5% and ±10°F on temperature measurements and by 0.5%–1.5% on steam flows depending on methods used. Hence, even if the actual exhaust gas flow differs from that estimated, one should apply these margins and see if the guaranteed steam flow is within the margin of error. Figure 5.20 shows the HRSG performance at guarantee point considering variations in exhaust gas flow and temperature. Then, if the steam generation and steam temperature are

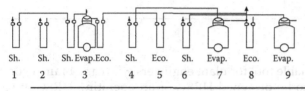

HRSG performance—Design case

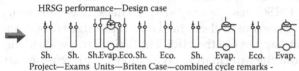

Sh. Sh. Sh.Evap.Eco.Sh. Eco. Sh. Evap. Eco. Evap.
Project—Exams Units—Briten Case—combined cycle remarks -

Amb. temp, °F= 80 Heat loss, % = 1 Gas temp. to HRSG C = 1,136 Gas flow, - Lb/h = 6,583,000
% vol CO_2 = 3. H_2O = 7. N_2 = 74.54 O_2 = 15. SO_2 =. ASME eff, % = 78.63 tot duty, MW Btu/h= 1496.2

Surf.	Gas temp. in/out °F		Wat./Stm. in/out °F		Duty MMB/h	Pres. Pisa	Flow kg/h	Pstm. %	Pinch. °F	Apprch. °F	US Btu/h °F	Module no.
Sh.	1,135	1,093	900	1,054	77.48	1825	796,305	100			599,975	1
Sh.	1,093	987	630	1,032	195.85	350	917,000	100			1,165,015	2
Sh.	987	871	625	900	212.19	1840	796,305	100			1,382,638	3
Evap.	871	640	610	625	413.48	1855	796,305	100	15	15	5,008,280	3
Eco.	640	607	558	610	57.21	1865	796,305	0			1,474,682	3
Sh.	607	599	439	575	13.45	375	148,982	100			168,690	4
Eco.	599	530	430	558	118.48	1900	796,305	0			1,775,908	5
Sh.	530	522	305	485	14.26	70	153,162	100			129,762	6
Evap.	522	454	430	439	118.53	380	148,916	100	15	9	2,974,070	7
Eco.	454	379	310	430	128.58	1900	1,021,000	0			2,991,608	8
Evap.	379	291	235	284	148.7	53	153,162	100	7	49	4,407,354	9

Gas–steam temperature profiles

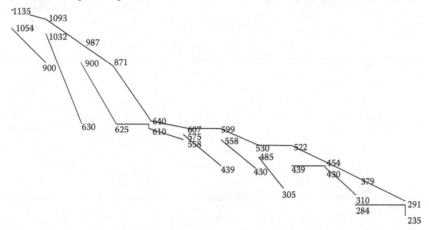

Suphtr 1 Suphtr 2 Suphtr 3 Evap 3   Econ 3 Suphtr 4 Econ 5 Suphtr 6 Evap 7   Econ 8   Evap 9

FIGURE 5.21
Multiple-pressure HRSG.

within the boundary limits, then they may agree that the HRSG performance is acceptable. For example, if 900,000 lb/h is the gas flow at 900°F used for guarantee purposes, performance curves may be developed applying margins as shown. If the steam generation or steam temperature falls outside these limits, then the plant engineer can challenge the HRSG design.

Conclusion

HRSG simulation is thus a valuable tool for plant engineers who can obtain a good idea of the performance of an yet-to-be-designed HRSG or an operating HRSG at various operating conditions. Complex HRSG designs may also be checked as shown in Figure 5.21. Plant engineers can use this as a tool to check field data, optimize HRSG configuration, or simply understand what happens to the HRSG in their plant when some steam or gas parameter changes without the information on its physical dimensions and tube geometry details and, most importantly, without the help of the HRSG supplier!

References

1. V. Ganapathy, Simulation aids cogeneration, *Chemical Engineering Progress*, Oct. 1993, p27.
2. V. Ganapathy, Simplify HRSG evaluation, *Hydrocarbon Processing*, March 1990, p77.
3. V. Ganapathy, Rethink planning for heat recovery systems, *Hydrocarbon Processing*, May 2009, p80.
4. V. Ganapathy, *Industrial Boilers and HRSGs*, CRC Press, Boca Raton, FL, 2003, p456.
5. J. Oakey, *Power Plant Life Management and Performance Improvement*, Woodhead Publishing, New York, 2011, Chapter 16, p621.

6

Miscellaneous Boiler Calculations

Condensing Economizers

As discussed earlier, one may permit condensation of acid vapor as well as water vapor on the heating surface and take advantage of the tremendous energy available in the flue gas in the form of latent heat of water vapor by condensing the flue gas to below water dew point. This is about 55°C for flue gases from combustion of natural gas, which has about 18% volume of water vapor. Boiler efficiency can approach 100% if exit gas temperature is reduced to values close to 55°C.

There would be significant benefits to cooling the flue gas to temperatures below water vapor and acid dew points, provided the acid corrosion problems can be overcome in a cost-effective way. With stack temperatures below the water vapor dew point, condensed water vapor would provide a source of water for use in power plant cooling; recovered latent and sensible heat from the flue gas improves the plant efficiency significantly. Steam for deaeration can be reduced if the condensate water is preheated in a condensing econo-mizer. Enormous environmental benefits will also be available as the amount of flue gas to be handled in the cleanup equipment is lesser along with a lower temperature (volume) and the cost of removing CO_2 at the back end of the boiler will be lower due to the lower vol-ume of flue gases to be handled. Fan size and cost used in pollution control system would be smaller. However, while estimating the cost of the system, one should also consider the cost of handling the dilute or concentrated liquid condensate from condensation of acid vapor and the treatment costs to meet local pollution control regulations. Sometimes, the cost of meeting the handling cost may outweigh the benefit of energy recovery. It has to be decided on a case-to-case basis.

Consider the case where the flue gas contains both the sulfuric acid vapor and water vapor. As flue gas is reduced in temperature below the sulfuric acid dew point, the acid first condenses as a highly concentrated liquid solution of sulfuric acid and water. If more heating surfaces are provided at lower than acid dew point temperature to, say, below the water dew point, more water is condensed, and the liquid mixture of water and sulfuric acid, which forms on low-temperature surfaces, is approximately a few orders of magnitude more dilute in sulfuric acid than the highly concentrated acid solu-tions, which form at temperatures above the water vapor dew point temperature but below the sulfuric acid dew point temperature. Economizer and air heater are prone to acid corrosion, and suppliers of these suggest minimum fluid temperature to avoid corrosion of back-end heating surfaces.

According to literature survey, 304 stainless steel is the best candidate for heat exchangers that operate at temperatures below the water vapor dew point temperature

FIGURE 6.1
Arrangement of a condensing economizer. (Courtesy of Condensing Heat Exchanger Corp., Timmins, Ontario, Canada.)

(handling dilute acid), while Teflon and Alloy 22 for heat exchangers that operate at temperatures above the water vapor dew point temperature but below the sulfuric acid dew point temperature (handling concentrated acid). The cost of Alloy 22 is about 12 times that of 304 stainless steel. Alloy 22 (containing 20%–22.5% chromium, 12.5%–14.5% Mo) is the preferred alloy for the high acid concentration due to its low corrosion rate, availability, and ability to be readily fabricated. The major attribute of Inconel Alloy 22 is outstanding resistance to a broad range of corrosive media. It resists oxidizing acids as well as reducing acids such as sulfuric and hydrochloric. Figure 6.1 shows the arrangement of a condensing economizer in a plant.

Water Dew Point of Flue Gases

Significant amount of energy can be recovered from flue gas containing water vapor if it is cooled below the water dew point. Water dew point is a function of the partial pressure of water vapor in the flue gas discussed in Chapter 1. For every kg of natural gas combusted, about 2.15 kg of water vapor is produced. This represents not only a significant amount of latent heat energy, but also a significant amount of water that can also be recovered. From combustion calculations of fuels, the volume fraction of various components such as CO_2, H_2O, N_2, O_2, and SO_2 is obtained. From the partial pressure of H_2O, one may estimate the water dew point, which is an important variable. For example, flue gas from the combustion of typical natural gas has about 17%–18% volume of water vapor while fuel oil has about 11%–12%. If flue gas pressure is atmospheric, the partial pressure of water vapor in natural gas products of combustion is 17.2 kPa or $0.17 \times 14.7 = 2.5$ psia. The water dew point corresponding to this pressure from steam tables is 134°F or 57°C. Oil products of combustion contain about 12% volume of water vapor, and the water dew point is about 50°C (122°F). If the flue gas is cooled below this temperature, then water in the flue gas starts condensing, and if a condensing economizer is used, the latent plus sensible heat is transferred to the water, which can be a significant amount. A large low-temperature heat sink is required for this purpose such as makeup water or condensate. The following example shows how much energy can be recovered in condensation process.

Energy Recoverable through Condensation

Example 6.1

Determine the amount of energy recovered below the water dew point temperature of 100,000 kg/h of flue gases from natural gas combustion that is cooled to 20°C. Flue gas analysis is % volume $CO_2 = 8$, $H_2O = 17$, $N_2 = 72$, $O_2 = 3$.

Solution

One can easily calculate the amount of energy that can be recovered while sensible heat is recovered from the flue gas by using the following equation: $Q = [W_g h_1 - (W_g - C)h_2]$, where W_g is the gas flow, C the amount of water condensed, and h_1, h_2 are the enthalpies of the flue gas corresponding to temperatures t_1 and t_2, the initial and final temperatures of the flue gas. Note that due to condensation, the flue gas quantity and analysis will change. The % volume of water vapor after condensation will be lesser than 17%. Estimating the duty below the dew point is quite involved as shown later, as C, the amount of water condensed must be first estimated.

At the inlet temperature, the partial pressure of water vapor $= 0.17 \times 14.7 = 2.5$ psia. This corresponds to 57°C or 134°F water dew point from steam tables. We have to estimate the energy recovered when 100,000 kg/h flue gases are cooled from 57°C to 20°C. Assume that a suitable heat sink is available.

The flue gas molecular weight $= 0.08 \times 44 + 0.17 \times 18 + 0.72 \times 28 + 0.03 \times 32 = 27.7$

The weight fraction or mass flow of each constituent is then

$$CO_2 = 0.08 \times 44/27.7 = 0.127 \text{ or } 12,700 \text{ kg/h.}$$
$$H_2O = 0.17 \times 18/27.7 = 0.11 \text{ or } 11,000 \text{ kg/h}$$
$$N_2 = 0.72 \times 28/27.7 = 0.728 \text{ or } 72,800 \text{ kg/h}$$
$$O_2 = 100,000 - 12,700 - 11,000 - 72,800 = 3,500 \text{ kg/h}$$

Let C kg/h of water condense when flue gas is cooled to 20°C.

Moles of flue gas after condensation at 20°C $= (12,700/44) + \{(11,000 - C)/18\} + (72,800/28) + (3,500/32) = 288.63 + 611.1 - (C/18) + 2,600 + 109.4 = 3,609.1 - (C/18)$

Volume fraction of H_2O after condensation $= (11,000 - C)/18/(3,609.1 - 0.0556C)$

Corresponding to 20°C from steam tables, the saturation pressure $= 0.34$ psia or fraction volume of $H_2O = 0.34/14.7 = 0.023$.

Solving for C,

$$0.023 = (11,000 - C)/18/(3,609.1 - 0.0556C) \text{ or } C = 9,730 \text{ kg/h}$$

TABLE 6.1

Results of Condensation Calculations

Item	Inlet	Exit
Gas flow, kg/h	100,000	90,270
% volume CO_2	8	9.4
H_2O	17	2.3
N_2	72	84.8
O_2	3	3.5
Gas temperature, °C	57	20
Energy recovered, kW	—	7,581

The total moles of the flue gases leaving the economizer: 3,609.1 − (9,730/18) = 3,068.5, and the gas flow after condensation = 100,000 − 9,730 = 90,270 kg/h. Let us obtain the flue gas analysis after condensation:

$$\% \text{ volume } CO_2 = 100 \times 12,700/44/3,068.5 = 9.4\%$$

$$\% \text{ volume } H_2O = 100 \times (11,000 - 9,730)/18/3,068.5 = 2.3\%$$

$$\% \text{ volume } N_2 = 100 \times 72,800/28/3,068.5 = 84.73\%$$

$$\% \text{ volume } O_2 = 100 \times 3500/32/3068.5 = 3.56\%$$

The sensible energy recovered from 57°C to 20°C = 100,000 × 10.8 − 90,270 × 1.07 = 983,411 kcal/h = 1,144 kW (3.9 MM Btu/h). (The enthalpy of flue gas corresponding to the flue gas analysis at 57°C is 10.8 kcal/kg, and 1.07 kcal/kg is the gas enthalpy at 20°C; see Appendix F on gas properties for the calculation of enthalpy of flue gases.) The latent energy recoverable = 9,730 × 569 = 5,536,370 kcal/h = 6,437 kW (21.9 MM Btu/h). An average latent heat of 569 kcal/kg (1024 Btu/lb) was used from steam tables. Total energy recovered = 1144 + 6437 = 7581 kW.

Table 6.1 summarizes the results. Figure 6.2 shows the amount of sensible and latent energy recovery from the flue gases cooled from water dew point temperature of 57°C to 20°C. The latent energy contribution is significant. However, there should be a large low temperature heat sink such as cooling water at about 5°C–10°C for condensation to occur.

One may estimate the total energy recovered and compute the boiler efficiency by using the following expression: efficiency = total energy recovered/fuel input in consistent units. Depending on final flue gas temperature, lower heating value efficiency can range from 95% to even 100%.

Condensation Heat Transfer Calculations

In the case of condensing economizers with horizontal plain or finned tubes, water vapor condenses outside while the cooling medium flows inside the tubes. Since the heat transfer coefficients on both inside and outside the tubes are of the same order, calculations for U must consider both the condensing and the tube-side coefficients.

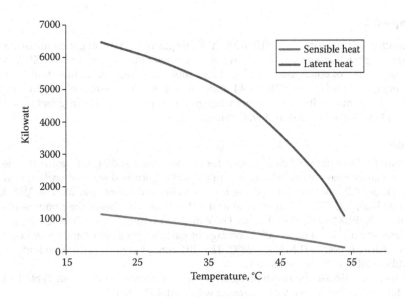

FIGURE 6.2
Amount of energy recovered from natural gas products of combustion.

The condensing heat transfer coefficient h_c in kcal/m² h°C is given by the following equation [1]:

$$h_c = 0.725 \left[\frac{k_l^3 \rho_l^2 \Delta H g}{\mu_l d \, \Delta T} \right]^{0.25}$$

(6.1)

where
  k_l is thermal conductivity of liquid, kcal/m h °C
  ρ_l is the density of liquid, kg/m³
  ΔH is the latent heat of condensing vapor, kcal/kg
  μ_l is the viscosity of liquid, kg/m h
  d is the tube OD, m
  ΔT is the saturation temperature – tube wall temperature, °C

$$g = 9.8 \times 3600^2 \text{ m/h}^2$$

Properties of the vapor may be estimated at the film temperature, which may be taken as the average of vapor and tube wall temperature. With a tube bundle, the presence of neighboring tubes adds to the complexity of the calculation of h_c. If condensate does not drain properly, it reduces h_c as the thickness of the water film increases. When these droplets strike the tubes below, splashing occurs, causing turbulence and stripping of the condensate; the heat transfer coefficient could be higher but difficult to predict. Staggered arrangement of tubes is recommended to enhance the condensation process, which is also impacted by the number of rows along the gas flow direction. Kern proposed a correction for the average condensing heat transfer coefficient h_{cm} considering the number of rows deep N_d as follows:

$$h_{cm} = h_c N_d^{-0.167}$$

(6.2)

Example 6.2

In a condensing economizer, 100,000 kg/h of flue gas from natural gas combustion with analysis as given earlier is cooled from the water dew point temperature of 57°C to 45°C by 250,000 kg/h of condensate at 20°C. Plain tubes are used. Determine the size of the exchanger; plain tube size = 50.8 × 44 mm. There are 30 tubes/row of length 3 m long. $S_T = 80$, $S_L = 70$ mm, with 30 streams in staggered arrangement. Fouling factor of 0.0002 m^2 h °C/kcal is used on gas and water sides.

Solution

One may perform detailed calculations for condensation at 45°C and estimate the sensible and latent heat energy transferred to the water both as discussed earlier; however, using Figure 6.2, one can see that the total sensible and latent heat duty is 3,976 kW = 3,419,000 kcal/h (3,542 kW latent heat and 434 kW sensible heat); the temperature rise of water = 3,419,000/250,000 = 13.6°C. The exit water temperature = 33.6°C. The average film temperature is taken as 40°C (average of gas and water temperatures). Log-mean temperature difference (LMTD) = 24°C. The difference between saturation temperature and tube wall may be taken as 20°C.

Let us first estimate the condensing heat transfer coefficient h_c. From Table F.13, one may determine the property of saturated water. At 40°C (104°F),

k_l = 0.3656 Btu/ft h °F = 0.545 kcal/m h °C; ρ_l = 61.9 lb/ft^3 = 990 kg/m^3; ΔH = 1035 Btu/lb = 575 kcal/kg; μ_l = 1.586 lb/ft h = 2.36 kg/m h; d = 0.0508 m; ΔT = 20°C

Substituting, h_c = 0.725[0.545^3 × 990^2 × 575 × 9.8 × 3600^2/(2.36 × 0.0508 × 20)]$^{0.25}$ = 6044 kcal/m^2 h °C (1233 Btu/ft^2 h °F). We may have to correct this for a number of rows deep. Assume the number of rows deep is 24 and correct it later if necessary. Correction factor = 24$^{-0.167}$ = 0.588 and corrected h_c = 0.588×6044 = 3554 kcal/m^2 h °C.

The tube-side convective heat transfer coefficient h_i at an average water temperature of 27°C is estimated as follows (from Table B.3). C = 250 − 0.0063 × 27 × 27 + 4.023 × 27 = 354. Hence, h_i = 0.0278 × (10,000/3,600)$^{0.8}$ × 354/.044$^{1.8}$ = 5,326 W/m^{2}K (4,580 kcal/m^{2}h °C).

The gas-side convective heat transfer coefficient may be estimated as follows using the methods discussed in Appendix C. At the gas film temperature of 40°C, gas properties are C_p = 0.2642 kcal/kg°C; μ = 0.066 kg/m h; and k = 0.0235 kcal/m h °C. Nonluminous coefficient is too small and neglected.

S_T/d = 80/50.8 = 1.57, S_L/d = 1.38. From Table C.2 by interpolation, B = 0.48, N = 0.565. Gas mass velocity G = 100,000/[30 × (0.080 − 0.050.8) × 30 × 3] = 38,051 kg/m^{2}h. Reynolds number Re = G d/μ = 38,051 × 0.0508/0.066 = 29,288

$$Nu = 0.48 \times 29288^{0.565} = 160 = h_o \times 0.0508/0.0235 \text{ or } h_o = 75 \text{ kcal/m}^2 \text{ h °C}$$

(neglect the effect of varying gas flow)

Weighted outside coefficient is estimated as follows = 3976/[(3542/3554) + (434/75)] = 586 kcal/m^{2}h °C. [The condensation duty divided by its heat transfer coefficient and the sensible heat duty divided by its coefficient are added, and the reciprocal gives an estimated weighted heat transfer coefficient as the condensation and sensible heat transfer occur in parallel.]

The overall U_o = 1/586 + 0.0002 + 0.0508 × ln(50.8/44)/2/37 + 0.0002 × 50.8/44 + (50.8/44) × (1/4632) = 0.001706 + 0.0002 + 0.000098 + 0.00023 + 0.000249 = 0.002483 or U_o = 402 kcal/m^{2}h °C

Required surface S = 3976 × 860/24/402 = 354 m^2 = 3.14 × 0.0508 × 30 × 3 × N or number of rows deep = 25. The correction for this is 0.584 versus 0.588. LMTD is 24°C.

Revised calculations are not necessary as the difference is marginal. Hence, one may use, say, 26 rows deep considering some margin; also an even number of rows brings the headers on the same side! The disadvantage of using plain tubes is obvious. It requires a large number of rows deep. One may use a smaller transverse pitch and improve the convective coefficient and see the effect on gas pressure drop. That is left as an exercise to the reader. The purpose of this exercise is to show the sizing procedure for a condensing economizer. Since condensation process is complex, field data from similar projects will help fine-tune or correct the U values.

Condensation over Finned Tubes

Finned tubes are used in condensing economizer service as a more compact tube bundle with lower gas pressure drop is obtained. The following equation is recommended by Beatty and Katz for condensation heat transfer [2,3]:

$$h_c = 0.689 \left[\frac{k_l^3 \rho_l^2 g \Delta H}{\mu_l d_h \Delta T} \right]^{0.25} \tag{6.3}$$

$$\left(\frac{1}{d_h} \right)^{0.25} = 1.3 \left[\frac{A_f \eta_f}{A_{ef} h^{0.25}} \right] + \frac{A_t}{(A_{ef} d^{0.25})} \tag{6.4}$$

where
A_f is the fin area, m²/m
ΔH is the latent heat, kcal/kg
η_f is the fin efficiency, fraction
h is the fin height, m
A_{ef} is the effective area, $\eta_f A_f + A_t$, m²/m
A_f, A_t are fin and tube areas, m²/m
d, d_h are tube diameter, effective diameter, m

Example 6.3

Solid finned tubes are used in the condensing economizer application provided earlier. Tube size = 50.8 × 44 mm, fins/m = 117, fin height = 15 mm, fin thickness = 1.5 mm. Transverse spacing S_T = 85 mm, S_L = 80 mm. Arrangement is staggered. Total external area = 0.901 m²/m, fin surface area = 0.77 m²/m. Plain tube area = 0.131 m²/m. Determine the size of the economizer for the same duty as Example 6.2.

Solution

Let us use the properties of the condensate at 40°C as before. Calculations for finned tubes are shown in Appendix E. Using the procedure described, the fin efficiency may be shown to be 0.8. The convective heat transfer coefficient = 82 kcal/m²h °C.

$$\left(\frac{1}{d_h} \right)^{0.25} = 1.3 \times \left(0.77 \times 0.80/0.901/0.015^{0.25} \right) + 0.131/0.901/0.0508^{0.25} = 1.953 + 0.305 = 2.258$$

The condensing heat transfer coefficient h_c

$$= 0.689 \left[\frac{0.545^3 \times 990^2 \times 9.8 \times 3600^2 \times 575}{(2.36 \times 20)} \right]^{0.25} \times 2.258 = 8937 \text{ kcal/m}^2\text{h °C}$$

Assume the number of rows deep is 12; the correction factor $= 12^{-0.167} = 0.66$ or $h_c = 8937 \times 0.66 = 5898$ kcal/m^{2}h °C. The tube-side coefficient is the same as before.

The weighted average outside heat transfer coefficient $= 3976/[(3542/5898) + (434/0.77/82)] = 531$ kcal/m^2 °C. [0.77 is the fin effectiveness as discussed in Appendix E.] The overall heat transfer coefficient is

$$U_o = 1/531 + 0.0002 + (0.901/0.1595) \times 0.0508 \times \ln(50.8/44)/2/37 + 0.0002 \times 50.8/44$$

$$+ (50.8/44) \times (1/4632) = 0.00188 + 0.0002 + 0.000553 + 0.00023 + 0.000249 = 0.00311$$

or $U_o = 321$ kcal/m^{2}h °C

Surface area required $= (3976 \times 860)/24/321 = 444$ m$^2 = 30 \times 3 \times 0.901 \times N_d$ or $N_d = 5.47$. Use six rows deep. The corrected outside coefficient $= 8937 \times 0.74 = 6613$ kcal/m^{2}h °C. The weighted average outside coefficient $= 3976/[(3542/6613) + (434/.77/82)] = 536$ kcal/m^{2}h °C (not much different from the earlier run). Hence, six rows should be fine.

Figure 6.3 shows the scheme of a condensing economizer. Tube and fin materials are of stainless steel as dilute acid is formed. Chapter 1 discusses the acid dew point correlations. More care should be taken when the flue gas contains sulfuric acid vapor as it will condense first and then the water vapor. Provision is made for proper drainage and handling of corrosive condensate while designing condensing economizers.

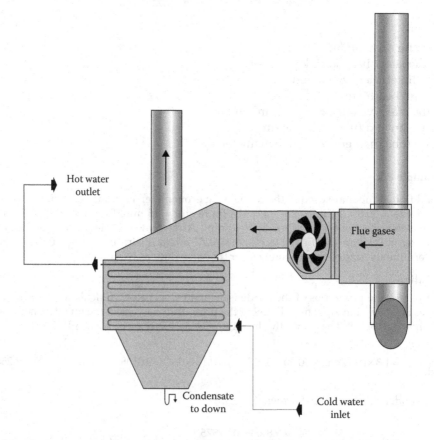

FIGURE 6.3
Scheme of condensing economizer. (Courtesy Condensing Heat Exchanger Corp., Timmins, Ontario, Canada.)

Wall Temperature of Uninsulated Duct, Stack

If in a steam plant, boiler ducts or stacks are not insulated, the heat loss from the casing can be substantial. However, a more important issue is the low stack wall temperature, which can cause acid dew point corrosion problems if the flue gas contains sulfur components. Water dew point temperature may also be reached at low loads of boiler, causing casing corrosion in the long run. Hence, plant engineers should have an idea of the casing wall temperatures when stacks are not insulated. These calculations assume significance particularly at low ambient and high wind velocity conditions. The calculations that follow show how one may estimate the stack wall temperature (Figure 6.4).

Let W kg/s of flue gas at a temperature of t_{g1} °C enter the stack. The heat loss from the casing to the environment in kW/m² is given by

$$q = 5.67 \times 10^{-11} \in \left[(273 + t_c)^4 - (273 + t_a)^4\right] + 0.00195 \times (t_c - t_a)^{1.25} \times \left[\frac{(V + 21)}{21}\right]^{0.5} \quad (6.5)$$

where
  q is the heat loss, kW/m²
  t_a, t_c are ambient and casing temperatures, °C
  V is the wind velocity, m/min
  $\in$ is the emissivity of casing

The temperature drop across the gas film is given by

$$t_g - t_{w1} = 1000q\,(d_o/d_i)/h_i \quad (6.6)$$

where
  h_i is the flue gas convective heat transfer coefficient inside the stack, W/m² K
  d_o, d_i are outside and inside stack diameters, m
  t_g, t_{w1} are average gas temperature and inner wall temperature, °C

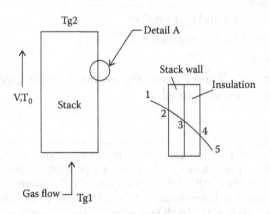

FIGURE 6.4
Stack wall temperature. *Note:* (1) T_g, (2) T_{w1}, (3) T_{wo}, (4) T_c, and (5) T_0.

As shown in Appendix B, the convective heat transfer coefficient h_i is given by

$$h_i = \frac{0.0278 C W^{0.8}}{d_i^{1.8}}$$

(6.7a)

where W = flue gas flow, kg/s

$$C = \left(\frac{C_p}{\mu}\right)^{0.4} k^{0.6}$$

(6.7b)

C_p, μ, k are the flue gas specific heat, viscosity, and thermal conductivity at the average film temperature of $0.5(t_g + t_{w1})$. C is given in Table B.2 for a few common gas streams. The temperature drop across the stack wall is given by

$$t_{w1} - t_c = \frac{q d_o \ln(d_o/d_i)}{(2 K_m)}$$

(6.8)

(This drop is small and can be neglected, or we can assume $t_{w1} = t_{w2} = t_c$)

The total heat loss $Q = 3.14 d_o H q$,

(6.9)

where H is the stack height, m.

The exit gas temperature from stack $t_{g2} = t_{g1} - Q/(W_g C_p)$

(6.10)

Average gas temperature $t_g = 0.5(t_{g1} + t_{g2})$.

(6.11)

The determination of stack wall temperature is an iterative process as explained later.

1. Assume an average gas temperature t_g.
2. Assume an average gas casing t_c.
3. Calculate q using (6.5).
4. Calculate Q using (6.9).
5. Calculate t_{g2} using (6.10) and t_g from (6.6).
6. Calculate h_i using a film temperature of $0.5(t_g + t_c)$.
7. Check t_g using (6.6).
8. If t_g calculated in steps 5 and 7 agree, the assumed casing temperature is correct; else, another iteration is warranted.

Example 6.4

13.86 kg/s (110,000 lb/h) of flue gases from the combustion of natural gas enter an uninsulated boiler stack with an outer diameter of 1.27 m (50 in.) and an inner diameter of 1.2195 m (48 in.) at 210°C (410°F). Flue gas analysis is % volume CO_2 = 8, H_2O = 18, N_2 = 71, O_2 = 3. If ambient temperature is 21°C (70°F) and wind velocity is 38 m/min (125 ft/min), determine the stack wall temperature, gas temperature leaving the stack, and the total heat loss. Casing emissivity is 0.9. Stack wall thermal conductivity is 43 W/m K (25 Btu/ft h °F). Stack height = 15 m (49 ft).

Solution

As the first trial value, let us assume that the casing temperature is the average of ambient and gas temperatures to start with = $0.5 \times (210 + 21) = 116°C$

Casing loss $q = 5.67 \times 10^{-11} \in [(273 + t_c)^4 - (273 + t_a)^4] + 0.00195 \times (t_c - t_a)^{1.25} \times [(V + 21)/21]^{0.5}$

$$q = 5.67 \times 10^{-11} \times 0.9 \times \left[(273 + 116)^4 - (273 + 21)^4 \right] + 0.00195 \times (116 - 21)^{1.25} \times (59/21)^{0.5}$$

$$= 0.787 + 0.969 = 1.756 \text{ kW/m}^2 \left(555 \text{ Btu/ft}^2 \text{ h} \right)$$

The drop across the stack wall can be shown to be = $1756 \times 1.27 \times \ln(1.27/1.2195)/2/43 = 1.05°C$ and can be neglected; hence, t_{w1}, t_{w2} need not be calculated.

Estimate h_i at a gas film temperature of $0.5 \times (117 + 210) = 163.5°C$. From Table B.2, C for natural gas products of combustion in the following equation is 159.4.

$h_i = 0.0278C \, W^{0.8}/d_i^{1.8} = 0.0278 \times 159.4 \times 13.86^{0.8}/1.2195^{1.8} = 25.4 \text{ W/m}^2 \text{ K} (4.47 \text{ Btu/ft}^2 \text{ h °F})$. Gas temperature drop = $1756/25.4 = 69°C$. Hence, $t_g = 117 + 69 = 186°C$.

The heat loss in the stack = $3.14 \times 1,756 \times 2.7 \times 15 = 223,310 \text{ W} = 223.31 \text{ kW} (0.762 \text{ MM Btu/h})$. The exit gas temperature = $210 - 223.310/13.86/1.153 = 196°C$ (using average gas specific heat of 1.153 kJ/kg°C from Appendix F). Average gas temperature t_g is 203°C. The difference in the computed value of t_g (203°C versus 186°C) is large, and hence, a few more iterations are required to arrive at t_g and hence t_c.

Use 121°C for casing wall temperature in the next iteration.

$q = $ Casing loss $= 5.67 \times 10^{-11} \in [(273 + t_c)^4 - (273 + t_a)^4] + 0.00195 \times (t_c - t_a)^{1.25} \times [(V + 21)/21]^{0.5} = 0.848 + 1.033 = 1.881 \text{ kW/m}^2 = 1881 \text{ W/m}^2$. Inner wall temperature = $121 + 1881 \times 1.27 \ln(1.27/1.2195)/2/43 = 122°C$. Hence, $(t_g - t_c) = 1881/25.36 = 74.2°C$ or $t_g = 122 + 74.2 = 196.2°C$. Correcting for film temperature of $(122 + 196.2)/2 = 159°C$, h_i will be unchanged at $25.36 \text{ W/m}^2 \text{ K}$. Heat loss in stack = $3.14 \times 1.881 \times 2.7 \times 15 = 239.2 \text{ kW}$. Gas temperature drop in stack = $239.2/13.86/1.153 = 15°C$. Hence, exit gas temperature = $210 - 15 = 195°C$. Average gas temperature = $0.5(195 + 210) = 202.5°C$. The difference has come down but is still large. Try 127°C as casing temperature.

$q = 2.0367 \text{ kW/m}^2$. Gas temperature drop = $2036.7/25.36 = 80°C$. $t_g = 127 + 80 = 207°C$.

Correcting for film temperature of $0.5 \times (127 + 207) = 167°C$, $h_i = 25.5 \text{ W/m}^2 \text{ K}$

Heat loss in stack = $3.14 \times 2.0367 \times 2.7 \times 15 = 259 \text{ kW}$. Gas temperature drop = $259/13.86/1.153 = 16°C$. Stack exit gas temperature = $210 - 16 = 194°C$. Hence, average gas temperature $t_g = 202°C$, which agrees somewhat closely with the estimated temperature of 207°C. Hence, casing temperature is about 124°C–126°C, and exit gas temperature from stack about 202°C. With a computer program, one can fine-tune the results considering variations in gas properties. Hence, plant engineers should be concerned about uninsulated stacks or ducts at gas temperatures particularly when the flue gas contains sulfur compounds. At low boiler loads, the casing temperature will be even lower as h_i decreases further; at low ambient and high wind velocity conditions, the condition worsens, and it may be better to insulate the stack.

Flue gas recirculation is commonly used in steam generators for NO_x reduction. Depending on the fuel analysis, combustion temperature and emission levels of NO_x anywhere from 10% to 25% of flue gases are recirculated from economizer exit to the forced draft fan suction in gas-fired units. If this duct is not insulated, the condensation of water vapor can occur in the recirculation duct causing corrosion and operational problems for the fan.

Insulation Calculations

Insulation and refractory performance may be evaluated using the following procedure. The heat loss from a surface is given by

$$q = 5.67 \times 10^{-11} \times \in \times \left[(273+t_c)^4 - (273+t_a)^4 \right] + 0.00195 \times (t_c - t_a)^{1.25} \times \left[(V+21)/21 \right]^{0.5} \quad (6.5)$$

$\in$, the emissivity of casing, may be taken as 0.9 for oxidized steel, 0.05 for polished aluminum, and 0.15 for oxidized aluminum.

$$q = K(t-t_c)/\left[\{(d+2L)/2\} \ln\{(d+2L)/d\} \right] = K(t-t_c)/L_e \quad (6.12)$$

where
t_a, t_c, t are the ambient, casing, and hot surface temperatures, °C
V is the wind velocity, m/min
K is the thermal conductivity of insulation, W/m K
d is the pipe outer diameter, m
L, L_e are thickness of insulation and its equivalent thickness, m (see Table 6.2)

Example 6.5

Determine the thickness of insulation to limit the casing temperature to 93°C when ambient is 27°C, wind velocity V = 80 m/min, casing emissivity = 0.15. K = thermal conductivity of insulation = 0.05 W/m K. The pipe diameter is 0.3 m and is at 426°C.

$$q = 5.67 \times 10^{-11} \times 0.15 \times \left[(273+93)^4 - (27+273)^4 \right] + 0.00195 \times (93-27)^{1.25} \times \{(80+21)/21\}^{0.5}$$

$$= 0.8837 \text{ kW/m}^2 = 883.7 \text{ W/m}^2$$

883.7 = 0.05(426 − 93)/L_e or L_e = 0.019 m (0.74 in.). One may use look-up tables to find out the thickness-based d and L. For large diameters, L_e will be nearly same as L. Here, L = 0.01905 m.

Example 6.6

What is the casing temperature when 0.025 m thick insulation is used in the earlier case?

$$L_e = \{(0.3 + 2 \times 0.025)\}/2 \ln[(0.3 + 2 \times 0.025)/0.3] = 0.0269 \text{ m}$$

Let t_c = 65°C. Use a K = 0.0486 W/m K for the revised average insulation temperature.

$$\text{LHS} = 5.67 \times 10^{-11} \times 0.15 \times \left[(273+65)^4 - (273+27)^4 \right] + 0.00195 \times (65-27)^{1.25} \times (101/21)^{0.5}$$

$$= 0.445 \text{ kW/m}^2 = 445 \text{ W/m}^2$$

RHS = 0.0486 × (426 − 65)/0.0269 = 652 W/m². Since there is a mismatch, try 77°C.
LHS = 0.6273 kW/m² and RHS = 0.630 kW/m². Hence, casing temperature = 77°C.
A computer program is ideal for these calculations particularly if multilayer insulation is involved.

TABLE 6.2

Equivalent Thickness of Insulation

Tube OD, in. (mm)	Thickness of Insulation, in. (mm)							
	0.5 (12.5)	1 (25.4)	1.5 (38)	2 (51)	3 (76)	4 (101)	5 (127)	6 (152)
1 (25.4)	0.69	1.65	2.77	4	6.8	9.9	13.2	16.7
2 (50.8)	0.61	1.39	2.29	3.3	5.5	8.05	10.75	13.62
3 (76.2)	0.57	1.28	2.08	2.97	4.94	7.15	9.53	12.07
4(101.8)	0.56	1.22	1.96	2.77	4.55	6.6	8.76	11.1
5 (127)	0.55	1.18	1.88	2.65	4.34	6.21	8.24	10.4
6 (152)	0.54	1.15	1.82	2.55	3.16	5.93	7.85	9.8
8 (203)	0.53	1.12	1.75	2.43	3.92	5.55	7.5	9.15
10 (254)	0.52	1.09	1.7	2.35	3.76	5.29	6.93	8.57
12 (304)	0.52	1.08	1.67	2.3	3.65	5.11	6.65	8.31
16 (406)	0.52	1.06	1.63	2.23	3.5	4.86	6.31	7.83
20 (508)	0.51	1.05	1.61	2.19	3.41	4.7	6.1	7.52

Example 6.7

A horizontal flat surface is at –12.2°C. Ambient temperature is 27°C and relative humidity is 80%. Determine the thickness of fibrous insulation that will prevent condensation of water vapor on the surface. Use K = 0.04 W/m K. Wind velocity may be neglected. Use emissivity of 0.9 for casing.

The surface must be above the water dew point to prevent the condensation of water vapor. From steam tables, the saturated vapor pressure is 0.51 psia. At 60% relative humidity, the vapor pressure = 0.6 × 0.51 = 0.408 psia. This corresponds to a saturation temperature of 73°F from steam tables. So we must select the thickness of insulation so that it is above 23°C (73°F).

$$q = 5.67 \times 10^{-11} \times 0.9 \times \left[(273 + 27)^4 - (23 + 273)^4 \right] + 0.00195 \times (27 - 23)^{1.25}$$

$$= 0.0326 \text{ kW/m}^2 = 32.6 \text{ W/m}^2$$

q = 0.04 × (23 + 12.2)/L = 32.6 or L = 0.043 m = 43 mm. Use slightly higher thickness considering variations in ambient conditions.

Example 6.8

A 1.5 in. Sch 40 pipe (1.9 in. OD, 1.61 in. ID), 300 m long carries hot water at 150°C. What is the heat loss from its surface if not insulated (case 1) or has 25, 50, and 75 mm thick insulation (case 2)? Assume K of insulation as 0.036 W/m K. Ambient temperature is 27°C and wind velocity is zero.

Case 1: For bare pipe,

$$q = 5.67 \times 10^{-11} \times 0.9 \times \left[(273 + 150)^4 - (273 + 27)^4 \right] + 0.00195 \times (150 - 27)^{1.25} = 2.0 \text{ kW/m}^2.$$

Case 2: Using a computer program, the heat loss was determined for various cases, which is shown later.

It can be shown that the heat losses are, respectively, 0.085, 0.044, 0.0284 kW/m² for the 25, 50, and 76 mm thick insulated pipes. If the 50 mm thick insulation pipe is considered, the total heat loss = 3.14 × .0.044 × 0.148 × 300 = 6.134 kW. If the specific heat of water is taken as 4.3 kJ/kg K and flow as 1 kg/s, the decrease in temperature of water is about 6.134/4.3 = 1.5°C.

Superheated steam pipes are large in diameter, and also, the specific heat of steam is lower. Hence, the temperature drop of steam can be as much as 5°C–7°C depending on insulation thickness and pipe sizes and length. Hence, such a calculation may be performed to estimate the drop in steam temperatures as lower temperature at steam turbine inlet reflects loss in power production. The pipe length can be split up into segments, and the effect of variations of K with temperature can be considered, and hence, more accurate results can be obtained. One may also arrive at the optimum thickness by evaluating the loss in energy in dollar terms and cost of installation for various thicknesses.

Hot Casing Design

Whenever hot gases are contained in an internally refractory-lined (or insulated) duct, the casing temperature can fall below the acid dew point in cold weather. The hot gases can seep through the refractory cracks over a period of time and can cause acid condensation with its associated problems. Some engineers prefer what is called a hot casing whereby the duct casing is made hot (above the acid dew point) by externally insulating the duct (Figure 6.5). This lowers the heat losses also. However, if calculations are not properly done and a thicker-than-required external insulation is used, the duct temperature can be made significantly hot and can cause expansion problems. If the gas temperature varies along the gas path (say, a boiler), then higher thicknesses have to be applied externally as the gas temperature reduces in order to keep the casing temperature within a range of temperature.

Example 6.9

A waste heat boiler has a gas inlet temperature of 704°C and cools to 250°C. The casing has 125 mm of Greencast 22 refractory followed by 50 mm of castable block mix refractory. The casing temperature based on 27°C ambient and zero wind velocity is estimated as 78°C as shown later. Since this is below the sulfuric acid dew point temperature of 120°C, external insulation should be applied to raise the casing temperature. The results from a computer program are shown in Table 6.3 for various cases. Let the acid dew point be 130 °C. (For simplicity and to illustrate the point, gas temperature is assumed to be the hot face temperature.)

In case 1, the hot gases are contained in a refractory-lined casing. The hot layer of refractory is Castolite 22. K ranges from 0.245 to 0.324 W/m K from 200°C to 871°C. Refractory 2 is castable block mix, and K ranges from 0.082 to 0.121 W/m K from 93°C to 427°C. Mineral fiber insulation K ranges from 0.052 to 0.095 W/m K from 93°C to 427°C.

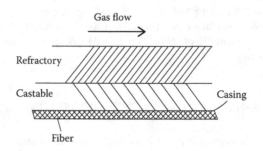

FIGURE 6.5
Hot casing arrangement.

TABLE 6.3

Summary of Refractory Calculations

Layer	Case 1 Temp. In–Out	Case 2 Temp. In–Out	Case 3 Temp. In–Out	Case 4 Temp. In–Out
Refractory 1	700°C–411°C	700°C–459°C	250°C–196°C	250°C–176°C
Refractory 2	411°C–78°C	459°C–204°C	196°C–134°C	176°C–89°C
Min fiber	—	204°C–72°C	134°C–38°C	89°C–42°C
Heat loss, kW/m²	0.640	0.548	0.107	0.150
Fiber thickness, mm	0	12	38	12

Note: Refractory 1: Castolite 22. Refractory 2: Castable block mix.

TABLE 6.4

Heat Loss Equation

SI Units	British Units
$q = 5.67 \times 10^{-11} \times \in (T_s^4 - T_a^4) + 0.00195(T_s - T_a)^{1.25} \times \{(V + 21)/21\}^{0.5}$	$Q = 0.1714 \times 10^{-8} \times \in (T_s^4 - T_a^4) + 0.296(T_s - T_a)^{1.25} \times \{(V + 69)/69\}^{0.5}$
q is the heat loss, kW/m²	Btu/ft² h
T_s is the hot face temperature, K	°R
T_a is the cold face temperature, K	°R
V is the wind velocity, m/min	ft/min
$\in$ is the emissivity	$\in$ is the emissivity

It can be seen that when we have only the two layers of refractory inside the casing, the casing temperature drops to 78°C. This may cause water and acid condensation if some flue gases seep through the refractory. When we add 12 mm insulation outside the casing, the casing temperature increases to 204°C, and heat loss is kept low as the casing temperature of the insulation is only 72°C. Thus, casing corrosion concerns are minimized. As the gas cools, we note that we have to use a higher external insulation thickness to keep the casing hot. If we use say 12 mm thick insulation when the gases are at 250°C (case 4), the casing temperature drops to 89°C and below the acid dew point temperature. If we use a higher thickness of 38 mm (case 3), the casing temperature increases to 134°C while the heat loss is also minimized.

The other option of minimizing corrosion is to use membrane wall casing; in this case, the tube wall temperature is close to the saturation steam temperature and always above the dew point. External mineral fiber insulation of say 50–75 mm will be adequate. The choice is left to the boiler supplier. Membrane wall casing may be more expensive for some boiler suppliers.

Table 6.4 shows the heat loss equations in both SI and British units.

Drum Coil Heater: Bath Heater Sizing

There are several applications where a coil immersed in hot water (such as in a boiler drum or a hot water tank) is used to heat up a thermic fluid or even cool steam for steam temperature control. Cold feed water from economizer inlet may also be preheated in a coil or exchanger tubes located inside the hot water bath to minimize acid dew point corrosion concerns. Drum heating coils using steam are also used to keep boiler warm in case

TABLE 6.5

Natural Convection Heat Transfer, hc

SI Units	Metric Units	British Units
$Nu = 0.54[d^3\rho^2gb\Delta TC_p/(\mu k)]^{0.25}$	$Nu = 0.54[d^3\rho^2gb\Delta TC_p/(\mu k)]^{0.25}$	$Nu = 0.54[d^3\rho^2gb\Delta TC_p/(\mu k)]^{0.25}$
$h_c = 0.953[k^3\rho^2b\Delta T\, C_p/(\mu d_o)]^{0.25}$	$h_c = 57.2[k^3\rho^2b\Delta TC_p/(\mu d_o)]^{0.25}$	$h_c = 144[k^3\rho^2b\Delta TC_p/(\mu d_o)]^{0.25}$
h_c, W/m²K	kcal/m²h °C	Btu/ft²h °F
d_o is the tube OD, m	m	ft
ρ is the density, kg/m³	kg/m³	lb/ft³
g is the acceleration gravity	9.8×3600^2 m/h²	32×3600^2 ft/h²
b is the expansion coefficient, 1/K	1/K	1/°R
ΔT, K	K	°R
k, W/m K	kcal/m h °C	Btu/ft h °F
μ, kg/m s	kg/m h	lb/ft h
C_p, J/kg K	kcal/kg °C	Btu/lb °F

of quick start applications. Superheated steam for steam temperature control purpose is also cooled in the boiler drum water using pipes immersed in the drum. In all of these applications, the heat transfer between the tubes and the liquid bath is through natural convection. The advantage of using this method for heating thermic fluids or viscous oils in a water bath rather than using a direct fired natural gas–fired heater is that the heat flux is much lower and is controlled unlike in a fired heater.

The outside natural convection heat transfer coefficient between the coil and the water or fluid in the drum may be obtained using the equation in Table 6.5.

$$Nu = 0.54\left[\frac{d^3\rho^2gb\Delta TC_p}{(\mu k)}\right]^{0.25} \tag{6.13}$$

where all the terms are in consistent units.

Simplifying, the convection heat transfer coefficient in kcal/m²h °C is given by

$$h_o = 57.2\times\left[\frac{k^3\rho^2b\Delta TC_p}{(\mu d_o)}\right]^{0.25} \tag{6.14}$$

where
    C_p is the specific heat, kcal/kg°C
    d_o is tube OD, m
    g is the acceleration due to gravity, m/h²
    k is the thermal conductivity, kcal/m h °C
    ρ is the fluid density, kg/m³
    b is the volumetric expansion coefficient, 1/K
    ΔT is the temperature difference between tubes and fluid, K
    μ is the viscosity, kg/m h

All fluid properties are evaluated at the mean temperature between fluid and tubes, while expansion coefficient is at fluid temperature.

Example 6.10

A coil made of 33.5 mm OD pipe (ID = 26.6 mm) is immersed in a drum, which is maintained at 38°C using steam at 110°C inside tubes. The purpose of the coil is to keep the water at the desired temperature. Tube temperature is assumed to be 93°C. Estimate the outside convection heat transfer coefficient, and the overall heat transfer coefficient assuming condensing coefficient is 11,000 kcal/m²h °C. Use a fouling factor of 0.0002 m²h °C/kcal (0.001 ft²h °F/Btu) on shell side and tube side.

First let us estimate the liquid properties at the film temperature of 66°C. We may have to correct the properties if the film temperature is different.

From Table F.13, at 66°C, k = 0.381 Btu/ft h°F = 0.5677 kcal/m h °C, b = 0.0002 1/°R = 0.00036 1/K, μ = 1.04 lb/ft h = 1.55 kg/m h, d_o = 0.0335 m, ρ = 61.2 lb/ft³ = 979 kg/m³, C_p = 1 kcal/kg °C, ΔT = 55°C

$$h_c = 57.2 \times [k^3 \rho^2 b \, \Delta T \, C_p/(\mu \, d_o)]^{0.25} = 57.2 \times [0.5677^3 \times 979^2 \times 0.0.00036 \times 55 \times 1/$$
$$(1.55 \times 0.0335)]^{0.25} = 920 \text{ kcal/m}^2\text{h °C (188 Btu/ft}^2\text{h °F)}.$$

The overall heat transfer coefficient is given by

$1/U_o$ = 1/920 + 0.0002 + (0.0335/2/35) × ln(33.5/26.6) + 0.0002 × 33.5/26.6 + 33.5/ (11,000 × 26.6) = 0.001085 + 0.0002 + 0.00011 + 0.000252 + 0.000114 = 0.00176 or U_o = 567 kcal/m²h °C (659 W/m²K)

$$\text{Heat flux} = 567 \times (110 - 38) = 40,824 \text{ kcal/m}^2 \text{ h} = 47.4 \text{ kW/m}^2$$

Tube wall temperature = 38 + 40,824 × (0.0002 + 0.001085) = 90°C. Properties may be corrected for the average temperature of 0.5 × (90 + 38) = 64°C. However, it is close to that assumed, and hence, we may proceed with the sizing of the coil.

If the duty and log-mean temperature difference are known, the required surface area may be determined. Say, the heat loss in the boiler or duty of the coil is 0.125 MM kcal/h. LMTD = (110 − 38) = 72°C. Hence, surface area required = 0.125 × 10⁶/72/567 = 3.1 m²

Bath heaters are often used to heat up viscous fluids or any fluid by immersing a coil or exchanger inside a bath of hot water or glycol maintained at the desired temperature by controlling the burner heat input. The heater looks like a fire tube boiler firing natural gas or oil and is of multipass design depending on the efficiency required. Since the bath is maintained at a low temperature, the fluid heated inside the tubes immersed in the bath is not subject to high film temperatures as in a fired heater. If a fired heater or a fire tube boiler is used for heating the liquid, then the high heat flux due to the high gas temperature profile in the boiler has to be considered, while with bath heaters, that concern is not present. Thermal expansion issues are also not a concern as both the hot and cold fluids are at reasonably low temperatures. Expansion of the hot water or bath fluid is handled by using an expansion tank built integral with the shell as shown in Figure 6.6. This type of heater is seen in oil fields.

As discussed in Chapter 3, superheated steam is sometimes cooled in a coil located inside the steam drum or mud drum. Its sizing may also be done in a similar manner.

Example 6.11

Saturated water temperature in a boiler steam drum is 190°C. For steam temperature control purposes as discussed in Chapter 3, 40,000 kg/h of superheated steam at 50 kg/cm²g and at 350°C is to be cooled to 300°C. Determine the coil size if drum length is about 8 m. Tubes of 50.8 × 43 mm are available for the exchanger.

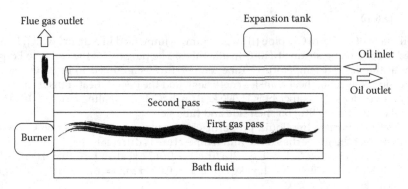

FIGURE 6.6
Bath heater fired by gas or oil.

Solution

First determine the tube-side heat transfer coefficient for superheated steam at 325°C average temperature from Appendix B. Let us start with 2000 kg/h per tube. This gives a steam velocity (see Table B.1), $V = 0.0003461 \, wv/d_i^2$. From steam tables, specific volume of steam is 0.04878 m³/kg. Steam velocity $V = 0.0003461 \times 2000 \times 0.04878/0.043^2 = 18.3$ m/s, which is a reasonable velocity.

Estimating h_i

From Table B.4, at 5000 kPa and 325°C, C = 314. Hence, $h_i = 0.0278 \times (2000/3600)^{0.8} \times 314/0.043^{1.8} = 1572$ W/m²K = 1352 kcal/m²h °C.

Estimating h_o

Assume that the tube OD is at 300°C. The average temperature of water is 245°C (473°F). From Table F.13, k = 0.3645 Btu/ft h°F, C_p = 1.125 Btu/lb °F, μ = 0.275 lb/ft h, d_o = 2 in., b = 0.00067/°R, ΔT = (300 − 190) = 110°C = 198°F. ρ = 50 lb/ft³

$$h_c = 144 \times \left[0.3645^3 \times 50^2 \times 0.00067 \times 198 \times 1.1125/(0.275 \times 2) \right]^{0.25} = 343 \, Btu/ft^2 \, h \, °F$$

$$= 1680 \, kcal/m^2 \, h \, °C.$$

$1/U_o = (1/1680) + 0.0002 + (.0508/2 \times 35)\ln (50.8/43) + (0.0002 \times 50.8/43) + 1/50.8/(43 \times 1352)$
$\quad = 0.000595 + 0.0002 + 0.000121 + 0.000236 + 0.000874 = 0.002026$ or $U_o = 493$ kcal/m²h °C

From steam table, the enthalpies of steam between 350°C and 300°C assuming a drop of 0.5 kg/cm² pressure are 733.3 and 698.7 kcal/kg. Hence, duty of exchanger = 40,000 × (733.3 − 698.77) = 1.381 MM kcal/h.

Using a tube length of 3.5 m and say 10 tubes across and 2 starts (20 streams), surface area required = $1.381 \times 10^6/[(325 − 190) \times 493]$ = 20.7 m² = 3.14 × 0.0508 × 3.5 × 10 × N or N = 3.71 rows. Use 4 rows of length 3.5 m with 20 streams in inlet and outlet headers. This provides some margin, and by controlling the steam flows, one may control the steam temperature. The 10 tubes in the header will need a width of about 700 mm, and the drum should be able to accommodate this coil. Figure 3.30 shows a similar coil with three starts, while this exchanger will have 10 tubes across with 2 starts, making the streams 20 as required with a total of four rows deep.

The steam pressure drop is (see Table B.6).

$\Delta P = 0.6375 \times 10^{-12} \text{ f } L_e \text{ v } w^2/d_i^5 = 0.6375 \times 10^{-12} \times 0.02 \times (3.5 \times 4 + 3 \times 32 \times 0.0508) \times 0.04877 \times 2000^2/0.043^5 = 0.32 \text{ kg/cm}^2$. The average steam velocity = 18.3 m/s. The inlet and exit losses will be about 1.5 times the velocity head = $1.5 \times 18.3 \times 18.3/(2 \times 9.8 \times 0.04877) = 525 \text{ kg/m}^2 = 0.0525 \text{ kg/cm}^2$. Total pressure drop is about 0.38 kg/cm^2.

Checking Heat Transfer Equipment for Noise and Vibration Problems

Whenever a fluid flows across a tube bundle such as boiler tubes in an economizer, evaporator, or air heater, vortices are formed and shed beyond the wake beyond the tubes. This shedding on alternate sides of the tubes causes a harmonically varying force on the tube perpendicular to the normal flow of the fluid and a self-excited vibration results. If the frequency of the von Karman vortices, as they are called, coincides with the natural frequency of vibration of the tubes, resonance occurs and tubes vibrate, leading to leakage and damage at tube supports. Vortex shedding is most prevalent in the Reynolds number from 300 to 2×10^5. Boiler components such as superheaters, evaporators, economizers, and air heaters operate very much within this range. There is some evidence from the literature that vortex shedding does not occur in two-phase flow and that it is a concern only in single-phase flows.

Another mechanism associated with vortex shedding is acoustic vibration, which is normal to both fluid flow and tube length. This is observed only with gases and vapors. Standing waves are formed inside the duct at a particular frequency, and the noise can be a nuisance. When the boiler load changes, the noise typically goes away.

Hence, in order to analyze tube bundle vibration and noise, three frequencies must be computed: natural frequency of vibration of tubes, vortex shedding frequency, and acoustic frequency. When these are apart by at least 20%, vibration and noise may be absent. There are instances when these frequencies matched but no vibration issues were noticed!

Determining Natural Frequency of Vibration of Tubes

The natural frequency of transverse vibrations of a uniform beam supported at each end is given by [4]

$$f_n = (C/2\pi)\left[EI/m/L^4\right]^{0.5} \tag{6.15}$$

where
  f_n is the natural frequency of vibration, 1/s
  C is a factor determined by end condition of tube support, Table 6.6
  E is Young's modulus of elasticity, N/m^2 (186 to 200×10^9 N/m^2 for carbon steel)
  $I = (\pi/64)(d^4 - d_i^4)$, m^4
  L is the length of tube between supports, m
  m is the tube weight, kg/m

TABLE 6.6

C Values for Various Modes of Vibration

End Support Type	Mode 1	Mode 2	Mode 3
Both ends clamped	22.37	61.67	120.9
One clamped, one hinged	15.42	49.97	104.2
Both hinged	9.87	39.48	88.8

Acoustic Frequency f_a

$$f_a \text{ is given by } V_s/\lambda \tag{6.16}$$

where V_s is the velocity of sound at the air temperature in the duct in m/s = $(\vartheta RT/MW)^{0.5}$. ϑ is the ratio of specific heats = 1.4, R = 8.314 J/mol K, MW = molecular weight, T is gas temperature in K. Simplifying and using an MW of 29, we have $V_s = 20\sqrt{T}$

Wavelength $\lambda = 2w/n$ where w is the width of duct, m, and n is the mode of vibration.

Vortex Shedding Frequency f_e

Strouhal number S is used to determine the vortex shedding frequency f_e.

$$S = \frac{f_e d}{V} \tag{6.17}$$

where
d is the tube OD, m
V is the gas velocity in the tube bundle, m/s

S is available in the form of charts (Figure 6.7 a through d). It typically ranges from 0.2 to 0.3. Note that these calculations give some idea whether vibrations are likely. One should rely more on field experience with certain tube spacing and geometry of the heat transfer equipment.

As mentioned earlier, if f_e and f_n are close by less than 20%, then bundle vibration is likely. If f_e and f_a are within 20% of each other, noise vibration is likely.

Example 6.12

A tubular air heater in a boiler plant is 3.57 m wide, 3.81 m deep, and 4.11 high. 50.8 × 46.7 mm carbon steel tubes are arranged in an inline fashion with a transverse pitch of 89 mm and longitudinal pitch of 76 mm. There are 40 tubes wide and 60 tubes deep. 136,000 kg/h of air at an average temperature of 104°C flows across the tubes. Tubes are fixed at both ends. Weight of tube = 2.485 kg/m. Check if tube bundle vibrations are likely.

First determine the natural frequency of vibration in the first and second modes.

$f_n = (22.37/2\pi)\left[200\times10^9 \times \pi \times (0.0508^4 - 0.0467^4)/2.485/4.11^4\right]^{0.5} = 18.3c/s$. In mode 2, it will be $(61.67/22.37) \times 18.3 = 50.3$ c/s

Let us compute the vortex shedding frequency f_e. From Figure 6.7a, for $S_T/d = 89/50.8 = 1.75$ and $S_L/d = 76/50.8 = 1.5$, S is 0.33.

Compute the actual air velocity. Density at 104°C and at atmospheric pressure = $\rho_g = 12.17$ MW/$(t_g + 273) = 12.17 \times 28.84/(104 + 273) = 0.932$ kg/m³.

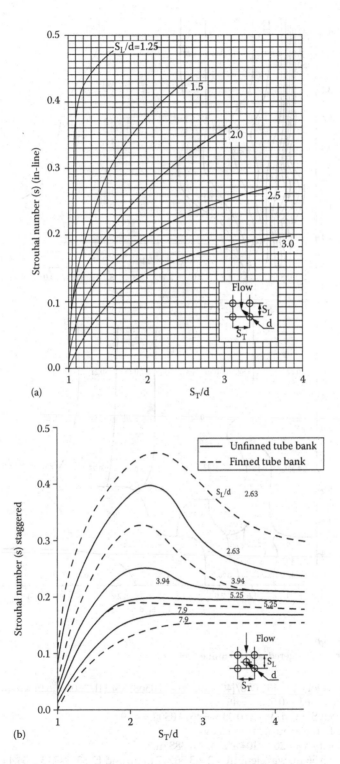

FIGURE 6.7
Strouhal number for (a, b) inline bank of tubes.

(Continued)

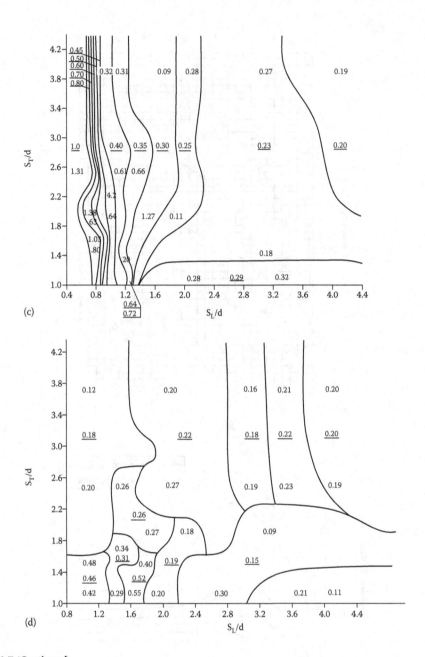

FIGURE 6.7 (Continued)
Strouhal number for (c) Staggered, and (d) inline tube banks.

Gas mass velocity = 136,000/[40 × (0.089 − 0.0508) × 4.11]/3600 = 6.01 kg/m²s

Air velocity = 6.01/0.932 = 6.45 m/s

Hence, from $S = f_e \, d/V$, $f_e = 0.33 \times 6.45/0.0508 = 41.9$ c/s

Compute f_a, the acoustic frequency

Sonic velocity $V_s = 20 \times (104 + 273)^{0.5} = 388$ m/s.

Width w = 3.56 m, wavelength = 2 × 3.56 = 7.13 m, and $f_a = 388/7.13 = 54.4$ c/s

The results are summarized in Table 6.7. It may be seen that without baffles, f_a and f_e are close. Hence, noise problems may arise, particularly as the load increases slightly. Then,

TABLE 6.7

Summary of Results

Mode of Vibration	1	2
f_n	18.3	50.3
f_e	41.9	41.9
f_a (Without baffle)	54.4	109
f_a (With 1 baffle)	109	218

the gas velocity and f_e will increase. If a baffle plate is inserted parallel to the tube length (dividing the duct width by half), then f_a will double thus keeping f_e and f_a far apart.

Vortex shedding is unlikely to cause damage. This is due to the large mass of the system compared to the low energy in the gas stream (the analogy is a tall pole that will flex in the wind, but if it is short, it may not). Thus, using tube supports at intervals of 2–2.5 m will help reduce the vibration.

The tube natural vibration frequencies will be small in the first mode of vibration. The amplitude of vibration will be smaller at higher modes, and hence, the first mode is generally considered in the analysis.

Damping Criterion

Often, the vortex shedding frequencies coincide with the acoustic frequency, but standing waves do not develop and the transverse gas column does not vibrate. Resonance is more the exception than the rule. Chen proposed a damping criterion y based on tube geometry as follows [1]:

$$y = Re\left\{(S_L/d - 1)/S_L/d\right\}^2 d/S_T \qquad (6.18)$$

where
 S_T, S_L are the transverse and longitudinal spacing
 d is the tube outer diameter

Strouhal number S is obtained as discussed earlier. For an inline bank of plain tubes, Chen stated that y must exceed 600 before a standing wave develops. In a study on spiral finned tube bank, y reached 15,000 before sonic vibration developed. if y is less than 20,000, vibrations due to vortex shedding may not occur. Vibration analysis is not an exact science, and a lot of it is based on experience with operating units of similar design. In a few cases, the vortex shedding and acoustic frequencies were close, but no damaging noise vibration was felt.

ASME Section 3 Appendix N1330, 1995, on flow-induced vibration suggests that if reduced damping factor C exceeds 64 where

$$C = \frac{4\pi mz}{\rho d^2} \qquad (6.19)$$

where
 m is the mass of tube, kg/m (the mass of ash deposits or fluid inside tube may be added to the tube weight)
 z is the damping factor, which varies significantly (typically 0.001 for systems without intermediate supports and 0.01 with intermediate supports)
 ρ is the gas density, kg/m³. d = tube OD, m

Table 6.9, later in the chapter, shows the results of calculation for a waste heat boiler with plain tubes at the gas inlet and finned tubes at the exit. Tube OD = 50.8 × 2.8 mm.

Fluid Elastic Instability

The need for intermediate tube supports is governed by fluid elastic instability consider-ations. ASME section 3 gives an idea of the stability of tube bundles. If the nondimensional flow velocity as a function of mass damping factor is above the curve shown in Figure 6.8, then intermediate supports are suggested; without them, fretting and wear of tubes due to vibration are possible. Basically, this criterion tells us that if we have a tall bundle without intermediate supports, it can oscillate due to gas flow; intermediate supports decrease the tube length between supports and hence increases the natural frequency of vibration. This reduces the nondimensional flow velocity $V/(f_n d)$, where V is the gas velocity and f_n the natural frequency of tubes and d the tube OD.

However, experience with similar units in operation gives better guidance than these criteria. Tube lengths of even 4 m have been used without intermediate supports with-out problems. Hence, experience based on field data is also a factor in using intermediate supports.

Example 6.13

A boiler has the following data as shown in Table 6.8. m = 4.66 kg/m, damping factor = 0.001. Gas velocity = 16.4 m/s. Gas density = 0.3 kg/m³, d = 0.0508 m, f_n = 8.8 c/s.

$$\text{Mass damping factor} = 2\pi mz/\rho d^2 = 2 \times 3.14 \times 4.66 \times 0.001/0.3/0.0505^2 = 37.8$$

From the chart, reduced flow velocity should be less than 20 for a stable operation.

Hence, gas velocity V should be 20 × 8.8 × 0.0505 = 8.9 m/s or lower to avoid any con-cerns about bundle stability. If the actual gas velocity at inlet is higher, intermediate supports are suggested. With a damping factor of 0.01, one can see that the mass damp-ing factor is much higher pushing the gas velocity for instable operation to much higher than 16.4 m/s. In practice, for tall bundles, intermediate supports at 3–4 m intervals are used. If the tube diameter is smaller, the spacing will be closer.

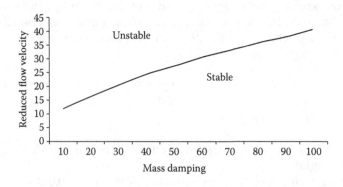

FIGURE 6.8
Damping factor versus nondimensional flow velocity.

TABLE 6.8

Damping Factor for Evaporator Tubes

Item	Bare Tube	Finned
Gas temperature, °C	871	265
Gas density, kg/m³	0.30	0.63
Gas velocity, m/s	16.4	7.9
Fins/m	Nil	78
Fin height, mm	—	19
Fin thickness, mm	—	1.9
Tube mass, kg/m	4.66	10.9
Strouhal number	0.25	0.25
Vortex shed frequency, c/s	80.8	38.9
Damping factor	0.001	0.01
Factor C	75.6	421
Natural frequency, c/s	8.8	5.7

Steam Drum Calculations

Steam Velocity in Drum

When sizing steam drum internals, the chevron steam separator plays a significant role in ensuring the final steam purity (Figures 6.9 and 6.10). The boiler designer must have some idea how the steam generation varies along the length of the drum. This depends on whether we have a longitudinal gas flow boiler or a cross-flow type of boiler. If the chevron separator is located such that the steam flow enters it uniformly from all sides, then the vanes in the chevron will not be overloaded. In a steam generator, based on furnace heat loading and gas temperature profile along the convection gas path, one can obtain a good estimate of the steam generated along every meter length of the drum and then locate the chevron separator in the region where the steam generation is nearly equal from each end. This location need not be at the middle of the drum length. Similarly, in a cross-flow boiler, a large percentage of steam will be generated in the first few rows of the evaporator (due to the higher gas temperature at the inlet section), and baffling of internals has to be done considering this.

There are two vapor velocities in the drum space, which should not be exceeded to ensure that water droplets are not carried along with steam ensuring the chevron driers or vanes can perform well. Adhering to these norms will help lower the burden on the chevron separator.

$$\text{The horizontal steam velocity } V_h = 0.2 \left[\frac{(\rho_l - \rho_v)}{\rho_v} \right]^{0.5} \tag{6.20}$$

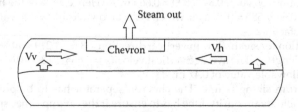

FIGURE 6.9
Steam drum showing the flow of steam to chevron separator.

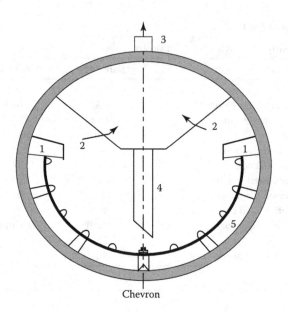

FIGURE 6.10
Steam drum internals. Note: (1) steam–water mixture, (2) wet steam, (3) dry stem, (4) drain pipe, and (5) belly pan.

$$\text{The vertical steam velocity } V_v = 0.0762\left[\frac{(\rho_l - \rho_v)}{\rho_v}\right]^{0.5} \tag{6.21}$$

where ρ_l, ρ_v are the densities of saturated vapor and liquid, kg/m³.

Example 6.14

A steam drum of diameter 1219 mm and length 8 m is generating 50 ton/h of steam at 40 barg. Determine the steam velocity in the vertical and horizontal directions assuming 100% of steam leaving the water line in the vertical direction and 50% enters the separator from each direction. Check if the drum cross section is reasonable.

The density of saturated water from steam tables is $\rho_l = 1/0.00125 = 800$ kg/m³ and that of saturated steam $\rho_v = 1/0.04792 = 20.87$ kg/m³.

Then, V_h should be less than $0.2 \times [(800 - 20.87)/20.87]^{0.5} = 1.22$ m/s.

V_v should be less than $(0.0762/0.2) \times 1.22 = 0.47$ m/s.

The horizontal cross section area of steam space $= \pi \times 1.219^2/8 = 0.583$ m².

Volume of steam $= 50,000/20.87/3,600 = 0.665$ m³/s. Hence, 50% of the horizontal flow velocity $= 0.5 \times 0.665/.583 = 0.57$ m/s, which is far below the allowable value of 1.22 m/s. Hence, it is satisfactory.

The cross section for steam flow in vertical direction $= 8.5 \times 1.219 = 10.36$ m² (including the dished end lengths). The average vertical velocity $= 0.665/10.36 = 0.064$ m/s, which is far below the allowable value of 0.47 m/s.

Hence, the drum sizing is fine. The chevron separator has to be properly selected based on allowable steam purity. One has to ensure if the holdup times specified, if any, have been met.

Blowdown Calculations

One should perform an energy balance around the deaerator to estimate the steam for deaeration and then perform the blowdown calculations based on the conductivity (or total dissolved solids) of the boiler water, makeup water, and the feed water.

Figure 6.11 shows a simple deaerator scheme with one condensate return line. Saturated steam for deaeration is taken from the drum. Makeup water is added in the deaerator from which the feed water for the boiler is taken. By doing a mass, energy, and conductivity (or TDS) balance, we can estimate the deaeration steam.

The allowable boiler water TDS (total dissolved solids) is given by both ASME and ABMA, which are shown in Tables 6.9 and 6.10, and based on practice, one may select the appropriate value for boiler water TDS.

Example 6.15

A boiler generates 50,000 kg/h saturated steam at 20 kg/cm²g of which 10,000 kg/h is taken for process and returns as condensate at 82°C. Makeup water is available at 21°C, and steam for deaeration is taken from the drum operating at 20 kg/cm² g. The deaerator operates at 1.76 kg/cm² a. Blowdown water has a TDS of 1500 ppm, and makeup has a TDS of 100 ppm. Evaluate the steam for deaeration and the blowdown.

From mass balance around the deaerator,

$$10,000 + D + M = F = 50,000 + B$$

Enthalpy of saturated steam in the drum from steam tables = 668.52 kcal/kg.
Enthalpy of makeup water at 21°C = 21.5 kcal/kg and that of condensate = 82.4 kcal/kg
Saturation temperature at deaerator = 115°C. Enthalpy of feed water = 115.8 kcal/kg
From energy balance around the deaerator,

$$10,000 \times 82.4 + D \times 668.52 + M \times 21.5 = 115.8 \times F$$

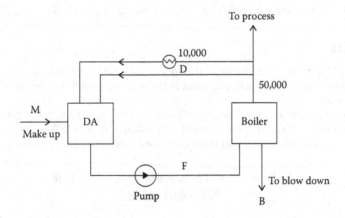

FIGURE 6.11
Blowdown scheme.

TABLE 6.9

ABMA Boiler Water limits and Associated Steam Purity at Steady-State Full Load Operation—Water Tube Drum-Type Boilers

Drum Pressure (psig)	TDS Range,[a] Boiler Water (ppm) (Max)	Range Total Alkalinity,[b] Boiler Water (ppm)	Suspended Solids Boiler Water (ppm) (Max)	TDS Range,[b,c] Steam (ppm) (Max Expected Value)
0–300	700–3500	140–700	15	0.2–1.0
301–450	600–3000	120–600	10	0.2–1.0
451–500	500–2500	100–500	8	0.2–1.0
601–750	200–1000	40–200	3	0.1–0.5
751–900	150–750	30–150	2	0.1–0.5
901–1000	125–625	25–125	1	0.1–0.5
1001–1800	100	_[d]	1	0.1
1801–2350	50		n.a.	0.1
2351–2600	25		n.a.	0.05
2601–2900	15	*Once-through boilers*	n.a.	0.05
1400 and above	0.05	n.a.	n.a.	0.05

Sources: American Boiler Manufacturers, 1982, Boiler water guidelines.
Note: n.a., not available.
[a] Actual values within the range reflect the TDS in the feed water. Higher values are for high solids in the feed water, and lower values for low solids.
[b] Actual values within the range are directly proportional to the actual value of TDS of boiler water. Higher values are for the high solid, and in the boiler water, lower values for low solids.
[c] These values are exclusive of silica.
[d] Dictated by boiler water treatment.

From solids balance,

$$100M = 1500B$$

Substituting these and solving, we get

$$M = 35{,}625 \text{ kg/h, } B = 2{,}375 \text{ kg/h, } D = 6{,}750 \text{ kg/h and } F = 52{,}375 \text{ kg/h}$$

Example 6.16

In a boiler plant if the conductivity of the condensate, makeup, and feed water are 800, 40, and 150 μmhos/cm, respectively, what is the approximate % of condensate returns in the feed water?

From mass balance around deaerator, $C + D + M = F$ (C, D, M, F, B refer to various flows as seen in Figure 6.11). From conductivity balance, $800C + 40M = 150F$.

We have only two equations and three unknowns. Simplifying, we have

$$M = F - C - D \text{ and } 800C + 40 \times (F - C - D) = 150F$$
$$760C - 40D = 110F$$

Since the question is regarding the ratio of C/F, make the practical assumption that D will be very small compared to F, and hence, neglecting the term with D, $C = (110/760)F =$ or about 15% of feed water. (This incidentally is a quiz question!)

TABLE 6.10

ASME Boiler and Feed Water Chemistry[a]

Boiler type: Industrial water tube, high duty, primary fuel fired, drum type

Makeup water percentage: Up to 100% of feed water

Conditions: Includes superheater, turbine drives, or process restriction on steam purity

Drum operating pressure[b], MPa (psig)	0–2.07 (0–300)	2.08–3.10 (301–450)	3.11–4.14 (451–600)	4.15–5.17 (601–750)	5.18–6.21 (751–900)	6.22–6.89 (901–1000)	6.90–10.34 (1001–1500)	10.35–13.79 (1501–2000)
Feed water[c]								
Dissolved oxygen (mg/L O_2) measured before oxygen scavenger addition[d]	<0.04	<0.04	<0.007	<0.007	<0.007	<0.007	<0.007	<0.007
Total iron (mg/L Fe)	≤0.100	≤0.050	≤0.030	≤0.025	≤0.020	≤0.020	≤0.010	≤0.010
Total copper (mg/L Cu)	≤0.050	≤0.025	≤0.020	≤0.020	≤0.015	≤0.015	≤0.010	≤0.010
Total hardness (mg/L $CaCO_3$)	<0.300	<0.300	<0.200	<0.200	<0.100	<0.050	n.d.	n.d.
pH range at 25°C	7.5–10.0	7.5–10.0	7.5–10.0	7.5–10.0	7.5–10.0	8.5–9.5	9.0–96	9.0–9 6
Chemicals for preboiler system protection	Use only volatile alkaline materials							
Nonvolatile TOCs (mg/L C)[e]	<1	<1	<0.5	<0.5	<0.5	As low as possible, < 0.2		
Oily matter (mg/L)	<1	<1	<0.5	<0.5	<0.5	As low as possible, < 0.2		
Boiler water								
Silica (mg/L SiO_2)	≤150	≤90	≤40	≤30	≤20	≤8	≤2	≤1
Total alkalinity (mg/L $CaCO_3$)	<350[f]	<300[f]	<250[f]	<200[f]	<150[f]	<100[f]	n.s.[f]	n.s.[f]
Free hydroxide alkalinity (mg/L $CaCO_3$)[g]	n.s.	n.s.	n.s.	n.s.	n.d.[h]	n.d.[h]	n.d.[h]	n.d.[h]
Specific conductance (μmho/cm) at 25°C without neutralization	<3500[i]	<3000[i]	2500[i]	<2000[i]	<1500[i]	<1000[i]	≤150	≤100

(Continued)

TABLE 6.10 (Continued)

ASME Boiler and Feed Water Chemistry[a]

Source: Adapted from ASME 1979 Consensus.

Note: n.d., not detectable; n.s., not specified.

[a] No values are given for saturated steam purity because steam purity achievable depends upon many variables, including boiler water total alkalinity and specific conductance as well as design of boiler, steam drum internals, and operating conditions (see footnote i). Because boilers in this category require a relatively high degree of steam purity, other operating parameters must be set as low as necessary to achieve this high purity for protection of the superheaters and turbines arid/ or to avoid process contamination.

[b] With local heal fluxes >473.2 kW/m² (>150,000 Btu/h ft²), use values for the nod higher pressure range.

[c] Boilers below 6 21 MPa (900 psig) with large furnaces, large steam release space, and internal chelant, polymer, and/or antifoam treatment can sometimes tolerate higher levels of feed water impurities than these in the table and still achieve adequate deposition control and steam purity. Removal of these impurities by external pretreatment is always a more positive solution. Alternatives must be evaluated as to practicality and economics in each case.

[d] Values in table assume the existence of a deaerator.

[e] Nonvolatile TOCs are the organic carbon not intentionally added as part of the water treatment regime.

[f] Maximum total alkalinity consistent with acceptable steam purity. If necessary, it should override conductance as blowdown control parameter. If makeup is demineralized water at 4.14–6.89 MPa (600–1000 psig), boiler water alkalinity and conductance should be that in table for 6.90–10.34 MPa (1001–1500 psig) range.

[g] Minimum level of OH⁻ alkalinity in boilers below 6.21 MPa (900 psig) must be individually specified with regard to silica solubility and other components of internal treatment.

[h] *Not detectable* in these cases refers to free sodium or potassium hydroxide alkalinity. Some small variable amount of total alkalinity will be present and measurable with the assumed congruent or coordinated phosphate pH control or volatile treatment employed at these high pressure ranges.

[i] Maximum values are often not achievable without exceeding suggested maximum total alkalinity values, especially in boilers below 6.21 MPa (900 psig) with >20% makeup of water whose total alkalinity is >20% of TDS naturally or after pretreatment with soda lime or sodium cycle ion-exchange softening. Actual permissible conductance values to achieve any desired steam purity must be established for each case by careful steam purity measurements. Relationship between conductance and steam purity is affected by too many variables to allow its reduction to a simple list of tabulated values.

Drum Holdup Calculations

Holdup time is an important parameter in steam drum design. Basically, it gives the time for all the water in the drum to evaporate at the rated steam capacity between certain levels without additional input of water from feed pump. Sometimes, distance between levels such as normal water level (NWL) and low water level (LWL) or high water level (HWL) is specified in terms of time. Then, based on these specifications, one determines the size of the steam drum.

Example 6.17

A boiler generates 50 t/h of steam at 40 kg/cm^{2}g (3921 kPa) (569 psig). The steam drum has an inner diameter of 1371 mm and length 10 m. Specifications call for 1.5 min of holdup time between NWL and LWL. The NWL is 50 mm below the drum centerline and the LWL is 150 mm below that. Is this adequate?

Solution

The volume of water in the drum is equal to the volume in the straight section V_s plus that in the dished ends V_e.

$$V_s = LR^2 (a/57.3 - \sin a \times \cos a) \tag{6.22}$$

and

$$V_e = 0.261H^2 (3R - H) \tag{6.23}$$

One must estimate the angle a as shown in Figure 6.12 for a given sector height. $R = 685$ mm and $H_1 = 635$ mm and $H_2 = 485$ mm.

$$a_1 = \cos^{-1}(50/685) = 85.8° \sin a_1 = 0.9973 \text{ and } \cos a_1 = 0.0732.$$

$$V_{s1} = 10 \times 0.685^2 \times (85.8/57.3 - 0.0732 \times 0.9973) = 6.68 \text{ m}^3$$

$V_{e1} = 0.261 \times 0.635^2 (3 \times 0.685 - 0.635) = 0.149$ m^3. With two ends, the dished end volume = 0.298 m^3.

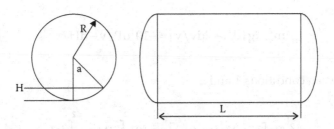

FIGURE 6.12
Partial volume of water in boiler drum.

$$a_2 = \cos^{-1}(200/685) = 73°. \sin a_2 = 0.956 \cos a_2 = 0.292$$

$$V_{s2} = 10 \times 0.685^2 \times (73/57.3 - 0.956 \times 0.292) = 4.67 \text{ m}^3.$$

$V_{e2} = 0.261 \times 0.0.485^2 \times (3 \times 0.685 - 0.485) = 0.096$ m³. With two ends, the volume = 0.192 m³.

Total volume between the NWL and LWL = (6.68 + 0.298) − (4.67 + 0.192) = 2.11 m³.

From steam tables, the specific volume of water at 40 kg/cm²g = 0.001253 m³/kg. Hence, the volume of water consumed at 50 tons/h steaming rate = 50,000 × 0.001253/60 = 1.044 m³/min.

Time available between the levels = 2.11/1.044 = 2.02 min. Hence, the size is adequate per specifications. In some waste heat boilers, the holdup time between NWL and empty could be as high as 5–7 min.

Estimating Flow in Blowdown Lines

The problem of estimating the discharge rates from a boiler drum or vessel to the atmosphere or to a vessel at low pressures involves two-phase flow calculations and is a lengthy procedure [5].

Presented in the following text is a simplified approach to the problem that can save considerable time for engineers who are involved in sizing or estimating discharge rates from boiler drums, vessels, or similar applications involving water. Several advantages are claimed for this approach as follows:

- No reference to steam tables is required.
- No trial and error procedure is involved for obtaining pipe size to discharge a desired rate of fluid.
- Effect of friction or equivalent length of piping can be checked.

Theory

The basic Bernoulli's equation can be written as follows for flow in a piping system:

$$10^4 v\,dp + V^2 dK/2g + v\,dv/g + dH = 0 \tag{6.24}$$

Substituting mass flow rate m = V/v,

$$m^2/2g(dK + 2dv/v) = -10^4 dP/v - dH/v^2 \tag{6.25}$$

Integrating between conditions 1 and 2,

$$m^2/2g\left[K + 2\ln(v_2/v_1)\right] = -10^4 \int_1^2 dP/v - \int_1^2 dH/v_2 \tag{6.26}$$

$$m = \left[2g/\{K+2\ln(v_2/v_1)\}\times\left\{-10^4\int_1^2 dP/v - \int_1^2 dH/v_2\right\}\right]^{0.5} \quad (6.27)$$

where
  V is the fluid velocity, m/s
  v is the specific volume, m³/kg
  K is the equivalent pipe resistance = fL_e/d where
  f is the friction factor
  L_e is the equivalent length of pipe, m
  d is the pipe inner diameter, m
  m is the mass flow rate, kg/m²s
  g is the acceleration due to gravity, m/s²
  P is the steam pressure, kg/cm² a
  H is the elevation, m

When the pressure of the vessel to which the blowdown pipe is connected is decreased, the flow rate increases until critical pressure is reached at the end of the pipe. Reducing the vessel pressure below critical pressure does not increase the flow rate. If the vessel pressure is less than critical pressure, critical flow conditions are reached resulting in sonic flow. From thermodynamics, the sonic velocity V_s and critical mass flow m_c can be shown to be

$$V_s = \left[-v^2 g\,(dP/dv)\times 10^4\right]^{0.5} \quad (6.28)$$

$$m_c = 100\left[-g\,(dP/dv)\right]^{0.5} \quad (6.29)$$

The term (dP/dv) refers to the change in pressure to volume ratio at critical flow conditions at constant entropy.

Hence, to estimate m_c, Equations 6.26 through 6.29 have to be solved. This is an iterative process. For the sake of simplicity, the gravity term may be neglected. Given K and P, one may also estimate P_c and m using Figure 6.14a and b later in the chapter. The procedure is as follows:

1. Assume a value for P_c, the critical pressure.
2. Calculate (dP/dv) at P_c for constant entropy conditions.
3. Calculate m_c using (6.29).
4. Solve (6.27) for m.

The term $-10^4\int_1^2 dP/v$ is computed as follows using Simpson's rule:

$$-10^4\int_1^2 dP/v = -10^4\int_1^2 \rho\,dP$$

$$\int_1^2 \rho\,dP = (P_s - P_c)/6\times(\rho_s + 4\rho_m + \rho_c)$$

where ρ_m = density at a mean pressure of $(P_s + P_c)/2$, kg/m³.

The second term in (6.27), namely, $[2\ln(v_2/v_1)] = 2\ln(\rho_s/\rho_c)$

Example 6.18

A boiler drum blowdown line is connected to a tank set at 8 atm. Drum pressure is 100 atm, and the resistance K of the blowdown line is 80. Estimate the critical flow m_c and critical pressure P_c.

Solution

Assume that critical pressure is 40 atm. From steam tables, $P_s = 100$ atm, $s = 0.7983$ kcal/kg °C, $h_l = 334$ kcal/kg, $v_l = 0.001445$ m³/kg, or $\rho = 692$ kg/m³ (s is the entropy, h_l, v_l are enthalpy and specific volume of saturated water, and ρ is the density).

At $P_c = 40$ atm, $h_l = 258.2$ kcal/kg, $h_v = 669$ kcal/kg, $s_l = 0.6649$, $s_v = 1.4513$, $v_l = 0.001249$, $v_v = 0.05078$ (subscripts v and l refer to vapor and liquid).

Steam quality $x = (0.7983 - 0.6649)/(1.4513 - 0.6649) = 0.17$

Specific volume $v = 0.0.001249 + 0.17 \times (0.05078 - 0.001249) = 0.009651$ m³/kg

Again, compute v at 41 atm. Using the same procedure as earlier, $v = 0.0093$ m³/kg

Hence, from (6.29), $m_c = 100 [9.8 \times 1/(0.00965 - 0.0093)]^{0.5} = 16733$ kg/m² s

We have to compute the actual mass flow rate m from (6.27) and see if it is equal to m_c. The density at initial pressure of 100 atm $\rho_s = 1/0.001445 = 692$ kg/m³

The dryness fraction corresponding to this $x = (334 - 258.2)/(669 - 258.2) = 0.1845$

$$v_c = 0.001249 + 0.1845 \times (0.05078 - 0.001249) = 0.010387 \text{ m}^3/\text{kg or } \rho_c = 96.3 \text{ kg/m}^3$$

Similarly, at $P_m = (100 + 40)/2 = 70$ atm, $v_m = 0.0378$ m³/kg or $\rho_m = 264$ kg/m³

$$-10^4 \int_1^2 dP/v = 10^4 \times \{(100 - 40)/6\} \times (692 + 4 \times 264 + 96.3) = 184.4 \times 10^6$$

$$m = \left[\{2 \times 9.8/(80 + 2 \ln(692/96.3)\} \times 184.4 \times 10^6 \right]^{0.5} = 6560 \text{ kg/m}^2 \text{ s}$$

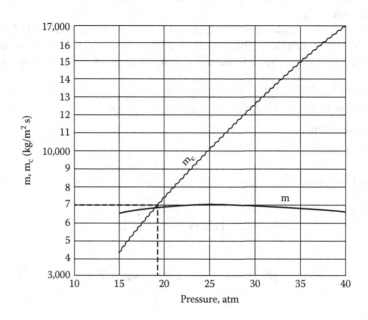

FIGURE 6.13
Results of calculation for m and m_c.

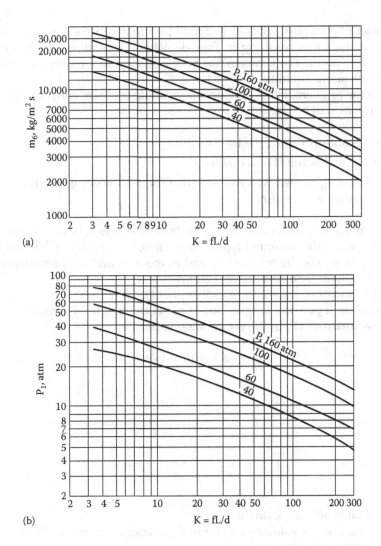

FIGURE 6.14
Charts for obtaining (a) flow rate and (b) critical pressure.

The two values m and m_c do not match. Hence, we have to assume another P_c and compute m and m_c. This has been done for $P_c = 30$ and 15 atm, and the results are shown in Figure 6.13. We see that at about 19 atm, the two curves intersect, and the mass flow rate is 7000 kg/m² s.

As seen earlier, the procedure is tedious. However, using Figure 6.14a and b, one may quickly check the mass flow rate and the critical pressure at various K values.

Flow Instability in Two-Phase Circuits

In once-through boilers or evaporators generating steam at high quality, the problem of flow instability is often a concern. This is due to the nature of the two-phase pressure drop characteristics inside tubes, which can have a negative slope with respect to flow under certain conditions. The problem is felt when multiple streams are connected to a common header in once-through boilers. Small perturbations can cause large changes in flow through a few tubes resulting in flow depletion, dry-out, or overheating of tubes. Vibration can also occur. The problem has been observed in a few low-pressure systems generating steam at high quality.

To illustrate the problem, let us take up the example of steam generation inside a tube. A few assumptions will be made:

1. Heat flux is uniform along the tube length.
2. Steam at the exit of tube has a quality x.
3. Some subcooling of feed water is present. (The feed water enters the tube at less than saturation temperature.)

If a tube is supplied with subcooled water, the boiling process starts after the enthalpy of the water has reached the saturated liquid state. Thus, the length of the boiler tube can be divided into two sections, the economizer and evaporator, and their lengths will be determined by heat input to their respective sections.

Let W be the flow of water entering in lb/h. Let Q = total heat input to the evaporator and Q_l the heat input per unit length, Btu/ft h. Let the length of economizer portion be L_1 ft. The pressure drop ΔP_1 in the economizer section is

$$\Delta P_1 = \frac{3.36 \times 10^{-6} f\, L_1 W^2 v_f}{d_i^5} \tag{6.30}$$

where

$$L_1 = \frac{W\Delta h}{Q_l} \tag{6.31}$$

where
Δh is the enthalpy absorbed, Btu/lb
v_f is the average specific volume of water in the economizer section, ft³/lb
d_i is the tube inner diameter, in.

The pressure drop in the evaporator section of length $(L - L_1)$ is given by

$$\Delta P_2 = \frac{3.36 \times 10^{-6} f\, (L - L_1) W^2 \left[v_f + .5 \times \left(v_g - v_f \right) \right]}{d_i^5} \tag{6.32}$$

Also,

$$x\, h_{fg} / \Delta h = (L - L_1)/L_1 \tag{6.33}$$

As the energy applied is uniform along the evaporator length, we are simply taking the ratio of energy absorbed in the evaporator and economizer, as being proportional to their lengths.

h_{fg} is the latent heat in Btu/lb
v_g, v_f are specific volumes of saturated liquid and vapor, ft³/lb

Substituting for x from (6.33) into Equation 6.32 and for L_1 from (6.31) and simplifying, we have

$$\Delta P = \Delta P_1 + \Delta P_2 = kW^3 \Delta h^2 \left(v_g - v_f\right)/2Q_l h_{fg} - kW^2 \left[\Delta h\left(v_g - v_f\right)/h_{fg} - v_f\right]$$
$$+ kWL^2 Q_1 \left(v_g - v_f\right)/2h_{fg} \tag{6.34}$$

or,

$$\Delta P = AW^3 - BW^2 + CW \tag{6.35}$$

Though this is a simplistic analysis, it may be used to show the effect of variables involved.

Equation 6.35 is depicted in Figure 6.15. It is seen that the curve of pressure drop versus flow is not monotonic but has a negative slope. This is more so if the steam pressure is low. Hence, it may lead to unstable conditions. For example, at some pressure drops, there could be three possible operating points, which may cause oscillations and large flow variations between circuits if there are multiple streams. (For the meaning of streams, refer to Appendix B.) Some tubes with low flows can reach departure from nucleate boiling conditions.

To improve the situation, one may place a restriction such as a control valve or orifice at the economizer inlet. The orifice increases the pressure drop in proportion to the square of the flow as shown by the term R in Equation 6.36. Figure 6.15a shows the effect of orifice, which makes the pressure drop monotonic:

$$\Delta P = AW^3 + (R - B)W^2 + CW \tag{6.36}$$

Due to the large specific volume and latent heat of steam at low steam pressures, the problem is more likely at low pressures as shown in Figure 6.15b. Decreasing the inlet subcooling by

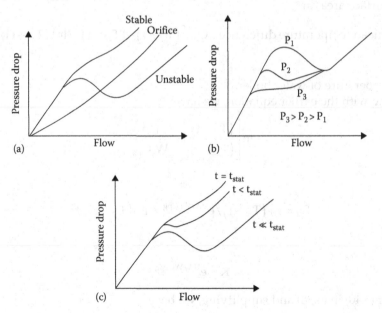

FIGURE 6.15
Effect of (a) orifice size, (b) pressure, and (c) inlet subcooling on stability of two-phase boiling circuits.

using a higher feed water temperature also helps as shown in Figure 6.15c. If inlet subcooling is eliminated, $\Delta h = 0$, and then Equation 6.35 becomes more stable as shown here:

$$\Delta P = BW^2 + CW \tag{6.37}$$

Transient Calculations

Often, one needs to estimate the time taken to heat up a boiler knowing the flue gas flow and temperature or the tube wall temperature attained by a superheater without any flow of steam inside the tubes (say, during startup of the boiler). Lumped mass system analysis is used, which gives good estimates sufficient for practical purposes.

Gas at a temperature T_{g1} enters an evaporator that is initially filled with water at a temperature of t_1. The following energy balance equation can be written neglecting the heat losses:

$$M_c dt/dz = W_g C_{pg} \left(T_{g1} - T_{g2} \right) = UA\Delta T \tag{6.38}$$

where

M_c is the water equivalent of the boiler = mass of steel × specific heat of steel + mass of water × specific heat of water (weight of boiler tubes, drums, and casing is included)
dt/dz is the rate of change of temperature °C/h
W_g is the gas flow, kg/h
C_{pg} is the gas specific heat, kcal/kg°C
T_{g1}, T_{g2} are gas temperatures entering and leaving the unit, °C.
U is the overall heat transfer coefficient, kcal/m²h °C
A is the surface area, m²

$$\Delta T = \text{log mean temperature difference, °C} = \left[\left(T_{g1} - t \right) - \left(T_{g2} - t \right) \right] / \ln \left[\left(T_{g1} - t \right) / \left(T_{g2} - t \right) \right] \tag{6.39}$$

t is the temperature of water boiler, °C
Combining with the earlier equation, we have

$$\ln \left[\frac{\left(T_{g1} - t \right)}{\left(T_{g2} - t \right)} \right] = \frac{UA}{W_g C_{pg}}$$

or

$$T_{g2} = t + \left(T_{g1} - t \right) / e^{UA/W_g C_{pg}} = t + \left(T_{g1} - t \right) / K \tag{6.40}$$

where

$$K = e^{UA/W_g C_{pg}} \tag{6.41}$$

Substituting (6.40) in (6.38) and simplifying, we have

$$M_c dt/dz = W_g C_{pg} \left(T_{g1} - t \right) (K - 1) / K \tag{6.42}$$

or

$$dt/(T_{g1}-t) = [W_g C_{pg}(K-1)/M_c K]dz \qquad (6.43)$$

Integrating from time t_1 to t_2, we have

$$\ln[[(T_{g1}-t_1)/(T_{g1}-t_2)] = [W_g C_{pg}(K-1)/(M_c K)]z \qquad (6.44)$$

This equation gives an idea of the time required to heat up the boiler from temperature t_1 to, say, the boiling temperature of 100°C. Once steam starts forming additional equations for latent heat, flow-through vent may be developed to obtain a more accurate startup curve.

Example 6.19

A waste heat boiler weighs 50,000 kg and contains 30,000 kg of water at 27°C. To the boiler enters 130,000 kg/h of flue gases at 800°C. Assuming that gas specific heat = 0.3 kcal/kg°C, steel specific heat = 0.12 kcal/kg°C, surface area for heat transfer = 2000 m², and overall adjusted U = 40 kcal/m²h °C, determine the time taken to heat the boiler with the water to 100°C.

Solution:

$$UA/W_g C_{pg} = 40 \times 2,000/(130,000 \times 0.3) = 2.05. \, K = e^{2.05} = 7.78$$
$$M_c = 50,000 \times 0.12 + 30,000 \times 1 = 36,000$$
$$\ln[(800-27)/(800-100)] = 0.099 = [130,000 \times 0.3 \times (7.78-1)/36,000/7.78]z$$
$$= 0.944z \text{ or } z = 0.1 \text{ h or } 6 \text{ min}$$

One has to then check the time for generating steam and ensure if the superheater gas inlet temperature is reasonable. In practice, some differences may be seen due to the assumptions made such as a uniform U for the complete evaporator; the drums may have a lower value of U compared to the tubes.

Example 6.20

The time required to heat up a superheater in a gas turbine exhaust plant from 27°C to 480°C has to be determined. At 550°C, 68,000 kg/h of turbine exhaust gases enter the HRSG. Assume the gas-side heat transfer coefficient is 60 kcal/m²h °C, gas specific heat = 0.286 kcal/kg°C, weight of the superheater is 2000 kg, and surface area = 180 m².

Solution

$$M_c = 2000 \times 0.12 = 240 \text{ kcal/°C}$$

Using Equations 6.41 and 6.44,

$$K = \exp[60 \times 180/(68,000 \times 0.286)] = \exp(0.555) = 1.74$$
$$\ln[(550-27)/(550-480)] = 68,000 \times 0.286 \times (1.74-1)z/1.74/240 \text{ or } z = 0.058 \text{ or } 3.6 \text{ min.}$$

This gives an idea of how fast the metal gets heated up. The steam generation based on the calculations done as in earlier example may take much longer than 3.6 min. If the startup is frequent, one can start up the gas turbine on low load when the exhaust gas temperature will be much lower. This is the reason that with high gas inlet temperatures, the superheater is shielded from the hot gases by a screen section that reduces the gas temperature entering it. Else, it can get oxidized without flow of steam. In fired boilers, the firing rate is controlled to keep the gas temperature lower at the superheater inlet if it is a radiant superheater. Here again, a convective superheater with a screen section helps.

Example 6.21

With 450°C exhaust gas temperature and the same gas flow (no bypass stack), what temperature will the tube metal reach in 10 min?

$\ln[(450 - 27)/(450 - t)] = 68{,}000 \times 0.286 \times 0.74 \times 0.167/(1.74 \times 240) = 5.75$ or $t = 449°C$. The superheater will reach the gas temperature in about 10 min. In 5 min, the tubes will attain 425°C. It is likely that steam will start flowing through the tubes within 10 min. As discussed in Appendix E, the finned evaporator weighs less and is compact, and hence, the water equivalent is lower compared to plain tube boilers, and hence, this helps to speed up the steam generation.

Analysis of an Evaporator

In an evaporator, the large mass of metal and water inventory results in a longer startup time, but the residual energy in the metal also helps to respond to load changes faster when the energy from flue gas or heat input to the boiler is cut off. Drum-level fluctuations are also smoothed out by a large water inventory.

The basic equation for energy transfer to the evaporator is

$$Q = W_s h_{fg} + (h_1 - h_f)W_f + (W_m C_p dt/dp + W_w dh/dp)dp/dz \qquad (6.45)$$

where
W_m is the mass of metal, kg
W_s, W_f are mass flow of steam, feed water, kg/h
W_w is the amount of water inventory in the boiler system including drums, tubes
dh/dp is the change of enthalpy to change of steam pressure, kcal/kg/kg/cm^2
dt/dp is the change of saturation temperature change to change in pressure, °C/kg/cm^2
Q is the energy transferred to evaporator, kcal/h
dp/dz is the rate of pressure change, kg/cm^2/h

Let the steam space between the drum level and the first valve = V m^3. The change in pressure may be written as
pv = C = pV/m where C is a constant
m is the mass of steam, kg in volume V or,

$$pV/m = C$$

or

$$dp/dz = (pv/V)(W_s - W_1) \qquad (6.46)$$

where
p is the steam pressure
W_s, W_1 are the steam generated and the steam withdrawn by process

When steam withdrawn is equal to steam generated, the steam pressure is unchanged. Pressure fluctuations occur when the steam generated and steam demand differ.

Example 6.22

A waste heat boiler evaporator has the following data:

Gas flow = 158,800 kg/h. Gas inlet temperature = 537°C. Gas exit temperature = 266°C. Steam pressure = 42 kg/cm² a. Feed water entering = 105°C.

Tubes: 50.8 × 2.7 mm, 30 tubes/row, 20 deep, 3.66 m with 177 fins/m, 19 mm high × 1.27 mm thick serrated fins.

Steam drum has an ID of 1371 mm, mud drum ID = 914 mm, 3.96 m long. Boiler generates 20,400 kg/h steam. Weight of steel tubes, drums = 38,000 kg. Weight of water in evaporator = 8600 kg. Volume of steam space = 3.26 m³. Feed water temperature = 105°C. Energy transferred to evaporator in normal operation = 11.67 mm kcal/h. What happens to the steam pressure and steam generation when the heat input and feed water supply are turned off?

Solution

For steam in the pressure range 42–44 kg/cm² a, from steam tables we have enthalpy of saturated liquid as 261.7 and 265 kcal/kg; saturation temperature: 252°C and 255°C; average latent heat h_{fg} = 405.3 kcal/kg; average specific volume of steam = 0.047 m³/kg

Hence,

$$dh/dp = (265 - 261.7)/2 = 1.65 \, kcal/kg/kg/cm^2$$
$$dt/dp = (255 - 252)/2 = 1.5h \, °C/kg/cm^2$$

From Equation 6.45, when Q = 0, W_f = 0,

$$W_s \times 405.3 + (38,000 \times 0.12 \times 1.5 + 8,600 \times 1.65) \times dp/dz = 0 \text{ or } 405.3W_s + 21,030dp/dz = 0$$

dp/dz from (6.46) = 43 × 0.047/3.26 ($W_s - W_l$) = 0.62 ($W_s - W_l$). Combining this with earlier equation,

$$W_s \times 405.3 + 21,030 \times 0.62 \times (W_s - W_l) = 0. \text{ If } W_l = 20,400 \text{ kg/h}, W_s = 19,784 \text{ kg/h}$$

From (6.46), pressure decay = dp/dz = 0.62 × (19,784 − 20,400) = −382 kg/cm²/h or −0.106 kg/cm²/s. This situation has often been experienced by plant engineers. If for some reason the gas flow to the HRSG is diverted and the feed water also cut off, there is decay in steam pressure till the heat input is restored. In the case of fresh air–fired HRSGs, there is a time lag between the bypassing of exhaust gases and the start of the fresh air fan. During this period, there is a pressure decay that may be correctly estimated by this method.

Example 6.23

Let us assume that the boiler is operating at 20,400 kg/h and suddenly the steam demand goes up to 22,686 kg/h.

Case 1: What happens to steam pressure if we maintain the heat input and feed water supply?

Case 2: What happens if feed water is cut off but heat input remains?

Case 1: Feed water enthalpy at 105°C = 105.8 kcal/kg, and feed water flow is the same as present steam flow assuming zero blowdown = 20,400 kg/h.

From (6.45), $20{,}400 \times (261.7 - 105.8) + W_s \times 405.3 + 21{,}030 dp/dz = 11.67 \times 10^6$ also, $dp/dz = 0.62 \times (W_s - 22{,}686)$

$$3.18 \times 10^6 + 405.3 W_s + 21{,}030 \times 0.62 \times (W_s - 22{,}686) = 11.67 \times 10^6$$
$$13{,}443.9 W_s = 295.79 \times 10^6 + 8.49 \times 10^6 \text{ or } W_s = 22{,}630 \text{ kg/h}$$

Thus, $dp/dz = 0.62 \times (22{,}630 - 22{,}686) = -35 \text{ kg/cm}^2/\text{h} = -0.01 \text{ kg/cm}^2/\text{s}$ (pressure falls slightly)

Case 2:

$$13{,}443.9 W_s = (295.79 + 11.67) \times 10^6 \text{ or } W_s = 22{,}869 \text{ kg/h}$$
$$dp/dz = 0.62 \times (22{,}869 - 22{,}686) = 113 \text{ kg/cm}^2/\text{h} = 0.0315 \text{ kg/cm}^2/\text{s}$$

The pressure actually increases because the cooling effect of the feed water is not sensed. In practice, controls respond fast and restore the balance among heat input, feed water flow, and steam flow. The preceding equations represent the worst-case scenario when controls do not act. If we do not adjust the heat input, the steam pressure will slide if we withdraw more steam than generated. The preceding equations may be used to show the trend and are simplistic models. Programs can be developed for more accurate modeling of transients in any steam generator.

Drum-Level Fluctuations

In cogeneration plants, the process conditions dictate the steam demand and could vary a lot from normal operating conditions. When the steam demand suddenly increases in a boiler, it may take a few milliseconds to a few seconds depending on the control system to adjust the flow and energy input to match the demand. When the evaporation rate in a boiler increases suddenly, water from the boiler tubes is displaced into the drum, which raises the water level momentarily or causes the *swell* effect. Momentarily, the pressure is also reduced. This swell occurs whether the increased evaporation is due to heat transfer in boiler surfaces or due to self-boiling as a result of falling steam pressure due to higher than ongoing steam generation. Similarly, a shrink effect occurs when the steam demand decreases and water level decreases momentarily. The change in water level is proportional to the change in evaporation. The effect of feed water flow will also have to be considered. The mass of water in the drum increases at the following rate:

$$[(W_i - W_s) + T(W_s - W_l)/dt]$$

where T is the mass of water displaced into drum by unit increase in evaporation = $3600 \times$ water weight $\times (h_l - h_{fw})/[W_s \times (h_v - h_{fw})]$, where W_i, W_s, W_l are the water flow in, normal steam generation, and steam withdrawn in kg/h, respectively.

If water inventory = 40,000 kg in a boiler and from steam tables $h_l = 198$, $h_{fw} = 105$, $h_v = 665$ kcal/kg, and $W_s = 18{,}000$ kg/h, then $T = 3{,}600 \times 40{,}000 \times (198 - 105)/[18{,}000 \times (665 - 105)] = 1{,}328$ s. In a steady-state situation, $W_s = W_l = W_i$ (neglecting blowdown). Then,

$$dh/dt = \text{level change/s} = v[(W_i - W_s + T(W_s - W_l)/t]/A$$

where
  A is the drum cross section, m^2
  v is the specific volume of saturated liquid, m^3/kg

Example 6.24

If in the boiler mentioned earlier, steam demand suddenly changes from 18,000 to 21,000 kg/h in 7 s, and if A = 10 m², v = 0.001125 m³/kg, then momentary change in level = 0.001125 × [1,328 × 3,000/3,600/7]/10 = 0.0177 m/s = 17.7 mm/s, which is significant assuming controls do not react immediately. In practice, the swell will be much smaller and will last for a much shorter period, say, for a few milliseconds with good controls such as three-element-level control system.

Fan Calculations

Density of air or flue gas has to be obtained first to perform fan-related calculations.

Density of Flue Gas, Air

The following formula may be used to estimate the density of air and flue gases:

$$\rho = MW \times P/8314/T \tag{6.47}$$

where
  MW is the molecular weight of flue gas
  P is the pressure of flue gas, air in Pascal
  T is the temperature, K
  ρ is the gas density, kg/m³
  For dry air, MW = 28.97. (8314 is the universal gas constant)

At 273 K, at sea level, and at atmospheric pressure (101,325 Pa), density of dry air is

$$\rho = 28.97 \times 101,325/273/8,314 = 1.293 \text{ kg/m}^3$$

In British units, from Table 6.11, $\rho = 0.0933 \times 28.97 \times 14.7/(460 + 32) = 0.08076$ lb/ft³ = 1.292 kg/m³. Effect of elevation on the density of air is given in Table 6.12, and Table 6.13 shows the molecular weight of some gases.

A simplified formula for the estimation of air density at atmospheric pressure, at sea level, and at a temperature of T K is

$$\rho = 353 / T \tag{6.48}$$

At 300 K, density of air = 353/300 = 1.177 kg/m³ = 0.0735 lb/ft³

Example 6.25

Flue gas at 100°C (212°F) having the following analysis % volume CO_2 = 15, H_2O = 8, N_2 = 75, O_2 = 2 at a 200 mm wc (1961 Pa) (7.9 in. wc) gauge pressure and at 1000 m (3280 ft) elevation flows inside a boiler plant duct. Determine the density.

Correction factor from Table 6.12 is 0.887 by interpolation. MW of flue gas = 0.75 × 28 + .02 × 32 + 0.15 × 44 + 0.08 × 18 = 29.68

$$\text{Gas pressure} = 101,325 \times 0.887 + 1,961 = 118,446 \text{ Pa}$$

$$\rho = 29.68 \times 118,446/8314/373 = 1.133 \text{ kg/m}^3 \left(0.0708 \text{ lb/ft}^3\right)$$

Hence, if 100,000 kg/h of flue gas flows in a duct at these conditions, the volume of flue gas = 100,000/1.133 = 88,261 m³/h.

TABLE 6.11

Density of Flue Gas, Air

SI	British	Metric
$\rho = 1.203 \times 10^{-4}$ MW P/T	$\rho = 0.0933$ MW P/T	$\rho = 11.792$ MW P/T
P is the absolute pressure, Pascal	psia	$kg/cm^2 a$
T, K	T, °R	T, K
ρ, kg/m^3	ρ, lb/ft^3	ρ, kg/m^3

Note: MW is the molecular weight of gas.

TABLE 6.12

Density Correction Factor

Elevation, m (ft)	Correction
0	1
305 (1000)	0.964
610 (2000)	0.930
915 (3000)	0.896
1220 (4000)	0.864
1524 (5000)	0.832
1829 (6000)	0.801
2134 (7000)	0.772
2439 (8000)	0.743

TABLE 6.13

Molecular Weight of Gases

Gas	MW	Gas	MW
Hydrogen	2.016	Ammonia	17.03
Oxygen	32.0	Carbon dioxide	44.01
Nitrogen	28.016	Carbon monoxide	28.01
Air (dry)	28.97	Nitrous oxide	44.02
Methane	16.04	Nitric oxide	30.01
Ethane	30.07	Nitrogen dioxide	46.01
Propane	44.09	Sulfur dioxide	74.06
n-Butane	58.12	Sulfur trioxide	80.06
Water vapor	18.02		

Pressure Drop in Ducts

In steam generators or waste heat boilers, one often has to estimate the air or flue gas pressure drop in ducts given the flue gas analysis and temperature. The Reynolds number has to be determined first from the following formula:

Re = $1.273w/(d_i\mu)$ where w is the flow in kg/h, μ is the viscosity in kg/m h, and d_i is the equivalent diameter of duct, m (see Table B.1).

Example 6.26

Determine the flue gas pressure drop in a rectangular duct with a cross section of 0.6 × 0.75 m. Flue gas flow is 11,300 kg/h at 150°C. Assume natural gas products of combustion and total developed length of 30 m. Flue gas from combustion of natural gas has the following analysis in % volume: $CO_2 = 8.29$, $H_2O = 18.17$. $N_2 = 71.08$, $O_2 = 2.46$. MW = 27.6

Solution

The duct equivalent diameter = $2ab/(a + b)$ = 2 × 0.6 × 0.75/1.35 = 0.666 m

Re = $1.273 \ w/d_i\mu$. The flue gas viscosity at 200°C from Table F.7 is 0.0789 kg/m h. Hence, Re = 1.273 × 11,300/(0.666 × 0.0789) = 276,240

Use the Blasius friction factor correlation. Friction factor f = $0.316/Re^{0.25}$ = 0.0138

Density of flue gas at atmospheric pressure = 12.17 × 27.6/(273 + 150) = 0.794 kg/m³

From Table B.6, $\Delta P = 6.382 \times 10^{-9} \ f \ L_e \ v \ w^2/d_i^5 = 6.382 \times 10^{-9} \ f \ L_e \ v \ w^2/d_i^5 = 6.382 \times 10^{-9} \times$ 0.0138 × 30 × $11,300^2/(0.794 \times 0.666^5)$ = 3.3 mm wc

Fan Selection

Package steam generators generating up to 120 tons/h of steam today use a single fan. The furnaces of oil- and gas-fired boilers are pressurized. Estimating the flow or head inaccurately can lead to operation of the fan in an unstable region or result in the horsepower being too high and the operation inefficient. The inlet density of air and the flow volume and temperature should be estimated accurately considering flue gas recirculation (FGR) if any. If flue gas at, say, 150°C is with ambient air and sent to the fan inlet, the density will be lower as the mixture temperature is higher than ambient temperature. The flow volume will also be higher considering the amount of flue gas recirculated. In addition, the elevation factor should be kept in mind. The fan is selected for the lowest density case as the fan always delivers the same volume, and hence, the mass flow is lower at lower density. Ensuring the required mass flow of air for combustion at the lowest density case will result in higher mass flow of air at higher density cases (say, lower ambient temperatures). The fan also delivers a lower head at lower density, while the steam generator may require a constant head due to a given mass flow of flue gas at a particular load. Hence, provision should be made to turn down the fan under such circumstances. The effect of air density on fan performance is shown in Figure 6.16.

One of the mistakes made by consultants or plant engineers is the use of large margins on flow and head. This leads to oversizing of the fan (considering that a single fan is often used) and operation close to the unstable region as shown in Figure 6.16 at part loads. Fan also will have a large motor, which is unnecessary. Typically, 10% margin on flow and a maximum of 15%–20% on head should be adequate for package boilers. Else, inlet vane control and variable speed drives may be required for the turndown making the fan system expensive. In a large utility boiler, two or even three fans are used, and when the turndown is low, say, 50%, a fan may be cut off. With a single fan system, this results in operation at low flows, and hence, the excess air control becomes difficult. Underestimating the fan head can also cause the fan to operate in the unstable region as shown in Figure 6.16a. The fan operating point should always be in the negatively sloping portion of the head versus flow curve; else, the fan could operate in the unstable region causing surges and vibration.

The fan inlet duct and downstream ductwork to the burners must have proper flow distribution. Pulsations and duct vibrations are likely if the inlet air flow to the fan blades is not smooth.

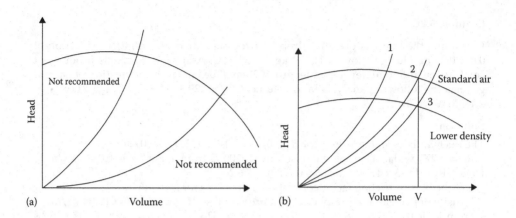

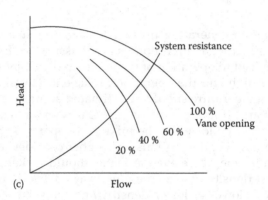

FIGURE 6.16
(a) Fan performance and range of operation. (b) Effect of system resistance on fan horsepower. Note: Curve 1 is the actual operating curve while curve 2 is the estimated. Operating at point 1 is not recommended. Also, a fan delivers a lower head at lower density. (c) Effect of vane position on flow reduction in fans.

Why Should the Fan Capacity Be Reviewed at the Lowest Density Condition?

As discussed in Chapter 1, the air flow in mass units required for combustion depends on the heat input to the steam generator, which is a function of boiler capacity and efficiency.

$W = 60\rho Q$ where ρ is the density in kg/m^3 and Q is the volumetric flow in m^3/s and W is the air flow in kg/s.

Fans discharge constant volumetric flow at any density. Hence, if the fan is sized for a particular volumetric flow, the mass flow will decrease when the density decreases. Hence, the minimum air flow in mass units should be obtained at the lowest density case. This means that the volumetric flow at higher density cases will be lower, and fan has to be modulated using devices such as inlet vanes or variable speed drive. Also the air pressure drop across the windbox is proportional to W^2/ρ. When the density decreases, the pressure drop across the windbox and in the duct to burner will be higher as W, the mass flow, will remain constant. Considering that H/ρ for a fan is a constant (H is the static head in mm wc), using the lowest density in fan selection ensures that the heat is available at the lowest density as seen in Figure 6.16b. At high densities, the fan has to be modulated to reduce the head.

Head Developed by Fan

A fan always develops the same volume and same head H_a in height of air column, but the pressure will vary depending on density. We can also write $H_a \rho_a = H_w \rho_w$, where H_w is the head developed in mm wc and ρ_w is the air density. It is a practice in the industry to work with H_w for fan as it is easy to measure. Hence, $H_a = H_w \rho_w / \rho_a$.

As density of water ρ_w is the same, we can use the following expression to understand how the fan head varies with air density:

$$H_{w1}/\rho_1 = H_{w2}/\rho_2$$

Hence, a fan will develop a lower head of water column at lower density. The flue gas pressure drop across boiler heating surfaces is calculated in mm wc (see Chapters 2, 3, and 5). Hence, we have to ensure that the fan can handle to flue gas pressure drop at the lowest air density case. Figure 6.16c also illustrates that.

Fan Power Consumption

$P = H_w Q/\eta_f \eta_m$ where H_w is the total pressure developed in Pa, P the power consumption in watts, η_f, η_m the efficiency of fan and motor in fractions, and Q is the volumetric flow in m³/s. Converting to mass units,

$$P = H_w W/\left(\rho \eta_f \eta_m \right)$$

If H_w is the head developed in mm wc (which is often used in boiler practice and in measurements), multiply it by 9.807 to obtain head in Pascal.

Example 6.27

A steam generator requires 8.5 m³/s (18,000 cfm) air at 30°C for combustion. Static head required is 457 mm wc (18 in. wc). Fan and motor efficiencies are 75% and 90%, respectively. Determine the motor power consumption.

$$P = 9.807 \times 457 \times 8.5/0.75/0.9 = 56{,}437 \text{ W} = 56.5 \text{ kW}$$

Density assuming atmospheric pressure is 1.165 kg/m³. Hence, mass flow is about 8.5 × 1.165 = 9.9 kg/s (78,560 lb/h).

Estimating Stack Effect

Whenever flue gas flows inside a tall vertical stack, a natural draft is created owing to the difference in density between the low density flue gas and ambient air, which has a higher density. Due to the friction loss inside the stack, the available draft is reduced. The estimation of available draft assumes significance in small boilers operating without forced or induced draft fans.

Example 6.28

Determine the stack effect when 50,000 kg/h of flue gas at 200°C from the combustion of fuel oil flows inside a 1.22 m diameter stack of height 16 m. Flue gas analysis is % volume $CO_2 = 11$, $H_2O = 12$, $N_2 = 74.5$, $O_2 = 2.5$. Ambient air is at 25°C.

Neglect the effect of elevation and gas temperature in the stack. Gas pressure is atmospheric. MW of flue gas = $28.66 \times 101{,}325/8{,}314/473 = 0.738$ kg/m³. Density of air = $353/298 = 1.185$ kg/m³

$$\text{Available draft} = (1.185 - 0.738) \times 16 = 7.15 \text{ kg/m}^2 = 7.15 \text{ mm wc}$$

Friction loss in stack:
$\Delta P = 0.08262 \text{ f } L_e \text{ v } w^2/d_i^5$ from Appendix B, where ΔP is in mm wc.

$$w = \text{flow in kg/s} = 50{,}000/3{,}600 = 13.9 \text{ kg/s}$$
$$v = \text{specific volume of fluegas at 200C} = 1/0.738 = 1.355 \text{ m}^3/\text{kg}$$

To estimate the friction factor, first the Reynolds number is computed. From Appendix F, viscosity of flue gas at 200°C is 0.0863 kg/m h = 2.397×10^{-5} kg/m s. Reynolds number is obtained from Appendix B as follows:

$$\text{Re} = 1.273 w/d/\mu = 1.273 \times 13.9/1.22/2.397 \times 10^{-5} = 6.05 \times 10^5$$

Friction factor in turbulent regime may be estimated using $f = 0.316/\text{Re}^{0.25} = 0.011$

$$\Delta P = 0.08262 \times 0.011 \times 16 \times 13.9^2/1.22^5 = 1.039 \text{ mm wc}$$
$$\text{Net draft available} = 7.15 - 1.039 = 6.1 \text{ mm wc}$$

References

1. D.Q. Kern, *Process Heat Transfer*, McGraw-Hill, New York, 1950, p263.
2. K.O. Beatty and D.L. Katz, Condensation of vapors outside of finned tubes, *Chemical Engineering Progress*, 44.1, 55–70, 1948.
3. Century Brass Products, *Heat Exchanger Manual*, Waterbury, CT, 1950.
4. V. Ganapathy, Avoid heat transfer equipment vibration, *Hydrocarbon Processing*, June 1987, p62.
5. F.J. Moody, Maximum two-phase vessel blowdown from pipes, *Transactions of ASME, Journal of Heat transfer*, August 1966, 285.

Appendix A: Boiler Design and Performance Calculations

Introduction

The key to evaluating the performance of a steam generator or a waste heat boiler is in understanding the thermal performance of each of its major heat transfer components such as furnace, superheater, evaporator, economizer, and air heater and then integrating them to arrive at the overall performance. In order to understand the performance of any component, U, its overall heat transfer coefficient must be evaluated, and then based on flue gas–fluid temperatures and surface area, its duty or energy transferred may be estimated. The component could have plain or finned tubes and could be arranged in an inline or staggered fashion. Since the details of the tube geometry and surface area details for any component are available to a plant engineer in existing boiler plants, the duty and performance may then be evaluated by the methods discussed in the following text. Then using the correlations for the estimation of tube inside and outside convective heat transfer and nonluminous heat transfer coefficients explained in the appendices that follow and numerous examples dispersed throughout the book on thermal design aspects, one may evaluate the maximum or minimum tube wall temperatures, gas- or steam-side pressure drops, boiler and heat recovery steam generator (HRSG) efficiency, and their off-design performance. Several tools for boiler and HRSG performance evaluation have been provided throughout the book using which plant and process engineers may independently analyze their performance.

This appendix gives an idea of how the overall performance of a boiler or HRSG may be evaluated in the design and off-design modes and the importance of gas-side heat transfer coefficient, which governs U. Plant engineers may sometimes be involved in the design of a component, say, when a new superheater has to be designed for a revamp project or replacement of a superheater for different duty. Often, they will also be involved with off-design performance calculations of the boiler which is in place and all they want to do is to check the field data from time to time and see if the performance is in line with the predicted performance or if there is any sign of problem. The need for the analysis of the performance of a boiler component arises when, for example, a superheater is not performing per expectations, tubes are failing due to overheating, the economizer exit gas temperature is higher than predicted, or the steam generation in an unfired HRSG is lower than that guaranteed.

The overall heat transfer coefficient U has to be evaluated either to design a boiler component or to check its off-design performance. For this purpose, one should be able to estimate the heat transfer coefficients of the fluids flowing inside and outside the tubes. The procedure for estimating convective heat transfer coefficient for flow inside tubes and outside plain and finned tubes is explained in Appendices B, C, and E, and the evaluation of nonluminous heat transfer coefficient is discussed in Appendix D. In all of these calculations, thermal and transport properties of flue gases play a vital role, and methods of estimating these are shown in Appendix F. Some practice with these calculations will help engineers evaluate the performance of their steam generators or waste heat boilers or any of the boiler components such as superheater, evaporator, or economizer or fire tube waste heat boiler or air heater with ease and see if the operating data are close to that predicted by the boiler supplier or not or whether a potential problem exists. Detailed calculations with real-life examples are given throughout the book so one may develop computer code for faster and more accurate calculations. Depending solely on the boiler supplier to perform these studies may not often be a good idea. Often, the reports from the boiler suppliers can be biased; the plant engineer can then challenge them if he can independently perform these studies. This is the main objective of this book; to educate plant and process engineers on boiler related applied heat transfer calculations so that they can arrive at the performance of various boiler components or that of the entire steam generator or HRSG independently. When there is a problem in a steam generator or HRSG, it is easy to shift the blame on poor operation or maintenance, while the real issue could be the poor design of the steam generator. Hence, plants with steam generators and waste heat boilers should develop in-house talent to perform thermal performance calculations, and this book provides tools for that very purpose. The information provided will also be pertinent to boiler suppliers.

Basic Equations for Energy Transfer

The total energy transferred Q in kW in a convective heat transfer equipment such as a superheater, evaporator, or economizer is given by the following basic equation:

$$Q = U \, A \, \Delta T \tag{A.1}$$

ΔT is the corrected log-mean temperature difference (LMTD), K. If surface area A is based on external surface area, then U is also based on external surface area. If A is based on tube inner diameter, then U is based on tube ID. The relation between the two is simple:

$$U_o A_o = U_i A_i \tag{A.2}$$

where
   U_o and U_i are the overall heat transfer coefficients based on external and internal surface areas
   A_o and A_i are the external and internal surface areas of the tubes

If external radiation Q_r from a cavity, flame, or furnace is received by the heat transfer surface, then the preceding equation is modified as

$$Q - Q_r = Q_c + Q_n = U \, A \, \Delta T \tag{A.3}$$

where Q_c, Q_n are the energy transferred by convection and nonluminous radiation, kW.

In addition to the above equation, energy balance equations should also be satisfied. The energy given by the hot fluid, namely, the flue gases, has to be absorbed by the colder fluid, say, water (in an economizer) or steam (in a superheater) or by air (air heater) or steam water mixture (boiler evaporator).

$$W_h \Delta h_h (1 - h_l) = W_c \Delta h_c = Q \tag{A.4}$$

where
   Q is the energy transferred in kW. Q_r is neglected when direct radiation from furnace or cavity is absent or when gas temperature is low, say, less than about 700°C. Chapter 2 discusses the method of estimating Q_r
   A = surface area, m^2
   W is the fluid flow or flue gas flow, kg/s (subscripts c and h stand for cold and hot fluids)
   Δh is the change in enthalpy, kJ/kg (subscripts h and c stand for hot and cold side fluids) = C_p × temperature change, where C_p is the specific heat, kJ/kg K

Correction factors for LMTD for different types of arrangement such as parallel-flow, cross-flow, and counter-flow are available in reference books. In pure counter-flow or parallel-flow, it is 1. For cross-flow arrangements, charts are available. See Figure A.1 [1].
   h_l is the heat loss, ranging from 0.2% to 1%. The method of evaluating this is explained in Chapter 6 on miscellaneous calculations; heat loss depends on the type of boiler casing, insulation thickness, ambient temperature, and wind velocity conditions.
   Table A.1 shows the different systems of units that have been used in this book. One can easily perform the calculations in any system of units.
   For extended surfaces, U_o may be obtained as follows [2]:

$$\frac{1}{U_o} = \frac{A_t}{(h_i A_i)} + ff_i \times \left(\frac{A_t}{A_i}\right) + \frac{A_t}{A_w} \times \left(\frac{d}{2K_m}\right) \times \ln\left(\frac{d}{d_i}\right) + ff_o + \frac{1}{\eta h_o} \tag{A.5}$$

where
   A_t is the total external surface area of finned tube per unit length, m^2/m
   A_i is the tube inner surface area per unit length = πd_i, m^2/m
   A_w is the average wall surface area = $\pi(d + d_i)/2$, m^2/m
   K_m is the thermal conductivity of tube wall (see Appendix E for properties of some boiler tube materials)
   d, d_i are the tube outer and inner diameters, m
   ff_i, ff_o are the fouling factors inside and outside tubes, m^2 K/W
   h_i, h_o are the tube inside and outside side heat transfer coefficients, W/m^2 K
   η is the fin effectiveness, fraction

If plain or bare tubes are used, the preceding equation may be simplified and written as follows:

$$\frac{1}{U_o} = \frac{d}{h_i d_i} + ff_i \times \left(\frac{d}{d_i}\right) + \left(\frac{d}{2K_m}\right) \times \ln\left(\frac{d}{d_i}\right) + ff_o + \frac{1}{h_o} \tag{A.6}$$

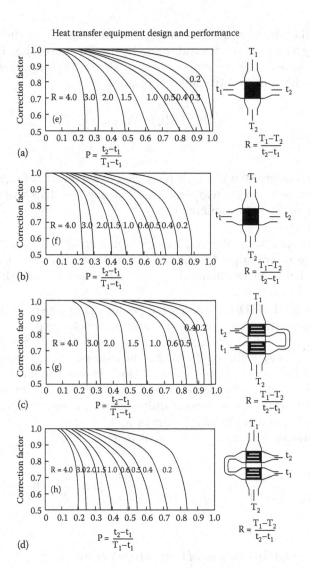

FIGURE A.1
Cross-flow correction factors.

h_o, the outside heat transfer coefficient, consists of a convective part h_c and a nonluminous part h_n, which is significant at gas temperatures above 700°C. In low gas temperature surfaces such as economizer or air heater, h_n may be neglected.

$$h_o = h_c + h_n \qquad (A.7)$$

Procedure for determining h_c, h_n, U, η, Q_r, Q are all given in Appendices B through E. There are basically five resistances (all in series) to heat transfer in a boiler component. To perform boiler calculations with ease, one must have some idea of the various thermal resistances as shown earlier and understand which are governing and hence important. This will save time when one wants to evaluate the duty or temperature profiles in a boiler quickly. Now let us see which side governs U in different types of boiler components.

TABLE A.1

Systems of Units

Item	SI	British	Metric
Gas flow	1 kg/s	7,934 lb/h	3,600 kg/h
Duty	1 kW	3,413 Btu/h	860 kcal/h
Specific heat	4187 J/kg K	1 Btu/lb °F	1 kcal/kg °C
Viscosity	1 kg/m s	2,419 lb/ft h	3600 kg/m h
Thermal conductivity	1 W/m K	0.5778 Btu/ft h °F	0.86 kcal/m h °C
Tube diameter	1 m	3.28 ft	1 m
Area	1 m²	10.76 ft²	1 m²
Enthalpy	4.187 kJ/kg	1.8 Btu/lb	1 kcal/kg
Fouling factor	0.0001759 m² K/W	0.001 ft² h °F/Btu	0.0002045 m² h °C/kcal
Heat transfer coefficient	100 W/m² K	17.59 Btu/ft² h °F	86 kcal/m² h °C
Heat flux or area heat release rate	100 kW/m²	31,719 Btu/ft² h	86,000 kcal/m² h
Volumetric heat release rate	100 kW/m³	9,670 Btu/ft³ h	86,000 kcal/m³ h
Velocity	1 m/s	3.28 ft/s	1 m/s
Specific volume	1 m³/kg	16 ft³/lb	1 m³/kg
Density	1 kg/m³	0.0625 lb/ft³	1 kg/m³
Fins density	39.36 fins/m	1 fins/in.	39.36 fins/m
Pressure	100 kPa (1 bar)	14.5 psi	1.02 kg/cm²

Note: How to read the table: 1 m² = 10.76 ft². 86 kcal/m² h °C = 100 W/m² K = 17.59 Btu/ft² h °F. 2 fins/in. = 78.7 fins/m.)

Water Tube Boilers, Economizers, Superheaters

The gas-side coefficient h_o governs U in water tube boilers, economizers, and superheaters. The other terms can be neglected. (In what follows, we are going to assume some typical values for h_i, h_o to explain the point, but Appendices B through E show how to estimate each of these terms. In a typical plain tube economizer, the tube side coefficient h_i will be in the range of 7000–8500 W/m² K (6020–7310 kcal/m² h °C or 1232–1496 Btu/ft² h °F); h_o, the gas-side coefficient will be about 80 W/m² K (68.8 kcal/m² h °C) (14 Btu/ft² h °F). Using d = 0.05 m (1.97 in.) and d_i = 0.042 m (1.847 in.), and K_m = 43 W/m K (24.8 Btu/ft h °F), ff_i, ff_o = 0.000175 m² K/W (0.0002 m² h °C/kcal) (0.001 ft² h °F/Btu) and substituting in (A.5) yields

$$\frac{1}{U_o} = 0.05/0.042/8000 + \frac{1}{80} + (0.05/43/2)\ \ln\left(\frac{0.05}{0.042}\right) + 0.000175 \times 0.05/0.042 + 0.000175$$

$$= 0.01323 \text{ or } U_o = 75.5 \text{ W/m}^2 \text{ K}\left(13.29 \text{ Btu/ft}^2 \text{ h °F}\right)$$

Thus, we see that the overall coefficient is close to the gas-side coefficient, which is the highest resistance in the system to energy transfer and hence governing. The metal thermal resistance and the tube-side resistance are not high enough to influence U. The purpose of this exercise is to let plant engineers know that for quick evaluation of performance, it is adequate to estimate the gas-side coefficient and use, say, 95% of it as U_o. This is also the situation in a boiler evaporator where h_i is on the order of 10,000–20,000 W/m² K and in the case of superheaters where h_i is in the range of 1500–2500 W/m² K.

Using $h_i = 1500$ W/m^{2}K (264 Btu/ft^{2}h °F), we can show that $U_o = 72$ W/m^{2}K (12.67 Btu/ft^{2}h °F), still close to h_o. If extended surfaces are used, then U will again be close to the corrected outside coefficient ηh_o.

Let us say we have an economizer that uses solid fins of density 78 fins/m (2 fins/in.), 19 mm high (0.75 in.), and 1.5 mm (0.059 in.) thick alloy steel fins. The ratio of external to internal surface can be shown to be 5.2. Assume $h_i = 8000$ W/m^{2}K (1408 Btu/ft^{2}h °F). The other data are as in previous example. Due to the use of fins, the gas-side heat transfer coefficient will decrease. Use, say, 60 W/m^{2}K for corrected gas-side heat transfer coefficient (ηh_o) and a fin effectiveness of 84% (h_o will be lower for finned tubes compared to plain tubes for the same gas velocity. This is discussed later in Appendix E). Then,

$$1/U_o = 5.2/8000 + 1/60 + 5.2 \times 0.05/2/43 \times \ln(0.05/0.042) + 0.000175 + 0.000175 \times 5.2$$
$$= 0.0394 \text{ or } U_o = 52.8 \text{ W/m}^2\text{K (9.29 Btu/ft}^2\text{h °F)}$$

which is again close to the corrected outside coefficient of 60 W/m^{2}K (10.56 Btu/ft^{2}h °F).

Fire Tube Boilers, Gas Coolers, Heat Exchangers

In these equipment, the flue gas flows inside the tubes while a fluid with high heat transfer coefficient is on the shell side.

Here let us use the same values for h_o and h_i and tube sizes, namely, $h_i = 50$ W/m^2 K (8.8 Btu/ft^{2}h °F), $h_o = 8000$ W/m^{2}K (1408 Btu/ft^{2}h °F). Tube OD × ID = 50 × 42 mm. From (A.6),

$$1/U_o = 0.05/0.042/50 + 1/8000 + (0.05/43/2) \ln(0.05/0.042) + 0.000175 + 0.000175 \times 0.05/0.042$$
$$= 0.0243 \text{ or } U_o = 41 \text{ W/m}^2\text{K (7.22 Btu/ft}^2\text{h °F) (based on outer surface area)}.$$

Converting to tube inner diameter basis, we have the following relation:

$$U_o \times d = U_i \times d_i \text{ or } U_i = \frac{41 \times 50}{42} = 48.8 \text{ W/m}^2 \text{ K or very close to } h_i.$$

Hence, in fire tube boiler also, the gas-side coefficient governs U. The gas-side coefficient governs the overall heat transfer coefficient in evaporators, superheaters, and economizers and fire tube boilers. One should also note that when we specify the U value, we should state whether it is based on inner or outer diameter of the tube.

Gas to Gas Exchangers: Air Heater

In the case of air heaters, both the tube side and outside coefficients have to be considered in determining U as both coefficients are comparable in value unlike in boilers or economizers where the gas-side coefficient was so small compared to the steam- or water-side coefficient.

If $h_i = 50$ W/m^2 K $\left(8.8 \text{ Btu/ft}^2\text{ h °F}\right)$ and $h_o = 60$ W/m^2 K $\left(10.56 \text{ Btu/ft}^2\text{ h °F}\right)$, then from (A.6)
$1/U_o = 50/42/50 + 1/60 + (0.05/2/43) \ln(50/42) + 0.000175 + 0.000175 \times 50/42 = 0.04096$ or
$U_o = 24.4$ W/m^{2}K (4.22 Btu/ft^{2}h °F). This is close to neither h_i nor h_o. The same conclusion

may be drawn in the case of liquid to liquid exchangers where both the tube inside and outside coefficients are large and comparable, and so *both* will influence U_o.

Thus, in boiler performance analysis, one has to keep in mind that the evaluation of the gas-side heat transfer coefficient assumes significance when it comes to evaluating the performance of superheaters, economizer, or boiler bank tubes or evaporator and greatly impacts their performance or duty. However, it is also essential to compute the tube-side coefficients to estimate the tube wall temperature accurately as in the case of superheaters. This helps in the selection of proper materials or warns us if low-temperature corrosion is likely in case of an economizer or air heater.

Design and Performance Calculations

Two types of calculations are done for any heat transfer equipment.

Design Calculation

In this process, the tube geometry (such as tube size, spacing of tubes, fin geometry, if any) is first assumed, then U is computed based on gas and steam–water parameters, and finally the surface area of the superheater or evaporator or economizer is arrived at based on the duty or energy required to be transferred and the LMTD. In design case, the mass flows of flue gas and steam or water and the temperatures at inlet and exit of both streams and the duty Q are known. U, the overall heat transfer coefficient, LMTD (ΔT), the log-mean temperature difference are computed, and then the required surface area A is obtained; then one checks if the geometry (tubes/row, length, fin configuration, rows deep) and other aspects such as tube wall temperatures, gas pressure drop, and space requirements are reasonable. If not, the configuration is changed, and the calculations are redone. The flue gas tube-side fluid parameters chosen for the design condition are important; it could be the maximum duty case or the load at which the equipment is likely to operate most of its lifetime. There is only one design condition. Note that in the design case, the LMTD (ΔT) is known, as we know from heat balance the flue gas and steam/water inlet and exit temperatures; U is computed based on gas and fluid data and tube geometry data, and A is the only unknown. This procedure is applicable to any heat transfer component whether flue gas flows inside or outside the tubes. The steps are as follows:

- Assume exit temperature of either stream, say, T_2, the hot gas stream.
- Compute the assumed duty $Q_a = W \times C_p \times (T_1 - T_2)$.
- Compute the exit temperature of the cold fluid using preceding duty. Since the mass flow is known, one can check the enthalpy difference and then the temperature change.
- Compute LMTD as now all four temperatures are known.
- Tube sizes, tube spacing, length, cross section of the equipment, and fin geometry, if any, have to be assumed. Using the procedure described in Appendices B through E, the various heat transfer coefficients are computed and then U.
- Estimate the surface area required using $A = Q_a/(U\Delta T) = N_w N_d L A_t$, where N_w, N_d, L, and A_t are the number of tubes wide, deep, tube length, and surface area per unit length.

- Check the number of rows deep required from the above step.
- Check the gas- and steam-side pressure drops, and if they are not reasonable, then repeat the exercise using different cross sections or tube geometries and finalize the tube geometry.

Off-Design or Performance Calculation

The off-design calculations are done to check how the given equipment performs at other gas or steam parameters, part loads, or how the boiler will perform with different fuels, steam pressures, excess air, at different flue gas recirculation rates, and so on. The tube geometry and surface area details are known. There can be numerous off- design conditions for a boiler component. In an HRSG, the design case may be the unfired 100% gas turbine load case and the off-design case could be the supplementary fired 70% gas turbine load case with import steam. Steam pressure or feed water temperature may also change. There are numerous possibilities, and the plant has to inform the boiler supplier of the possible operating modes. For example, if a boiler is fired with natural gas and then switched to a low BTU fuel, then the flue gas flow through the boiler can increase by about 60%–70%, which results in a different gas and steam temperature profile along the gas path and high back pressure through the boiler. Fans have to be checked to see if they can handle the high flue gas quantity. Off-design calculations are done in advance to check these issues. Plant engineers have a large role to play here as they know about the needs of the plant. If they miss something and the boiler designer does not consider that piece of information, then the design may be in jeopardy.

Though design calculations are performed by the boiler supplier, the procedure is discussed in case the plant engineer has to come up with the redesign of a boiler component due to changes in some parameters. Or the procedure may also be used to check the sizing done by the boiler vendor.

Superheater Design and Off-Design Calculation

We will now illustrate the design and off-design calculations for a finned tube super-heater (Figure A.2). One may refer to Appendix E, where the U value is evaluated for this superheater.

Example A.1

A 50.8 × 44 mm (2 × 1.732 in.) solid finned superheater with 78 fins/m (2 fins/in.), 12.7 mm (0.5 in.) high fins, and 1.52 mm (0.06 in.) thick is arranged in inline fashion with 101.8 mm (4 in.) square pitch: 18 tubes/row, 6 rows deep, 3.1 m (10.2 ft) long tubes with 9 streams. Gas flow is 100,000 kg/h (220,400 lb/h) with the following exhaust gas analysis by volume% $CO_2 = 3$, $H_2O = 7$, $N_2 = 75$, $O_2 = 15$. The bundle is designed with inlet and exit exhaust gas temperatures of 550°C (1022°F) and 479°C (892°F) and with 25,000 kg/h (55,100 lb/h) of saturated steam at 51 kg/cm² g (725 psig) raised from saturation temperature of 265°C (509°F) to 369°C (696°F). Surface area is 188 m² (2030 ft²). Check if the design is reasonable. Flue gas data from Appendix F at the average gas temperature of 514°C have been estimated as follows: $C_p = 0.2755$ kcal/kg °C (0.2755 Btu/lb °F), $\mu_g = 0.127$ kg/m h (0.085 Btu/ft h), $k_g = 0.0468$ kcal/m h °C (0.0314 Btu/ft h °F). Assume 1% heat loss.

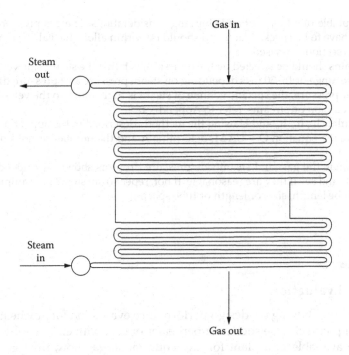

Gas in

Steam out

Steam in

Gas out

FIGURE A.2
Typical convective superheater.

Solution

First let us check the design.

The exit gas temperature is 479°C. The duty is then

$$Q = W_g \times C_{pg} \times (T_{g1} - T_{g2}) \times (1 - h_l) = 100{,}000 \times 0.2755 \times (550 - 479) \times .99$$
$$= 1.936 \times 10^6 \text{ kcal/h (7.68 MM Btu/h) (2.25 MW)}.$$

Enthalpy absorbed by steam = $1.936 \times 10^6/25{,}000 = 77.4$ kcal/kg (139.3 Btu/lb). From steam tables at 51.5 kg/cm² of steam pressure, saturated steam enthalpy from steam tables is 667.3 kcal/kg (steam-side pressure drop of 1.5 kg/cm² was assumed); exit enthalpy = 667 + 77.4 = 744.7 kcal/kg. Hence, exit steam temperature = 368°C (694°F).

For counter-flow arrangement,

$$\Delta T = [(479 - 265) - (550 - 368)]/\ln [(479 - 265)/(550 - 368)] = 197.6°C \text{ (355°F)}.$$

U_o from Appendix E is 51.9 kcal/m²h °C (10.6 Btu/ft²h °F).

Surface area required A = $1.936 \times 10^6/197.6/51.9 = 189$ m² (2033 ft²). Surface area provided is 188 m² (2023 ft²). Hence, the duty is close to that assumed, and the design is reasonable. If U is not given, then one may use the methods discussed in Appendices B through E to arrive at U.

What if the plant engineer is asked to design a superheater from scratch for a specific duty or parameters? In such cases, the following points may be kept in mind.

1. Calculate the duty of the superheater as we know the mass flow of steam and its inlet and desired exit temperatures; evaluate the flue gas exit temperature. The flue gas mass flow should be known or will be available. Then, the LMTD is computed as all the four temperatures are known.
2. The cross section of the superheater should be selected such that the gas velocity is in the range of 20–35 m/s (for dust-laden gases, the velocity will be much lower, say, around 10–20 m/s). Values beyond the range may sometimes be

acceptable due to layout and shipping considerations. The gas pressure drop will have to be checked later and should be within allowable values; if not, the cross section is revised.

3. Streams should be selected on the tube side such that the steam-side velocity is in the range of 15–30 m/s depending on steam pressure and pressure drop. At lower pressure, the specific volume of steam is higher and so the velocity. For a discussion on streams, refer to Appendix B.

3. Calculate h_c, h_n, h_i, and U using the methods described in the appendices.

4. Using the equation $Q = UA\Delta T$, determine A and the number of rows deep as before.

5. Compute the gas- and steam-side pressure drops as shown in Appendices B and C and see if they are reasonable. If not, repeat from step 2 by manipulating the tube length, size, or length or tube spacing.

Performance Evaluation

There are two ways of doing off-design performance evaluation for any heat transfer component without phase change such as superheater or economizer. The following information should be available in a plant for any equipment: gas flow, inlet gas temperature, surface area of exchanger, and steam or water flow and its inlet temperature. The performance has to be evaluated. By performance, we mean the duty, exit temperatures of both fluids, pressure drops on gas and steam side, and tube wall temperatures.

Conventional Method

- Assume exit temperature of either stream, say, the hot gas stream, T_2.
- Compute the assumed duty, $Q_a = W \times C_p \times (T_1 - T_2)$.
- Compute the exit temperature of the cold fluid using preceding duty from energy balance. Since the mass flow is known, one can compute the enthalpy difference and then the exit temperature of the cold fluid.
- Compute LMTD, ΔT as now all four temperatures are known.
- Compute U using average fluid or film temperatures as appropriate.
- Compute the transferred duty using $Q_t = UA\Delta T$.
- If both Q_a and Q_t are within a desired margin, say, 0.5%, then iterations stop, and all the values of duty and temperatures are frozen; if not, another assumption of exit temperature of the hot stream is made, and all of the preceding steps are repeated. There are quick convergence techniques to speed up the process.

NTU Method of Performance Evaluation (Number of Transfer Units)

The NTU method is more elegant. Textbooks on heat transfer will explain the basis of this method for various arrangements of the heat transfer surface such as counter-flow, parallel-flow, single- or multi-pass cross-flow, and so on [3–5].

Using this method, one can estimate the duty of heat transfer equipment with no phase change (and with little or no external radiation) in one step, given the flue gas and fluid inlet temperatures, surface area, and U. Though we say it is one-step method, it is in reality not so as thermal and transport properties used in the heat transfer evaluation depend on the average gas and fluid temperatures and also U. When one has only the inlet temperature of both the fluids, the assumption of gas properties cannot be accurate. An evaluation has to be made of the average gas and steam or water temperature and the specific heats, and U may be fine-tuned after a few runs. With a computer program, one may achieve the final results in short order. In boiler practice, we use it for predicting the performance of a superheater or economizer or the air heater. A slightly different procedure is used for an evaporator with phase change at constant temperature, which will be discussed later. The steps are shown in the following text for clarity.

The duty of any heat transfer equipment is given by

$$Q = \epsilon \, C_{min} \left(T_1 - t_1\right) \tag{A.8}$$

where ϵ is the effectiveness factor, which depends on the heat exchanger configuration (parallel- or counter-flow or multi-pass cross-flow, etc.) See Table A.2 for ϵ values for some configurations. T_1, t_1 are inlet gas and inlet fluid temperatures, °C.

$$C = \frac{\left(WC_p\right)_{min}}{\left(WC_p\right)_{max}} = \frac{C_{min}}{C_{max}} \tag{A.9}$$

and

$$C_{min} = \left(WC_p\right)_{min} \tag{A.10a}$$

$$C_{max} = \left(WC_p\right)_{max} \tag{A.10b}$$

TABLE A.2

Effectiveness Factors for Exchangers

Exchanger Type	Effectiveness ϵ
Parallel-flow, single-pass	$\varepsilon = \dfrac{1 - \exp[-NTU \times (1+C)]}{1+C}$
Counter-flow, single-pass	$\varepsilon = \dfrac{1 - \exp[-NTU \times (1-C)]}{1 - C\exp[-NTU \times (1-C)]}$
Shell-and-tube (one shell pass; 2, 4, 6, etc., tube passes)	$\varepsilon_1 = 2\left[1 + C + \dfrac{1 + \exp[-NTU \times (1+C^2)^{1/2}]}{1 - \exp[-NTU \times (1+C^2)^{1/2}]} \times (1+C^2)^{1/2}\right]^{-1}$
Shell-and-tube (n shell pass; 2n, 4n, 6n, etc., tube passes)	$\varepsilon_n = \left[\left(\dfrac{1 - \varepsilon_1 C}{1 - \varepsilon_1}\right)^n - 1\right]\left[\left(\dfrac{1 - \varepsilon_1 C}{1 - \varepsilon_1}\right)^n - C\right]^{-1}$
Cross-flow, both streams unmixed	$\varepsilon \approx 1 - \exp\{C \times NTU^{0.22}[\exp(-C \times NTU^{0.78}) - 1]\}$
Cross-flow, both streams mixed	$\varepsilon = NTU\left[\dfrac{NTU}{1 - \exp(-NTU)} + \dfrac{NTU \times C}{1 - \exp(-NTU \times C)} - 1\right]^{-1}$
Cross-flow, streams C_{min} unmixed	$\varepsilon = \{1 - \exp[-C[1 - \exp(-NTU)]]\}/C$
Cross-flow, streams C_{max} unmixed	$\varepsilon = 1 - \exp\{-1/C[1 - \exp(-NTU \times C)]\}$

(WC$_p$ (mass flow × specific heat) of the flue gas and the other fluid such as steam or water or air is computed, and the lower of the two values is (WC$_p$)$_{min}$. Typically, in a steam generator, the tube side (WC$_p$) will be larger while in gas turbine exhaust boilers (HRSGs), the gas side (WC$_p$) will be larger.)

$$\text{NTU is the number of transfer units} = \frac{UA}{\left(WC_p\right)_{min}} \qquad \text{(A.11)}$$

Note that NTU is nondimensional. Hence, we can use kcal/m^2 h °C (Btu/ft^2 h °F)(W/m^2 K) for U, m^2 (ft^2) for A, kg/h (kg/s) (lb/h) for W, kcal/kg °C (kJ/kg K) (Btu/lb °F) for C$_p$.

Using the effectiveness factor from Table A.2, one computes the duty and then arrives at the exit flue gas and fluid temperatures. One may note that though NTU method appears to be a direct calculation procedure for the duty and exit temperatures of a heat transfer equipment, a few iterations are required to perform this calculation accurately as U and specific heats of hot and cold fluids vary with the average gas and fluid temperatures, and the average gas and fluid temperatures depend on the inlet and exit temperatures of both the fluids! A well-written program is helpful.

Example A.2

70,000 kg/h of flue gases with the same analysis as in Example A.1 enters this superheater at 500°C while 19,000 kg/h of saturated steam flows inside the tubes at 51 kg/cm^2 a. What will be the duty and the flue gas and steam exit temperatures? Assume that the flue gas average specific heat is 0.273 kcal/kg °C and that of steam 0.768 kcal/kg °C (specific heat is obtained by dividing the enthalpy difference by the temperature difference. These may also be checked later based on actual temperatures if they differ significantly from those assumed.) Surface area is 188 m^2.

Solution

Flue gas WC$_p$ = 70,000 × 0.273 = 19,110 and for steam, WC$_p$ = 19,000 × 0.768 = 14,592. Hence, (WC$_p$)$_{min}$ is 14,592. (Heat loss factor on gas side is neglected here.) From Table A.2 for counter-flow arrangement,

$$\in = \frac{[1 - \exp\{-NTU \times (1-C)\}]}{[1 - C\exp\{-NTU \times (1-C)\}]} \qquad \text{(A.12a)}$$

where

$$NTU = \frac{UA}{C_{min}} \quad \text{and} \quad C = \frac{\left(WC_p\right)_{min}}{\left(WC_p\right)_{max}}$$

$$\left(WC_p\right)_{min} = 14{,}592 \quad \text{and} \quad C = \frac{14{,}592}{19{,}110} = 0.7636$$

U for the full load case was 51.9 kcal/m^2 h °C (10.6 Btu/ft^2 h °F), and for this case, we may assume that U varies as a function of gas flow to the power of 0.65 and use an approximate value of 51.9 × (70,000/100,000)$^{0.65}$ = 41.1 kcal/m^2 h °C (8.42 Btu/ft^2 h °F). This can be checked later if a computer program is used. We made use of the fact that gas-side heat transfer coefficient is governing U. NTU = 41.1 × 188/14,592 = 0.53.

$$\in = \frac{\left[1 - \exp\{-0.53 \times (1 - 0.7636)\}\right]}{\left[1 - 0.7636 \ \exp\{-0.53 \times (1 - 0.7636)\}\right]} = \frac{0.1178}{0.3263} = 0.36$$

Hence, Q = 0.36 × 14,592 × (500 − 264) = 1.24 × 10^6 kcal/h (1,442 kW) (4.92 MM Btu/h)

$$\text{Exit gas temperature} = 500 - 1.24 \times 10^6/70000/.273 \ = \ 435\,^\circ\text{C}\left(815\,^\circ\text{F}\right)$$

$$\text{Exit steam temperature} = 264 + 1.24 \times 10^6/19,000/0.768 = 349\,^\circ\text{C}\left(660\,^\circ\text{F}\right)$$

These are good estimates. With a computer program, one may fine-tune the values of U, C_p of steam and flue gas and estimate the duty more accurately. The procedure explains how off-design calculation may be done for an existing equipment. One can then proceed to check the gas- and steam-side pressure drops and tube and fin tip temperatures as detailed in Appendix E.

One may also check these results as follows. Compute ΔT = [(500 − 349) − (435 − 264)]/ ln[(500 − 349/((435 − 264)] = 160.7°C. Q_t = UAΔT = 41.1 × 188 × 160.7 = 1.242 MM kcal/h.

Estimating Performance of Evaporator

A boiler evaporator generates steam at constant temperature. Hence, it is easy to predict its duty or exit gas temperature using the following procedure. This is true for condensers also where one fluid is at constant temperature, and only the other has varying inlet and exit temperatures.

Let T_1, T_2 be the gas inlet and exit temperatures and T_s the saturation temperature.

$$W_g \times C_{pg} \times (T_1 - T_2) = UA \ [(T_1 - T_s) - (T_2 - T_s)]/\ln[(T_1 - T_s)/(T_2 - T_s)]$$
$$= UA \ (T_1 - T_2)/\ln[(T_1 - T_s)/(T_2 - T_s)] \qquad (A.13a)$$

ignoring heat losses. Simplifying,

$$\text{Ln}\left[\frac{(T_1 - T_s)}{(T_2 - T_s)}\right] = \frac{UA}{(W_g C_{pg})} \qquad (A.13b)$$

(Note that this equation is valid for both fire tube and water tube boilers.)

If T_1, T_s, U, A are known, T_2 is easily solved for from which duty Q may be obtained.

Once T_2 is known, the duty is estimated using Q = $W_g \times C_{pg} \times (T_1 - T_2)$ from which one may estimate the steam production.

Again it should be noted that U is a function of the average gas temperature in the evaporator (if finned tubes are used or the average film temperature if plain tubes are used), and hence, a few iterations are required to accurately estimate the duty or exit gas temperature.

The application of the NTU method is shown later for a complete steam generator.

Evaluating Boiler Performance from Exit Gas and Fluid Inlet Temperatures

Field data are available in any operating plant using which one can evaluate the complete boiler performance of the steam generator or HRSG and verify the data given by the supplier; in case they are significantly different from the estimated values, one can challenge the boiler supplier or try to find out why they are different. It is a fact that the gas temperature leaving the economizer T_2 and the feed water entering temperature t_1 can measured more reliably than the gas temperature near the furnace exit; hence, these data at the boiler exit region may be used to arrive at the complete boiler performance by the methodology discussed later.

Figure A.3 shows the economizer temperature profiles. T_1, T_2 are gas temperatures entering and leaving, and t_1, t_2 are the water temperatures entering and leaving.

The methodology is very much like the NTU method discussed earlier. For a counterflow configuration, which is usually used and neglecting heat loss, we have

$$Q = W_h C_{ph}\left(T_1 - T_2\right) = W_c C_{pc}\left(t_2 - t_1\right) = \frac{UA\left[\left(T_1 - t_2\right) - \left(T_2 - t_1\right)\right]}{\ln\left[\left(T_1 - t_2\right)/\left(T_2 - t_1\right)\right]} \qquad \text{(A.14a)}$$

$$\text{Let}\left(\frac{W_h C_{ph}}{W_c C_{pc}}\right) = C; \text{ Hence, } t_2 - t_1 = C\left(T_1 - T_2\right) \text{ or } t_2 = t_1 + C\left(T_1 - T_2\right) \qquad \text{(A.14b)}$$

(subscripts c and h refer to cold and hot fluids)

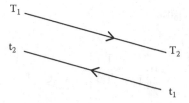

FIGURE A.3
Gas–water temperature profiles at the economizer.

Substituting for t_2 in (A.14a) and simplifying,

$$Q = \frac{UA\left[T_1 - t_1 - C(T_1 - T_2) - T_2 + t_1\right]}{\ln\left[\{T_1 - t_1 - C(T_1 - T_2)\}/(T_2 - t_1)\right]} \tag{A.15}$$

$$W_h C_{ph}(T_1 - T_2) = Q = \frac{UA(T_1 - T_2)(1 - C)}{\ln\left[\{T_1 - t_1 - C(T_1 - T_2)\}/(T_2 - t_1)\right]} \tag{A.16}$$

$$\ln\left[\frac{\{T_1 - t_1 - C(T_1 - T_2)\}}{(T_2 - t_1)}\right] = \frac{UA(1 - C)}{W_h C_{ph}} \tag{A.17}$$

$$\left[\frac{\{T_1 - t_1 - C(T_1 - T_2)\}}{(T_2 - t_1)}\right] = \exp\left[\frac{UA(1 - C)}{W_h C_{ph}}\right] = K \tag{A.18}$$

$$T_1(1 - C) - t_1 + CT_2 = K(T_2 - t_1) \tag{A.19}$$

$$T_1 = [K(T_2 - t_1) + t_1 - CT_2]/(1 - C) \tag{A.20}$$

From (A.14a), it may be shown that

$$Q = \frac{(K - 1)W_h C_{ph}(T_2 - t_1)}{(1 - C)} \tag{A.21}$$

Now K, T_2, t_1 are known or can be reasonably estimated as the tube geometry, and surface areas are known. Hence, T_1 may be estimated from this and then t_2 and then Q.

Example A.3

Let us apply this procedure to the boiler discussed in Example 3.7 and check the overall boiler performance and efficiency. Start from the economizer end as the field measurements are generally more accurate at the cold end.

Solution

Economizer

T_2 = economizer exit gas temperature = 154°C and t_1 = feed water in = 116°C

$$\frac{W_h C_{ph}}{W_c C_{pc}} = C = \frac{123{,}000 \times 0.2815}{(97{,}740 \times 1.04)} = 0.3393$$

U = 40.6 kcal/m² h °C and A = 2395 m² (from Table 3.13)

$$K = \exp\left[\frac{40.6 \times 2395(1 - 0.3393)}{(123,000 \times 0.2815)}\right] = 6.391$$

From (A.20), T_1 = [6.391 × (154 − 116) + 116 − .34 × 154]/(1 − 0.34) = 464°C, which agrees with the calculations in Chapter 3. One may also compute Q and then t_2 from T_1, T_2 and then obtain the LMTD and check the Q.

$$Q = \text{duty of economizer} = 123,000 \times 0.2815 \times (464 - 154) = 10.73 \times 10^6 \text{ kcal/h}$$

The specific heat of water = 1.044 kcal/kg °C from the earlier text. Hence, t_2 = 116 + 10,730,000/97,740/1.044 = 221°C

Evaporator Performance

Let us use Equation A.13b.

$$\ln\left[\frac{(T_1 - t)}{(T_2 - t)}\right] = \frac{UA}{(W_g C_{pg})}$$

In our case, U = 105 kcal/m² h °C, A = 322 m², C_{pg} = 0.3024, t_s = 256°C. Hence,

$$\ln\left[\frac{(T_1 - 256)}{(464 - 256)}\right] = \frac{105 \times 322}{(123,000 \times 0.3024)} = 0.909$$

$$\left[\frac{(T_1 - 256)}{(464 - 256)}\right] = \exp(0.909) = 2.482 \text{ or } T_1 = 777\,^\circ C$$

Duty of evaporator = 123,000 × (777 − 466) × 0.3024 = 11.57 × 10⁶ kcal/h

Primary Superheater

$W_h C_{ph}/W_c C_{pc}$ = C = 123,000 × 0.3163/(96,740 × 0.8) = 0.502 (C_{pg} and C_{pc} values taken from Example 3.10). About 3260 kg/h spray was used.

$$U = 105 \text{ kcal/m}^2 \text{h}\,^\circ C, A = 77 \text{ m}^2, K = \exp\left[(105 \times 77 \times (1 - 0.502)/123,000 \times 0.3163)\right] = 1.109$$

$$T_1 = \left[1.109 \times (777 - 256) + 256 - 0.502 \times 777\right]/(1 - 0.502) = 891\,^\circ C$$

$$Q = \text{duty} = 123,000 \times 0.3163 \times (891 - 777) = 4.435 \times 10^6 \text{kcal/h}$$

Steam temperature at exit = 256 + 4.435 × 10⁶/(96,740 × 0.8) = 313°C

Final Superheater

Gas exit temperature = 891°C and the steam inlet temperature after spray = 286°C.

$$W_h C_{ph}/W_c C_{pc} = C = 123,000 \times 0.3244/100,000 \times 0.629 = 0.634$$

U = 109.3 kcal/m² h °C, A = 103 m²; K = exp[(109.3 × 103 × (1 − 0.634)/(123,000 × 0.3244)] = exp(0.103) = 1.108

$$T_1 = [1.108 \times (891 - 286) + 286 - .634 \times 891]/(1 - 0.634) = 1070°C$$

$$Q = 123,000 \times 0.3244 \times (1070 - 891) = 7.14 \times 10^6 \text{ kcal/h}$$

Exit steam temperature $= 286 + 7.14 \times 10^6/100,000/0.629 = 400°C$

Screen Section

The screen section is nothing but an evaporator. Due to its location and function, namely, to shield or protect the superheater, it is called the screen section. The calculation procedure is the same as that for the evaporator.

$$\ln[(t_{g1} - t_s)/(t_{g2} - t_s)] = UA/(W_g C_{pg})$$

$$U = 117.8 \text{ kcal/m}^2\text{h °C}; A = 105 \text{ m}^2, C_{pg} = 0.3328 \text{ kcal/kg °C}$$

$$\ln[(t_{g1} - 256)/(1,071 - 256)] = 117.8 \times 105/(123,000 \times 0.3328) = 0.302 \text{ or } t_{g1} = 1359°C$$

Screen duty $Q = 123,000 \times 0.3328 \times (1,359 - 1,071) = 11.79 \times 10^6$ kcal/h
The total energy absorbed by steam = 64.68 MM kcal/h (Table 3.13). Hence, furnace duty = $(64.68 - 11.79 - 7.14 - 4.435 - 11.57 - 10.73)10^6 = 19.01 \times 10^6$ kcal/h
From Chapter 2 on furnace calculations, the furnace duty may be estimated as follows:

$$Q_f = W_f \times LHV \times (1 - \text{casing loss} - \text{unaccounted loss}) - W_g \times h_e$$

$$W_g = \text{total flue gas flow} = 123,000 \text{ kg/h}$$

Casing loss and unaccounted loss = 0.01

$$LHV = 11,653 \text{ kcal/kg}$$

Hence,

$$\text{Furnace duty} = W_f \times 11,653 \times 0.99 - 123,000 \times (412 - 1.4) + 117,000 \times 1.3 = 19.01 \times 10^6$$

(412 and 1.4 kcal/kg refer to enthalpies of the flue gas at the furnace exit gas temperature and at the reference or ambient temperature. Enthalpy values of flue gas are taken from Appendix F.)

Hence, $W_f = 6,012$ kg/h (air flow = 117,000 kg/h and flue gas flow = 123,000 kg/h). This fuel flow agrees with the data provided by the boiler supplier of 6024 kg/h.
Hence, efficiency on LHV basis = $64.68 \times 10^6/(6,012 \times 11,653) = 92.32\%$ agrees closely with the data provided earlier. Hence, the boiler performance is reasonable.
The plant engineer may also apply the methods discussed in Chapter 2 on the estimation of furnace exit gas temperature and see based on the heat input and effective furnace area whether the furnace exit gas temperature is close to that estimated as earlier.
Thus, we have arrived at the furnace exit gas temperature from the boiler rear end measurements. This method may also be used to model the furnace exit gas temperature as a function of boiler load and net effective heating surface. As furnace modeling is a complex

process, substantiating the model with field data and such back calculations will ensure that the model for that type of furnace, fuel, excess air, burner location is reasonably accurate. One may also evaluate the efficiency from the oxygen measurements and exit gas temperature and see how it compares with design or proposal values. Thus, there are several tools for a process engineer to evaluate a boiler performance independently and challenge the supplier if significant differences arise. For example, if the boiler can operate only at, say, 75% initially for various plant restrictions, then how can one prove that the boiler performance is satisfactory? The aforementioned procedure can be applied. The fuel gas flow may be measured and compared with the calculated. One may work from the burner end as discussed in Example 3.7, or from the economizer exit as described earlier. If there are large variations, then the design may be challenged.

HRSG Performance

Example 4.2 shows how the performance of a complete HRSG (Figure A.4) may be evaluated at off-design conditions. Example 5.10 on simulation also evaluates the off-design performance of the same HRSG using the simulation concept. These are good exercises and learning tools for the plant engineers who want to evaluate the performance of their HRSGs or steam generators at any operating conditions and ascertain if the operating data are reasonable.

We have now a fairly good idea of basic design and performance calculations of boiler components from the earlier discussions. Appendix B explains the procedure for the estimation of heat transfer coefficient inside tubes, which will be useful for estimating the heat transfer coefficients in superheater, air heater, or economizer tubes, while Appendices C

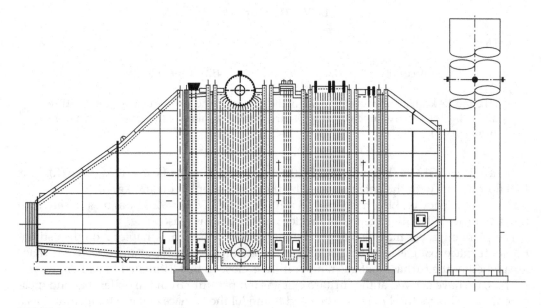

FIGURE A.4
Unfired HRSG in gas turbine plant.

through E show how one may compute the gas-side coefficients outside tubes. Plant engineers can go through the various examples in the book and apply the calculation procedures to critically evaluate their steam generator or HRSG or any component.

References

1. V. Ganapathy, *Industrial Boilers and HRSGs*, CRC Press, Boca Raton, FL, 2003, p. 445.
2. V. Ganapathy, Evaluated extended surfaces carefully, *Hydrocarbon Processing*, Oct. 1990, p. 65.
3. V. Ganapathy, Evaluate the performance of waste heat boilers, *Chemical Engineering*, Nov. 16, 1981, p. 291.
4. D.Q. Kern, *Process Heat Transfer*, McGraw-Hill, New York, 1950, p. 570.
5. V. Ganapathy, *Applied Heat Transfer*, Pennwell Books, Tulsa, OK, 1982, p. 621.

Appendix B: Tube-Side Heat Transfer Coefficients and Pressure Drop

Estimating the tube-side convective heat transfer coefficient h_c for single-phase fluids is an important task in boiler performance evaluation whether it is a fire tube or water tube boiler or a fluid heater. It is also important as h_c impacts the tube wall temperatures in the case of superheaters, economizers, and air heaters. Simplified procedures for estimating h_c for steam, water, air, flue gas, or any fluid are presented in the following text. Appendix D shows how the nonluminous heat transfer coefficient inside the tubes may be estimated. Then the total inside heat transfer coefficient h_i is the sum of h_n and h_c. From h_i, one may estimate the overall heat transfer coefficient U as shown in Appendix A.

Tube-side coefficients may be estimated using the following correlation:

$$Nu = 0.023 \, Re^{0.8} Pr^{0.4} \tag{B.1}$$

where

Nusselt number $Nu = h_c \, d_i / k$
Reynolds number $Re = \rho V d_i / \mu = \rho \, (4w/\rho/\pi/d_i^2) \, d_i/\mu = 1.273 \, w/(d_i \mu)$
Prandtl number $= \mu C_p / k$
w is the fluid flow per tube = W/number of streams, kg/s
W is the total flow, kg/s
C_p is the specific heat, J/kg K
μ is the viscosity of fluid, kg/m s
V is the fluid velocity, m/s
k is the thermal conductivity, W/m K
ρ is the density, kg/m³
d_i is the tube inner diameter, m
h_c is the tube-side convective heat transfer coefficient in W/m² K

(All fluid properties are estimated at the average fluid temperature.) Simplifying equation (B.1), we have

$$h_c = 0.0278 \, w^{0.8} \left(\frac{C_p}{\mu} \right)^{0.4} \frac{k^{0.6}}{d_i^{1.8}} = 0.0278 \, C \, \frac{w^{0.8}}{d_i^{1.8}} \tag{B.2}$$

where

$$C = \left(\frac{C_p}{\mu} \right)^{0.4} k^{0.6} \tag{B.3}$$

Table B.1 shows this equation in SI, metric, and British units.

Tables B.2 through B.4 show the values of C for air, flue gas, water, and steam. Plant engineers as an exercise may develop C values as a function of fluid temperature for other fluids such as thermic fluids, oils, and waste flue gas in fire tube boilers in their plant.

TABLE B.1

Equations for Tube-Side Heat Transfer Coefficient and Velocity

SI Units	British Units	Metric Units
$h_c = 0.0278 \, w^{0.8}(C_p/\mu)^{0.4}k^{0.6}/d_i^{1.8}$	$h_c = 2.44 \, w^{0.8}(C_p/\mu)^{0.4}k^{0.6}/d_i^{1.8}$	$h_c = 0.0278 \, (C_p/\mu)^{0.4}k^{0.6}w^{0.8}/d_i^{1.8}$
$h_c = 0.0278 \, Cw^{0.8}/d_i^{1.8}$	$h_c = 2.44 \, Cw^{0.8}/d_i^{1.8}$	$h_c = 0.0278 \, Cw^{0.8}/d_i^{1.8}$
h_c, W/m^2 K	h_c, Btu/ft^2 h °F	h_c, kcal/m^2 h °C
w, kg/s	w, lb/h	w, kg/h
d_i, m	d_i, in.	d_i, m
C_p, J/kg K	C_p, Btu/lb° F	C_p, kcal/kg °C
μ, kg/ms	μ, lb/ft h	μ, kg/m h
k, W/m K	k, Btu/ft h °F	k, kcal/m h °C
$C_{SI} = (C_p/\mu)^{0.4}k^{0.6}$	C, multiply C_{SI} by 0.001134	C, multiply C_{SI} by 0.001229
$V = 1.246 \, wv/d_i^2$	$V = 0.05 \, wv/d_i^2$	$V = 3.461 \times 10^{-4} \, wv/d_i^2$
v, m^3/kg	ft^3/lb	m^3/kg
V, velocity in m/s	ft/s	m/s
Re = 1.273 w/(d$_i\mu$)	Re = 15.2 w/(d$_i\mu$)	Re = 1.273 w/(d$_i\mu$)

Note: C is defined in equation B.3.

TABLE B.2

C for Dry Air, Flue Gases from Combustion of Natural Gas, Fuel Oil

Temperature, °C	Dry Air	Natural Gas (Flue Gas)	Fuel Oil (Flue Gas)
30	145.1	148.3	145.2
50	146.3	150.1	147.0
100	149.1	154.5	151.2
200	154.6	162.7	159.3
300	159.9	170.5	166.8
400	165.2	178.0	173.9
500	170.3	185.1	180.8
600	175.4	192.2	187.4
700	180.3	199.0	193.8
800	185.1	205.7	200.0
900	189.9	212.1	205.9
1000	194.5	218.5	211.7

Notes: This table shows C values using SI units. To obtain h_c in British units, multiply C by 0.0011343, and in metric units, multiply by 0.001229; Natural gas products of combustion: % volume $CO_2 = 8$, $H_2O = 18$, $N_2 = 71.5$, $O_2 = 2.5$. Fuel oil products of Combustion: % volume $CO_2 = 12$, $H_2O = 12$, $N_2 = 73.5$, $O_2 = 2.5$. Gas pressure is atmospheric. Due to the presence of water vapor, the specific heat and thermal conductivity increases while the viscosity decreases, and hence, h_c is higher for natural gas products of combustion than air or for flue gas from combustion of oil.

TABLE B.3

C for Compressed Water

Temperature, °C	C
20	327.9
50	432.5
100	587.6
150	716.5
200	803.2
250	864
300	919.3

Notes: This table shows C values using SI units. To obtain h_c in British units, multiply C by 0.0011343, and in metric units, multiply by 0.001229; The equation $C = 250 + 4.023t - 0.0063t^2$ describes this trend, where T is the temperature of water in °C.

TABLE B.4

C Values for Saturated and Superheated Steam

Pressure, KPa	1,000	2,000	3,500	5,000	7,000	10,000
C-sat steam	266.5	305.3	354.6	400.1	462.3	571.7
200°C	255.4	—	—	—	—	—
250°C	248.0	274.0	337.0	—	—	—
300°C	250.7	264.5	291.5	328.4	404.9	
350°C	256.8	265.6	281.1	300.0	332.0	402.5
400°C	264.5	270.6	281.0	292.7	310.8	344.5
450°C	272.8	277.5	285.0	293.3	305.4	326.3
500°C	281.5	285.3	291.1	297.4	306.3	321.0

Note: This table shows C values using SI units. To obtain h_c in British units, multiply C by 0.0011343, and in metric units, multiply by 0.001229.

Heat Transfer at Tube Entrance

Correlation (B.2) gives the heat transfer coefficient after the flow has fully developed inside the tube. The heat transfer coefficient at entrance to tube has been found by research to be significantly higher than that for fully developed flow (Figure B.1) [1].

Figure B.1 shows typical variation of local heat transfer coefficient at tube entrance compared to that obtained after the flow has stabilized. This is due to the formation of boundary layer as the flow develops. At the tube entrance, the boundary layer is yet to form, and hence, h_c is high. This information is of interest when tube sheet temperature in any fire tube boiler has to be computed as the heat flux at the tube sheet will be much higher than that downstream inside the tubes. This is discussed in Chapter 4 on waste heat boilers.

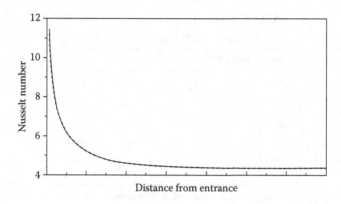

FIGURE B.1
Variation of Nusselt number with gas flow direction.

Example B.1

0.0252 kg/s (200 lb/h) of air at 427°C (800°F) flows inside an air heater tube of inner diameter 44.4 mm (1.75 in.). Determine h_c.

Solution

C_p from Appendix F = 0.26 kcal/kg °C = 0.26 Btu/lb °F; μ = 0.1203 kg/m h = 0.0807 lb/ft h; k = 0.0428 kcal/m h °C = 0.0287 Btu/ft h °F.

From Table B.2, C at 427°C is 167.2 for air (in SI units). Hence, from (B.2),

$$h_c = 0.0278 \times 167.2 \times \frac{0.0252^{0.8}}{0.0444^{1.8}} = 66.57 \text{ W/m}^2 \text{ K}$$

In British units, from Table B.1, h_c = 2.44 × (200^{0.8}/1.75^{1.8}) × (0.26/0.0807)^{0.4} × 0.0287^{0.6} = 11.71 Btu/ft² h °F. (Also, using C = 0.0011343 × 167.2 = 0.1896, h_c = 2.44 × 200^{0.8} × 0.1896/1.75^{1.8} = 11.71 Btu/ft² h °F).

   Using metric units, from Table B.1, h_c = 0.0278 × (0.26/0.1203)^{0.4} × 0.0428^{0.6} × 90.72^{0.8}/0.0444^{1.8} = 57.2 kcal/m² h °C. (Using C = 0.001229 × 167.2 = 0.2055, h_c = 0.0278 × 90.72^{0.8} × 0.2055/0.0444^{1.8} = 57.24 kcal/m² h °C).

   Thus we may use equations from Table B.1 to determine h_c in appropriate units.

Effect of Gas Pressure

The effect of pressure on C is not very significant for common gases and may be neglected up to 40 bar pressure (see Figure B.2) [2,3].

Example B.2

Determine the tube-side heat transfer coefficient when 6 kg/s (47,606 lb/h) of feed water in a boiler economizer flows at an average temperature of 150°C (302°F). Tube ID = 73 mm (2.87 in.).

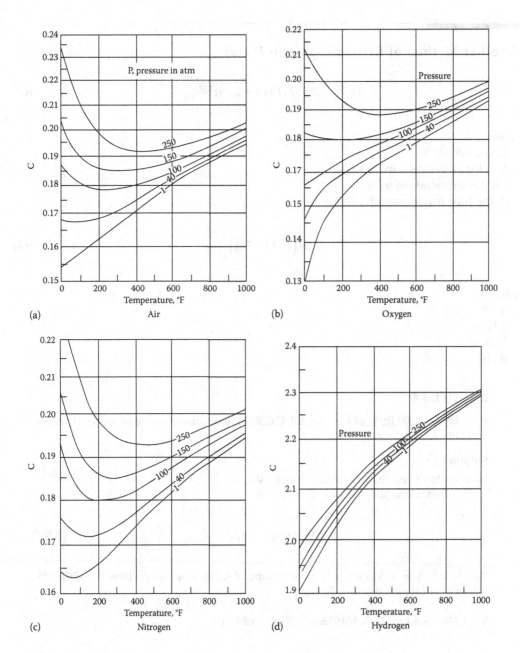

FIGURE B.2
Effect of gas pressure on heat transfer inside tubes.

Solution

Using this equation and C = 716.5 from Table B.3,

$$h_c = 0.0278 \times 6^{0.8} \times \frac{716.5}{0.073^{1.8}} = 9278 \text{ W/m}^2 \text{ K } \left(1635 \text{ Btu/ft}^2\text{h}°\text{F}\right) \left(7995 \text{ kcal/m}^2\text{h}°\text{C}\right)$$

For British units, use C = 0.001134 × 716.5 = 0.8127. h_c = 2.44 × 0.8127 × 47,606$^{0.8}$/2.87$^{1.8}$ = 1,641 Btu/ft² h °F.

Another Method of Estimating h_c for Water

$$h_c = 7.037 (199.6 + 2.79t) \frac{V^{0.8}}{d_i^{0.2}} \tag{B.4a}$$

where
  h_c is in W/m^2 K
  V is the water velocity in m/s
  t is the temperature in °C
  d_i the tube inner diameter in m

$$h_c = (150 + 1.55t) \frac{V^{0.8}}{d_i^{0.2}} \tag{B.4b}$$

where
  h_c, Btu/ft ^{2}h °F
  t, °F
  V, ft/s
  d_i, in.

Example B.3

If 6.3 kg/s (50,000 lb/h) of water at 121°C (250°F) flows inside a pipe of inside diameter 73.6 mm (2.9 in.), determine h_c.

Solution

First one has to estimate the water velocity V.
    V is given by the simple formula in Table B.1.

$$\text{In SI units, } V = 1.246 \frac{wv}{d_i^2} \tag{B.5a}$$

V-m/s; w = flow in kg/s; v = specific volume of water from steam tables = 0.0010625 m^3/kg (0.017 ft^3/lb); d = tube ID in m.

V = 1.246 × 6.3 × 0.0010625/0.07366/0.07366 = 1.54 m/s

$$\text{In British units, } V = 0.05 \frac{wv}{d_i^2} \tag{B.5b}$$

V, ft/s; w, flow in lb/h; v, specific volume in ft^3/lb;

$$V = 0.05 \times 50,000 \times \frac{0.017}{2.9^2} = 5.05 \text{ ft/s}$$

Then using (B.4), we have

$$h_c = 7.037 \times (199.6 + 2.79 \times 121) \times \frac{1.54^{0.8}}{0.0736^{0.2}}$$

$$= 8996 \text{ W/m}^2 \text{ K} \left(1580 \text{ Btu/ft}^2\text{h}°\text{F}\right)\left(7737 \text{ kcal/m}^2\text{h}°\text{C}\right)$$

(As discussed earlier, the performance of an economizer is impacted mainly by the gas-side heat transfer coefficient, and hence, small variations in h_c of water will not affect the overall heat transfer coefficient U; hence, will not cause significant error in evaluating their thermal performance or duty. However, h_c is important for evaluating tube wall temperature that suggests the possibility of low-temperature corrosion.)

Example B.4

0.5 kg/s (3967 lb/h) of saturated steam at 35 bara (507 psia) flows inside a superheater tube of inner diameter 38 mm (1.5 in.). What is h_c?

Solution

C value from Table B.4 for saturated steam at 3500 kPa or 35 bara is 354.6. Using (B.4),

$$h_c = 0.0278 \times 354.6 \times 0.50^{0.8}/0.038^{1.8} = 2038 \text{ W/m}^2 \text{ K}$$

In British units, C = 0.0011343 × 354.6 = 0.4022. h_c = 2.44 × 3967$^{0.8}$ × 0.4022/1.5$^{1.8}$ = 358 Btu/ft²h °F.

In Metric units, C = 0.001229 × 354.6 = 0.4358.

$$h_c = 0.0278 \times 0.4358 \times \frac{18000.8}{0.0381.8} = 1753 \text{ kcal/m}^2\text{h}°\text{C}$$

Example B.5

In this case, if the steam temperature is 400°C, what is h_c?

Solution

C = 281; h_c = 0.0278 × 281 × 0.50$^{0.8}$/0.038$^{1.8}$ = 1615 W/m² K.

In British units, C = 0.001229 × 281 = 0.3187. h_c = 2.44 × 0.3187 × 3967$^{0.8}$/1.5$^{1.8}$ = 284 Btu/ft²h °F.

In Metric units, C = 0.001229 × 281 = 0.3453. hc = 0.0278 × 18000.8/0.0381.8 = 1389 kcal/m²h °C.

By performing such calculations manually, plant engineers can understand how h_c values change with the temperature and pressure of steam. We note that h_c for super-heated steam in the preceding example is lower than that of saturated steam. Also one may note that C decreases first and then increases at low pressures. Hence, depending on pressure and temperature, the tube-side coefficient will vary and affect the tube wall temperature. At very high pressures (>7000 kPa), the heat transfer coefficient decreases as steam temperature increases, while at lower pressure, a dip in the value is seen. Figure B.3 shows the trend at 30–70 kg/cm² a steam pressures as a function of temperature for a flow of 1500 kg/h inside 38 mm tube.

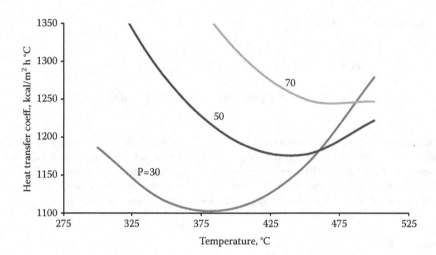

FIGURE B.3
Heat transfer coefficient of steam inside tubes from 30 to 70 kg/cm² a.

Importance of Streams in Superheater, Economizer

The number of streams in a superheater or economizer is a very important piece of information, and plant engineers should know how many streams are used in their superheater and economizer. Often, when there is a tube failure or a problem with superheater tube wall or temperature, the first parameter one estimates is the tube-side heat transfer coefficient and velocity. In order to do that, information on streams is essential. Streams are the number of tubes through which the entire tube-side fluid flows. From this, one may estimate the flow per tube (w = flow per tube = total tube-side flow/number of streams).

Flow per tube determines the fluid velocity, heat transfer coefficient, as well as pressure drop, and hence, one should understand the importance of this term. Plant engineers should also review the drawings provided by boiler supplier and find out how many streams are used in their superheater and economizer. Sometimes, this information is not clearly indicated or provided by the boiler suppliers, and plant engineers also don't understand the significance of this. Sometimes baffles may be used in the superheater headers, which may decrease the streams by half or even a third; having these data handy will help plant engineers evaluate their superheater or economizer performance when occasion arises such as when there is a superheater tube failure or overheating of tubes. They may also use this information to estimate the tube-side pressure drop and cross-check with field data to ensure that the steam-side pressure drop inside the superheater is reasonable.

The author has seen many superheaters fail in operation due to wrong selection of streams by the boiler supplier or lack of understanding of streams while designing the superheater leading to overheating of tubes, stagnation of flow, or even reverse flow. Selection of streams is done based on many considerations in superheaters such as velocity, pressure drop, and height of superheater (if tubes are vertical and superheater steam flow is in downward direction). Chapter 3 discusses a superheater tube failure problem due to low steam-side pressure drop. Since gas flow to steam flow ratio may vary depending upon the application, the number of tubes across the boiler width need not be the same as the number of streams. It can be a fraction or a multiple of the number of tubes wide. Some examples of superheater (and economizer) geometry are presented as follows to illustrate the concept of streams.

1. *Superheater with horizontal tubes*: In this case (Figure B.4a), we have a superheater in staggered arrangement. If there are, say, 30 tubes across the header in each row, then there are a total of $30 \times 2 = 60$ streams. Some engineers call the number of tubes in the same plane as *starts*. So there are 30 tubes across and 2 starts. The bottom line is that if 200,000 kg/h is the total flow of steam in this superheater, then $200,000/60 = 3,333$ kg/h is the average flow per tube. This is w to be used in the calculation of h_c or ΔP. Figure B.4a shows only four streams (2×2).

2. *Vertical tube superheater*: There are 18 tubes across, but by reviewing the header drawing (Figure B.4b), the plant engineer notices a baffle plate in the inlet header, and hence, there are nine streams only. The steam inlet and exit nozzles are on the same header. The other header is just a turnaround header, and so there is no baffle. Steam flows through nine tubes or nine streams in counter-flow direction and then makes a 180° turn and returns to the same header in parallel-flow fashion. This is done sometimes to lower the superheater tube wall temperatures (by increasing h_c). Hence, if 27,000 kg/h is the total steam flow, the average flow per tube will be 3,000 kg/h.

3. *Economizer (or superheater) with vertical tubes*: In this drawing of a gas turbine HRSG economizer with vertical tubes (Figure B.4c), there are 12 tubes carrying water (across the width), but 24 tubes/row across the boiler width. All tubes are in counter-flow arrangement. Twenty-four tubes across are required from gas velocity considerations, but steam velocity consideration requires only half that number. This arrangement differs from that in Figure B.4b in that we have complete counter-flow in this case, while it is party counter- and partly parallel-flow in earlier case. The LMTD values will accordingly be different and have to be evaluated. One may note that a collection header is required at the bottom for each row of tubes as there are incoming and outgoing tubes in each header. It is a more complicated arrangement to have streams as a third of the number of tubes wide. Hence, this arrangement is suitable for the number of streams half the number of tubes wide. If streams are the same as the number of tubes wide, then we can eliminate the collection header at the top of each row of tubes.

4. *Economizer with low number of streams*: In the case of Figure B.4d, there are eight tubes per row in the economizer with vertical downward gas flow, but the water flow requires only four streams. Hence, each tube is bent in the horizontal as well as in the vertical plane. Similarly, there can be 30 tubes across while streams can be 15 or 10 or 5 or even 6 or even 30 streams depending on water velocity required. Typically, 1–2 m/s is a good starting value for the velocity of water at 100% load. This arrangement is seen in gas turbines where the ratio of gas flow to steam flow is so large that the gas velocity demands a large number of tubes in the cross section while the water velocity requirement dictates a few tubes only. Figure shows eight tubes across with four streams. Note that in this type of arrangement, the transverse and longitudinal spacing are determined by the bend radius available for a given tube size.

5. *Economizer or superheater with multiple streams*: In this example of an economizer (Figure B.4e) or superheater, there are more streams than the tubes across. If the number of tubes wide is say 10, the streams are a multiple of it, 20. There can be 30 or even 60 streams depending on the ratio of gas to water flow. In a few waste

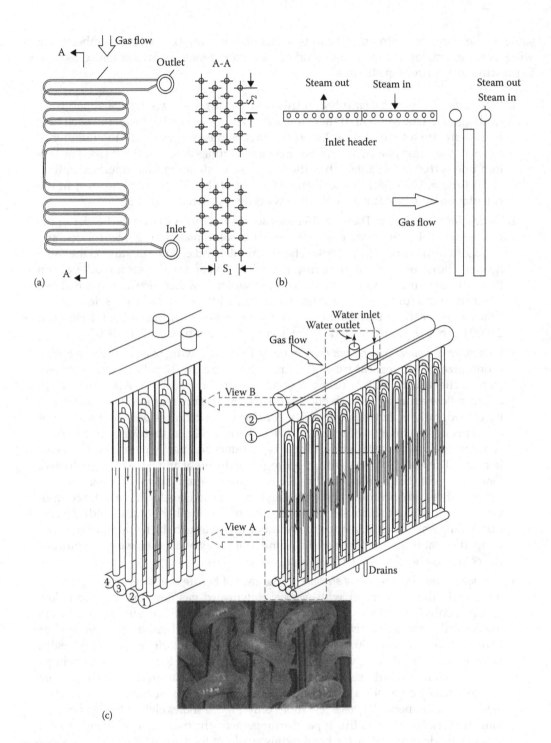

FIGURE B.4
(a) Horizontal tube superheater. (b) Vertical tube superheater with baffle in header. (c) Vertical tube superheater in a waste heat boiler.

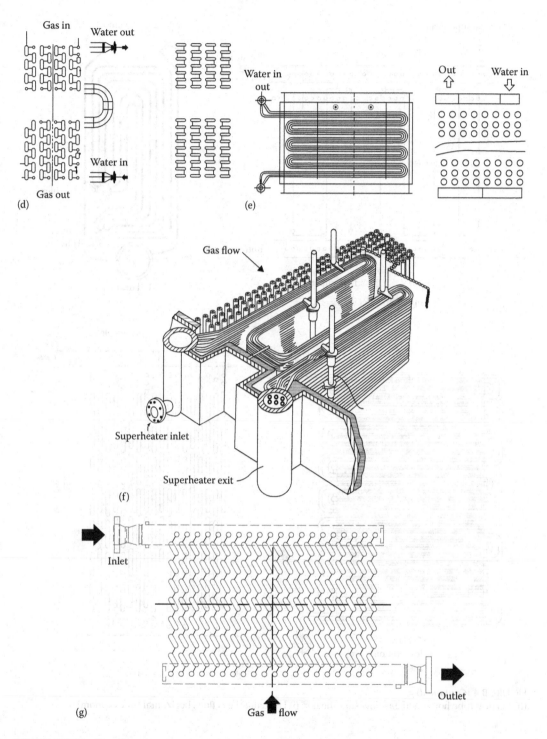

FIGURE B.4 (*Continued*)
(d) Streams are a fraction of number of tubes wide. (e) Horizontal gas flow, horizontal tube superheater.
(f) Horizontal gas flow, horizontal tube superheater. (g) Counter-flow staggered finned tube economizer.

(*Continued*)

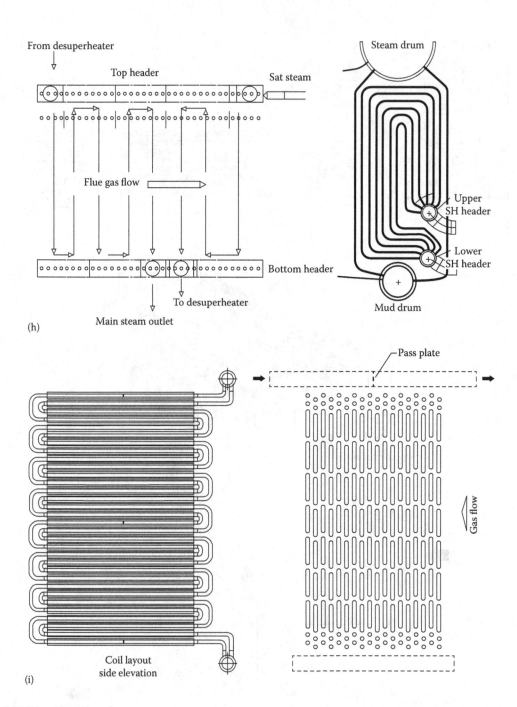

FIGURE B.4 (*Continued*)
(h) Vertical tube horizontal gas flow superheater. (i) Horizontal gas flow, horizontal tube economizer.

heat boilers, the flue gas flow will be quite small while the steam generation or water flowing in the economizer or condensate heater will be very large requiring multiple streams. Figure shows four streams (2 × 2).

6. *Horizontal gas flow superheater with multiple tubes on each header*: In this horizontal gas flow, horizontal tube superheater header (Figure B.4f), there are 2 tubes in each horizontal plane, and if there are 30 tubes along the header length, the number of streams is then 60. (Two starts on the header and 30 tubes across.) This could be arranged in parallel- or counter-flow fashion depending on steam and tube wall temperatures. If there are four starts on the header, then the streams will be 30 × 4 = 120.

7. *Vertical gas flow counter-flow economizer*: In this counter-flow economizer (or superheater) (Figure B.4g), there are 20 tubes/row and 20 streams in staggered arrangement of tubes.

8. *Inverted loop superheater*: This is a common arrangement in package boilers (Figure B.4h). We have an inlet and exit header with multiple starts. The number of streams is decided based on steam velocity, pressure drop, tube wall temperature considerations, gravity versus friction loss at low loads, and height of the tubes. At the steam inlet, we have 6 starts with 5 tubes along the gas flow for a total of 30 streams. There is a baffle in the header after the 10th tube, so the 30 streams can reverse and flow up and down again. This is a commonly used arrangement in package boilers. The steam from the desuperheater has only 24 streams for illustration purposes. Here we have a baffle after the eighth row in the header.

Note that due to tube wall temperature limitations, the first stage is counter-flow while the second stage is in parallel-flow. Thus, plant engineers should spend some time understanding their superheater flow configuration, which will help them evaluate its performance and those like this. As discussed in Chapter 3, at low loads, if the gravity loss in downward flow direction is larger than the steam pressure drop, then stagnation or reverse flow may occur. Hence, this is also a consideration for selecting the steam flow per tube or the number of streams.

9. *Horizontal gas flow, horizontal tube economizer*: In this economizer, the arrangement is uncommon (Figure B.4i). It has horizontal tubes and horizontal gas flow with a two-pass design! Sometimes due to layout considerations, this arrangement is warranted. There are 18 streams (9 × 2) in staggered arrangement. The water flows down from top-to-bottom header and then from bottom-to-top header. There is a baffle in the top header. Note that the total length along gas flow direction is small. We may have any number of such passes depending on the duty.

Pressure Drop inside Tubes

Pressure drop inside the tubes is also an important piece of information while designing superheaters, air heaters, economizers. Often, consideration of pressure drop and heat transfer coefficient determines the initial cost and cost of operation. Hence, pressure drop data along with the heat transfer coefficient help to optimize a design as will be shown in Chapter 2. The basic equation for pressure drop is

$$\Delta P = 810 \times 10^{-6} f L_e v w^2 / d_i^5 \tag{B.6}$$

where ΔP is the pressure drop in KPa (gas-steam-liquids).

Friction factor f varies with Reynolds number inside the tube and the tube roughness factor. Typically, flow in boilers, superheaters, economizers, and air heaters is turbulent, and hence, the friction factor is taken as a function of tube inner diameter as shown in Table B.5. Table B.6 shows the pressure drop equations in all systems of units.

TABLE B.5

Friction Factor as a Function of Tube Inner Diameter

Tube inner diameter (mm)	19	25	38	51	63.5	76	101	126	200	250
Friction factor f		0.0245	0.023	0.021	0.0195	0.018	0.0175	0.0165	0.016	0.014	0.013

Note: Standard pipe sizes are shown in Table B8.

TABLE B.6

Pressure Drop Equations

SI Units	British Units	Metric Units
$\Delta P = 810 \times 10^{-6} f L_e v w^2 / d_i^5$	$\Delta P = 3.36 \times 10^{-6} f L_e v w^2 / d_i^5$	$\Delta P = 0.6375 \times 10^{-12} f L_e v w^2 / d_i^5$
w, kg/s	w, lb/h	w, kg/h
ΔP, kPa	ΔP, psi	ΔP, kg/cm^2
L_e, m	L_e, ft	L_e, m
v, m^3/kg	v, ft^3/lb	v, m^3/kg
d_i, m	d_i, in.	d_i, m
$\Delta P = 0.08262 f L_e v w^2 / d_i^5$	$\Delta P = 93 \times 10^{-6} f L_e v w^2 / d_i^5$	$\Delta P = 6.382 \times 10^{-9} f L_e v w^2 / d_i^5$
ΔP, mm wc in above equation	ΔP, in wc	ΔP, mm wc

Notes: f = friction factor inside the tubes. L_e = effective length of tube, m. One may use Table B.7 to determine the effective lengths in a coil or piping with bends and fittings. w = flow per tube = total flow of fluid/streams, kg/s (if there are multiple streams, then based on effective length of each stream and tube diameter, flow and pressure drop equations must be solved to arrive at the flow in each stream. d_i = tube inner diameter, m. v = specific volume of fluid, m^3/kg.

TABLE B.7

Effective Lengths for Bends, Valves, Fitting

Fitting	L_e/d_i
90° square bend	57
45° elbows	15
90° elbows, standard radius	32
90° elbows, medium radius	26
90° elbows, long sweep	20
180° elbows, close return bends	75
180° medium radius return bends	50
Gate valves open	7
Gate valves, one-fourth closed	40
Gate valves, half closed	200
Gate valves, three-fourths closed	800
Globe valves open	300

Example B.6

Determine the pressure drop in a 76 mm schedule 80 line carrying water at 38°C and 9000 kPa if the total equivalent length is 305 m (1000 ft). Flow is 4.8 kg/s (38,085 lb/h) and specific volume of water is 0.001 m³/kg (0.016 ft³/lb) from steam tables. Inner diameter of pipe = 76 mm (3 in.).

Solution

$\Delta P = 810 f L_e v w^2 / d_i^5$
$= 810 \times 10^{-6} \times 0.0175 \times 305 \times 0.001 \times 4.8^2 / 0.076^5 = 39.3$ kPa

In British units, $\Delta P = 3.36 \times 10^{-6} f L_e v w^2 / d_i^5 = 3.36 \times 10^{-6} \times 0.0175 \times 1,000 \times 0.016 \times 38,085^2 / 3^5 = 5.6$ psi.

In metric units, $\Delta P = 0.6375 \times 10^{-12} f L_e v w^2 / d_i^5 = 0.6375 \times 10^{-12} \times 0.0175 \times 305 \times 0.001 \times 17,280^2 / 0.076^5 = 0.4$ kg/cm².

Example B.7

Determine the gas pressure drop of 0.021 kg/s (166.6 lb/h) of flue gas with a molecular weight of 29 at atmospheric pressure and at an average gas temperature of 500°C, which flows inside a tube of inner diameter 45 mm (1.77 in.). Tube total effective length = 10 m. (It is common practice in boiler industry to express pressure drop of air or flue gas in mm wc as it can be easily measured in this unit.)

The density of gas at atmospheric pressure = MW × 273/(273 + 500)/22.4 = 0.457 kg/m³. Hence, specific volume = 1/density = 2.19 m³/kg (35 ft³/lb). Friction factor from Table B.5 may be taken as 0.02.

$$\Delta P = 0.08262 \ f \ L_e v \ w^2 / d_i^5 = 0.08262 \times 0.02 \times 10 \times 2.19 \times 0.021 \times \frac{0.021}{0.045^5} = 86.5 \text{ mm wc}$$

In British units, $\Delta P = 93 \times 10^{-6} f L_e v w^2 / d_i^5 = 93 \times 10^{-6} \times 0.02 \times 35 \times 32.8 \times 166.6^2 / 1.77^5 = 3.41$ in. wc (Table B.8).

TABLE B.8

Thickness of Standard Steel Pipe

Pipe Size (in.)	OD (in.)	Sch 40	Sch 80
1/8	0.405	0.068	0.095
1/4	0.540	0.088	0.119
3/8	0.675	0.091	0.126
1/2	0.840	0.109	0.147
3/4	1.05	0.113	0.154
1.0	1.315	0.133	0.179
1.25	1.66	0.140	0.191
1.5	1.9	0.145	0.200
2	2.375	0.154	0.218
2.5	2.875	0.203	0.276
3	3.5	0.216	0.300
4	4.5	0.237	0.337
5	5.563	0.258	0.375
6	6.625	0.280	0.432
8	8.625	0.322	0.500
10	10.75	0.365	0.594
12	12.75	0.406	0.688
16	16	0.500	0.844

References

1. P. Abbrecht and S. Churchill, The thermal entrance region in fully developed turbulent flow, *AIChE Journal*, 6, 268, June 1960.
2. V. Ganapathy, *Applied Heat Transfer*, Pennwell Books, Tulsa, OK, 1982, p600.
3. V. Ganapathy, *Industrial Boilers and HRSGs*, CRC Press, 2003, Boca Raton, FL, p531.

Appendix C: Heat Transfer Coefficients Outside Plain Tubes

In order to determine the performance of a plain tube bundle such as boiler bank tubes or superheater tube bundle (Figure C.1), one must determine the outside heat transfer coefficient h_o, which consists of h_c, the convective heat transfer coefficient, and h_n, the nonluminous heat transfer coefficient, which is significant above 700°C. This appendix explains how h_c and the gas side pressure drop ΔP_g may be estimated. There are numerous correlations, but the few commonly used are cited in the following text. (For finned tubes, see Appendix E.)

Fishenden and Saunders correlation for h_c for cross-flow of gases over plain tube banks takes the following form:

$$Nu = 0.35 \, F_h Re^{0.6} Pr^{0.3} \tag{C.1}$$

F_h depends on the tube geometry, whether staggered or inline, and is shown in Table C.1 (Figure C.2).

If one reviews this table, the correction factor for inline and staggered arrangement for industrial boilers will be nearly the same for tube spacing typically seen in boiler practice, namely, S_T/d of 2–3 and S_L/d of 2–3. In the case of shell and tube heat exchangers where tubes sizes are very small (on the order of 0.375–1 in. versus 1.5–2.5 in. in boilers), smaller longitudinal spacing is used in which case staggered arrangement may give a slightly higher h_c value over inline arrangement. Hence, from heat transfer point of view, not much is gained by using staggered arrangement in boilers or heaters. On the other hand, as shown later, the gas pressure drop across the staggered tube bundle is much larger than an inline bundle, and hence, staggered plain tubes are generally avoided in boiler practice. One may also note from Table C.1 that as we reduce S_L in inline arrangement, the correction factor or the heat transfer coefficient decreases, while in staggered arrangement, it increases! One cannot decrease S_L as it will affect the ligament efficiency and increase the thickness of the drum or headers.

A conservative correlation for both inline and staggered arrangements is as follows [1]:

$$Nu = 0.33 \, Re^{0.6} Pr^{0.33} \tag{C.2}$$

where
Reynolds number $Re = Gd/\mu$
G is the gas mass velocity, kg/m^2 s
$G = W_g/[N_w L(S_T - d)]$
W_g is the flow over tubes, kg/s
d is the tube outer diameter, m
N_w is the number of tubes/row or tubes wide
L is the effective length of tube, m
S_T, S_L are transverse and longitudinal pitch, m
μ is the viscosity of gas, kg/m s or Pa s
Nusselt number $Nu = h_c d/k$, where h_c, the convective heat transfer coefficient, is in W/m^2 K; d is in m; and k, the thermal conductivity of the gas, is in W/m K
Prandtl number $Pr = \mu C_p/k$, where C_p, the gas specific heat, is in J/kg K

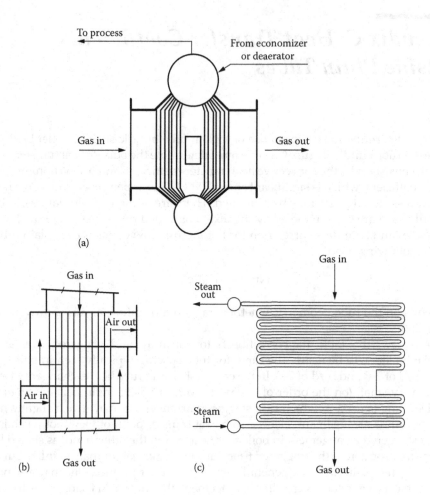

FIGURE C.1
Typical tube banks in boilers. (a) Evaporator; (b) tubular air heater; (c) superheater.

Note that all thermal and transport properties for heat transfer coefficient for plain tubes are estimated at the gas film temperature, which is approximately the average of the tube wall and gas temperature.

Substituting for Nu, Re, and Pr in (C.2) and simplifying, we have

$$h_c = 0.33F\left(G^{0.6}/d^{0.4}\right) \quad \text{where} \quad F = k^{0.67}C_p^{0.33}/\mu^{0.27} \tag{C.3}$$

Grimson's correlation is widely used in boiler design practice:

$$Nu = B Re^N \tag{C.4}$$

TABLE C.1

F_h for Inline and Staggered Arrangements

S_t/d		1.25	1.5	2	3	1.25	1.5	2	3
S_T/d	Re		Inline Bank				Staggered Bank		
1.25	2,000	1.06	1.06	1.07	1	1.21	1.16	1.06	0.96
1.25	8,000	1.04	1.05	1.03	0.98	1.11	0.99	0.92	0.95
1.25	20,000	1	1	1	0.95	1.04	1.02	0.98	0.94
1.5	2,000	0.95	0.95	1.03	1.03	1.17	1.15	1.08	1.02
1.5	8,000	0.96	0.96	1.01	1.01	1.1	1.06	1.00	0.96
1.5	20,000	0.95	0.95	1.0	0.98	1.04	1.02	0.98	0.94
2	2,000	0.73	0.73	0.98	1.08	1.22	1.18	1.12	1.08
2	8,000	0.83	0.83	1	1.02	1.12	1.1	1.04	1.02
2	20,000	0.90	1	1	1	1.09	1.07	1.01	0.97
3	2,000	0.66	0.66	0.95	1	1.26	1.26	1.16	1.13
3	8,000	0.81	0.81	1.02	1.02	1.16	1.15	1.11	1.06
3	20,000	0.91	0.91	1.01	1	1.14	1.13	1.1	1.02

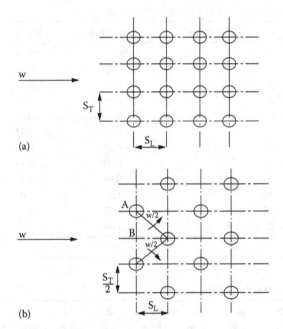

FIGURE C.2
(a) Inline and (b) staggered tube banks.

B and N have been arrived at for Reynolds number varying from 2,000 to 40,000. Table C.2 shows the B and N factors, while Table C.3 shows the correction factor for the number of rows deep while Table C.4 shows the correction factor for angle of attack.

NOTE: Based on his experience and analysis of field data of numerous boilers, the author has found that these correlations underpredict the heat transfer coefficient by about 10%. Hence, a correction factor of 1.1 may be used if one wants to be less conservative.

TABLE C.2

Grimson's Values of B and N

S_T/d	1.25		1.5		2.0		3.0	
S_L/d	B	N	B	N	B	N	B	N
Stagger								
1.25	0.518	0.556	0.505	0.554	0.519	0.556	0.522	0.562
1.5	0.451	0.568	0.460	0.562	0.452	0.568	0.488	0.568
2.0	0.404	0.572	0.416	0.568	0.482	0.556	0.449	0.570
3.0	0.310	0.592	0.356	0.580	0.440	0.562	0.421	0.574
Inline								
1.25	0.348	0.592	0.275	0.608	0.100	0.704	0.0633	0.752
1.5	0.367	0.586	0.250	0.620	0.101	0.702	0.0678	0.744
2.0	0.418	0.570	0.299	0.602	0.229	0.632	0.198	0.648
3.0	0.290	0.601	0.357	0.584	0.374	0.581	0.286	0.608

TABLE C.3

Correction F_n Factor for Rows Deep

No. of Rows Deep	Staggered	Inline
1	0.68	0.64
2	0.75	0.80
3	0.83	0.87
4	0.89	0.90
5	0.92	0.92
6	0.95	0.94
7	0.97	0.96
8	0.98	0.98
9	0.99	0.99
10	1.0	1.0
15	1.02	1.0
25	1.043	1.0

TABLE C.4

Correction for Angle of Attack

Degree	90	80	70	60	50	40	30	20	10
F_n	1.0	1.0	0.98	0.94	0.88	0.78	0.67	0.52	0.42

Example C.1

Flue gases of 20 kg/s (158,688 lb/h) from the combustion of natural gas (analysis: % volume CO_2 = 8, H_2O = 18, N_2 = 71, O_2 = 3) at 400°C (752°F) average film temperature flow over a tube bundle with more than 10 rows deep. Tube OD = 50.8 mm (2.0 in.), and transverse and longitudinal pitch 101.8 and 101.8 mm (4.0 in.), respectively. Effective tube length is 3.5 m (11.5 ft), and there are 12 tubes/row. What is h_c for inline and staggered arrangements? From the table of gas properties in Appendix F,

We can see that at 400°C, specific heat C_p = 0.289 kcal/kg °C (1210 J/kg K), viscosity μ = 0.1088 kg/m h (0.0000302 kg/m s), and thermal conductivity k = 0.0414 kcal/m h °C (0.0481 W/m K).

$$G = \frac{20}{\left[12 \times 3.5 \times (0.1018 - 0.0508)\right]} = 9.39 \text{ kg/m}^2\text{s} \left(6{,}898 \text{ lb/ft}^2\text{h}\right)$$

$$Re = Gd/\mu = 9.39 \times 0.0508/0.0000302 = 15{,}795$$

Using Grimson's correlation, B = 0.482, N = 0.556 for staggered, and B = 0.229, N = 0.632 for inline arrangement for $S_T/d = S_L/d = 2$.

For inline arrangement, Nu = 0.229 × 15,795$^{0.632}$ = 103.1 or h_c = 103.1 × 0.0481/.0508 = 97.6 W/m^2 K (83.9 kcal/m^2 h °C) (17.2 Btu/ft^2 h °F).

For staggered arrangement, Nu = 0.482 × 15,795$^{0.556}$ = 104 or h_c = 98.56 W/m^2 K (84.9 kcal/m^2 h °C) (17.36 Btu/ft^2 h °F).

Using Fishenden and Saunders equation,

$$Nu = 0.35\, F_h Re^{0.6} Pr^{0.3}$$

$$Re = 15{,}795$$

$$Pr = \frac{\mu C_p}{k} = 0.0000302 \times \frac{1{,}210}{0.0481} = 0.759$$

F_h for inline arrangement is 1.0, while for staggered, it is about 1.02.

$$Nu = 0.35 \times 15{,}795^{0.6} \times .759^{0.3} = 106.4 = h_c \times \frac{0.0508}{0.0481}$$

or

$$h_c = 100.8 \text{ W/m}^2\text{K} \left(86.7 \text{ kcal/m}^2\text{h}°\text{C}\right)\left(17.75 \text{ Btu/ft}^2\text{h}°\text{F}\right)$$

h_c is about 2% more for staggered arrangement.

Hence, it may be seen that the difference in h_c between staggered and inline arrangements is not significant. Hence, gas pressure drop over plain tubes should be estimated to see which arrangement is better. Gas pressure drop gives an idea of the incremental addition to the fan power consumption. This is discussed later.

Variation of Convective Heat Transfer Coefficients around Tube Periphery

The correlations discussed earlier for heat transfer provide an estimate for the average value of h_c for the tube bundle. h_c varies around the tube periphery and also along the tube rows as shown in Figure C.3a through c. Figure C.3b is for inline, while Figure C.3c is for staggered arrangement. While determining the maximum tube wall temperature for selecting materials for a superheater, which operates at high gas temperature, one should estimate the maximum value of h_c in a given tube row. The maximum value of h_c occurs at the tube front and decreases to a minimum at about 90° and again increases to a maximum at 180°. The ratio of hc'/hc, the ratio of local heat transfer coefficient to

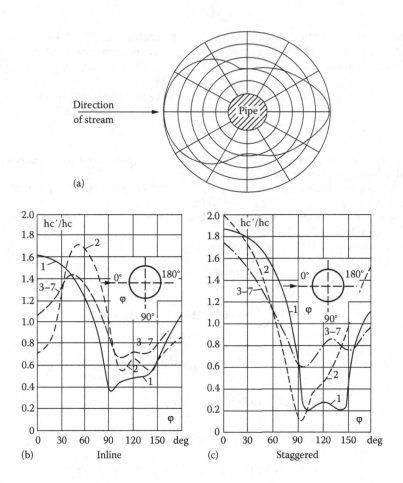

FIGURE C.3
Variation of heat transfer coefficient around tube periphery. The numbers 1–7 (in b and c) refer to row along gas flow.

the average, depends on the Reynolds number and tube arrangement, whether inline or staggered. It can be seen that the second tube shows a maxima larger than the first for inline arrangement (Figure C.3b). In staggered arrangement, a similar effect is seen (Figure C.3c) but at a different point in the periphery. In inline arrangement, hc′ can be 1.6 times the average hc in row 1, while in row 2, it is about 1.7. Beyond two rows, hc′ is about 1.4 times the average value. Hence, for estimating tube wall temperature and selecting materials in counter-flow arrangements, the first two to three rows assume significance.

Though the figures stress the fact that there is variation in heat transfer coefficient around the tube periphery and along the gas flow direction, it is better to do actual tube wall temperature measurements in the first few rows of high temperature surface such as a superheater and confirm the correction factors or correlate it with the calculated tube wall temperatures. This will help the engineer select appropriate materials for these critical heating surfaces. The charts provided earlier show results from limited tests and are shown for informative purposes.

Gas Pressure Drop in Tube Bundles

One may arrive at the optimum tube spacing for the boiler components if, in addition to heat transfer coefficient, the flue gas pressure drop is known.

The gas pressure drop across the bank tubes having N_d rows deep is given by

$$\Delta P_g = \frac{0.204\, f\, G^2 N_d}{\rho_g} \tag{C.5}$$

where

G is the gas mass velocity in kg/m² s

ΔP_g is the gas pressure drop in mm wc

ρ_g is the gas density at atmospheric pressure and at average gas temperature t_g, kg/m³

$$\rho_g = \frac{12.17\, MW}{\left(t_g + 273\right)} \tag{C.6}$$

f is the friction factor given by Equations C.7a and C.7b:

For inline arrangement, for $S_T/d = 1.5$–4 and for 2,000 < Re < 40,000,

$$f = Re^{-0.15}\left[0.044 + \frac{0.08\, S_L/d}{\left(S_T/d - 1\right)^{(0.43 + 1.13\, d/S_L)}} \right] \tag{C.7a}$$

For staggered arrangement,

$$f = Re^{-0.16}\left[0.25 + \frac{0.1175}{\left(S_T/d - 1\right)^{1.08}} \right] \tag{C.7b}$$

where

S_T/d is the transverse pitch to diameter ratio

S_L/d is the longitudinal pitch to diameter ratio

The viscosity of the gas and density are estimated at the average gas temperature for pressure drop calculations. However, even if one uses the gas film temperature at low gas temperatures, the error will be small as Reynolds number is raised to the power of −0.15 (Table C.5).

TABLE C.5

Pressure Drop Equations

SI Units	British	Metric
$\Delta P_g = 0.204 f G^2 N_d / \rho_g$	$\Delta P_g = 9.3 \times 10^{-10}\, f\, G^2 N_d / \rho_g$	$\Delta P_g = 1.574 \times 10^{-8}\, f\, G^2 N_d / \rho_g$
ΔP_g, mm wc	ΔP_g, in. wc	ΔP_g, mm wc
G, kg/m² s	G, lb/ft² h	G, kg/m² h
ρ_g, kg/m³	ρ_g, lb/ft³	ρ_g, kg/m³
$\rho_g = 12.17\, MW/(t_g + 273)$	$\rho_g = 1.37\, MW/(t_g + 460)$	$\rho_g = 12.17\, MW/(t_g + 273)$
t_g, °C	t_g, °F	t_g, °C

Note: MW, molecular weight. Gas density at atmospheric pressure.

Example C.2

For the data in Example C.1 (with Re = 15,795), determine the gas pressure drop for both inline and staggered arrangements for 10 rows deep. Gas pressure is atmospheric. G = 9.39 kg/m² s.

Solution

$S_T/d = S_L/d = 2$. If the average gas temperature is, say, 500°C, the density of flue gas = 12.17 × 27.6/773 = 0.4345 kg/m³, where 27.6 is the flue gas molecular weight.

For inline arrangement, f = 15,795$^{-0.15}$[0.044 + 0.08 × 2/1] = 0.0478
  In SI units, ΔP_g = 0.204 × 10 × 0.0478 × 9.39²/0.4345 = 20 mm wc
  In British units, G = 6,898 lb/ft² h, ρ_g = 1.37 × 27.6/(932 + 460) = 0.02716 lb/ft³
  Hence, ΔP_g = 9.3 × 10^{-10} × 6,898² × 0.0478 × 10/0.02716 = 0.78 in wc
  In metric units, G = 9.39 × 3,600 = 33,804 kg/m² h. Hence, ΔP_g = 1.574 × 10^{-8} × 10 × 0.0478 × 33,804²/0.4345 = 20 mm wc

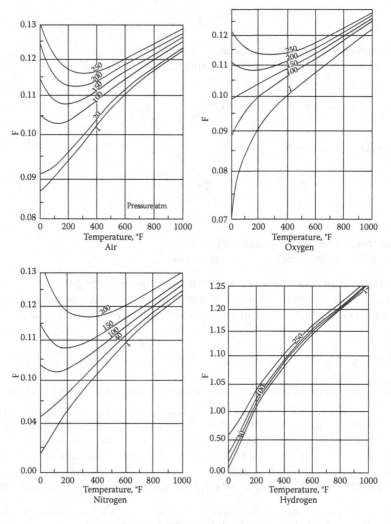

FIGURE C.4
Effect of gas pressure on convective heat transfer coefficient outside tubes (pressure in atm).

For staggered arrangement, $f = 15{,}795^{-0.16}[0.25 + 0.1175/1] = 0.078$
Hence, gas pressure drop will be $0.078/0.0478 \times 20 = 33$ mm wc

Hence, for nearly the same heat transfer coefficient or size of tube bundle, the gas pressure drop is about 60% higher, which means more operating cost. *Hence, inline arrangement of plain tubes is generally preferred over staggered arrangement in boiler practice for plain tubes.*

Effect of Gas Pressure on Heat Transfer

The effect of gas pressure is similar to that of flow inside tubes. Figure C.4 shows results for common gases. The effect is not significant till 40 bar pressure, and hence, for boiler, heater, or HRSG applications, the effect of gas pressure may be neglected.

Effect of Tube Size on Boiler Design

It may be shown that when small-diameter tubes are used in the boiler for the same gas pressure drop and duty, the smaller-diameter option will have lower weight and hence easier handling and lower shipping costs. However, the number of tubes to be welded to the headers (or drums) will be more, increasing the cost of labor. One also has to check the cost of materials including the cost of longer drums or headers. In any event, effect of tube size on heat transfer must be understood by plant engineers as different boiler suppliers around the world may come up with boiler designs with different tube sizes, and an idea of how this affects the size, surface area, and weight is important while evaluating the designs. It is the author's view that in locations where labor costs are low and material costs are high, smaller-diameter tubes may be used.

Example C.3

Flue gases of 200,000 kg/h from natural gas products of combustion has to be cooled from 600°C to 311°C in a bare tube boiler generating saturated steam at 40 kg/cm² a (saturated steam temperature = 249°C). Tubes measuring 50.8 mm and 38 mm are available. For the same gas pressure drop, determine which option weighs less. Flue gas analysis % volume $CO_2 = 8$, $H_2O = 18$, $N_2 = 71$, $O_2 = 3$. Use inline arrangement. Fouling factors are 0.0002 m² h °C/kcal for both gas and steam. (Assume boiling heat transfer coefficient as 12250 kcal/m²h°C.)

Let us use 50.8 × 43 mm plain tubes 24 tubes/row, 8 m long with transverse and longitudinal pitch = 78 mm. For the 38 mm tubes, use a rectangular spacing of 66 mm. The number of rows deep has to be estimated from U_o, the overall heat transfer coefficient.

Let us determine the convective heat transfer coefficient. Gas properties at average film temperature have to be obtained. The tube wall temperature will be close to that of steam temperature due to the high boiling heat transfer coefficient. Hence, let us assume that tube outer wall temperature = 260°C. Average gas temperature is $(600 + 311)/2 = 455$°C. Average film temperature = $(455 + 260)/2 = 357$°C.

From Appendix F at film temperature of 357°C, $C_p = 0.2860$ kcal/kg °C, $\mu = 0.1038$ kg/m h, $k = 0.03925$ kcal/m h °C. At an average gas temperature of 455°C, $C_p = 0.2924$, $\mu = 0.1143$ kg/m h, $k = 0.04395$ kcal/m h°C.

Gas mass velocity $G = 200,000/[24 \times 8 \times (0.078 - 0.0508)] = 38,297$ kg/m² h
$Re = Gd/\mu = 38,297 \times 0.0508/0.1038 = 19,232$. Use Grimson's correlations. Obtain factors B and N from Table C.2. $B = 0.25$ $N = 0.62$
$Nu = 0.25 \times 19232^{0.62} = 113.2 = h_c \times 0.0508/0.0.03925$ or $h_c = 87.46$ kcal/m² h °C (17.85 Btu/ft² h °F). (We may also obtain h_c in W/m² K using appropriate properties as done earlier. Let us not use any correction factors as the purpose is to see the effect of tube size only.)

Nonluminous Heat Transfer Coefficient h_n (See Appendix D)

Beam length $L = 1.08 \times (0.078 \times 0.078 - 0.785 \times 0.0508 \times 0.0508)/0.0508 = 0.086$ m

$$K = (0.8 + 1.8 \times 0.18) \times (1 - 0.38 \times 0.728) \times \frac{(0.08 + 0.18)}{(0.26 \times 0.086)^{0.5}} = 1.413$$

Gas emissivity $\in_g = 0.9 \times (1 - e^{-1.413 \times 0.086}) = 0.102$

$$h_n = 5.67 \times 10^{-8} \times 0.102 \times \left[\frac{(7.28^4 - 5.33^4)}{(728 - 533)} \right] = 5.93 \text{ W/m}^2 \text{ K} = 5.1 \text{ kcal/m}^2\text{h°C}$$

$$U_o = \frac{1}{(5.1 + 87.46)} + 0.0002 + 0.0002\left(\frac{50.8}{43}\right) + \left(\frac{1}{12,250}\right) \times \left(\frac{0.0508}{043}\right) + \left(\frac{0.0508/2}{37}\right) \ln\left(\frac{50.8}{43}\right)$$

$$= 0.01144 \quad \text{or} \quad U_o = 87.36 \text{ kcal/m}^2\text{h°C}\left(17.83 \text{ Btu/ft}^2\text{h°F}\right)$$

Duty $= 200,000 \times .99 \times (600 - 311) \times 0.2924 = 16.73 \times 10^6$ kcal/h (use a heat loss of 1%)

LMTD $= (600 - 311)/[(600 - 249)/(311 - 249)] = 167$°C. Surface area required $= 16.73 \times 10^6/167/87.36 = 1151$ m² $= 3.14 \times 0.0508 \times 24 \times 8 \times N_d$, where N_d is the number of rows deep. $N_d = 37.5$. 38 rows were used as headers have to be brought to the same side of the bundle.

Gas Pressure Drop

$$\Delta P_g = 0.204 \text{ f } G^2 N_d/\rho_g$$

Reynolds number for pressure drop is computed at the average gas temperature. $Re = 38,297 \times 0.0508/0.1143 = 17,020$. ρ_g at average gas temperature of 455°C $= (27.6/22.4) \times 273/(273 + 455) = 0.4614$ kg/m³ (MW of flue gas $= 27.6$).

$$f = Re^{-0.15}\left[0.044 + \frac{0.08 \text{ } S_L/d}{(S_T/d - 1)^{(0.43 + 1.13 d/S_T)}} \right] = 0.0693$$

$$\Delta P_g = 0.204 \times 0.0693 \times \left(\frac{38,297}{3,600}\right)^2 \times \frac{38}{0.4614} = 132 \text{ mm wc}$$

Similar calculations may be done for the 38 mm tubes. Results are shown in Table C.6. It can be seen that the overall heat transfer coefficient is higher with the smaller tube. It may be also noted that the weight of the 38 mm tube design is lower for the same gas pressure drop and duty. However, this has to be balanced by the cost of manufacturing as we have

TABLE C.6

Comparison of 50.8 and 38 mm Tubes

Item	50.8 mm	38 mm
Gas flow, kg/h	200,000	200,000
Gas inlet temp., °C	600	600
Gas exit temp., °C	311	311
Duty, MM kcal/h	16.7	16.9
Gas press. drop, mm wc	132	130
Tubes/row	24	24
Length, m	8	8
No. of rows deep	38	47
Transverse pitch, mm	78	66
Longitudinal pitch, mm	78	66
U_o, kcal/m² h °C	86.8	94.0
Surface area, m²	1,160	1,074
Weight, kg	34,310	25,840

Note: Large longitudinal spacing is not recommended.

more tubes to be welded or rolled on to the drums or headers. Hence, one may review options with smaller-diameter tubes if weight of the boiler is a concern. As mentioned earlier, the cost of fabrication may differ depending on local manufacturing costs.

Effect of Longitudinal Spacing

With plain tubes, the longitudinal spacing in inline arrangement impacts the gas pressure drop significantly while contributing slightly to the overall heat transfer coefficient. Tube spacing is generally based on tube bend radius available for a given tube size and ligament efficiency in boiler drums and also on the boiler manufacturer's practice. However, choosing a large longitudinal spacing increases the gas pressure drop and thus increases the operating cost. The nonluminous heat transfer coefficient is increased due to the increase in beam length. It also increases the drum or header lengths and increases the cost of the boiler.

Example C.4

Flue gases of 100,000 kg/h in a boiler are cooled from 700°C to 326°C. Flue gas analysis is % volume $CO_2 = 8$, $H_2O = 18$, $N_2 = 71$, $O_2 = 3$. Tubes are 50.8 mm OD × 43 mm. Steam is generated at 40 kg/am² a.Compare the design with a transverse tube spacing of 76 mm with longitudinal spacing of 76,101 and 152 mm, respectively, in inline arrangement.

Using Grimson's correlations, calculations were performed using a computer program, and the results are presented in Table C.7. While the convective heat transfer coefficient changes within 1%, the nonluminous coefficient increases due to the higher beam length (see Appendix D). Hence, U_o values are slightly higher when S_L increases. However, the gas pressure drop increases much more, by about 40% when $S_L/d = 3$ compared to when $S_L/d = 1.5$.

So the boiler designer should consider using smaller bend radius while fabricating the bundle or a smaller longitudinal pitch, while the plant engineer should be aware of the implications of S_L.

TABLE C.7

Effect of S_L/d on Boiler Performance

S_L/d	1.5	2	3
Gas temp in, °C	700	700	700
Gas temp out, °C	328	327	325
U_o, kcal/m² h °C	88.79	89.94	91.4
h_c, kcal/m² h °C	89.49	90.00	90.30
h_n, kcal/m² h °C	5.13	6.22	7.9
Gas pressure drop, mm wc	101	115	146
Friction factor f	0.0748	0.0860	0.1096

Note: 30 tubes/row; 30 deep; length = 4 m; t_{sat} = 249°C; S_T/d = 1.5.

Simplified Procedure for Evaluating Performance of Plain Tube Bundles

There are numerous parameters involved in the evaluation of design and performance of plain tube bundles. A computer program would be the best tool for this exercise as numerous cases can be evaluated and the optimum chosen. However, as an academic exercise, a simplified procedure for convective plain tube bundles is presented later using which one can arrive at the geometry of a tube bundle for a specified duty and gas pressure drop. It is assumed that the nonluminous heat transfer coefficient is negligible, and only convective heat transfer is predominant. Such derivations also help plant engineers quickly visualize the effect of tube geometry or other factors on boiler performance. Keep in mind that in the following derivation, the effect of nonluminous heat transfer coefficient is neglected; this gives reasonable results up to 600°C–700°C.

From Appendix A, it was shown that $U = 0.95h_o$ for boilers as other resistances may be neglected.

h_o is obtained from $Nu = 0.35Re^{0.6}Pr^{0.3}$

$Nu = h_o d/k$, $Re = Gd/\mu$, $Pr = \mu Cp/k$; all the properties in consistent units.

$$G = \frac{W_g}{\left[N_w L(S_T - d)\right]} \tag{C.8a}$$

(W_g, kg/h; h_o, kcal/m²h °C; d, m; k, kcal/m h °C; G, kg/m² h; μ, kg/m h; C_p, kcal/kg °C).
Substituting for the various terms and simplifying, we have

$$h_o = \frac{0.35G^{0.6}F_2}{d^{0.4}} \quad \text{where } F_2 = \left(\frac{C_p}{\mu}\right)^{0.3} k^{0.7} \tag{C.8b}$$

$$U = \frac{0.3325\ G^{0.6}F_2}{d^{0.4}} \tag{C.9}$$

(gas properties are computed at the average film temperature)

$$\frac{Q}{\Delta T} = UA = \left(\frac{0.3325\ G^{0.6}F_2}{d^{0.4}}\right)\pi dN_w N_d L = 1.044\ F_2 G^{0.6}N_w N_d L \tag{C.10}$$

Substituting for G from (C.8a), we have

$$\frac{Q}{\Delta T} = UA = \frac{1.044\, F_2 W_g^{0.6} N_w N_d L}{N_w^{0.6} L^{0.6} \left(S_T/d-1\right)^{0.6}} \tag{C.11}$$

$$= \frac{1.044 F_2 W_g^{0.6} Nw^{0.4} L^{0.4} N_d}{\left(S_T/d-1\right)^{0.6}} \tag{C.12}$$

This equation may be used to relate the tube geometry with duty. However, when there is a phase change as in a boiler evaporator, the equation may be simplified further.

$$\ln\left[(T_1 - T_s)/(T_2 - T_s)\right] = UA/(W_g C_p) \text{ (as seen in Appendix A)}$$

$$= \frac{1.044 F_2 W_g^{0.6} Nw^{0.4} L^{0.4} N_d / \left(S_T/d-1\right)^{0.6}}{\left(W_g C_p\right)}$$

$$= \frac{1.044 F_2 N_w^{0.4} L^{0.4} N_d \left(S_T/d-1\right)^{0.6}}{\left(W_g^{0.4} C_p\right)}$$

$$= \frac{1.044\left(F_2/C_p\right) N_d}{\left[G^{0.4}\left(S_T/d-1\right)d^{0.4}\right]} \tag{C.13a}$$

If the tube geometry is known, one can estimate N_d or G for a specific thermal performance. Now introducing the equation for gas pressure drop (inline which is widely used),

$$\Delta P_g = \frac{0.204\, f\, G^2 N_d}{\rho_g} \ \left(\text{if G is in kg/m}^2\text{s}\right) \quad \text{or} \quad \Delta P_g = \frac{1.574 \times 10^{-8} f\, G^2 N_d}{\rho_g} \left(\text{G in kg/m}^2\text{h}\right) \tag{C.13b}$$

$$f = Re^{-0.15} X \tag{C.13c}$$

where

$$X = 0.044 + \frac{0.08 S_L/d}{\left(S_T/d-1\right)^{(0.43+1.13d/S_L)}}$$

Substituting for Re = Gd/μ in above equation C.13c and for N_d from (C.13), we have

$$\Delta P_g = 1.5077 \times 10^{-8} \times G^{2.25} \times \ln\left[\frac{(T_1 - T_s)}{(T_2 - T_s)}\right] \times \frac{(S_T - d) X}{\left[F_3 d^{0.75} \rho_g\right]} \tag{C.14}$$

where

$$F_3 = \frac{k^{0.7}}{\left(Cp^{0.7} \mu^{0.45}\right)} \tag{C.15}$$

This is a very important equation linking gas pressure drop with thermal performance and duty. While a computer program would yield close results, equations such as this help one visualize the impact of variables on boiler design and performance.

Example C.5

Flue gases of 29,945 kg/h from a steam generator are cooled from 627°C to 227°C in an evaporator generating saturated steam at 177°C. Flue gas analysis is CO_2 = 8.29, H_2O = 18.1, N_2 = 71.08, O_2 = 2.46% volume. $S_T = S_L$ = 101.6 mm and d = 50.8 mm. Gas film temperature is about 300°C, and density at average gas temperature of 427°C = 0.48 kg/m³. From Appendix F, C_p = 0.2821 kcal/kg°C, μ = 0.0972 kg/m h, and k = 0.0364 kcal/m h°C at film temperature of 300°C. Determine the boiler geometry for a gas pressure drop of 76 mm wc.

First one has to determine G from (C.14). X = 0.044 + 0.08 × 2 = 0.204. F_3 = $0.0364^{0.7}/.2821^{0.7}/0.0972^{0.7}$ = 0.681 Density ρ_g = 27.58 × 273/22.4/(273 + 427) = 0.48 kg/m³. Hence,

$$76 = 1.5077 \times 10^{-8} \times G^{2.25} \times \mathrm{Ln}\left[\frac{(627-177)}{(227-177)}\right] \times 0.0508 \times \frac{0.204}{[0.681 \times 0.0508^{0.75} \times 0.48]} \quad \text{or}$$

$$G = 24{,}812 \text{ kg/m}^2 \text{ h}$$

$$24{,}812 = G = \frac{29{,}945}{[N_w L \times 0.0508]} \quad \text{or} \quad N_w L = 23.75.$$

If N_w = 8, then L = 2.97 m.
Let us get N_d from (C.13). F_2 = $(Cp/\mu)^{0.3} k^{0.7}$ = $(0.2821/.0972)^{0.3} \times 0.0364^{0.7}$ = 0.135

$$\mathrm{Ln}[[(T_1 - T_s)/(T_2 - T_s)] = 1.044(F_2/C_p)N_d/[G^{0.4}(S_T/d - 1)d^{0.4}] \quad (\text{C.16})$$

$$2.197 = 1.044 \times \left(\frac{0.135}{0.2821}\right)\frac{N_d}{(24{,}812^{0.4} \times 0.0508^{0.4})} \quad \text{or} \quad N_d = 76.$$

Thus, the entire geometry is arrived at. One can also see the effect of changing tube spacing, tube diameter, and so on.

Example C.6

If the number of rows deep is limited to 50, what happens to the gas pressure drop and exit gas temperature? Assume same cross section and tube spacing and gas inlet conditions.
Substituting in (C.16),

$$\mathrm{Ln}[[(T_1 - T_s)/(T_2 - T_s)] = 1.044 \times (0.135/.2821) \times 50/(24812^{0.4} \times 0.0508^{0.4}) = 1.441$$

or

$$[(T_1 - T_s)/(T_2 - T_s)] = 4.22 = (627 - 177)/(T2 - 177) \text{ or } T_2 = 284°C$$

Using (C.14), we have

$$\Delta P_g = 1.5077 \times 10^{-8} \times G^{2.25} \times \mathrm{ln}[[(T_1 - T_s)/(T_2 - T_s)] \times (S_T - d)X/[F_3 d^{0.75}\rho_g]$$

$$= 1.5077 \times 10^{-8} \times 24{,}812^{2.25} \times 1.441 \times 0.0508 \times \frac{0.204}{\left(0.681 \times 0.0508^{0.75} \times 0.48\right)} = 50 \text{ mm wc}$$

One can fine-tune for the lower density due to the higher exit gas temperature; however, these equations may be used to get some quick ideas of the variables involved in design or performance of convective tube plain tube bundles.

Reference

1. V. Ganapathy, *Applied Heat Transfer*, Pennwell Books, Tulsa, OK, 1982, p130.

Appendix D: Nonluminous Heat Transfer Calculations

Nonluminous heat transfer plays a significant role in high gas temperature heat transfer equipment such as superheaters or boiler banks or unfired furnaces of waste heat boilers. During combustion of fossil, fuels like natural gas, fuel oil, or coal triatomic gases such as CO_2, H_2O, and SO_2 are generated, which contribute to radiation. The emissivity pattern of these gases has been studied by Hottel, who has developed charts to estimate the emissivity of these gases if the partial pressure and beam length of the tube bank are known.

Net interchange of radiation between gases and surroundings like a water-cooled wall or a tube bundle or a cavity may be written as

$$\frac{Q}{A} = \sigma\left(\epsilon_g\, T_g^4 - a_g T_o^4\right) \tag{D.1}$$

where
ϵ_g is the emissivity of gases at average gas temperature T_g
T_o is the tube wall temperature
a_g is the absorptivity at T_o

$$\epsilon_g = \epsilon_c + \eta\,\epsilon_w - \Delta\epsilon \tag{D.2}$$

Computing absorptivity is tedious; also, as T_o^4 is typically much smaller than T_g^4, without much loss of accuracy, one can write

$$\frac{Q}{A} = \sigma\epsilon_g\left[T_g^4 - T_o^4\right] = h_n\left(T_g - T_o\right) \tag{D.3}$$

or

$$h_n = \frac{\sigma\epsilon_g\left[T_g^4 - T_o^4\right]}{\left(T_g - T_o\right)} \tag{D.4a}$$

where
σ is the Steffan–Boltzman constant $= 5.67 \times 10^{-8}\,W/m^2\,K^4$
T_g, T_o are gas and wall temperatures, K
h_n is the nonluminous coefficient, $W/m^2\,K$

In British units,

$$h_n = \frac{\sigma \epsilon_g \left[T_g^4 - T_o^4 \right]}{\left(T_g - T_o \right)} \tag{D.4b}$$

where
  σ is the Steffan–Boltzman constant = 0.1714×10^{-8}
  T_g, T_o in °R
  h_n in Btu/ft²h °F

To estimate h_n, partial pressures of triatomic gases and beam length L are required. L is a characteristic dimension that depends on the bundle arrangement and shape of enclosure.
  For a plain tube bundle,

$$L = \frac{1.08 \times \left(S_T S_L - 0.785 d^2 \right)}{d} \tag{D.5}$$

For a cavity, L is approximately 3.4–3.6 times the volume of the space divided by the surface area of the heat-receiving surface. If a, b, c are the dimensions of the cavity,

$$L = \frac{3.4abc}{2(ab + bc + ca)} = \frac{1.7}{(1/a + 1/b + 1/c)} \tag{D.6}$$

For flow inside tubes, the tube inner diameter may be taken as L. As d_i is generally small, the beam length and hence the emissivity is small, and h_n inside tubes may be neglected if the convective coefficient is large.
  For a finned tube bundle,

$$\text{Beam length } L_b = (d + 2h) \left[\frac{(S_T \times S_L)}{(d + 2h)^2} - 0.85 \right] \tag{D.7}$$

Example D.1

Estimate the beam length when 50.8 mm tubes in a boiler bank are arranged with a transverse pitch of 101 mm and longitudinal pitch of 127 mm.

Solution

L = 1.08 × (0.1018 × 0.127 – 0.785 × 0.0508 × 0.0508)/0.0508 = 0.23 m

Example D.2

Flue gases with CO_2 = 12% and H_2O = 16% by volume are flowing across a superheater tube bundle. If 50.8 mm tubes are arranged with S_T = 101 mm and S_L = 127 mm, determine h_n. Flue gas pressure is atmospheric. Average gas temperature T_g = 900°C and average wall temperature T_o = 300°C.

Solution

$L = 0.23$ m from earlier example. P_c, the partial pressure of $CO_2 = 0.12$ and P_w, the partial pressure of water vapor $= 0.16$

$$P_cL = 0.23 \times 0.12 = 0.0276 \text{ atm m} = 0.09 \text{ atm ft}$$
$$P_wL = 0.23 \times 0.16 = 0.0368 \text{ atm m} = 0.12 \text{ atm ft}$$

From Figure D.1a and b at $T_g = 900 \times 1.8 + 32 + 460 = 2112°R$ and $T_o = 300 \times 1.8 + 32 + 460 = 1032°R$,

$\epsilon_c = 0.07$ and $\epsilon_w = 0.057$. In Figure D.1c, corresponding to $(P + P_w)/2 = 1.16/2 = 0.58$ and $P_wL = 0.12$, $\eta = 1.05$ and corresponding to $P_w/(P_w + P_c) = 0.16/0.28 = 0.57$ and $(P_c + P_w)L = 0.21$, $\Delta v = 0.002$

Hence, $\epsilon_g = 0.07 + 1.05 \times 0.057 - 0.002 = 0.128$

Using (D.2),

$$h_n = 5.67 \times 10^{-8} \times 0.128 \times (1173^4 - 573^4)/(1173 - 573) = 21.60 \text{ W/m}^2\text{ K}$$

Using (D.4b),

$$h_n = 0.1714 \times 10^{-8} \times 0.128 \times (2112^4 - 1032^4)/(2112 - 1032) = 3.81 \text{ Btu/ft}^2\text{ h °F}$$

While charts are often used, equations are also available for estimating h_n.

$$\epsilon_g = 0.9 \times \left(1 - e^{-KL}\right) \tag{D.8}$$

where attenuation factor

$$K = \frac{(0.8 + 1.6\, P_w) \times (1 - 0.00038\, T_g) \times (P_c + P_w)}{\left[(P_c + P_w)L\right]^{0.5}} \tag{D.9}$$

T_g is in K, L is in m.

Example D.3

Compute h_n in Example D.2 using Equations D.8 and D.9.

$$K = \frac{(0.8 + 1.6 \times 0.16) \times (1 - .00038 \times 1173) \times 0.28}{(0.28 \times 0.23)^{0.5}} = 0.645$$

$$\epsilon_g = 0.9 \times \left(1 - e^{-0.645 \times 0.23}\right) = 0.124$$

$$h_n = \frac{5.67 \times 10^{-8} \times 0.124 \times \left(1173^4 - 573^4\right)}{(1173 - 573)} = 20.9 \text{ W/m}^2 \text{ K} \left(3.68 \text{ Btu/ft}^2 \text{ h °F}\right)$$

Now, let us see how the performance of an unfired furnace may be evaluated. This feature is seen in several waste heat boilers (Figure D.2).

Example D.4

How is the duty of an unfired waste heat boiler furnace determined? Waste gases of 27.8 kg/s at 1000°C enter a waste heat boiler furnace that is completely water cooled and is generating steam at 16 barg. Saturation temperature is 202°C. Furnace dimensions are 3 m wide × 4 m deep × 8 m long. Flue gas analysis is % vol $CO_2 = 12$, $H_2O = 15$, $N_2 = 67$, $O_2 = 6$. Effective cooling area $= 2 \times (3 \times 4 + 4 \times 8 + 3 \times 8) = 136 \text{ m}^2$.

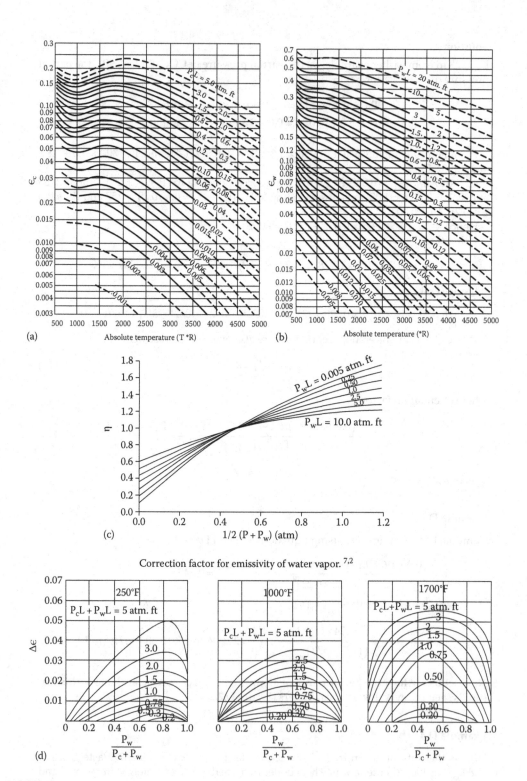

FIGURE D.1
Emissivity of (a) carbon dioxide and (b) water vapor. Correction factor for (c) the emissivity of water vapor and (d) for the presence of water vapor and carbon dioxide.

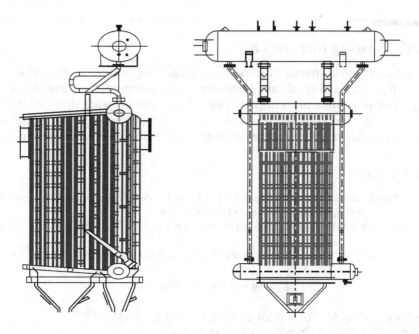

FIGURE D.2
Waste heat boiler with furnace section.

Solution

Assume the exit gas temperature of 800°C = 1073 K. Use saturation temperature plus 18°C for tube outer wall temperature = 220°C = 493 K. Average specific heat of the gas is 1323 J/kg K between 1000°C and 800°C.

At 900°C = 1173 K average gas temperature, the emissivity of the flue gases may be computed.

The beam length of the furnace is 1.7/(1/3 + 1/4 + 1/8) = 2.41 m

$$K = \frac{(0.8 + 1.6 \times .15)(1 - .00038 \times 1173) \times 0.27}{(0.27 \times 2.41)^{0.5}} = 0.193$$

Gas emissivity = $0.9 \times (1 - e^{-0.193 \times 2.41})$ = 0.334
Hence, radiation to furnace walls = $136 \times 0.334 \times 0.9 \times 5.67 \times 10^{-8} \times (1173^4 - 493^4)$ = 4,251 kW
Energy from flue gases = $27.8 \times 1323 \times (1,000 - 800) \times .99$ = 7,282,300 W = 7,282 kW
So the assumed exit gas temperature is too low.
Try exit gas temperature = 900°C or average gas temperature = 950°C = 1,223 K
Gas emissivity = $(0.8 + 1.6 \times 0.15) \times (1 - 0.00038 \times 1223) \times 0.27/(0.27 \times 2.41)0.5$ = 0.186
Gas emissivity = $0.9 \times (1 - e^{-0.186 \times 2.41})$ = 0.325
Hence, radiation to furnace walls = $0.325 \times 0.9 \times 5.67 \times 10^{-8} \times (1223^4 - 493^4) \times 136$ = 4912 kW
Energy from flue gases = $27.8 \times 1,323 \times (1,000 - 900)$ = 7,355,880 W = 3,677 kW
Try exit gas temperature = 870°C average gas temperature = 935°C = 1,208 K
Using the same emissivity,
Radiation to furnace = $0.325 \times 0.9 \times 5.67 \times 10^{-8} \times (1208^4 - 494^4) \times 136$ = 4,668 kW
Energy from flue gases = $27.8 \times 1323 \times 130 \times 0.99$ = 4,733 kW.

Hence, flue gas exit temperature is approximately 870°C. With a computer program, one can study the effect of varying the furnace dimensions on the furnace duty and exit gas temperature. For improved accuracy, one may split up the furnace into zones and arrive at the exit gas temperature and duty more accurately.

Large Tube Spacing Increases h_n

In boiler practice, some manufacturers use a large longitudinal spacing due to considerations such as availability of tube bends, manufacturing practice, and drum ligament efficiency. Transverse tube pitch impacts gas velocity and pressure drop, while longitudinal pitch increases the beam length and hence h_n. Longitudinal pitch in plain tubes also increases the gas-side pressure drop as shown in Appendix C and hence should be avoided.

Example D.5

A tube bundle uses an S_T/d of 1.5 and S_L/d of 1.5 and 3. What is the impact on h_n at 700°C gas temperature? $P_c = 0.08$, $P_w = 0.18$ and $d = 50.8$ mm.
Beam length L when $S_T/d = 1.5$, $S_L/d = 1.5$ is $1.08 \times (1.5 \times 1.5 - 0.785) \times 0.0508 = 0.08$ m

$$K = (0.8 + 1.8 \times 0.18) \times (1 - 0.00038 \times 973) \times 0.26/(0.26 \times 0.08)^{0.5} = 1.236$$

$$\epsilon_g = 0.9 \times (1 - e^{-1.236 \times 0.08}) = 0.085$$

Similarly, when $S_L/d = 3$, $L = 0.204$ m, $K = 0.774$, and $\epsilon_g = 0.131$.
Hence, gas emissivity increases, which increases h_n.
However, as shown in Appendix C, the gas pressure drop increases significantly as S_L/d increases, and hence, this is not a good design practice. Use as small an S_L/d as possible. This will also conserve space and reduce the length of the bundle, drums, and hence cost of boiler. Using a large S_L increases the U and heat flux and the tube wall temperature. Effect of S_L is also discussed in Appendix C, Example C.4.

Example D.6

In a fire tube boiler, the average gas temperature of flue gases inside the tubes is 537°C (810 K) and tube inner diameter is 45 mm. Tube wall temperature is about 195°C (468 K). $P_c = P_w = 0.12$. Flue gas is at atmospheric pressure. Estimate h_n.
The beam length of the tube ID = 0.045 m

$$K = \frac{(0.8 + 1.6 \times 0.12) \times (1 - 0.00038 \times 810) \times 0.24}{(0.24 \times 0.045)^{0.5}} = 1.586$$

$$\epsilon_g = 0.9(1 - e^{-1.586 \times 0.045}) = 0.062$$

$$h_n = \frac{5.67 \times 10^{-8} \times 0.062 \times (810^4 - 468^4)}{(810 - 468)} = 3.93 \, W/m^2 K$$

h_n inside fire tube boilers is usually small and neglected as the beam length is small.

Bibliography

1. W. Roshenow, J.P., Hartnett, *Handbook of Heat Transfer*, McGraw-Hill, New York, 1972.
2. V. Ganapathy, *Applied Heat Transfer*, Pennwell Books, Tulsa, OK, 1982.

Appendix E: Calculations with Finned Tubes

Introduction

Solid and serrated finned tubes, Figure E.1, are extensively used in boiler economizers, HRSG evaporators, superheaters, feed water heaters, air-cooled condensers, fluid heaters, and similar equipment for recovering energy from clean gas streams such as gas turbine exhaust, flue gas from combustion of clean fuels, and clean incinerator exhaust gas, to name a few applications.

The choice of fin geometry and arrangement whether inline or staggered is determined by several factors such as the tube-side coefficient, gas pressure drop, space limitations, and cost. Finned surfaces are effective when the ratio of the tube-side heat transfer coefficient to outside (or gas-side) coefficient is very large as in the case of boiler evaporators or economizers where the gas-side coefficient is in the range of 50–75 W/m² K, while the tube-side coefficient is in the range of 5,000–12,000 W/m² K. If the outside and inside coefficients are in the same range, then finned tubes are not effective and can lead to unnecessary increase in heat flux inside the tubes, overheating of tubes, and increased gas-side pressure drop. That is why we do not see finned tubes in tubular air heaters as both the gas-side and air-side heat transfer coefficients are in the range of 40–50 W/m² K. In the case of gas turbine HRSG, finned tubes are widely used as multiple modules are required and space occupied by the HRSG should be as small as possible. Choosing fin geometry is also done with care. The fin density should be chosen with care for the superheater or in thermal fluid heaters (such as therminol or fuel oils) as the tube-side coefficient is only in the range of 1,200–1,500 W/m² K compared to, say, 10,000–15,000 W/m² K for the evaporator and economizer. A very large ratio of external to internal surface areas will be counterproductive when tube-side coefficient is small as will be shown later. For a steam superheater or thermal fluid heater, 78–117 fins/m (2–3 fins/in.) will suffice, while in HRSG evaporators or economizers, higher fin density of 197–236 fins/m (5–6 fins/in.) may be justified [1]. These points are explained with the following examples.

There are a few myths with finned tubes that will also be dispelled in this appendix. Many engineers are of the view that more the surface area for a given duty, better the design and more energy transfer. In the case of finned tubes, this view is very wrong. This is due to the decrease in heat transfer coefficient with increase in fin surface area. The variation in surface areas can be 50%–100% for the same duty and gas pressure drop depending on fin geometry chosen. Besides, a large fin surface can result in high heat flux inside the tubes and consequent high tube wall temperatures particularly in the case of superheaters or thermal fluid heaters and lower their life or result in overheating and consequent failure.

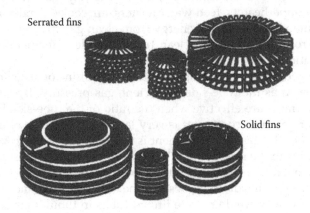

FIGURE E.1
Solid and serrated fins.

Why Finned Tubes?

First let us understand why finned tubes are used. Finned tubes make boiler components compact and light and reduce the gas pressure drop compared to a plain tube design for the same parameters. In addition, the space occupied by the finned bundle will be smaller compared to a plain tube bundle. Given in Table E.1 is a summary of design details for a boiler evaporator with plain and finned tubes for the same duty. The procedure for sizing with plain tubes is discussed in Appendix C. We will discuss the calculation procedure for finned tubes a little later.

A study was made for a boiler evaporator with plain tubes and finned tubes. The steam and gas parameters are shown in Table E.1. The first column shows a plain tube design with a given cross section, while column 2 shows a finned tube bundle design using serrated fins with the same cross section. The third column shows a plain tube bundle designed to match the gas pressure drop of the finned tube bundle.

It may be seen that for the same cross section, the number of rows deep for the plain tube boiler is 57, while for the finned tube option, it is only 13. As a result, the length of the headers will be small. This means a large savings in space if finned tubes are used. In addition, the gas pressure drop and weight of the finned tube bundle are lower. However, the surface area is much more and so is the heat flux inside the tubes, which will increase the tube wall temperature. If the choice of fins is properly done, the boiler can be compact

TABLE E.1

Plain Tube versus Finned Tube Evaporator

Boiler Design	Plain Tubes	Finned Tube	Plain Tubes
Tubes/row	21	22	21
Number of rows deep	57	13	52
Effective length, m	4	4	5
Fins/m	0	197	0
Fin height, mm	0	19	0
Fin thickness, mm	0	1.5	0
Serration, mm	0	4	0
Transverse pitch, mm	101	101	101
Longitudinal pitch, mm	101	101	101
Surface area, m²	762	1,588	868
U, kcal/m² h °C (W/m² K)	62.9 (73)	30.3 (35.2)	55.2 (64.1)
Tube wall temperature, °C	259	293	259
Gas pressure drop, mm wc	55	32	33
Weight of tubes, kg	20,600	12,150	23,500
Heat flux inside tubes, kW/m²	26.3	110	26

Notes: Tubes: 50.8 × 44 mm at 101 mm square pitch. Steam pressure = 39 kg/cm² g, steam generation = 3.39 kg/s. Feed water at 105°C. Gas flow = 27.8 kg/s at 560°C in and 302°C out. Duty = 8 MW. Gas analysis: % volume $CO_2 = 3$, $H_2O = 7$, $N_2 = 75$, $O_2 = 15$. Fouling factors on gas and steam sides—0.0002 m² h °C/kcal (0.000172 m² K/W).

and operate well within its allowable temperature limits. That is why we see finned tube bundles used in all sections of a gas turbine HRSG or in any clean gas heat recovery application. Finned tubes have also been used in gas-fired package boilers for the superheater as well as the evaporator as discussed in Chapter 3. If one had used plain tubes, it would be nearly impossible to build HRSGs with multiple pressure modules. Also one may note that the weight of the finned evaporator is lower due to the fewer number of rows deep. If we try to match the gas pressure drop of finned tubes by manipulating the tube spacing and geometry, the cross section of the plain tube bundle will be larger and also its weight. This is shown in the third column of Table E.1. In spite of a bigger cross section, the length occupied by the evaporator will be much more than the finned tube bundle. At 101 mm longitudinal spacing, the minimum length for plain tube bundle is 5.2 m, while for the finned bundle, it is only 1.3 m.

Heat Transfer Calculations

Sizing or performance evaluation of finned tubes is a tedious process. The determination of heat transfer coefficient itself requires inputs on tube wall and fin tip temperatures and number of rows of tubes deep. Hence, the tube-side as well as the overall heat transfer coefficient has to be evaluated to correct the gas-side coefficient, which implies that this is an iterative process. This is aptly done using a computer program written for the purpose. One has to assume a cross section and the number of rows deep for a given duty and check if that configuration meets the duty or try another tube geometry that meets the duty. This is the general procedure for design as well as off-design performance evaluation of a finned tube bundle. With a well-written program, it takes a few seconds to arrive at an optimum tube bundle size or check the performance of a given bundle for varying gas inlet or fluid conditions.

Determination of Heat Transfer Coefficient h_c

Convective Heat Transfer Coefficient

The ESCOA correlations that were revised in 1993 [2,3] for solid and serrated fins are widely used in the industry (Table E.2). For an inline bundle,

$$h_c = C_1 C_3 C_5 \left\{ \frac{(d+2h)}{d} \right\}^{0.5} \left(\frac{T_g}{T_f} \right)^{0.5} GC_p \left(k/\mu/C_p \right)^{0.67} \tag{E.1}$$

For a staggered bundle,

$$h_c = C_1 C_3 C_5 \left\{ \frac{(d+2h)}{d} \right\}^{0.5} \left(\frac{T_g}{T_f} \right)^{0.25} GC_p \left(k/\mu/C_p \right)^{0.67} \tag{E.2}$$

$$G = \frac{W_g}{\left[N_w L (S_T - A_o) \right]} \tag{E.3}$$

where

$$A_o = d + 2nbh \tag{E.4}$$

A_o is the obstruction area, m^2/m
C_p is the gas specific heat at average temperature, J/kg K
h_c is the convective heat transfer coefficient, $W/m^2 K$
k is the thermal conductivity of gas, W/m K
μ is the viscosity, kg/m s
(Gas properties are computed at the average gas temperature.)
T_g, T_f are absolute temperature of gas and fin, K
G is the gas mass velocity, kg/m^2 s
d is the outer diameter of tube, m
N_w is the tubes/row or number of tubes wide
L is the effective length, m
h is the fin height, m
b is the fin thickness, m
n is the fin density, fins/m
W_g is the gas flow, kg/s
S_T, S_L are transverse and longitudinal pitch, m
S is the spacing between fins = 1/n-b, m

TABLE E.2

ESCOA Revised Correlations: Factors C_1–C_6 for Solid and Serrated Fins in Inline and Staggered Arrangements

Solid fins

Inline

$$C_1 = 0.053[1.45 - (2.9S_L/d)^{-2.3}]Re^{-0.21} \quad C_2 = 0.11 + 1.4Re^{-0.4}$$

$$C_3 = 0.20 + 0.65e^{-0.25h/s} \quad C_4 = 0.08(0.15S_T/d)^{-1.1(h/s)^{0.15}}$$

$$C_5 = 1.1 - (0.75 - 1.5e^{-0.7N_d})e^{-2.0SL/S_T} \quad C_6 = 1.6 - (0.75 - 1.5e^{-0.7N_d})e^{-2.0(SL/S_T)^2}$$

$$J = C_1C_3C_5[(d + 2h)/d]^{0.5}[(t_g + 460)/(t_a + 460)]^{0.5}$$

$$f = C_2C_4C_6[(d + 2h)/d][(t_g + 460)/(t_a + 460)]^l$$

Staggered

$$C_1 = 0.091Re^{-0.25} \quad C_2 = 0.75 + 1.85Re^{-0.3}$$

$$C_3 = 0.35 + 0.65e^{-0.25h/s} \quad C_4 = 0.11(0.05S_T/d)^{-0.7(h/s)^{0.20}}$$

$$C_5 = 0.7 + (0.7 - 0.8e^{-0.15N_d^2})[e^{-1.0SL/S_T}] \quad C_6 = 1.1 + (1.8 - 2.1e^{-0.15N_d^2})e^{-2.0(SL/S_T)}$$
$$- [0.7 - 0.8e^{-0.15N_d^2}]e^{-0.6(SL/S_T)}$$

$$J = C_1C_3C_5[(d + 2h)/d]^{0.5}[(t_g + 460)/(t_a + 460)]^{0.25}$$

$$f = C_2C_4C_6[(d + 2h)/d]^{0.5}[(t_g + 460)/(t_a + 460)]^{-0.25}$$

Serrated fins

Inline

$$C_1 = 0.053[1.45 - (2.9S_L/d)^{-2.3}]Re^{-0.21} \quad C_2 = 0.11 + 1.4Re^{-0.4}$$

$$C_3 = 0.25 + 0.6e^{-0.26h/s} \quad C_4 = 0.08(0.15S_T/d)^{-1.1(h/s)^{0.15}}$$

$$C_5 = 1.1 - (0.75 - 1.5e^{-0.7N_d})e^{-2.0SL/S_T} \quad C_6 = 1.6 - (0.75 - 1.5e^{-0.7N_d})e^{-2.0(SL/S_T)^2}$$

$$f = C_2C_4C_6[(d + 2h)/d][(t_g + 460)/(t_a + 460)]^{0.25}$$

Staggered

$$C_1 = 0.091Re^{-0.25} \quad C_2 = 0.75 + 1.85Re^{-0.3}$$

$$C_3 = 0.35 + 0.65e^{-0.17h/s} \quad C_4 = 0.11(0.05S_T/d)^{-0.7(h/s)^{0.2}}$$

$$C_5 = 0.7 + (0.7 - 0.8e^{-0.15N_d^2})[e^{-1.0SL/S_T}] \quad C_6 = 1.1 + (1.8 - 2.1e^{-0.15N_d^2})e^{-2.0(SL/S_T)}$$
$$- [0.7 - 0.8e^{-0.15N_d^2}]e^{-0.6(SL/S_T)}$$

$$J = C_1C_3C_5[(d + 2h)/d]^{0.5}[(t_g + 460)/(t_a + 460)]^{0.25}$$

$$f = C_2C_4C_6[(d + 2h)/d]^{0.5}[(t_g + 460)/(t_a + 460)]^{-0.25}$$

Source: Fintube Technologies, Tulsa, OK.

Fin Efficiency and Effectiveness

For Both Solid and Serrated Fins

$$\eta = \frac{1-(1-E)A_f}{A_t} \tag{E.5}$$

where A_f, A_t are area of fins and total area per unit length, m²/m.

For Solid Fins

$$A_f = \pi n \times \left(2dh + 2h^2 + bd + 2bh\right) \tag{E.6}$$

$$A_t = A_f + \pi d\left(1 - nb\right) \tag{E.7}$$

Fin efficiency E is given by the following formula using Bessel functions:

$$E = \left[2r_0 / m / \left(r_e^2 - r_0^2\right)\right] \times \frac{\left[I_1\left(mr_e\right)K_1\left(mr_o\right) - K_1\left(mr_e\right)I_1\left(mr_o\right)\right]}{\left[I_0\left(mr_o\right)K_1\left(mr_e\right) + I_1\left(mr_e\right)K_0\left(mr_o\right)\right]} \tag{E.8}$$

where
  Bessel functions I_1, I_o, K_1, K_o are shown in Table E.4
  $r_e = (d + 2h)/2$ and $r_o = d/2$, m

  A simpler formula for E is

$$E = \frac{1}{\left[1 + .33\ m^2 h^2 \left\{(d + 2h)/d\right\}^{0.5}\right]} \tag{E.9}$$

where

$$m = \sqrt{\left(2h_o / K_m / b\right)} \tag{E.10}$$

K_m is the thermal conductivity of fin, W/m K.

For Serrated Fins

$$A_f = \frac{\pi dn\left[2h \times \left(ws + b\right) + b \times ws\right]}{ws} \tag{E.11}$$

$$A_t = A_f + \pi d\left(1 - nb\right) \tag{E.12}$$

$$E = \frac{\tanh{(mh)}}{mh} \tag{E.13}$$

where

$$m = \sqrt{\left[2h_o\left(b+ws\right)/K_m/b/ws\right]} \tag{E.14}$$

ws is the width of serration, m.

Nonluminous Heat Transfer Coefficient

With finned tubes, due to the large surface area for a given volume, the beam length will be much smaller than a comparable plain tube bundle at the same tube spacing, and hence, the nonluminous heat transfer coefficient is not high. The beam length may be approximated as

$$L = \left(d+2h\right)\left[\frac{\left(S_T \times S_L\right)}{\left(d+2h\right)^2} - 0.85\right] \tag{E.15}$$

Using the procedure discussed in Appendix D, the gas emissivity and the nonluminous heat transfer coefficient h_n may be estimated. Generally, this is quite small compared to plain tubes due to the larger surface area in a given volume.

Gas Pressure Drop across Finned Tube Bundles

$$\Delta P_g = \frac{0.205\left(f+a\right)G^2 N_d}{\rho_g} \tag{E.16}$$

where
ΔP_g is the gas pressure drop, mm wc
N_d is the number of rows deep
ρ_g is the density of gas at average gas temperature, kg/m³

For staggered arrangement,

$$f = \text{friction factor} = C_2 C_4 C_6 \left[\frac{\left(d+2h\right)}{d}\right]^{0.5} \times \left(\frac{T_g}{T_f}\right)^{-0.25} \tag{E.18}$$

For inline arrangement,

$$f = C_2 C_4 C_6 \left[\frac{\left(d+2h\right)}{d}\right] \times \left(\frac{T_g}{T_f}\right)^{0.25} \tag{E.19}$$

$$a = \frac{\left(1+B^2\right)\times\left(t_{g2}-t_{g1}\right)}{\left[4N_{dx}\left(t_g+273\right)\right]}$$ (E.20)

$$B = \left(\frac{\text{Free gas area}}{\text{Total flow area}}\right) = \left[\frac{\left(S_T-A_o\right)}{S_T}\right]^2$$ (E.21)

Factors C_1–C_6 are given in Table E.2 for solid and serrated fins. These are the revised ESCOA correlations.

Tube Wall and Fin Tip Temperatures

For solid fins, the relation between the tube wall temperature t_b and fin tip temperature t_f is given by

$$\frac{\left(t_g-t_f\right)}{\left(t_g-t_b\right)} = \frac{\left[K_1\left(mr_e\right)\times I_o\left(mr_e\right)+I_1\left(mr_e\right)\times K_o\left(mr_e\right)\right]}{\left[K_1\left(mr_e\right)\times I_o\left(mr_o\right)+K_o\left(mr_o\right)\times I_1\left(mr_e\right)\right]}$$ (E.22)

For serrated fins,

$$\frac{\left(t_g-t_f\right)}{\left(t_g-t_b\right)} = \frac{1}{\cosh\left(mb\right)}$$ (E.23)

A good estimate of t_f can be obtained for either type of fin using

$$t_f = t_b +\left(t_g-t_b\right)\times\left(1.42-1.4E\right)$$ (E.24)

The fin base temperature t_b is estimated as follows:

$$t_b = t_i +q_o\left(R_3+R_4+R_5\right)$$ (E.25)

where R_3, R_4, and R_5 are the resistances of the inside fluid, fouling layer, and tube wall, respectively.

$$\text{Heat flux } q_o = U_o\left(t_g-t_i\right), \quad W/m^2$$ (E.26)

(Note that depending on the location in the gas path, one may obtain average heat flux or maximum heat flux by using appropriate t_g and t_i values.)
    The significance of various resistances and U_o was shown in Appendix A.

$$R_3 = \frac{\left(A_t/A_i\right)}{h_i}.$$

$$R_4 = \frac{ff_i \times A_t}{A_i}.$$

$$R_5 = \left(\frac{A_t}{A_w}\right)\left(\frac{d}{2K_m}\right)\ln\left(\frac{d}{d_i}\right)$$

where $A_i = \pi d_i$.

TABLE E.3

F Factors

Material	F
304, 316, 221 alloys	1.024
Carbon steel	1
409, 410, 430 alloys	0.978
Nickel 200	1.133
Inconel 600, 625	1.073
Inconel 800	1.013
Inconel 825	1.038
Hastelloy B	1.179

Note that these correlations have a margin of error, and hence, based on experience and field measurements, suitable correction factors are applied. The author's experience is that these correlations are conservative and underpredict actual h_c (obtained in the field) by about 10%.

Weight of Finned Tubes

$$\text{Solid fins}: w = 10.68 \times Fbn(d+h) \times (h+0.03) \tag{E.27}$$

$$\text{Serrated fins}: w = 10.68 \times Fbnd \times (h+0.12) \tag{E.28}$$

w is the weight in lb/ft. Factor F is given in Table E.3. Note that d, h, b are in in. in (E.27) and (E.28) and n is fins/in.

Example E.1

A 50.8 × 44 mm solid finned superheater having a fin density of 78.7 fins/m of height 12.7 mm and thickness 1.52 mm is arranged in inline fashion at 101.6 mm square pitch. The supplier has used 18 tubes/row and 6 rows deep, 3.1 m long tubes with 9 streams. Gas flow is 100,000 kg/h with the following analysis by volume% $CO_2 = 3$, $H_2O = 7$, $N_2 = 75$, $O_2 = 15$. The bundle is designed with inlet and exit gas temperatures of 550°C and 479°C; steam flow is 25,000 kg/h (6.94 kg/s) at 369°C at 49 kg/cm² g (4,804 kPa) at superheater exit, and saturated steam enters the superheater in counter-flow fashion at 265°C. Surface area is 188 m². Check the suitability of the design.

Solution

Flue gas data from Appendix F at the average gas temperature of 514°C is

$$C_{pg} = 0.2755 \text{ kcal/kg °C}, \ \mu_g = 0.127 \text{ kg/m h}, \ k_g = 0.0468 \text{ kcal/m h °C}$$

$$A_o = 0.0508 + 2 \times 78.7 \times 0.00152 \times 0.0127 = 0.0538 \text{ m}^2/\text{m}$$

$$A_f = \pi \times 78.7 \times (2 \times .0508 \times .0127 + 2 \times .0127 \times 0.0127 + .00152 \times 0.0508 + 2 \times .00152 \times .0127) = 0.427 \text{ m}^2/\text{m}$$

$$A_t = 0.427 + \pi \times .0508 \times (1 - 78.7 \times .00152) = 0.567 \text{ m}^2/\text{m}$$

$$S = \text{clearance between fins} = (1/78.7) - .00152 = 0.0112 \text{ m}$$

Convective Heat Transfer Coefficient

$$G = W_g/[N_wL(S_T - A_o)] = 100,000/[18 \times 3.1 \times (0.1016 - 0.0538)] = 37491 \text{ kg/m}^2 \text{ h}$$
$$= 10.414 \text{ kg/m}^2 \text{ s}$$

$$Re = Gd/\mu = 37,491 \times .0508/0.127 = 14,996$$

$$C_1 = 0.053 \times [1.45 - 2.9 \times (0.1016/0.0508)^{-2.3}] \times 14,996^{-0.21} = 0.006059$$

$$C_3 = 0.2 + 0.65e^{-.25 \times 12.7/11.2} = 0.6900$$

$$C_5 = 1.1 - [(0.75 - 1.5e^{-0.7 \times 6})]e^{(-2SL/ST)} = 1.0016$$

Assume that average fin temperature is 400°C. After performing a round of calculations, we will be able to fine-tune this.

$$h_c = 0.006059 \times 0.690 \times 1.0016 \times 37,491 \times 0.2755 \times (0.0468/0.127/0.2755)^{0.67} \times [(514 + 273)/$$
$$(400 + 273)]^{0.5} \times [(0.0508 + 0.0254)/0.0508]^{0.5} = 69.6 \text{ kcal/m}^2 \text{ h °C } (80.93 \text{ W/m}^2 \text{ K})$$

Nonluminous Heat Transfer Coefficient

Beam length $L_b = (d + 2h)[(S_T \times S_L)/(d + 2h)^2 - 0.85] = (0.0508 + 0.0254) (0.1016 \times 0.1016/0.0762^2 - 0.85) = 0.0707$ m. Note that beam length is smaller with finned tubes, and hence, the nonluminous heat transfer coefficient will be small for similar gas conditions.

From Appendix D, for $p_c = 0.03$, $p_w = 0.07$, $L_b = 0.0707$,

$$K = (0.8 + 1.6 \times .07) (1 - .38 \times 0.787) \times 0.1/(0.1 \times 0.0707)^{0.5} = 0.76$$

$$\epsilon_g = 0.9 \times (1 - e^{-0.76 \times 0.0707}) = 0.047. \text{ From Appendix D,}$$

$$h_n = 5.67 \times 10^{-8} \times 0.047 \times (787^4 - 673^4)/(787 - 673) = 4.17 \text{ W/m}^2 \text{ K} = 3.58 \text{ kcal/m}^2 \text{ h °C}$$

(Average gas temperature is 514°C, and average fin temperature is assumed as 400°C.)

$$h_o = 69.55 + 3.58 = 73.13 \text{ kcal/m}^2 \text{ h °C } (85 \text{ W/m}^2 \text{ K}) (14.96 \text{ Btu/ft}^2 \text{ h °F})$$

$$m = [2 \times (73.13)/30/0.00152]^{0.5} = 56.6 \text{ 1/m (fin thermal conductivity of 30 kcal/m h °C}$$
$$(25.8 \text{ W/m K}) (20.15 \text{ Btu/ft h °F}) \text{ is used.}$$

$$mh = 56.6 \times 0.0127 = 0.719$$

Using the simplified approach,

$$E = 1/[1 + 0.33 \times 0.719^2 \times (0.0762/0.0508)^{0.5}] = 0.827$$

Effectiveness $\eta = 1 - [(1 - 0.827) \times 0.427/0.567] = 0.87$
Using the Bessel functions (Table E.4),

$$r_e = 0.0381, r_o = 0.0254, mr_e = 56.6 \times 0.0381 = 2.156, mr_o = 56.6 \times 0.0254 = 1.437$$

$$E = [2r_o/m/(r_e^2 - r_o^2)] \times [I_1(mr_e)K_1(mr_o) - K_1(mr_e)I1(mr_o)]/[I_o(mr_o)K_1(mr_e) + I_1(mr_e)K_o(mr_o)]$$

$$I_{1\,R4}(2.156) = 1.843, K_1(2.156) = 0.115, K_1(1.437) = 0.306, I_1(1.437) = 0.849, I_o(1.437)$$
$$= 1.516, K_o(1.437) = 0.234$$

TABLE E.4

I_o, I_1, K_o, K_1 Values for Various Arguments

X	$I_0(X)$	$I_1(X)$	$K_0(X)$	$K_1(X)$
0	1.0	0	8	8
0.1	1.002	0.05	2.427	9.854
0.2	1.010	0.10	1.753	4.776
0.3	1.023	0.153	1.372	3.056
0.4	1.040	0.204	1.114	2.184
0.5	1.063	0.258	0.924	1.656
0.6	1.092	0.314	0.778	1.303
0.7	1.126	0.372	0.66	1.05
0.8	1.166	0.433	0.565	0.862
0.9	1.213	0.497	0.487	0.716
1.0	1.266	0.565	0.421	0.602
1.2	1.394	0.715	0.318	0.434
1.4	1.553	0.886	0.244	0.321
1.6	1.75	1.085	0.188	0.241
1.8	1.99	1.317	0.146	0.183
2.0	2.28	1.591	0.114	0.140
2.2	2.629	1.914	0.0893	0.108
2.4	3.049	2.298	0.0702	0.0837
2.6	3.553	2.755	0.554	0.0653
2.8	4.157	3.301	0.0438	0.0511
3.0	4.881	3.953	0.0347	0.0402
3.2	5.747	4.734	0.0276	0.0316
3.4	6.785	5.670	0.0220	0.0250
3.6	8.028	6.793	0.0175	0.0198
3.8	9.517	8.140	0.0140	0.0157
4.0	11.30	9.759	0.0112	0.0125
4.2	13.44	11.70	0.0089	0.0099
4.4	16.01	14.04	0.0071	0.0079
4.6	19.09	16.86	0.0057	0.0063
4.8	22.79	20.25	0.0046	0.0050
5.0	27.24	24.34	0.0037	0.0040

Source: Ganapathy, V., *Applied Heat Transfer*, Pennwell Books, Tulsa, OK, 1982, p501.

$E = [2 \times .0254/56.6/(0.0381^2 - 0.0254^2)] \times [(1.843 \times 0.306 - 0.115 \times 0.849)/(1.516 \times 0.115 + 1.843 \times 0.234)] = 0.856$

$$\eta = 1 - [1 - (1.856) \times 0.427/0.567] = 0.89$$

Ratio of external to internal area $A_t/A_i = 0.567/\pi/0.044 = 4.1$

Tube-side coefficient h_i from Appendix B: $C = 320$ from Table B.4 at 5000 kPa (51 kg/cm²) and 318°C.

Mass flow/tube = 25,000/9 = 2778 kg/h = 0.772 kg/s (9 streams used)

$$h_i = \frac{0.0278 \times 320 \times 0.772^{0.8}}{0.044^{1.8}} = 2000 \text{ W / m}^2 \text{ K} \left(1720 \text{ kcal/m}^2 \text{ h } ^\circ\text{C}\right)\left(351.7 \text{ Btu/ft}^2 \text{ h } ^\circ\text{F}\right)$$

$1/U_o = 1/(0.89 \times 73.13) + 0.0002 + (0.0508/2 \times 30) \times \ln(0.0508/0.044) \times (0.567/\pi/0.044)$
   $+ 0.0002 \times 4.1 + 4.1/1720 = 0.01536 + 0.0002 + 0.0005 + 0.00082 + 0.00238$
   $= 0.01926$ or $U_o = 51.9$ kcal/m²h °C (60.35 W/m²K) (10.61 Btu/ft²h °F)

Superheater Duty

$Q = 100{,}000 \times 0.99 \times 0.2755 \times (550 - 479) = 1.936$ MM kcal/h $(2.25$ MW$)(7.682$ MM Btu/h$)$

Enthalpy absorbed by steam = $1.936 \times 10^6/25{,}000 = 77.45$ kcal/kg (139.4 Btu/lb) (324.3 kJ/kg).
   Exit steam enthalpy = 744.65 kcal/kg or temperature = 369°C from steam tables.

$$\text{LMTD} = [(550 - 369) - (479 - 265)]/\ln[(550 - 369)/(479 - 265)] = 197°C.$$

Surface area required = $1.936 \times 10^6/(197 \times 51.9) = 189$ m² and matches the value shown by the boiler supplier. Hence, the design seems reasonable.
   Estimation of heat flux $q_o = 51.9 \times (550 - 369) = 9394$ kcal/m²h (10.92 kW/m²) (3464 Btu/ft²h) (the U_o value should have computed at the gas inlet temperature and steam exit temperature conditions and will show a higher value. However, here we are interested in showing the methodology for computing heat flux and the tube and fin tip temperatures.)
   Fin base or maximum tube wall temperature $t_b = 369 + 9394 \times (0.00238 + 0.00082 + 0.0005) = 404°C$. Suitable correction factors or margins may be used for tube and fin material selection.
   One may use Equation E.22 to estimate the fin tip temperature.

$(t_g - t_f)/(t_g - t_b) = [0.115 \times 2.28 + 1.591 \times 0.114]/[0.14 \times 1.513 + 0.2625 \times 1.591] = 0.795$

$(550 - t_f)/(550 - 404) = 0.795$ or $t_f = 433°C$. Average fin temperature at the hot end = $(404 + 433)/2 = 418C$. Similarly, at the cold end, one has to check the average fin temperature and use that value for t_a in the equation for h_c. However, the difference will be small and hence neglected in this exercise.
   Using (E.24), $t_f = 405 + (550 - 405) \times (1.42 - 1.4 \times .856) = 432°C$
   If local heat transfer coefficients were used, t_f and t_b values would be higher, and a computer program would help perform these calculations more accurately. Note that the tube-side coefficient at higher steam temperature will also be lower increasing the tube wall and fin tip temperatures. A computer program will be helpful in performing these iterations.

Gas Pressure Drop

$$C_2 = 0.11 + 1.4 \text{Re}^{-0.4} = 0.14.$$

$$C_4 = 0.08(0.15 \times 0.1016 / 0.0508)^{(-1.1 \times 0.0127/.0112)^{0.15}} = 0.03084$$

$$C_6 = 1.6 - (0.75 - 1.5e^{-4.2})e^{-0.2} = 1.0044$$

$f = 0.14 \times 0.0384 \times 1.0044 \times [(0.0508 + 2 \times 0.0127)/0.0508] \times [(514 + 273)/(673)]^{0.25} = 0.0676$

$$B = \left[(S_T - A_o)/S_T\right]^2 = \left[(0.1016 - 0.0538)/0.1016\right]^2 = 0.221$$

$a = (1 + B^2)(t_{g1} - t_{g2})/(4N_d\,T_g) = (1 + 0.221^2)\,(479 - 550)/(4 \times 6 \times 787) = -0.00394$

Density at an average gas temperature of 514°C is $\rho_g = 12.17 \, MW/T_g = 12.17 \times 28.38/787 = 0.4388 \, kg/m^3$

$$\Delta P_g = \frac{0.205 \times (0.0676 - 0.00394) \times 10.414^2 \times 6}{0.4388} = 19.3 \, mm \, wc.$$

Steam-Side Pressure Drop

Let us estimate the total developed length. From Appendix B, $L_e = 3.1 \times 6 \times 18/9 + 32 \times .044 \times 18/9$ (9 streams) = 54 m. (Standard bend data from Table B.7 were used). Friction factor is 0.02. From steam tables, the specific volume of steam at 50 kg/cm² g and 315°C is obtained as 0.04868 m³/kg.

Hence, using the equation from Table B.1, $\Delta P = 0.6375 \times 10^{-12} \times 0.02 \times 54 \times (25,000/9)^2 \times 0.04868/0.044^5 = 1.57 \, kg/cm^2$ (22.3 psi) (154 kPa).

$$\text{Inlet velocity of steam} = \frac{3.46 \times 10^{-4} \times (25000/9) \times 0.0394}{0.044^2} = 19.6 \, m/s$$

Exit velocity = $19.6 \times 0.0554/0.0394 = 27.5$ m/s (specific volumes of steam at inlet and exit are 0.0394 and 0.0554 m³/kg, respectively)

Inlet velocity head = $19.6 \times 19.6/(2 \times 9.8 \times 0.0394) = 497 \, kg/cm^2$. Loss = 0.5×497 (half the velocity head) = 248 kg/m² = 0.025 kg/cm².

Exit velocity head = $27.5 \times 27.5/(2 \times 9.8 \times 0.0554) = 696 \, kg/m^2$. Exit loss = 0.07 kg/cm² (one velocity head).

Total pressure drop = 1.57 + 0.025 + 0.07 = 1.67 kg/cm². It may be seen that these calculations are tedious. Hence, a computer program developed for the purpose is used.

High Fin Density in Superheaters Is Not Recommended

Many HRSG suppliers, in order to reduce the cost of fabrication, use high-density fins in superheaters; high fin density reduces the number of rows along gas flow path for the same cross section of the HRSG, and hence, some labor may be saved; however, it is not a good practice when gas and steam temperatures are high. When the tube-side heat transfer coefficient is low as in superheaters (compared to evaporators or economizers), high fin density not only increases the heat flux inside the tubes but also increases the tube wall temperature and gas pressure drop. This increases the operating cost as well as cost of materials or reduces the life of the superheater. It may be fine with low steam temperature applications where tube wall temperature is not in severe creep range. This point is illustrated in the following example [5,6].

Example E.2

A simple finned tube superheater design with T11 material and alloy steel fin is presented with several solid fin density options. Tube size is 50.8 × 3.1 mm. There are 24 tubes/row, and effective length is 3.1 m. Fin height is 15 mm, and thickness is 1.52 mm. Fin density is varied to study the superheater performance for the same duty. The tubes are in staggered fashion and in counter-flow direction. Steam pressure is 35 kg/cm² a (3431 kPa) at the superheater inlet.

TABLE E.5

Summary of Results on Finned Tubes in Superheater

Fins/m	78	118	157	216
Gas exit temperature, °C	412	413	412	412
Duty, MM kcal/h	6.80	6.78	6.80	6.79
Gas press. drop, mm wc	30	32	35	46
Steam exit temp., °C	425	424	425	424
Heat flux, kcal/m² h	69,124	82,900	92,250	106,430
Tube wall temp., °C	510	524	535	54.7
U, kcal/m² h °C	54.1	46.4	40.8	32.7
Surface area, m²	582	673	771	983
No rows deep	12	10	9	8

Note: Gas flow = 80,000 kg/h, 700°C in, steam flow = 60,000 kg/h at 244°C in.

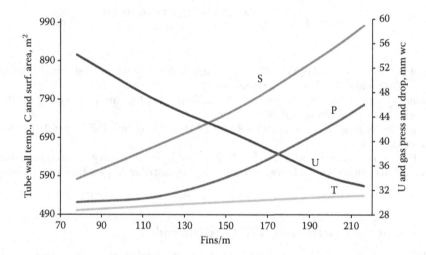

FIGURE E.2

Effect of fin geometry on surface area (S), gas pressure drop (P), overall heat transfer coefficient (U), and tube wall temperature (T).

Solution

The results obtained from a computer program are shown in the Table E.5. Figure E.2 shows the performance as a function of fin density. One can see that due to the higher heat flux with higher fin density, the tube wall temperature is much higher, 547°C versus 510°C. Remember that all the options are for the same duty. Though T11 material may be adequate for all the options, the life of the superheater is severely reduced when the tube wall temperature increases. The gas pressure drop is also higher even though the number of rows deep is fewer.

How Is Life of Superheater Affected by High Tube Wall Temperatures?

Many plant engineers and consultants ignore the fact that though the material selected may be acceptable for the temperature in consideration, the life of the tubes can be lower with higher tube wall temperature. The life of the superheater tubes is often estimated using Larsen Miller chart (Figure E.3).

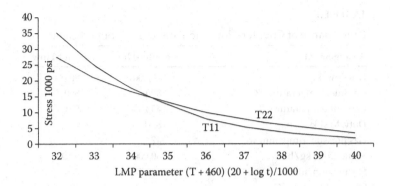

FIGURE E.3
Larsen Miller parameters for T11 and T22 materials.

Let the maximum operating pressure be 39 kg/cm² g (555 psig).The stress inside the tubes due to the steam pressure = Pr/t = 555 × 1/0.12 = 4625 psig. The LMP factor from Figure E.3 is 37,300. Using the equation, LMP = (460 + T_w) × (20 + log t), where t is the number of hours to failure and T_w is the tube wall temperature, °F.

With 78 fins/m, the operating tube wall temperature = 510°C. Use 15°C margin. T_w = 977°F = 1437°R. Hence, (20 + log t) = 37,300/1437 = 25.95 or t = 912,000 h = 114 years!

When the fin density is 216 fins/m, the maximum tube wall temperature is, say, 555°C (1031°F). Then, (20 + log t) = 37,300/(1,031 + 460) = 25.0 or t = 100,000 h or less than 12.5 years, using 8,000 h of operation in a year. One can see the concern with using higher than required fin density in a superheater. These may be estimates but are given to show the effect of tube wall temperature on the life of tubes. Even a 10°C change can bring about a large decrease in the operating life of the tubes. If the temperature of tubes had been 545°C maximum, then (20 + log t) = 37,300/1,473 = 25.32 or t = 210,100 h = 26 years. That is, a 10°C decrease has doubled the life of the tubes! Hence, boiler designers should not flippantly select the fin geometry in superheaters, and plant engineers should understand the implications of the high fin density and challenge the boiler supplier, who may not even furnish the tube wall temperature calculations at the time of purchase unless asked!

There is another important point to be made here. If the tube wall temperature data and the overall heat transfer coefficient U are not shown in the table, unsuspecting customers or plant engineers who are purchasing the boiler would assume that they are getting a lot of surface area with the high fin density design, 908 m² versus 582 m², and may even reject the option with the 78 fins/m design assuming it will not be adequate for the duty! It will be shown later that U decreases as the fin density and the external surface area increases. Plant engineers should look at the product of U × A (overall heat transfer coefficient and surface area) rather than A in isolation for a given duty!

The heat flux inside tubes is an important parameter. For example, for the 216 fins/m case, the heat flux inside the tubes = 32.7 × (700 − 424) × 983/(π × 0.0446 × 24 × 3.1 × 8) = 106,430 kcal/m² h. Table E.5 shows the summary of results with varying fin densities.

Example E.3

Plant engineers are noticing high tube wall temperatures in a superheater. The design and performance details of the superheater supplied are shown in Table E.6 in column 1. Exit steam pressure is 38 kg/cm² g. The present design is counter-flow. Plant engineers want to know what will happen to the performance with same gas and steam inlet

TABLE E.6

Performance of Counter-Flow and Parallel-Flow Superheaters

Arrangement	Counter-Flow	Parallel-Flow
Gas flow, kg/h	80,000	80,000
Gas inlet temperature, °C	800	800
Gas exit temperature, °C	453	501
Duty, MM kcal/h	8.31	7.21
Gas pressure drop, mm wc	47	48
Steam flow, kg/h	60,000	60,000
Steam temp. in, °C	249	249
Steam exit temp., °C	473	439
Max heat flux, kcal/m²h	129,168	218,000
Estimated max tube wall temp., °C	600	475
U, kcal/m²h °C	35.5	35.8
Surface area, m²	908	908
No rows deep	8	8
Steam pressure drop, kg/cm²	1.3	1.2
Steam-side coefficient, kcal/m²h °C	1,333	1,363

conditions but with parallel-flow configuration. Flue gas is from combustion of natural gas.% volume $CO_2 = 8$, $H_2O = 18$, $N_2 = 71.5$, $O_2 = 2.5$. Tube size: 50.8 × 44.6,24 tubes/row, 8 rows deep, 3.1m long, 101.8 mm spacing, staggered.

Solution

These are typical problems in an operating plant. One has access to lot of valuable field data and should know how to use the information to predict the performance at other operating cases. Here, since we have the data on gas and steam temperatures and hence the duty, we can estimate the U value. Surface area is 908 m². Ratio of external to internal area = 908/(3.14 × 0.0446 × 24 × 8 × 3.1) = 10.9

Q = 60,000 × (808.7 − 669) = 8.38 × 10⁶ kcal/h. LMTD = [(800 − 473) − (453 − 249)]/ ln[800 − 473)/(453 − 249)] = 261°C. Hence, U = 8.38 × 10⁶/(261 × 908) = 35.3 kcal/m²h °C. We can use these data for starters.

There are two ways to solve this problem without doing elaborate calculations. The U value in both cases will remain nearly the same, though slightly higher in the parallel-flow case due to the higher average gas temperature.

Method 1

This is the traditional method. We assume a certain steam temperature and compute the duty and the exit gas temperature and LMTD and then check the transferred duty. Assume that in the parallel-flow case, the steam temperature is 445°C. The enthalpy of steam at 38 kg/cm² g and at 450°C is 793.4 kcal/kg from steam tables. Using the same steam-side pressure drop of 1.3 kg/cm², the saturation temperature at 39.5 kg/cm² g is 249°C and enthalpy is 669 kcal/kg.

Q_a = energy absorbed by steam is 60,000 × (793.4 − 669) = 7.46 × 10⁶ kcal/h. Average gas specific in previous case is 0.3057 kcal/kg from Appendix F. Gas temperature drop = 7.46 × 10⁶/(80,000 × 0.99 × 0.3057) = 308°C. Hence, exit gas temperature is 800°C − 308°C = 492°C.

LMTD = [(800 − 249) − (492 − 445)]/ln[(800 − 249)/(492 − 445)] = 205°C.

Q_t= energy transferred = UAΔT = 35.3 × 908 × 205 = 6.57 MM kcal/h. Hence, the assumption we made is incorrect as we cannot transfer the assumed duty with this surface area, and actual duty will be lesser than 7.46 MM kcal/h.

Let us try 435°C as steam temperature. $Q_a = (787.5 - 669) \times 60,000 = 7.1 \times 10^6$ kcal/h. Gas temperature drop = 234°C. Hence, exit gas temperature = 800°C − 294°C = 506°C.

LMTD = 234°C. $Q_t = 35.3 \times 908 \times 234 = 7.5$ MM kcal/h. Hence, the assumed steam temperature is lower than what it should be. More iterations such as these are required, but assuming 439°C, we get $Q_a = 60,000 \times (789.7 - 669) = 7.24$ MM kcal/h. Gas temperature drop = 299°C. Exit gas temperature is 501°C. LMTD = 224°C. $Q_t = 908 \times 35.3.2 \times 224 = 7.18$ MM kcal/h. Now Q_a and Q_t agree closely. One should now perform detailed calculations as shown in Example E.1 to arrive at the duty more accurately. The U value will be slightly higher due to the higher average gas temperature and can be shown to be 35.8 kcal/m²h °C. Also the steam-side coefficient will be slightly higher due to the lower steam temperature (see Appendix B).

Method 2

The NTU method discussed in Appendix A is another way of handling this problem. The specific heat of steam may be estimated as $C_{ps} = (789.7 - 669)/(439 - 249) = 0.635$ kcal/kg °C using the steam temperatures obtained earlier. Let us use 0.3061 as specific heat of gas from Appendix F at 650°C for natural gas products of combustion.

$$Q_t = \epsilon C_{min} \left(t_{g1} - t_{s1} \right)$$

$\epsilon = [1 - \exp \{-NTU (1 + C)\}]/(1 + C)$ for parallel-flow exchanger.

$C_{min} = 80,000 \times .99 \times 0.3061 = 24,243$. $C_{max} = 60,000 \times 0.635 = 38,100$ C = 24,243/38,580 = 0.636. $(1 - C) = 0.364$. NTU = $UA/C_{min} = 35.8 \times 908/24,243 = 1.34$

Hence, $\epsilon = [1 - \exp(-1.34 \times 1.636)]/1.6284 = 0.543$.

Hence, $Q_t = 0.543 \times 24,243 \times (800 - 249) = 7.25$ MM kcal/h

Exit gas temperature = $800 - 7.25 \times 10^6/(80,000 \times 0.3061 \times 0.99) = 501$°C

Exit steam temperature = $249 + 7.25 \times 10^6/(60,000 \times 0.636) = 439$°C.

The predicted performance is shown in column 2. The other values may be calculated as shown in Example E.1.

In counter-flow arrangement, the steam exit end will always see the higher tube wall temperature. However, in parallel-flow arrangement, one has to calculate the tube wall temperatures at both the steam inlet end and exit end.

Heat flux inside the tube at the exit in parallel-flow mode = $35.8 \times (501 - 439) \times 10.9 = 24193$ kcal/m² h. Inner wall temperature = $439 + 24,193 (0.0002 + 1/1,363) = 462$°C.

At the gas inlet end, heat flux = $35.8 \times (800 - 249) \times 10.9 = 215,011$ kcal/m²h. Inner wall temperature = $249 + 215,011/(0.0002 + 1/1,363) = 450$°C.

With counter-flow arrangement, the steam exit end will see the hottest temperature. Heat flux = $35.3 \times (800 - 472) \times 10.9 = 125,820$ kcal/m² h. Tube wall temperature = 473 + 125,820/(0.0002 + 1/1,333) = 593°C. If we consider the actual heat transfer coefficients at the inlet and exit conditions, the heat flux will be higher and hence the tube wall temperatures. Also, some margins over calculated values are used in practice. These are shown in the Table E.6. One can see that the wall temperatures are much lower with parallel-flow arrangement.

How Does Gas-Side Heat Transfer Coefficient Vary with Fin Density?

With finned tubes, the surface area variations with fin geometry can be significant. The following data were used in generating Figure E.4 for a boiler evaporator. Gas flow is 100,000 kg/h at 560°C in and 300°C out, 51 × 44 mm tubes, serrated fins of height 19 mm, 1.5 mm thick and 4 mm serration, with fin density varying from 78 to 236 fins/m and staggered tubes at 101 mm transverse and longitudinal spacing.

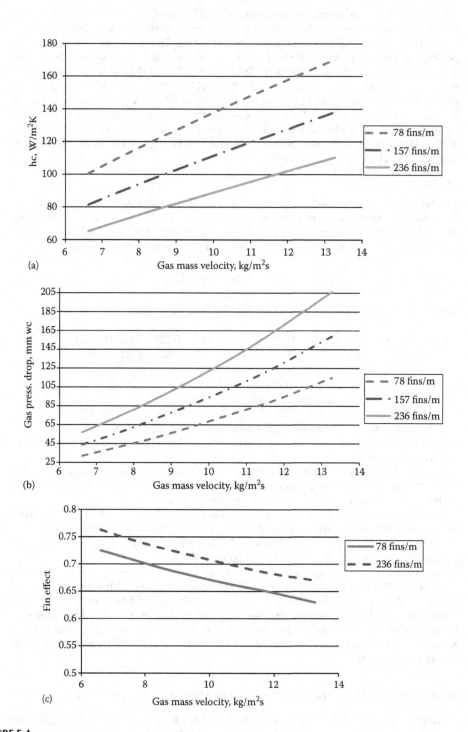

FIGURE E.4
Effect of gas mass velocity and fins/m on (a) convective heat transfer coefficient, (b) gas pressure drop, and (c) fin effectiveness.

Figure E.4a through c show the gas pressure drop, convective heat transfer coefficient, and fin effectiveness as a function of gas mass velocity for various fin densities ranging from 78 to 236 fins/m. It is seen that for a given gas mass velocity,

1. The heat transfer coefficient increases as the fin density decreases
2. The gas-side pressure drop for the same number of rows deep increases as the fin density increases
3. Fin effectiveness is slightly higher for the higher fin density; however, the U value will be lower as will be shown later

Hence, one may see a wide variation in surface areas for a given duty depending on fin geometry used and should not be carried away by a large surface area in finned tube components as ηh_c varies and hence the U value, which is close to h_c.

Effect of h_i on Fin Selection

The tube-side coefficient plays a dominant role in the selection of fin geometry. If the tube-side heat transfer coefficient is small, then it does not pay to use a large fin surface on the tube to take the advantage of finning. For example, in a tubular air heater, both the gas-side and air-side heat transfer coefficients are comparable, and hence, it does not make sense to use fins on the tubes. In a boiler superheater, the tube-side coefficient is reasonably high but not as much as that in the economizer or evaporator. Hence, we should use low fin density on a superheater. Using 5 or 6 fins/in. does not make sense on a superheater as it unnecessarily increases the heat flux inside the tubes and increases the tube wall temperature and reduces the superheater life as shown earlier. In an economizer or evaporator, the tube-side coefficient is very high compared to that on the gas side, and hence, we can afford to use 5 or even 6 fins/in.

The following example shows why the finning is not effective when the tube-side coefficient is small.

Example E.4

Over a tube bundle, 150,000 kg/h of flue gas from the combustion of natural gas at 600°C flows. The tube-side coefficient is 150 kcal/m² h °C (174 W/m² K) in one case and 7000 kcal/m²h °C (8140 W/m² K) in another case. For both cases, study the performance using fins of density 216 fins/m versus 78 fins/m. Fin height = 19 mm, serration = 4, thickness = 1.5 mm, staggered tubes.

The following conclusions may be drawn (Table E.7):

1. When the tube-side coefficient is very low, it is not productive to use a high fin density. The ratio of U_i values between 216 and 78 fins/m (tube-side coefficient) is only 1.19 with $h_i = 150$ kcal/m²h °C when we have used 2.12 times the surface area. That is, if we have 2.12 times increase in surface area, we should get at least some increase in U_i close to 2.0. With higher tube-side coefficient, for the surface area increase ratio of 1.46, we get about 1.71 times increase in U_i, which is much better. That is, the use of high fin density is justified when tube-side coefficient is high. Else, it is counterproductive.

TABLE E.7

Effect of Tube-Side Coefficient on Finned Bundle Performance

Item	Case 1	Case 2	Case 3	Case 4
Tube-side coefficient	7000	7000	150	150
Duty, MM kcal/h (MW)	11.64 (13.53)	11.50	11.55	11.5
Gas press. drop, mm wc	43	31	220	106
Tube wall temp., °C	344	305	550	513
Number of rows deep	6	10	28	32
U_o, kcal/m²h °C	40.1	56.9	8.45	17.8
Heat flux inside tubes	175,235	117,235	42,265	36,675
Fin density, fins/m	216	78	216	78
U_i, kcal/m²h °C	571	334	120.4	104.4
Ratio U_i	1.71		1.19	
Ratio ΔP_g	1.39		2.07	
Surface area, m²	1,573	1,080	7,340	3,456
Ratio of surface area	1.456		2.12	

Note: 36 tubes/row, 3.65 m length, $S_T = S_L = 101$ mm. Heat flux in kcal/m²h.

2. The gas pressure drop increases by only 1.39 times when h_i is 7000 kcal/m²h °C, compared to 2.07 times when we have $h_i = 150$ kcal/m²h °C. The gas pressure drop increase is too much.
3. The tube wall temperature is also very high when we use fins with $h_i = 150$ kcal/m²h °C, which is obvious as the temperature drop across the low tube-side film is much higher.

Tube-side film temperature drop with $h_i = 150$ kcal/m² h °C and fins/m of 216 = 42,265/150 = 282°C, while with $h_i = 7,000$ kcal/m² h °C, the tube-side film drop = 175,235/7,000 = 25°C only. *Thus, tubes also run much hotter when high fin densities are used in low tube-side heat transfer coefficient cases.*

The bottom line is as follows: Avoid fins when tube-side coefficients are very low. In superheaters where h_i can be in the range of 1200–1600 kcal/m²h °C, use 78–119 fins/m and not more; otherwise, the heat flux inside the tubes will shoot up, increasing the tube wall temperature and decreasing the life of the superheater. The author has seen several designs of superheaters using 5 fins/in. or even 6. This is a very poor design. Some boiler suppliers do this either without the knowledge of heat flux and tube wall temperatures (whose ill effects will be seen after the warranty period usually!) or with an eye on reducing the number of tubes deep, which reduces the manufacturing costs. The end user is stuck with such poor designs as the life is reduced when superheater tubes operate continuously at a higher temperature.

Effect of Fin Geometry on Heat Transfer Coefficients

When the tube-side heat transfer coefficient is large as in evaporator or economizer, then a high density makes sense. Here is an example of an evaporator designed with different fin densities. The tube length and spacing were adjusted slightly to obtain the same duty and gas pressure drop.

TABLE E.8

Effect of Fin Geometry in Evaporator

Item	Case 1	Case 2	Case 3	Case 4
Gas temperature in, °C	600	600	600	600
Gas temperature out, °C	281	281	281	281
Duty, MM kcal/h (MW)	8.53 (9.92)	8.54	8.53	8.52
Gas pressure drop, mm wc	50	48	50	50
Steam flow, kg/h	15,230	15,235	15,230	15,215
Fins/m, height, thickness, serration	78 × 15 × 1.5 × 4	117 × 15 × 1.5 × 5	157 × 15 × 1.5 × 4	197 × 15 × 1.5 × 4
S_T × S_L (staggered)	90 × 90	92 × 90	92 × 90	90 × 90
Tubes/row	20	20	20	20
Number of rows deep	16	13	11	9
Length, m	5	5	5	5.5
Weight of tubes, kg	10,750	10,850	10,700	11,000
U_o, kcal/m² h °C (W/m² K)	60 (69.8)	53.5 (62.2)	49.2 (57.2)	44.6 (51.9)
Tube wall temperature, °C	293	303	313	320
Surface area, m²	1,075	1,207	1,310	1,440
Heat flux inside tubes, kcal/m² h	102,500	126,200	148,900	164,600

Note: Reduce weight of tube bundles using smaller tubes.

Example E.5

In an HRSG evaporator, 100,000 kg/h of gas turbine exhaust at 600°C is cooled to 281°C. Tubes are 50.8 × 44 mm. Arrangement is staggered. Steam pressure is 39 kg/cm² g (saturation temperature is 249°C) and feed water is at 105°C. Fouling factors are 0.0002 m² h °C/kcal on both gas and steam sides.

Analysis: All the options have nearly the same weight, duty, and gas pressure drop and cross section (Table E.8).

- The number of rows deep is lesser when the fin density is higher. Hence, one can evaluate the cost of the bundle considering material and labor costs and choose the best option. In some locations, cost of labor may be low, and hence, lower fin density option may work out to lower cost as cost of finned tubes will be lower, and more rows can be assembled at an overall attractive cost.
- Surface areas are higher with higher fin density as expected due to the lower heat transfer coefficients with higher fin densities.
- Heat flux inside the tubes is higher with higher fin densities. However, considering the steam pressure, these heat fluxes are acceptable as discussed in Chapter 2.
- The tube wall temperature is higher with higher fin density due to higher heat flux. Hence, plant engineers should ask for this information also when obtaining quotes from various vendors. It is likely that some vendors may have offered designs with much higher gas velocity and high gas pressure drop with higher heat flux inside the tubes and consequent higher tube wall temperatures.

Reduce Weight of Tube Bundles Using Smaller Tubes

It is possible to reduce the weight of a boiler using smaller tubes for the same duty and gas pressure drop values. This is true for plain as well as finned tubes. The reason is the higher heat transfer coefficient with smaller-diameter tube. The weight of the tube bundle will also be less. However, the number of tubes used will be more with smaller tube option, which may increase the labor cost or cost of fabrication of the boiler as the number of tubes to be welded increases as the tube diameter decreases. Hence, the decision must be made based on not only the cost of material and weight of boiler but also the cost of fabrication.

Example E.6

Here is an example of an HRSG with an evaporator and economizer generating saturated steam at 40 kg/cm² g using feed water at 111°C. Exhaust gas flow = 150,000 kg/h at 580°C and analysis is % volume $CO_2 = 3$, $H_2O = 7$, $N_2 = 75$, $O_2 = 15$. Two options were considered in the design of the HRSG. One uses 50.8 tubes for the evaporator and economizer, while the other uses 38 mm tubes for evaporator and economizer. Tubes are all in staggered arrangement. Tables E.9 and E.10 show the geometry for both options, while Table E.10 shows the performance results.

1. Total gas pressure drop and duty are same for both the options. Steam production is the same.
2. It is seen that the 38 mm tube option has higher heat transfer coefficients about 3%–8%.
3. The 38 mm tube option has lower weight; about 60,000 kg versus 67,500 kg for the 51 mm tube option. With large boilers, this may be a deciding factor as shipping costs can be lowered.
4. The evaporator has 720 tubes and economizer has 561 tubes to be welded in the 38 mm option, while the 51 mm tube option has only 540 tubes in the evaporator and 420 in the economizer. In countries where labor costs are high, this may be a deciding factor.

Hence, based on local conditions, shipping, cost of labor or materials, and cost of fabrication, one may make suitable decisions. This exercise shows that with smaller tubes, it is possible to lower the weight of boilers and perhaps the shipping and manufacturing costs also.

TABLE E.9

Geometrical Data for 38 and 50.8 mm OD Tubes

Item	Evaporator	Economizer	Evaporator	Economizer
Tube OD, mm	38	38	50.8	50.8
Tube ID, mm	31.8	31.8	43	43
Fins/m	197	197	197	197
Fins height, mm	19	19	19	19
Fin thickness, mm	1.5	1.5	1.5	1.5
Serration, mm	4	4	4	4
Tubes/row	36	33	30	30
Number of rows deep	20	17	18	14
Length, m	5	5	5.3	5.3
Transverse pitch, mm	88	88	101	101
Longitudinal pitch	88	88	101	101
Streams	—	11	—	6
Arrangement	Staggered	Staggered	Staggered	Staggered
Flow direction		Counter-flow		Counter-flow

TABLE E.10

Performance Results for 38 and 50.8 mm OD Tubes

Item	Evaporator-38	Economizer-38	Evaporator-51	Economizer-51
Gas temp. in, ±5°C	580	253	580	253
Gas temp. out, ±5°C	253	149	253	150
Gas spec. heat, kcal/kg °C	0.2715	0.2584	0.2715	0.2584
Duty, MM kcal/h	13.18	3.99	13.18	3.95
Surface area, m²	5,023	3,914	5,190	4,036
Gas pressure drop, mm wc	65	46	73	40
Fouling factor—gas	0.0002	0.0002	0.0002	0.0002
U, kcal/m²h °C	39.51	36.64	38.01	33.80
LMTD, °C	67	28	67	29
Gas velocity, m/s	15	10	16	10
Weight, kg	33,662	26,228	37,962	29,526
Fluid temp. in, ±5°C	233	111	232	111
Fluid temp. out, ±5°C	251	233	251	232
Pressure drop, kg/cm²	—	1.39	—	1.47
Fouling factor—fluid	0.0002	0.0002	0.0002	0.0002
Fluid velocity, m/s	—	1.1	—	1.1
Steam flow, kg/h	30,760		30,687	—

Inline versus Staggered Arrangement

In the case of plain tubes, using staggered arrangement is not helpful as the increase in overall heat transfer coefficient is not significant while the increase in gas pressure drop is. This fact was illustrated with an example in Appendix C. Hence, plain tubes are often used in inline arrangement only. However, with finned tubes, staggered arrangement may help reduce weight and size for the same duty and gas pressure drop.

Example E.7

Over a finned evaporator bundle, 200,000 kg/h of flue gas from natural gas combustion flows. Gases are cooled from 700°C to 263°C. Using 51 × 43 mm tubes and 216 fins/m × 19 mm high, 1.5 mm thick with 4 mm serration, determine how inline arrangement fares against staggered arrangement. Duty and gas pressure drop should be approximately the same for both options. Steam pressure is 40 kg/cm² a, and feed water temperature is 120°C.

Solution

Table E.11 shows the results of calculations. Tube length and pitch were slightly manipulated to obtain the same duty and gas pressure drop in column 3 as in column 1 for inline arrangement.

It is seen that due to the higher overall heat transfer coefficient U_o, the staggered arrangement requires fewer tubes and lesser surface area. However, if we do not manipulate the tube spacing or length, the gas pressure drop will be higher. Column 3 shows that when we manipulate the tube pitch and length, the gas pressure drop becomes comparable to the inline arrangement. (Some manufacturers prefer inline arrangement as drilling patterns are simpler or have been streamlined.) Due to the higher heat transfer coefficients in staggered arrangement, the tube wall and fin tip temperatures will be higher than those in inline arrangement.

TABLE E.11

Inline versus Staggered Finned Evaporator

Item	Inline	Staggered	Staggered
Gas flow, kg/h	200,000	200,000	200,000
Inlet gas temperature, °C	700	700	700
Exit gas temperature, °C	264	263	264
Duty, MM kcal/h	25.29	25.37	25.31
Steam generation, kg/h	46,390	46,540	46,420
Tubes/row	32	32	32
Number of rows deep	20	15	14
Effective length, m	5.0	5.0	5.5
Surface area, m^2	6,384	4,788	4,915
U_o, kcal/m^{2}h °C	31	42.6	40.5
Gas pressure drop, mm wc	93	126	89
Tube wall temperature, °C	340	374	368
Fin tip temperature, °C	470	543	530
Transverse pitch S_T, mm	101	101	103
Longitudinal pitch, S_L, mm	101	101	101

Effect of Tube-Side Fouling

When tube-side fouling is likely, finned tubes are not attractive. Higher the ratio of fin to tube inner surface, poorer its performance will be. (This is similar to the use of high fin density when tube-side coefficient is low.) To illustrate this point, an evaporator has been designed for the same duty conditions in clean conditions, and then its performance is checked when tube-side fouling builds up. Results are shown in Table E.12.

TABLE E.12

Effect of Tube-Side Fouling

Item	Case 1	Case 2	Case 3	Case 4	Case 5	Case 6
Gas temp. in, °C	600	600	600	600	600	600
Gas out, °C	339	356	339	379	340	398
Duty, MM kcal/h	7.56	7.15	7.56	6.43	7.51	5.88
Gas press. drop, mm wc	30	30.5	11	11.5	12.5	13.5
Steam flow, kg/h	13,860	13,115	13,840	11,788	13,780	10,780
Fins—n × h × t	None	None	78 × 19 × 1.5	78 × 19 × 1.5	216 × 19 × 1.5	216 × 19 × 1.5
S_T × S_L	76 × 76	76 × 76	95 × 95	95 × 95	95 × 95	95 × 95
Tubes/row	30	30	28	28	28	28
Number of rows deep	20	20	10	10	6	6
Length, m	6	6	5	5	5.3	5.3
Tube-side fouling	0.0002	0.002	0.0002	0.002	0.0002	0.002
U_o, kcal/m^{2}h °C	69.3	60.6	34.39	25.27	22.1	14.1
Ratio A_t/A_i	1.18	1.18	6.08	6.08	14.75	14.75
Heat flux inside tubes	28,703	25,128	73,390	53,928	114,417	73,000
Tube wall temp., °C	262	310	281	378	299	421

Notes: Gas flow = 100,000 kg/h, 50.8 × 43 mm tubes, steam pressure = 40 kg/cm^2 a, feed water = 120°C. Gas-side fouling = 0.0002 m^{2}h °C/kcal (0.000172 m^{2}K/W) (heat flux in kcal/m^{2}h). Inline arrangement of tubes is used.

Cases 1 and 2 are with plain tubes in clean and dirty conditions.
Cases 3 and 4 are for 78 fins/m design in clean and dirty conditions.
Cases 5 and 6 are for 216 fins/m design in clean and dirty conditions.

1. The cases 1, 3, and 5 are the design cases with plain tubes, tubes with 78 fins/m and tubes with 216 fins/m. Cases 2, 4, and 6 show their off-design performance when tube-side fouling increases from 0.0002 m² h °C/kcal to 0.002 m² h °C/kcal (0.000172–0.00172 m² K/W).

2. In the design cases 1, 3, and 5, the boiler duty and exit gas temperatures are the same, namely, 340°C and hence the steam generation. However, when the tube-side fouling is high, the off-design performance shows significant differences.

3. The loss of duty or steam production is minimal when tubes are plain. The exit gas temperature goes up to only 353°C from 339°C, and steam output drops from 13,860 to 13,115 kg/h, a decrease of 5.5% only. However, when we used 216 fins/m, though the design is for the same steam output, the duty decreases drastically when tube-side fouling increases. The exit gas temperature goes up to nearly 400°C with steam output dropping by 22%! Thus, one should use caution when tube-side fouling is high.

4. The heat flux inside the tubes is also shown. Though U_o and heat flux decrease with higher fouling, the drop across the fouling layer will be much higher. Hence, there is an increase in tube wall temperatures. The tube wall temperature goes up to 421°C when fins/m is 216 compared to 310°C with plain tubes.

Effect of Outside Fouling Factor

The effect of outside fouling factor is not significant. The decrease in duty is marginal with gas-side fouling (Table E.13).

TABLE E.13

Effect of Gas-Side Fouling

Item	Case 1	Case 2	Case 3	Case 4
Gas temp. in, °C	600	600	600	600
Gas out, °C	339	347	340	345
Duty, MM kcal/h (MW)	7.56 (8.79)	7.33 (8.52)	7.51 (8.73)	7.37 (8.56)
Gas pressure drop, mm wc	11	11	12.5	12.5
Steam flow, kg/h	13,840	13,450	13,780	13,500
Fins—n × h × b	78 × 19 × 1.5	78 × 19 × 1.5	216 × 19 × 1.5	216 × 19 × 1.5
$S_T \times S_L$	95 × 95	95 × 95	95 × 95	95 × 95
Tubes/row	28	28	28	28
Number of rows deep	10	10	6	6
Length, m	5	5	5.3	5.3
Gas-side fouling, m² h °C/kcal	0.0002	0.002	0.0002	0.002
U_o, kcal/m² h °C	34.39	32.36	22.1	21.23
Ratio A_t/A_i	6.08	6.08	14.75	14.75
Heat flux inside tubes, kcal/m² h	73,390	69,059	114,417	109,930
Tube wall temp., °C	281	279	299	297

For the evaporator used in the earlier example, the effect of gas-side fouling was studied for two fin geometries, namely, 78 and 216 fins/m. It is seen that the effect is marginal only, about 2%–3% loss in duty. Hence, tube-side fouling is more serious compared to gas-side fouling.

Optimizing Finned Surface Design

Example E.8

An evaporator has to be designed for the following parameters. Turbine exhaust gas flow = 100,000 kg/h at 600°C in and 315°C out. Duty = 7.65 MM kcal/h. Steam pressure is 40 kg/cm² g. Tube size 51 × 43 mm and 95 mm square pitch. Steam flow = 14,000 kg/h at 40 kg/cm² a, and 120°C feed water. How does one arrive at the optimum design?

Solution

Though this problem is often encountered by the boiler supplier, plant engineers should know that several design options are possible with finned tubes. Here is a way to review options for a finned evaporator bundle (one could extend this concept to superheaters and economizers also). It gives the effect of fin geometry and gas velocity on gas pressure drop for the same duty at one glance. Curves are shown for just two fin geometries, but one can work out for different tube sizes and fin geometries and plot the gas pressure drop on one scale and surface area on the other. Now for a desired gas pressure drop, one can select various tube or fin geometry as described later.

It is seen that as fin density increases, the surface area required for the same duty is more. The gas pressure drop is higher for the smaller fin density for the same duty. One can see that the variation in surface area can be 50% for the same duty and gas pressure drop. Hence, while evaluating bids for HRSGs, one should not get carried away by surface areas but understand why there is so much difference. The smaller fins have a higher U as discussed earlier and hence lesser surface area requirements. However, as more rows are required with smaller fins, the gas pressure drop goes up. The results are shown in Figure E.5 as well as in Table E.14.

Let us, say, limit the gas pressure drop to 30 mm wc. As we move from 30 mm wc on the right-hand scale to the left, the line intersects the pressure drop curve. For the 78 fins/m design, the gas velocity to meet this pressure drop is about 19 m/s, while for the 216 fins/m design, it is about 21.5 m/s. Then go up to intersect the corresponding

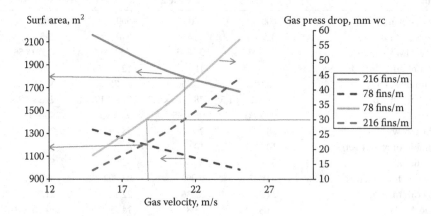

FIGURE E.5
Chart relates gas velocity, pressure drop, and surface area.

TABLE E.14

Evaporator Design with Different Gas Velocities and Fin Geometry

Tubes/row	20	20	20	20	20	20	20	20	
Number of rows deep	13	15	16	7	8	9	8	14	
Length	6.22	4.67	3.73	7.71	5.78	4.62	5.5	5.0	m
Fins/m	78	78	78	216	216	216	216	78	
Fin height	19	19	19	19	19	19	19	19	mm
Fin thick	1.5	1.5	1.5	1.5	1.5	1.5	1.5	1.5	mm
Serration	4	4	4	4	4	4	4	4	mm
Surf. area	1,334	1,155	985	2,161	1,851	1,666	1,762	1,155	m^2
U_o	34.3	40.4	45.7	20.9	24.5	27.5	25.1	38.9	$kcal/m^2\,h\,°C$
Gas press. drop	18	35	57	13	26	45	28	28	mm wc
Gas velocity	15	20	25	15	20	25	21	19	m/s
Weight	13,500	11,691	9,960	15,370	13,169	11,842	12,533	11,683	kg

lines for surface areas. It is seen that for 78 fins/m design, the surface area is about 1200 m^2, while for the 216 fins/m design, it is about 1800 m^2. Table E.14 shows the actual results from a computer program. The last two columns show the desired design for both the fin geometries.

Developing such charts helps one to understand the effect of fin geometry and also how an optimum design can be arrived at. Based on tubes deep and length and fin geometry, one can estimate the material and labor cost and arrive at an optimum design. Say for 28 mm wc gas pressure and same duty, we have two options; one with 78 fins/m, but it has 14 rows deep, while the 216 fins/m option has 8 rows deep, and though the surface area is more, the labor cost could be less. One can also study the effect of using 38 mm tubes and so on as discussed earlier. Plant engineers need to be aware that a wide difference can exist in tube or fin geometry or surface area for the same performance with finned tube exchangers, and hence, this study explains the reason.

Effect of Fin Thickness and Conductivity

The thickness of fins and the thermal conductivity of fins used in boiler tubes affect the heat transfer, duty, and tube wall and fin tip temperatures and hence should be chosen with care. Thermal conductivity of alloy steels is lower than that of carbon steels as seen in Table E.15. Hence, when we use alloy fins in carbon steel evaporators, then the K value of fin material reduces decreasing the U value and increases the fin tip temperature as shown later. Note that the K value of stainless fins will be much lower than that of carbon steel. When gas inlet temperature is high, say, above 700°C, the boiler is designed with a few rows of plain tubes, followed by a few rows of tubes with alloy steel fins and then followed by several rows of carbon steel fins. As the gas cools, carbon steel fins are used. In finned superheaters, it may be necessary to use the alloy steel fins with alloy steel tubes.

Example E.9

A boiler evaporator uses the following geometry: 50.8 × 44.6 tubes, 36 tubes/row, 15 deep, 5 m long, 101 mm S_T, S_L, staggered arrangement, 216 fins/m, 19 mm high, 1.5 mm thick, 4 mm serration, 40 kg/cm² a steam pressure, 105°C feed water. Gas flow = 150,000 kg/h. Study the effect of using alloy steel material for fins and thicker fins.

TABLE E.15

Effect of Fin Thickness and Thermal Conductivity

Gas inlet temperature, °C	800	800	800	800
Exit gas temperature, °C	255	251	254	260
Duty	23.96	24.1	24.0	23.73
Gas pressure drop, mm wc	66	140	141	61
U_o, kcal/mm²h °C	37.74	38.9	34.0	31.9
Steam generation, kg/h	42,770	43,000	42,840	42,350
Surface area, m²	5,386	6,228	6,228	5,386
Fin conductivity, kcal/m h °C	30	30	15	15
Fin thickness, mm	1.5	2.67	2.67	1.5
Max tube wall temperature, °C	372	396	377	354
Max fin tip temperature, °C	557	532	596	635

Solution

The results from the computer program are shown as follows. The differences in surface areas are due to the fin thickness though the number of tubes remains unchanged.

The following points may be noted.

1. As fin conductivity decreases for the same fin thickness, duty decreases slightly.
2. As the fin conductivity decreases, the tube wall temperature decreases while the fin tip temperature increases.
3. Flue gas pressure drop increases significantly if fin thickness is increased due to an increase in obstruction area A_o.
4. As the fin thickness increases, the fin tip temperature decreases while the tube wall temperature increases.

Hence, it is not prudent to increase the fin thickness as the operating cost in the form of gas pressure drop increases significantly. If we can keep the tube wall and fin tip temperatures within limits, using a nominal fin thickness of 1.25–1.5 mm is fine; else, we are unnecessarily increasing the weight and cost of the boiler tubes. The boiler duty also does not improve much due to the use of thicker fins. One should also note the impact of using alloy steel fins with low thermal conductivity.

The purpose of this example is to make plant engineers aware of the effect of fin thickness on performance and discuss intelligently with the HRSG suppliers if they use thick fins in their design. The operating costs in the form of gas pressure drop can be reduced by using nominal fin thickness. Thick fins may be required to lower the fin tip temperature in some instances to avoid the use of alloy steel fins.

Why Are Fins Not Used in Gas–Gas Exchangers?

Finned surfaces are not used in tubular air heaters as both the air-side and flue gas–side heat transfer coefficients are in the same range. The effect of adding finned surface will not be reflected in the duty.

Example E.10

Tubes of size 50.8 × 46.4 mm are used in inline arrangement in an air–flue gas exchanger. Three options are reviewed, one with plain tubes and two others with 78 and 197 serrated fins with a fin height of 15 mm and thickness of 1.5 mm and serration width of 4 mm. Calculations for the plain tube exchanger are worked out in Chapter 4 on waste heat boilers. The finned exchanger performance was obtained by using the procedure discussed earlier. Calculations for the 197 fins/m option are shown later, while results from computer program are shown for the 78 fins/m case. Fouling factors are 0.0002 m² h °C/kcal on air and flue gas sides. $S_T = 107$ mm, $S_L = 90$ mm. Tubes/row = 50 and number deep = 56. Effective length = 2.4 m. See Equations E.1 through E.20 for the calculation procedure. Note that the following is for serrated fins, while calculations for solid fins were shown in Example E.1 (Table E.16).

Solution

$A_o = d + 2nbh = 0.0508 + 2 \times 197 \times 0.015 \times 0.0015 = 0.05967$ m²/m

$$A_f = \frac{3.14 \times 0.0508 \times 197 \times \left[2 \times 0.015 \times 0.0055 + 0.0015 \times 0.004\right]}{0.004} = 1.343 \text{ m}^2/\text{m}$$

$$A_t = 1.343 + 3.14 \times 0.0508 \times \left(1 - 197 \times 0.0015\right) = 1.456 \text{ m}^2/\text{m}$$

Ratio of external to tube inner surface = 1.456/(3.14 × 0.0464) = 10

$$G = \frac{200{,}000}{\left[50 \times 2.4 \times \left(0.107 - 0.05967\right)\right]} = 35{,}213 \text{ kg/m}^2\text{h} = 9.78 \text{ kg/m}^2 \text{ s}$$

The overall thermal performance, duty, and air–gas exit temperatures with the finned tubes are the same as that with plain tubes. Hence, let us use gas properties at average air temperature of 103°C as follows: $C_p = 0.2455$ kcal/kg °C, $\mu = 0.0806$ kg/m h, k = 0.0282 kcal/m h °C from Appendix F.

Calculation of Heat Transfer Coefficient

Re = 35,213 × 0.0508/0.0806 = 22,193. Compute the constants C_1–C_6.

$$C_1 = 0.053\left[1.45 - 2.9 \times \left(\frac{90}{50.8}\right)^{-2.3}\right] \times 22193^{-0.21} = 0.00435$$

$$C_3 = 0.25 + 0.6 \text{ exp}\left[\frac{(-0.26 \times 0.015)}{\{(1/197) - 0.0015\}}\right] = 0.4516$$

$$C_5 = 1.1 - \left[0.75 - 1.5 \text{ exp}(-0.7 \times 56)\right] \text{exp}(-2 \times 90 / 107) = 0.9605$$

Average air temperature = 103°C. Assume average fin temperature = 135°C. (These values may be corrected later if necessary, but their effect is small). Convective heat transfer coefficient

$h_c = 0.00435 \times 0.4516 \times 0.9605 \times [(50.8 + 30)/50.8]^{0.5}$ (376/408)$^{0.5}$ × 35,213 × 0.2455 × (0.0282/0.0806/0.02455)$^{0.67}$ = 24.93 kcal/m²h °C

m = [2 × 24.93 × (0.0015 + 0.004)/(30 × 0.0015 × 0.004)]$^{0.5}$ = 39.0 mh = 39.0 × 0.015 = 0.585

tanh (mh) = [(1.796 − .557)/(1.796 + 0.557)] = 0.5265. Fin efficiency E = 0.5265/0.585 = 0.896

Fin effectiveness η = 1 − (1 − 0.896) × 1.343/1.456 = 0.904.

TABLE E.16

Thermal Conductivity of Metals, Btu/ft h °F

Material	\multicolumn{14}{c}{Temperature (°F)}													
	200	300	400	500	600	700	800	900	1000	1100	1200	1300	1400	1500
Aluminum (annealed)														
Type 1100-0	126	124	123	122	121	120	118							
Type 3003-0	111	111	111	1111	111	111	111							
Type 3004-0	97	98	99	100	102	103	104							
Type 6061-0	102	103	104	105	106	106	106							
Aluminum (tempered)														
Type 1100 (all tempers)	123	122	121	120	118	118	118							
Type 3003 (all tempers)	96	97	98	99	100	102	104							
Type 3004 (all tempers)	97	98	99	100	102	103	104							
Type 6061-T4 and T6	95	96	97	98	99	100	102							
Type 6063-T5 and T6	116	116	116	116	116	115	114							
Type 6063 T42	111	111	111	111	111	111	111							
Cast icon	31	31	30	29	28	27	26	25						
Carbon steel	30	29	28	27	26	25	24	23						
Carbon moly steel	29	28	27	26	25	25	24	23						
Chrome moly steels														
1% Cr, $\frac{1}{2}$% Mo	27	27	26	25	24	24	23	21	21					
$2\frac{1}{4}$% Cr: 1% Ma	25	24	23	23	22	22	21	21	21	20				
5% Cr, $\frac{1}{2}$% Mo	21	21	21	20	20	20	20	19	19	19				
12% Cr	14	15	15	15	16	16	16	16	17	17	17	18		

Austenitic stainless steels															
18% Cr, 8% Ni	9.3	9.8	10	11	11	12	12	13	13	14	14	14	14	15	15
25% Cr, 20% Ni	7.6	3.4	5.9	9.5	10	11	11	12	12	13	13	14	14	15	15
Admiralty metal	70	75	79	84	89										
Naval brass	71	74	77	80	83										
Cupper (electrolytic)	225	225	224	224	223										
Copper and nickel alloys															
90% Cu, 10% Ni	30	31	34	37	42	47	49	51	53						
80% Cu, 20% Ni	22	23	25	27	29	31	34	37	40						
70% Cu, 30% Ni	18	19	21	23	25	27	30	33	37						
30% Cu, 70% Ni (Monel)	15	16	16	16	17	18	18	19	20	20					
Nickel	30	36	33	31	29	28	28	29	31	33					
Nickel–chrome–iron	9.4	9.7	9.9	10	10	11	11	12	12	12	13	13			
Titanium (or B)	10.9	10.4									10.5				

Note: Multiply by 1.489 to obtain values in kcal/m h °C or by 1.732 to obtain values in W/m K.

Neglect nonluminous radiation due to low air temperatures and absence of water vapor and carbon dioxide in air. Using the same h_i as in the example shown in Chapter 4 as tube-side flow and tube count are very close, we have

$1/U_o = 1/(0.904 \times 24.93) + 0.0002 + 0.0508 \times 10 \times \ln(50.8/46.4) \times /(2 \times 30) + 0.0002 \times 10 + (10/49) = 0.04437 + 0.0002 + 0.000767 + 0.002 + 0.204 = 0.2513$ or $U_o = 3.98$ kcal/$m^2 h$ °C

Gas Pressure Drop

See Equations E.16 through E.21.

$$C_2 = 0.11 + 1.4 \times 22193^{-0.4} = 0.1355.$$

Spacing between fins $S = 1/n - b = (1/197) - 0.0015 = 0.003576$

$$C_4 = 0.08 \times (0.15 \times 107/50.8)^{[-0.11 \times 0.015/0.003576]^{0.15}} = 0.3851$$

$$C_6 = 1.6 - \left[0.75 - 1.5\,\exp(-0.7 \times 57)\right]\exp\left[-0.2 \times (90/107)^2\right] = 0.949$$

$f = 0.1355 \times 0.3851 \times 0.949 \times (80.8/50.8) \times (376/408)^{0.25} = 0.0771$ (average air temperature is 103°C (376 K) and average fin temperature is 135°C (408 K).

$$B = \left[(0.107 - 0.05967)/0.107\right]^2 = 0.1957$$

$$a = (1 + 0.1957 \times 0.1957) \times (181 - 25)/(4 \times 57)/(376) = 0.00189.$$

$$\text{Gas density} = \frac{12.17 \times 28.74}{376} = 0.93 \text{ kg/m}^3$$

$$\Delta P_g = \frac{0.205 \times (0.0771 + 0.00189) \times 9.78^2 \times 56}{0.93} = 93 \text{ mm wc}$$

Checking the Duty

Apply the NTU method for the cross-flow exchanger performance. See Appendix A for equations for various types of exchangers. The air side is mixed while the flue gas is not.

$$C_{max} = 225,000 \times 0.279 = 62,775. \quad C_{min} = 200,000 \times 0.246 \times 0.99 = 48,708$$

$$C = 48,708/62,775 = 0.7759.$$

$$A = 1.456 \times 50 \times 56 = 9784 \text{ m}^2.$$

$$NTU = 3.98 \times 9,784/(200,000 \times .99 \times 0.2455) = 0.801$$

$$\epsilon = 1 - \exp\left[\frac{-\{1 - \exp(-NTU \times C)\}}{C}\right] = 0.449.$$

Hence, $Q_T = 0.449 \times 48,708 \times (375 - 25) = 7.65$ MM kcal/h (8.9 MW).
Results for all the cases are shown in Table E.17.
The following observations may be made.

TABLE E.17

Performance of Air Heater with and without Fins

Air Flow	200,000	200,000	200,000
Inlet air temperature, °C	25	25	25
Exit air temperature, °C	182	182	181
Duty, MM kcal/h	7.69	7.67	7.64
Air-side pressure drop, mm wc	97	94	95
U_o, kcal/m² h °C	28.93	8.02	3.98
Surface area, m²	1,360	4,883	9,772
Flue gas flow, kg/h	225,000	225,000	225,000
Flue gas temperature in, °C	375	375	375
Flue gas temperature out, °C	253	253	253
Flue gas pressure drop, mm wc	30	28	26
Fins/m	0	78	197
Tubes/row	50	50	50
Number of rows deep	57	57	56
Length, m	3	2.55	2.4
Transverse pitch, mm	82	97	107
Longitudinal pitch, mm	70	90	90
Ratio of external to internal surface area	1.09	4.48	10
Weight, kg	24,300	40,300	64,900

The cross section of the exchanger, duty, and pressure drops on air and gas sides are same. However, the surface area of the finned exchangers are much more and also their weight. We are adding weight and cost to the exchanger for the same duty, which is foolish. Hence, the cost of the finned bundle will also be higher. One can see how low the U_o value is reduced as the fin density increases. Hence, finned tubes are not used when the tube-side heat transfer coefficient is comparable with the tube outside coefficient.

A summary of important points about finned tubes is given in Table E.18.

TABLE E.18

What You Should Know about Finned Tubes

The higher the fin surface area, the lower will be the heat transfer coefficient. One should not evaluate finned surfaces by surface area alone. Product of (UA) should be compared and not A alone, which is what many do!

Surface areas in boiler components can vary by 50%–100% for the same duty due to improper selection of fin geometry!

Higher fin surface also results in higher heat flux inside tubes and hence higher tube wall and fin tip temperatures. Life of superheaters is affected by poor selection of fin geometry.

Use smaller fin density when tube-side coefficient is small. For superheaters, 78–117 fins/m is adequate, while for evaporators and economizers, 157–255 fins/m may be fine.

Staggered arrangement is slightly better than inline for finned tubes.

Smaller-diameter tubes with same fin configuration results in a less heavy tube bundle compared to a larger tube diameter bundle.

Fouling inside the tubes impacts finned tubes more than plain tubes through lower duty and higher tube wall temperatures.

Serrated fins give a slightly higher heat transfer coefficient and also higher gas pressure drop.

Using thick fins increases the tube wall temperature while decreasing the fin tip temperature. The duty increases slightly. Hence, fin thickness may be limited to 1.5–1.9 mm. The use of 3 mm or thicker fins may be permitted for a row or two to limit the fin tip temperature but not for the entire HRSG.

References

1. V. Ganapathy, *Industrial Boilers and HRSGs*, CRC Press, Boca Raton, FL, 2003.
2. ESCOA Corp., *ESCOA Fin Tube Manual*, Tulsa, OK, 1979.
3. Fin Tube Technologies, Tulsa, OK.
4. V. Ganapathy, *Applied Heat Transfer*, Pennwell Books, Tulsa, OK. 1982, p501.
5. V. Ganapathy, Evaluated extended surfaces carefully, *Hydrocarbon Processing*, Oct. 1990, p65.
6. V. Ganapathy, Understand finned heat exchangers, *Chemical Engineering*, Sept. 2013, p62.

Appendix F: Properties of Gases

In boiler performance estimation, one of the important calculations is the determination of thermal and transport properties of a flue gas mixture such as products of combustion of fuel oil or natural gas or any waste gas stream. The specific heat, viscosity, and thermal conductivity data for individual gas such as CO_2, H_2O, N_2, O_2, and SO_2 are available in literature (see Tables F.1 through F.3); mixture properties have to be evaluated in order to determine the tube-side or outside heat transfer coefficient.

For a gaseous mixture, the following equations may be used:

$$\mu_m = \Sigma y_i \; {}_i \sqrt{MW_i} / \Sigma y_i \sqrt{MW_i} \tag{F.1}$$

$$k_m = \Sigma y_i k_i (MWi)^{.33} / \Sigma y_i (MW)^{0.33} \tag{F.2}$$

$$C_{pm} = \Sigma y_i C_{pi} (MW_i) / \Sigma y_i (MW_i) \tag{F.3}$$

where
  μ_m, k_m, C_{pm} are the viscosity, thermal conductivity, and specific heat of the mixture, respectively
  y_i is the volume fraction of the particular constituent
  MW is the molecular weight of that constituent

Example F.1

Turbine exhaust gases contain % volume of $CO_2 = 3$, $H_2O = 7$, $N_2 = 75$, $O_2 = 15$. Determine the C_{pm}, k_m, μ_m at 300°C. Table F.4 gives the data for the individual gases.

Solution

$C_{pm} = (0.2521 \times 0.03 \times 44 + 0.4795 \times 0.07 \times 18 + 0.2569 \times 0.75 \times 28 + 0.2395 \times 0.15 \times 32)/$
$\quad (0.03 \times 44 + 0.07 \times 18 + 0.75 \times 28 + 0.15 \times 32) = 0.2636$ kcal/kg °C = 1.103 KJ/
$\quad$ kg K = 0.2636 Btu/lb °F.

$\mu_m = (0.0941 \times 0.03 \times \sqrt{44} + 0.0734 \times 0.07 \times \sqrt{18} + 0.1017 \times 0.75 \times \sqrt{28} + 0.1204 \times 0.15$
$\quad \times \sqrt{32})/(0.03 \times \sqrt{44} + 0.07 \times \sqrt{18} + 0.75 \times \sqrt{28} + 0.15 \times \sqrt{32}) = 0.1028$ kg/m h
$\quad = 2.86 \times 10^{-5}$ Pa s = 0.0689 lb/ft h

$k_m = [0.0328 \times 0.03 \times (44)^{0.33} + 0.0375 \times 0.07 \times (18)^{0.33} + 0.0365 \times 0.75 \times (28)^{0.33} + 0.0397$
$\quad \times 0.15 \times (32)^{0.33}]/[0.03 \times (44)^{0.33} + 0.07 \times (18)^{0.33} + 0.75 \times (28)^{0.33} + 0.15 \times (32)^{0.33}]$
$\quad = 0.0369$ kcal/m h °C = 0.0429 W/m K [0.02478 Btu/ft h °F]

Enthalpy calculation is often required while doing heat balances (Tables F.1 through F.3 and F.5 through F.9). Tables F.10 and F.11 shows the enthalpy of a few common gases with a reference of 15°C.

TABLE F.1

Specific Heats for Various Gases, kcal/kg °C

Temp., °C	100	200	300	400	500	600	700	800	900	1000
CO_2	0.2176	0.2362	0.2521	0.2656	0.2770	0.2864	0.2942	0.3005	0.3056	0.3098
H_2O	0.4543	0.4662	0.4795	0.4939	0.5092	0.5251	0.5413	0.5575	0.5735	0.5889
N_2	0.2499	0.2530	0.2569	0.2613	0.2661	0.2711	0.2762	0.2813	0.2861	0.2906
O_2	0.2256	0.2329	0.2395	0.2453	0.2505	0.2550	0.2590	0.2625	0.2656	0.2683
SO_2	0.1588	0.1701	0.1795	0.1871	0.1932	0.1980	0.2016	0.2042	0.2060	0.2073
SO_3	0.1697	0.1900	0.2062	0.2189	0.2287	0.2361	0.2417	0.2461	0.2497	0.2532
CO	0.2509	0.2552	0.2599	0.2648	0.2699	0.2750	0.2801	0.2849	0.2894	0.2936
CH_4	0.5947	0.6769	0.7581	0.8376	0.9141	0.9868	1.0547	1.1168	1.1721	1.2196
H_2	3.4518	3.4638	3.4824	3.507	3.5369	3.5716	3.6104	3.6527	3.6979	3.7453
HCl	0.1931	0.1940	0.1958	0.1983	0.2015	0.2051	0.2089	0.2128	0.2166	0.2202
Air	0.2455	0.2497	0.2542	0.2590	0.2639	0.2689	0.2738	0.2786	0.2831	0.2873

Note: Multiply by 4.187 to convert to kJ/kg K and by 1 to obtain C_p in Btu/lb °F.

TABLE F.2

Viscosity of Gases, kg/m h

Temp., °C	100	200	300	400	500	600	700	800	900	1000
CO_2	0.0662	0.0805	0.0941	0.1070	0.1192	0.1307	0.1415	0.1517	0.1611	0.1699
H_2O	0.0478	0.0608	0.0734	0.0858	0.0978	0.1096	0.1210	0.1322	0.1431	0.1536
N_2	0.0780	0.0901	0.1017	0.1128	0.1234	0.1334	0.1430	0.1520	0.1606	0.1686
O_2	0.0910	0.1060	0.1204	0.1342	0.1473	0.1598	0.1716	0.1828	0.1934	0.2034
SO_2	0.0585	0.0728	0.0866	0.0998	0.1123	0.1242	0.1356	0.1463	0.1564	0.1659
SO_3	0.0556	0.0694	0.0829	0.0959	0.1086	0.1208	0.1327	0.1441	0.1551	0.1658
CO	0.0783	0.0904	0.1020	0.1132	0.1239	0.1340	0.1437	0.1529	0.1616	0.1698
CH_4	0.0500	0.0607	0.0717	0.0830	0.0947	0.1066	0.1189	0.1315	0.1444	0.1577
H_2	0.0371	0.0440	0.0505	0.0565	0.0620	0.0671	0.0717	0.0759	0.0796	0.0828
HCl	0.0625	0.0789	0.0951	0.1111	0.1270	0.1427	0.1583	0.1737	0.1890	0.2041
Air	0.0806	0.0934	0.1056	0.1173	0.1285	0.1391	0.1492	0.1587	0.1677	0.1762

Note: Multiply by 0.0002778 to convert to Pa s (kg/m s) or divide by 1.489 to obtain lb/fth.

TABLE F.3

Thermal Conductivity of Gases, kcal/m h °C

Temp., °C	100	200	300	400	500	600	700	800	900	1000
CO_2	0.0190	0.0261	0.0328	0.0392	0.0453	0.0510	0.0564	0.0615	0.0662	0.0706
H_2O	0.0206	0.0289	0.0375	0.0462	0.0552	0.0643	0.0737	0.0832	0.0929	0.1028
N_2	0.0283	0.0324	0.0365	0.0405	0.0446	0.0486	0.0526	0.0567	0.0607	0.0647
O_2	0.0281	0.0340	0.0397	0.0452	0.0505	0.0556	0.0605	0.0652	0.0698	0.0741
SO_2	0.0114	0.0161	0.0206	0.0249	0.0291	0.0330	0.0367	0.0403	0.0436	0.0468
SO_3	0.0120	0.0186	0.025	0.0310	0.0368	0.0423	0.0475	0.0525	0.0571	0.0616
CO	0.0268	0.0318	0.0367	0.0414	0.0459	0.0503	0.0545	0.0586	0.0625	0.0662
CH_4	0.0386	0.0534	0.0689	0.0854	0.1026	0.1208	0.1398	0.1596	0.1803	0.2019
H_2	0.1838	0.2132	0.2425	0.2719	0.3012	0.3306	0.3600	0.3894	0.4188	0.4483
HCl	0.0171	0.0212	0.0255	0.0300	0.0347	0.0397	0.0449	0.0503	0.0559	0.0617
Air	0.0282	0.0327	0.0372	0.0416	0.0460	0.0503	0.0545	0.0588	0.0629	0.0670

Note: Divide by 0.86 to convert to W/m K or divide by 1.489 to obtain Btu/ft h °F.

TABLE F.4

Gas Properties at 300°C

	C_p	μ	k	MW
CO_2	0.2521	0.0941	0.0328	44
H_2O	0.4795	0.0734	0.0375	18
N_2	0.2569	0.1017	0.0365	28
O_2	0.2395	0.1204	0.0397	32

Note: C_p in kcal/kg °C, μ in kg/m h, k in kcal/m h °C.

TABLE F.5

Analysis of Important Flue Gases

Gas	Temp., °C	Press. bara	CO_2	CO	H_2O	N_2	O_2	SO_2	SO_3	HCl	H_2	CH_4	H_2S
1	300–1000	1				80	10	10					
2	250–500	1				81	11	1	7				
3	300–1100	40	6	8	37	0.5					43	5.5	
4	100–600	1	3		7	75	15						
5	200–1000	1	12		10	72	6			Trace			Trace
6	300–1450	1.2	6	3	23	55		6			4	3	
7	120–450	1	28		7	60	5						
8	250–1350	1	7		8	76	4		5				
9	100–1500	1	8		18	71.5	2.5						
10	100–1500	1	12		12	73.5	2.5						

Notes: (1) Raw sulfur gases, (2) SO_3 gases after converter, (3) reformed gas in hydrogen plant, (4) gas turbine exhaust, (5) MSW incinerator exhaust, (6) sulfur condenser effluent, (7) preheater gas in cement plants, (8) incineration of plastics, (9) flue gas from natural gas combustion, (10) flue gas from fuel oil combustion (sulfur neglected).

TABLE F.6

Specific Heat of Some Common Flue Gases, kcal/kg °C

Temp, °C	100	200	300	400	500	600	700	800	900	1000
Gas 1	0.2691	0.2755	0.2821	0.2890	0.2959	0.3027	0.3095	0.3160	0.3222	0.3279
Gas 2	0.2592	0.2659	0.2727	0.2795	0.2863	0.2929	0.2993	0.3054	0.3111	0.3164
Gas 3	0.2533	0.2583	0.2636	0.2691	0.2747	0.2804				
Gas 4	0.573	0.5894	0.6065	0.6240	0.6416	0.6593	0.6767	0.6936	0.7099	0.7252
Gas 5	0.2292	0.2344	0.2397	0.2449	0.2499	0.2549	0.2596	0.2640	0.2681	0.2717

Notes: Multiply by 4.187 to convert to kJ/kg K and by 1 to obtain C_p in Btu/lb °F.
Gas 1: Flue gases from combustion of natural gas—% volume CO_2 = 8.29, H_2O = 18.17, N_2 = 71.08, O_2 = 2.46.
Gas 2: Flue gases from combustion of fuel oil—% volume CO_2 = 11.57, H_2O = 12.29, N_2 = 73.63, O_2 = 2.51.
Gas 3: Gas turbine exhaust—% volume CO_2 = 3, H_2O = 7, N_2 = 75, O_2 = 15.
Gas 4: Reformed gas in hydrogen plant—% volume CO_2 = 6, CO = 8, H_2O = 37, N_2 = 0.5, CH_4 = 5.5, H_2 = 43.
Gas 5: Sulfur combustion—% volume SO_2 = 10, N_2 = 80, O_2 = 10.

TABLE F.7

Viscosity of Flue Gases, kg/m h

Temp, °C	100	200	300	400	500	600	700	800	900	1000
Gas 1	0.0726	0.0852	0.0972	0.1088	0.1198	0.1303	0.1404	0.1499	0.1589	0.1674
Gas 2	0.0737	0.0863	0.0983	0.1099	0.1209	0.1314	0.1414	0.1509	0.1598	0.1682
Gas 3	0.0779	0.0906	0.1028	0.1145	0.1256	0.1362				
Gas 4	0.0524	0.0641	0.0755	0.0865	0.0971	0.1074	0.1173	0.1269	0.1360	0.1449
Gas 5	0.0765	0.0892	0.1014	0.1131	0.1242	0.1348	0.1448	0.1543	0.1633	0.1717

Note: Multiply by 0.0002778 to convert to Pa s (kg/m s) or divide by 1.489 to obtain lb/ft h.

TABLE F.8

Thermal Conductivity of Flue Gases, kcal/m h °C

Temp., °C	100	200	300	400	500	600	700	800	900	1000
Gas 1	0.0262	0.0313	0.0364	0.0414	0.0465	0.0515	0.0565	0.0616	0.0665	0.0715
Gas 2	0.0263	0.0312	0.0362	0.0411	0.0460	0.0508	0.0556	0.0603	0.0651	0.0697
Gas 3	0.0275	0.0322	0.0369	0.0416	0.0462	0.0507				
Gas 4	0.0641	0.0777	0.0914	0.1052	0.1190	0.1330	0.1471	0.1613	0.1755	0.1900
Gas 5	0.0261	0.0305	0.0348	0.0390	0.0432	0.0473	0.0514	0.0554	0.0594	0.0633

Note: Divide by 0.86 to convert to W/m K or divide by 1.489 to obtain Btu/ft h °F.

TABLE F.9

F_g Factor for Various Gases = $k^{0.67} C_p^{0.33}/\mu^{0.32}$

Temp., °C	100	200	300	400	500	600	700	800
Gas A	0.1295	0.1380	0.1462	0.1541	0.1616	0.1687	—	—
Gas B	0.1298	0.1387	0.1472	0.1553	0.1631	0.1707	0.1781	0.1852

Note: Gas A: GT exhaust. % volume CO_2 = 3, H_2O = 7, N_2 = 75, O_2 = 15;
Gas B: Fired GT exhaust. % volume CO_2 = 4.28, H_2O = 9.51, N_2 = 74.0, O_2 = 12.17.

Example F.2

Analysis of a flue gas is as follows: % volume CO_2 = 8, H_2O = 18, N_2 = 71.5, O_2 = 2.5. Determine the energy recovered when 100,000 kg/h of this flue gas is cooled from 500°C to 100°C.

First let us convert the % volume analysis to % weight basis.

   Molecular weight MW = 0.08 × 44 + 0.18 × 18 + 0.715 × 28 + 0.025 × 32 = 27.58

   % weight CO_2 = 8 × 44/27.58 = 12.76

   % weight H_2O = 18 × 18/27.58 = 11.74

   % weight N_2 = 71.5 × 28/27.58 = 72.59

   % weight O_2 = 2.5 × 32/27.58 = 2.9

   Enthalpy of gas at 100°C = 0.1276 × 18 + 0.1174 × 37.7 + 0.7259 × 20.9 + 0.029 × 18.7 = 22.43 kcal/kg

   Enthalpy at 500°C = 0.1276 × 118.2 + 0.1174 × 231 + 0.7259 × 124.5 + 0.029 × 113.8 = 135.87 kcal/kg

   Energy recoverable = 100,000 × (135.87 − 22.42) = 11.34 MM kcal/h = 13.19 MW (45 MM Btu/h) (Tables F.10 and F.11)

TABLE F.10

Enthalpy of Products of Combustion, kcal/kg

Temperature, °C	Nat. Gas	Fuel Oil	Dry Air
1900	606.9	583.9	529
1800	570.4	549.2	498
1700	534.3	514.8	467.2
1600	498.4	4480.6	436.6
1400	428.0	413.1	376.3
1200	359.3	347.0	317.1
1000	292.7	282.8	259.4
800	228.2	220.5	203.2
600	166.1	160.4	148.6
400	106.5	102.8	95.9
200	49.7	47.9	45.0

Note: Natural gas is Gas 1 and fuel oil refers to Gas 2 as shown in Table F.6. Multiply enthalpy given by 4.187 to obtain in kJ/kg and by 1.8 to obtain in Btu/lb.

TABLE F.11

Enthalpy of Some Common Gases, kcal/kg

Temp., °C	100	200	300	400	500	600	700	800	900	1000
CO_2	18	40.8	65.2	91.1	118.2	146.4	175.5	205.4	235.9	267
H_2O	37.7	83.7	131.3	180.4	231	283	336.6	391.6	448.1	506.2
N_2	20.9	46.1	71.8	97.9	124.5	151.6	179.1	207	235.4	264.2
O_2	18.7	41.5	65	89.1	113.8	139	164.7	190.7	217	243.6
SO_2	13	29.5	47	65.3	84.5	104.2	124.5	145.1	166.1	187.2

Note: Multiply by 1.8 to obtain enthalpy in Btu/lb and by 4.187 to obtain enthalpy in kJ/kg.

Estimating Energy Recovered from Turbine Exhaust

Using the enthalpy of exhaust gases shown in Table F.12, one may estimate the energy from typical unfired or fired turbine exhaust gases. For example, the energy recovered when 100,000 kg/h of turbine exhaust gas is cooled from 470°C to 120°C in an unfired HRSG is 100,000 × (118.9 − 26.2) = 9,268,000 kcal/h = 10,776 kW. Enthalpy values are interpolated to 470°C and 120°C.

Turbine exhaust gases are fired to 1000°C and cooled to 400°C. Determine the energy recoverable. The correct method is to compute the exhaust gas analysis when fired to 1000°C and then compute the enthalpy. As an approximation, we may use the 1200°C gas and see that the change in enthalpy = (283.2 − 103.5) = 179.7 kcal/kg. By extrapolating the 800°C gas, we see that at 1000°C, the enthalpy is 274.3 kcal/kg, and at 400°C, it is 101.2, and the difference = 173.1 kcal/kg. Hence, the average difference is 0.5(179.7 + 173.1) = 176.4 kcal/kg (738.59 kJ/kg) (317.52 Btu/lb).

TABLE F.12

Enthalpy of Turbine Exhaust Gases

Temperature, °C	A—1600	B—1200	C—800	D—500
1600	495.2			
1500	460.1			
1400	425.3			
1300	391.0			
1200	357.1	347.3		
1100	323.8	315.0		
1000	290.9	283.2		
900	258.6	251.9		
800	226.9	221.1	215.9	
700	195.7	190.8	186.4	
600	165.1	161.1	157.5	
500	135.2	132.0	129.4	127.1
400	105.9	103.5	101.2	99.8
300	77.3	75.6	74	73.0
200	49.4	48.4	47.4	46.7
100	22.3	21.8	21.4	21.1

Notes: Multiply by 1.8 to obtain enthalpy in Btu/lb and by 4.187 to obtain enthalpy in kJ/kg.

A—turbine exhaust fired from 500°C to 1600°C: % volume $CO_2 = 8.27$, $H_2O = 17.28$, $N_2 = 71$, $O_2 = 3.4$.

B—turbine exhaust fired from 500°C to 1200°C: % volume $CO_2 = 6.2$, $H_2O = 13.24$, $N_2 = 72.56$, $O_2 = 7.98$.

C—turbine exhaust fired from 500°C to 800°C: % volume $CO_2 = 4.28$, $H_2O = 9.51$, $N_2 = 74.0$, $O_2 = 12.17$.

D—unfired turbine exhaust: % volume $CO_2 = 3$, $H_2O = 7$, $N_2 = 75$, $O_2 = 15$

Flue Gas Mixture Properties Calculation

Flue gas streams with different gas analysis often mix, and the final gas analysis is required to be determined to evaluate the mixture gas properties. This simple procedure is illustrated in the following example.

Example F.3

Determine the mixture volumetric analysis of 50,000 kg/h of air (with analysis % volume $H_2O = 1$, $N_2 = 78$, $O_2 = 21$) that mixes with 10,000 kg/h of flue gases having an analysis % volume $CO_2 = 8$, $H_2O = 18$, $N_2 = 71$, $O_2 = 3$.

Solution

In steam generators, flue gas recirculation is widely adopted to lower NO_x emissions. The duct carrying the flue gases (often from economizer exit) should be insulated; else, water dew point temperature could be reached in the duct itself or as soon as it is mixed with cold air. In order to estimate the water dew point after mixing, the flue gas analysis of the mixture must be known.

First convert the % volume to % weight basis for both the streams.

MW of air $= 0.01 \times 18 + 0.78 \times 28 + 0.21 \times 32 = 28.74$. % weight $H_2O = 100 \times 0.01 \times 18/28.74 = 0.626$.% weight of $N_2 = 100 \times 0.78 \times 28/28.74 = 76$. % weight $O_2 = 100 \times 0.21 \times 32/28.74 = 23.38$

Hence, amount of water vapor in air $= 50,000 \times 0.00626 = 313$ kg/h
Amount of $N_2 = 50,000 \times 0.76 = 38,000$ kg/h. Amount of $O_2 = 11,687$ kg/h
Similarly, MW of flue gas $= 0.08 \times 44 + 0.18 \times 18 + 0.71 \times 28 + 0.03 \times 32 = 27.6$
% weight of $CO_2 = 8 \times 44/27.6 = 12.75$. % weight $H_2O = 18 \times 18/27.6 = 11.74$
% weight of $N_2 = 71 \times 28/27.6 = 72$. % weight $O_2 = 3 \times 32/27.6 = 3.48$
Amount of $CO_2 = 10,000 \times 0.1275 = 1,275$ kg/h

$$H_2O = 0.1174 \times 10,000 = 1,174 \text{ kg/h } N_2 = 0.72 \times 10,000 = 7,200 \text{ kg/h}$$

$$O_2 = 0.0348 \times 10,000 = 348 \text{ kg/h}$$

Total CO_2 in mixture $= 1,275$ kg/h. Amount of $H_2O = 1,174 + 313 = 1,487$ kg/h
Amount of $N_2 = 38,000 + 7,200 = 45,200$ kg/h. Amount of $O_2 = 11,687 + 348 = 12,035$ kg/h
Converting to % volume basis: total moles $= 1,275/44 + 1,487/18 + 45,200/28 + 12,035/32 = 2,102$
% volume $CO_2 = 100 \times (1275/44)/2102 = 1.38$, % volume $H_2O = 100 \times (1487/18)/2102 = 3.93$
% volume $N_2 = 100 \times 45,200/28/2,102 = 76.8$ and % volume $O_2 = 17.89$
The water vapor dew point based on 0.0393 kg/cm^2 a gas partial pressure is 29°C.
The effect of gas pressure on properties is discussed in Appendices B and C.
Properties of saturated water are required for estimating the natural convection or condensation heat transfer coefficients. This is given in Table F.13.
Here is a compilation of important formulas used for heat transfer and pressure drop calculations inside and outside tubes (Table F.14).

TABLE F.13

Properties of Saturated Water

t (°F)	C_p (Btu/lb °F)	ρ (lb/ft³)	μ (lb/ft h)	ν (ft²/h)	κ (Btu/h ft °F)	α (ft²/h)	β (R⁻¹)	N
32	1.009	62.42	4.33	0.0694	0.327	0.0052	0.03×10^{-3}	13.37
40	1.005	62.42	3.75	0.0601	0.332	0.0053	0.045	11.36
50	1.002	62.38	3.17	0.0508	0.338	0.0054	0.070	9.41
60	1.000	62.34	2.71	0.0435	0.344	0.0055	0.10	7.88
70	0.998	62.27	2.37	0.0381	0.349	0.0056	0.13	6.78
80	0.998	62.17	2.08	0.0334	0.355	0.0057	0.15	5.85
90	0.997	62.11	1.85	0.0298	0.360	0.0058	0.18	5.13
100	0.997	61.99	1.65	0.0266	0.364	0.0059	0.20	4.52
110	0.997	61.84	1.49	0.0241	0.368	0.0060	0.22	4.04
120	0.997	61.73	1.36	0.0220	0.372	0.0060	0.24	3.65
130	0.998	61.54	1.24	0.0202	0.375	0.0061	0.27	3.30
140	0.998	61.39	1.14	0.0186	0.378	0.0062	0.29	3.01
150	0.999	61.20	1.04	0.0170	0.381	0.0063	0.31	2.72
160	1.000	61.01	0.97	0.0159	0.384	0.0063	0.33	2.53
170	1.001	60.79	0.90	0.0148	0.386	0.0064	0.35	2.33
180	1.002	60.57	0.84	0.0139	0.389	0.9064	0.37	2.16
190	1.003	60.35	0.79	0.0131	0.390	0.0065	0.39	2.03
200	1.004	60.13	0.74	0.0123	0.392	0.0065	0.41	1.90
210	1.005	59.88	0.69	0.0115	0.393	0.0065	0.43	1.76
220	1.007	59.63	0.65	0.0109	0.395	0.0066	0.45	1.66
230	1.009	59.38	0.62	0.0104	0.395	0.0066	0.47	1.58
240	1.011	59.10	0.59	0.0100	0.396	0.0066	0.48	1.51
250	1.013	58.82	0.56	0.0095	0.396	0.0066	0.50	1.43
260	1.015	58.51	0.53	0.0091	0.396	0.0067	0.51	1.36
270	1.017	58.24	0.50	0.0086	0.396	0.0067	0.53	1.28
280	1.020	57.94	0.48	0.0083	0.396	0.0067	0.55	1.24
290	1.023	57.64	0.46	0.0080	0.396	0.0067	0.56	1.19
300	1.026	57.31	0.45	0.0079	0.395	0.0067	0.58	1.17
350	1.044	55.59	0.38	0.0068	0.391	0.0067	0.62	1.01
400	1.067	53.65	0.33	0.0062	0.384	0.0068	0.72	0.91
450	1.095	51.55	0.29	0.0056	0.373	0,0066	0.93	0.85
500	1.130	49.02	0.26	0.0053	0.356	0.0064	1.18	0.83
550	1.200	45.92	0.23	0.0050	0.330	0.0060	1.63	0.84
600	1.362	42.37	0.21	0.0050	0.298	0.0052	—	0.96

TABLE F.14

Important Formulae in All Three Units

SI Units	British Units	Metric Units
Heat transfer inside tubes		
$h_c = 0.0278w^{0.8} (C_p/\mu)^{0.4} k^{0.6}/d_i^{1.8}$	$h_c = 2.44w^{0.8} (C_p/\mu)^{0.4}k^{0.6}/d_i^{1.8}$	$h_c = 0.0278(C_p/\mu)^{0.4}k^{0.6}w^{0.8}/d_i^{1.8}$
$h_c = 0.0278Cw^{0.8}/d_i^{1.8}$	$h_c = 2.44Cw^{0.8}/d_i^{1.8}$	$h_c = 0.0278Cw^{0.8}/d_i^{1.8}$
h_c, W/m²K	h_c, Btu/ft²h °F	h_c, kcal/m²h °C
w, kg/s	w, lb/h	w, kg/h
d_i, m	d_i, in.	d_i, m
C_p, J/kg K	C_p, Btu/lb °F	C_p, kcal/kg °C
μ, kg/m s	μ, lb/ft h	μ, kg/m h
k, W/m K	k, Btu/ft h °F	k, kcal/m h °C
$C_{SI} = (C_p/\mu)^{0.4}k^{0.6}$	C, multiply C_{SI} by 0.001134	C, multiply C_{SI} by 0.001229
$V = 1.246wv/d_i^2$	$V = 0.05wv/d_i^2$	$V = 3.461 \times 10^{-4}wv/d_i^2$
v, m³/kg	ft³/lb	m³/kg
V, velocity in m/s	ft/s	m/s
$Re = 1.273w/(d_i\mu)$	$Re = 15.2w/(d_i\mu)$	$Re = 1.273w/(d_i\mu)$
Density of gas, air		
$\rho = 1.203 \times 10^{-4}$ MW P/T	$\rho = 0.0933$MW P/T	$\rho = 11.792$ MW P/T
P abs pressure, Pa	Psia	kg/cm² a
T, K	T, °R	T, K
ρ, kg/m³	ρ, lb/ft³	ρ, kg/m³
Pressure drop inside tubes, duct		
$\Delta P = 810 \times 10^{-6}fL_evw^2/di^5$	$\Delta P = 3.36 \times 10^{-6}fL_evw^2/d_i^5$	$\Delta P = 0.6375 \times 10^{-12}fL_evw^2/d_i^5$
w, kg/s	w, lb/h	w, kg/h
ΔP, kPa	ΔP, psi	ΔP, kg/cm²
L_e, m	L_e, ft	L_e, m
v, m³/kg	v, ft³/kg	v, m³/kg
d_i, m	d_i, in.	d_i, m
Pressure drop of flue gas, air		
$\Delta P = 0.08262fL_evw^2/d_i^5$	$\Delta P = 93 \times 10^{-6}fL_evw^2/d_i^5$	$\Delta P = 6.382 \times 10^{-9}fL_evw^2/d_i^5$
ΔP, mm wc in above equation	ΔP, in wc	ΔP, mm wc
Natural convection coefficient		
$Nu = 0.54[d^3\rho^2gb\Delta TC_p/(\mu k)]^{0.25}$	$Nu = 0.54[d^3\rho^2gb\Delta TC_p/(\mu k)]^{0.25}$	$Nu = 0.54[d^3\rho^2gb\Delta TC_p/(\mu k)]^{0.25}$
$h_c = 0.953[k^3\rho^2b\Delta TC_p/(\mu d_o)]^{0.25}$	$h_c = 144[k^3\rho^2b\Delta TC_p/(\mu d_o)]^{0.25}$	$h_c = 57.2[k^3\rho^2b\Delta TC_p/(\mu d_o)]^{0.25}$
h_c, W/m²K	Btu/ft²h °F	kcal/m²h °C
d_o, tube OD, m	ft	M
ρ, density, kg/m³	lb/ft³	kg/m³
g, acceleration gravity	32×3600^2 ft/h²	9.8×3600^2 m/h²
b, expansion coefficient, 1/K	1/°R	1/K
ΔT, K	°R	K
k, W/m K	Btu/ft h °F	kcal/m h °C
μ, kg/m s	lb/ft h	kg/m h
C_p, J/kg K	Btu/lb °F	kcal/kg °C

(Continued)

TABLE F.14 (*Continued*)

Important Formulae in All Three Units

SI Units	British Units	Metric Units
Gas pressure drop outside tubes		
$\Delta P_g = 0.204 f G^2 N_d / \rho_g$	$\Delta P_g = 9.3 \times 10^{-10} f G^2 N_d / \rho_g$	$\Delta P_g = 1.574 \times 10^{-8} f G^2 N_d / \rho_g$
ΔP_g, mm wc	ΔP_g, in wc	ΔP_g, mm wc
G, kg/m² s	G, lb/ft² h	G, kg/m² h
ρ_g, kg/m³	ρ_g, lb/ft³	ρ_g, kg/m³
N_d, number of rows deep		
Pressure drop in two-phase flow		
$\Delta P_a = 7.65 \times 10^{-11} v_f G i^2 r2$	$1.664 \times 10^{-11} v_f G^2 r2$	$\Delta P_a = 7.8 \times 10^{-13} v_f G i^2\ r2$
$\Delta P_f = 38 \times 10^{-12} f v_f L G_i^2 r3 / d_i$	$4 \times 10^{-10} f v_f L G^2 r3 / d_i$	$\Delta P_f = 0.388 \times 10^{-12} f v_f L G_i^2 r3 / d_i$
$\Delta P_g = 0.00981 L r4 / v_f$	$0.00695 L r4 / v_f$	$\Delta P_g = 0.0001 L r4 / v_f$
G, kg/m² h	lb/ft² h	G, kg/m² h
w, kg/s	lb/h	w, kg/s
L, m	ft	L, m
f, Moody's friction factor	Fanning friction factor = Moody's/4	f, Moody's friction factor
d_i, m	in.	d_i, m
v_f, m³/kg	ft³/lb	v_f, m³/kg
ΔP, kPa	psi	kPa

Note: This table is a compilation of important formulas used for heat transfer and pressure drop calculations inside and outside tubes.

Properties of steam and water are provided in Tables F.15 through F.25.

TABLE F.15

Thermodynamic Properties of Dry Saturated Steam—Pressure Table

Abs Press., psi p	Temp. °Fi	Specific Volume		Enthalpy			Entropy			Internal Energy		Abs Press., psi p
		Sat. Liquid v_f	Sat. Vapor h_e	Sat. Liquid h_f	Evap. h_{fe}	Sat. Vapor h_e	Sat. Liquid s_f	Evap. s_{fe}	Sat. Vapor s_e	Sat. Liquid u_e	Sat. Vapor u_e	
1.0	101.74	0.01614	333.6	69.70	1036.3	1106.0	0.1326	1.8456	1.9782	69.70	1044.3	1.0
2.0	126.08	0.01623	173.73	93.99	1022.2	1116.2	0.1749	1.7451	1.9200	93.98	1051.9	2.0
3.0	141.48	0.01630	118.71	109.37	1013.2	1122.6	0.2008	1.6855	1.8863	109.36	1056.7	3.0
4.0	152.97	0.01636	90.63	120.86	1006.4	1127.3	0.2198	1.6427	1.8625	120.85	1060.2	4.0
5.0	162.24	0.01640	73.52	130.13	1001.0	1131.1	0.2347	1.6094	1.8441	130.12	1063.1	5.0
6.0	170.06	0.01645	61.98	137.96	996.2	1134.2	0.2472	1.5820	1.8292	137.94	1065.4	6.0
7.0	176.85	0.01649	53.64	144.76	992.1	1136.9	0.2581	1.5586	1.8167	144.74	1067.4	7.0
8.0	182.86	0.01653	47.34	150.79	988.5	1139.3	0.2674	1.5383	1.8057	150.77	1069.2	8.0
9.0	188.28	0.01656	42.40	156.22	985.2	1141.4	0.2759	1.5203	1.7962	156.19	1070.8	9.0
10	193.21	0.01659	38.42	161.17	982.1	1143.3	0.2835	1.5041	1.7876	161.14	1072.2	10
14.696	212.00	0.01672	26.80	180.07	970.3	1150.4	0.3120	1.4446	1.7566	180.02	1077.5	14.696
15	213.03	0.01672	26.29	181.11	969.7	1150.8	0.3135	1.4415	1.7549	181.06	1077.8	15
20	227.96	0.01683	20.089	196.16	960.1	1156.3	0.3356	1.3962	1.7319	196.10	1081.9	20
25	240.07	0.01692	16.303	208.42	952.1	1160.6	0.3533	1.3606	1.7139	208.34	1085.1	25
30	250.33	0.01701	13.746	218.52	945.3	1164.1	0.3680	1.3313	1.6993	218.73	1087.8	30
35	259.28	0.01708	11.898	227.91	939.2	1167.1	0.3807	1.3063	1.6870	227.80	1090.1	35
40	267.25	0.01715	10.498	236.03	933.7	1169.7	0.3919	1.2844	1.6763	235.90	1092.0	40
45	274.44	0.01721	9.401	243.36	928.6	1172.0	0.4019	1.2650	1.6669	243.22	1093.7	45
50	281.01	0.01727	8.515	250.09	924.0	1174.1	0.4110	1.2474	1.6585	249.93	1095.3	50
55	287.07	0.01732	7.787	256.30	919.6	1175.9	0.4193	1.2316	1.6509	256.12	1096.7	55
60	292.71	0.01738	7.175	262.09	915.5	1177.6	0.4270	1.2168	1.6438	261.90	1097.9	60
65	297.97	0.01743	6.655	267.50	911.6	1179.1	0.4342	1.2032	1.6374	267.29	1099.1	65
70	302.92	0.01748	6.206	272.61	907.9	1180.6	0.4409	1.1906	1.6315	272.38	1100.2	70
75	307.60	0.01753	5.816	277.43	904.5	1181.9	0.4472	1.1787	1.6259	277.19	1101.2	75
80	312.03	0.01757	5.472	282.02	901.1	1183.1	0.4531	1.1676	1.6207	281.76	1102.1	80
85	316.25	0.01761	5.168	286.39	807.8	1184.2	0.4587	1.1571	1.6158	286.11	1102.9	85
90	320.27	0.01766	4.896	290.56	894.7	1185.3	0.1641	1.1471	1.6112	290.27	1103.7	90

(*Continued*)

TABLE F.15 (*Continued*)

Thermodynamic Properties of Dry Saturated Steam—Pressure Table

Abs Press., psi p	Temp. °Fi	Specific Volume		Enthalpy			Entropy			Internal Energy		Abs Press., psi p
		Sat. Liquid v_f	Sat. Vapor h_e	Sat. Liquid h_f	Evap. h_{fe}	Sat. Vapor h_e	Sat. Liquid s_f	Evap. s_{fe}	Sat. Vapor s_e	Sat. Liquid u_e	Sat. Vapor u_e	
95	324.12	0.01770	4.652	294.56	891.7	1186.2	0.4692	1.1376	1.6068	294.25	1104.5	95
100	327.81	0.01774	4.432	298.40	888.8	1187.2	0.4740	1.1286	1.6026	298.08	1105.2	100
110	334.77	0.01782	4.049	303.66	883.2	1188.9	0.4832	1.1117	1.5948	305.80	1106.5	110
120	341.25	0.01789	3.728	312.44	877.9	1190.4	0.4916	1.0962	1.5878	312.05	1107.6	120
130	347.32	0.01796	3.455	318.81	879.9	1191.7	0.4995	1.0817	1.5812	318.38	1108.6	130
140	353.02	0.01802	3.220	324.82	868.2	1193.0	0.6069	1.0682	1.5751	324.35	1109.6	140
150	358.42	0.01809	3.015	330.51	863.6	1194.1	0.5138	1.0556	1.5094	330.01	1110.5	150
160	363.53	0.01815	2.834	333.93	859.2	1195.1	0.5204	1.0436	1.5640	335.39	1111.2	160
170	368.41	0.01822	2.675	341.09	854.9	1196.0	0.5266	1.0324	1.5590	340.52	1111.9	170
180	373.06	0.01827	2.532	346.03	850.8	1196.9	0.5325	1.0217	1.5542	345.42	1112.5	180
190	377.51	0.01833	2.404	350.79	846.8	1197.6	0.5381	1.0116	1.5497	350.15	1113.1	190
200	381.79	0.01839	2.288	355.36	843.0	1198.4	0.5435	1.0018	1.5453	354.68	1113.7	200
250	400.95	0.01865	1.8438	376.00	825.1	1201.1	0.5675	0.9588	1.5263	375.14	1115.8	250
300	417.33	0.01890	1.5433	393.84	809.0	1202.8	0.5879	0.9225	1.5104	392.79	1117.1	300
350	431.72	0.01913	1.3260	409.69	794.2	1203.9	0.6056	0.8910	1.4966	408.45	1118.0	350
400	444.59	0.0193	1.1613	424.0	780.5	1204.5	0.6214	0.8630	1.4844	422.6	1118.5	400
450	456.28	0.0195	1.0320	437.2	767.4	1204.6	0.6356	0.8378	1.4734	435.5	1118.7	450
500	467.01	0.0197	0.9278	449.4	755.0	1204.4	0.6487	0.8147	1.4634	447.6	1118.6	500
550	476.94	0.0199	0.8424	460.8	743.1	1203.9	0.6608	0.7934	1.4542	458.8	1118.2	550
600	486.21	0.0201	0.7698	471.6	731.6	1203.2	0.6720	0.7734	1.4454	469.4	1117.7	600
650	494.90	0.0203	0.7083	481.8	720.5	1202.3	0.6826	0.7548	1.4374	479.4	1117.1	650
700	503.10	0.0205	0.6554	491.5	709.7	1201.2	0.6925	0.7371	1.4296	488.8	1116.3	700
750	510.86	0.0207	0.6092	500.8	699.2	1200.0	0.7019	0.7204	1.4223	598.0	1115.4	750
800	518.23	0.0209	0.5687	509.7	688.9	1198.6	0.7108	0.7045	1.4153	506.6	1114.4	800
850	525.26	0.0210	0.5327	518.3	678.8	1197.1	0.7194	0.6891	1.4085	515.0	1113.3	850

900	531.98	0.0212	0.5006	526.6	668.8	1195.4	0.7275	0.6744	1.4020	523.1	1112.1	900
950	536.43	0.0214	0.4717	534.6	659.1	1193.7	0.7355	0.6602	1.3957	530.9	1110.8	950
1000	544.61	0.0216	0.4456	542.4	649.4	1191.8	0.7430	0.6467	1.3897	538.4	1109.4	1000
1100	556.31	0.0220	0.4001	557.4	630.4	1187.8	0.7575	0.6205	1.3780	552.9	1106.4	1100
1200	567.22	0.0223	0.3619	571.7	611.7	1183.4	0.7711	0.5956	1.3667	566.7	1103.0	1200
1300	577.46	0.0227	0.3293	585.4	593.2	1178.6	0.7840	0.5719	1.3559	580.0	1099.4	1300
1400	587.10	0.0231	0.3012	598.7	574.7	1173.4	0.7963	0.5491	1.3454	592.7	1095.4	1400
1500	596.23	0.0235	0.2765	611.6	556.3	1167.9	0.8082	0.5269	1.3351	605.1	1091.2	1500
2000	635.82	0.0257	0.1878	671.7	463.4	1135.1	0.8619	0.4230	1.2849	662.2	1065.6	2000
2500	668.13	0.0287	0.1307	730.6	360.5	1091.1	0.9126	0.3197	1.2322	717.3	1030.6	2500
3000	695.36	0.0346	0.0858	802.5	217.8	1020.3	0.9731	0.1885	1.1615	783.4	972.7	3000
3206.2	705.40	0.0503	0.0503	902.7	0	902.7	1.0580	0	1.0580	872.9	872.9	3206

Source: Abridged from Keenan, J.H. and Keyes, F.G., *Thermodynamic Properties of Steam*, John Wiley & Sons, Inc., New York, 1937.

TABLE F.16

Thermodynamic Properties of Dry Saturated Steam—Temperature Table

Temp., °Fi	Abs Press., psi p	Specific Volume			Enthalpy			Entropy			Temp., °Ft
		Sat. Liquid v_f	Evap. v_{fe}	Sat. Vapor v_e	Sat. Liquid h_f	Evap. h_{fe}	Sat. Vapor h_e	Sat. Liquid s_f	Evap. a_{fe}	Sat. Vapor s_e	
32	0.08854	0.01602	3306	3306	0.00	1075.8	1075.8	0.0000	2.1877	2.1877	32
35	0.09995	0.01602	2947	2947	3.02	1074.1	1077.1	0.0061	2.1709	2.1770	35
40	0.12170	0.01602	2444	2444	8.05	1071.3	1079.3	0.0162	2.1435	2.1597	40
45	0.14752	0.01602	2036.4	2036.4	13.06	1068.4	1081.5	0.0262	2.1167	2.1429	45
50	0.17811	0.01603	1703.2	1703.2	18.07	1065.6	1083.7	0.0361	2.0903	2.1264	50
60	0.2563	0.01604	1206.6	1206.7	28.06	1059.9	1088.0	0.0555	2.0393	2.0948	60
70	0.3631	0.01606	867.8	867.9	38.04	1054.3	1092.3	0.0745	1.9902	2.0647	70
80	0.5069	0.01608	633.1	633.1	48.02	1048.6	1096.6	0.0932	1.9428	2.0360	80
90	0.6982	0.01610	468.0	468.0	57.99	1042.9	1100.0	0.1115	1.8972	2.0087	90
100	0.9492	0.01613	350.3	350.4	67.97	1037.2	1105.2	0.1295	1.8531	1.9826	100
110	1.2748	0.01617	265.3	265.4	77.94	1031.6	1109.5	0.1471	1.8106	1.9577	110
120	1.6924	0.01620	203.25	203.27	87.92	1025.8	1113.7	0.1645	1.7694	1.9339	120
130	2.2225	0.01625	157.32	157.34	97.90	1020.0	1117.9	0.1816	1.7296	1.9112	130
140	2.8886	0.01629	122.99	123.01	107.89	1014.1	1122.0	0.1984	1.6910	1.8894	140
150	3.718	0.01634	98.06	97.07	117.89	1008.2	1126.1	0.2149	1.6537	1.8685	150
160	4.741	0.01639	77.27	77.29	127.89	1002.3	1130.2	0.2311	1.6174	1.8485	160
170	5.992	0.01645	62.04	62.06	137.90	996.3	1134.2	0.2472	1.5822	1.8293	170
180	7.510	0.01651	50.21	50.23	147.92	990.2	1138.1	0.2630	1.5480	1.8109	180
190	9.339	0.01657	40.94	40.96	157.95	984.1	1142.0	0.2785	1.5147	1.7932	190
200	11.526	0.01663	33.62	33.64	167.99	977.9	1145.9	0.2938	1.4824	1.7762	200
210	14.123	0.01670	27.80	27.82	178.05	971.6	1149.7	0.3090	1.4508	1.7598	210
312	14.696	0.01672	26.78	26.80	180.07	970.3	1150.4	0.3120	1.4446	1.7566	212
220	17.186	0.01677	23.13	23.15	188.13	965.2	1153.4	0.3239	1.4201	1.7440	220
230	20.780	0.01684	19.365	19.382	198.23	958.8	1157.0	0.3387	1.3901	1.7288	230

240	24.969	0.01692	16.306	16.323	208.34	952.2	1160.5	0.3531	1.3609	1.7140	240
250	29.825	0.01700	13.804	13.821	216.48	945.5	1164.0	0.3675	1.3323	1.6998	250
260	35.429	0.01709	11.746	11.763	228.64	938.7	1167.3	0.3817	1.3043	1.6860	260
270	41.858	0.01717	10.044	10.061	238.84	931.8	1170.6	0.3958	1.2769	1.6727	270
280	49.203	0.01726	8.628	8.645	249.06	924.7	1173.8	0.4096	1.2501	1.6597	280
290	57.556	0.01735	7.444	7.461	259.31	917.5	1176.8	0.4234	1.2238	1.6472	290
300	67.013	0.01745	6.449	6.466	269.59	910.1	1179.7	0.4369	1.1980	1.6350	300
310	77.68	0.01755	5.609	5.626	279.92	902.6	1182.5	0.4504	1.1727	1.6231	310
320	89.66	0.01765	4.896	4.914	290.28	894.9	1185.2	0.4637	1.1478	1.6115	320
330	103.06	0.01776	4.289	4.307	300.68	887.0	1187.7	0.4769	1.1233	1.6002	330
340	118.01	0.01787	3.770	3.788	311.13	879.0	1190.1	0.4900	1.0992	1.5891	340
350	134.63	0.01799	3.324	3.342	321.63	870.7	1192.3	0.5029	1.0754	1.5783	350
360	153.04	0.01811	2.939	2.957	332.18	862.2	1194.0	0.5158	1.0519	1.5677	360
370	173.37	0.01823	2.606	2.625	342.79	853.5	1196.3	0.5286	1.0287	1.5573	370
380	195.77	0.01836	2.317	2.335	353.45	844.6	1198.1	0.5413	1.0059	1.5471	380
390	220.37	0.01850	2.0651	2.0836	364.17	835.4	1199.6	0.5539	0.9832	1.5371	390
400	247.31	0.01864	1.8447	1.8633	374.97	826.0	1201.0	0.5664	0.9608	1.5272	400
410	276.75	0.01878	1.6312	1.6700	385.83	816.2	1202.1	0.5788	0.9386	1.5174	410
420	308.83	0.01894	1.4811	1.5000	396.77	806.3	1203.1	0.5912	0.9166	1.5078	420
430	343.72	0.01910	1.3308	1.3499	407.79	796.0	1203.8	0.6035	0.8947	1.4982	430
440	381.59	0.01926	1.1979	1.2171	418.90	785.4	1204.3	0.6158	0.8730	1.4887	440
450	422.6	0.0194	1.0799	1.0993	430.1	774.5	1204.6	0.6280	0.8513	1.4793	450
460	466.9	0.0196	0.9748	0.9944	441.4	763.2	1204.6	0.6402	0.8298	1.4700	460
470	514.7	0.0198	0.8811	0.9009	452.8	751.5	1204.3	0.6523	0.8083	1.4606	470
480	566.1	0.0200	0.7972	0.8172	464.4	739.4	1203.7	0.6645	0.7868	1.4513	480
490	621.4	0.0202	0.7221	0.7423	476.0	726.8	1202.8	0.6766	0.7653	1.4419	490
500	680.8	0.0204	0.6545	0.6749	487.8	713.9	1201.7	0.6887	0.7438	1.4325	500
520	812.4	0.0209	0.5385	0.5594	511.9	686.4	1198.2	0.7130	0.7006	1.4136	520
540	962.5	0.0215	0.4434	0.4649	536.6	636.6	1193.2	0.7374	0.6568	1.3942	540
560	1133.1	0.0221	0.3647	0.3868	562.2	624.2	1186.4	0.7621	0.6121	1.3742	560

(Continued)

TABLE F.16 (*Continued*)

Thermodynamic Properties of Dry Saturated Steam—Temperature Table

Temp., °Fi	Abs Press., psi p	Specific Volume			Enthalpy			Entropy			Temp., °Ft
		Sat. Liquid v_f	Evap. v_{fe}	Sat. Vapor v_e	Sat. Liquid h_f	Evap. h_{fe}	Sat. Vapor h_e	Sat. Liquid s_f	Evap. a_{fe}	Sat. Vapor s_e	
580	1325.8	0.0228	0.2989	0.3217	588.9	588.4	1177.3	0.7872	0.5659	1.3532	580
600	1542.9	0.0236	0.2432	0.2668	617.0	548.5	1165.5	0.8131	0.5176	1.3307	600
620	1786.6	0.0247	0.1955	0.2201	646.7	503.6	1130.3	0.8398	0.4664	1.3062	620
640	2059.7	0.0260	0.1538	0.1798	678.6	452.0	1130.5	0.8679	0.4110	1.2789	640
660	2365.4	0.0278	0.1165	0.1442	714.2	390.2	1104.4	0.8987	0.3485	1.2472	660
680	2708.1	0.0305	0.0810	0.1115	757.3	309.9	1067.2	0.9351	0.2719	1.2071	680
700	3093.7	0.0369	0.0392	0.0761	823.3	172.1	993.4	0.9905	0.1484	1.1359	700
705.4	3206.2	0.0503	0	0.0503	902.7	0	902.1	1.0580	0	1.0580	706.1

Source: Abridged from Keenan, J.H. and Frederick, G., Keyes, *Thermodynamic Properties of Steam*, John Wiley & Sons, New York, 1937.

TABLE F.17

Thermodynamic Properties of Superheated Steam

Abs. Press., psi (Sat. Temp.)		Temp. °F												
		200	300	400	500	600	700	800	900	1000	1100	1200	1400	1600
1 (101.74)	v	392.6	452.3	512.0	571.6	631.2	690.8	750.4	809.9	869.5	929.1	988.7	1107.8	1227.0
	h	1150.4	1195.8	1241.7	1288.3	1335.7	1383.8	1432.8	1482.7	1533.5	1585.2	1637.7	1745.7	1857.5
	s	2.0612	2.1153	2.1720	2.2233	2.2702	2.3137	2.3542	2.3923	2.4283	2.4625	2.4952	2.5566	2.6137
5 (163.24)	v	78.16	90.25	102.26	114.22	126.16	138.10	150.03	161.95	173.87	185.79	197.71	221.6	245.4
	h	1148.8	1195.0	1241.2	1288.0	1335.4	1383.6	1432.7	1482.6	1533.4	1585.1	1637.7	1745.7	1857.5
	s	1.8718	1.9370	1.9942	2.0456	2.0927	2.1361	2.1767	2.2148	2.2509	2.2851	2.3178	2.3792	2.4363
10 (193.21)	v	38.85	45.00	51.04	57.05	63.03	69.01	74.98	80.95	86.92	92.88	98.84	110.77	122.69
	h	1146.6	1193.9	1240.6	1287.5	1335.1	1383.4	1432.5	1482.4	1533.2	1585.0	1637.6	1745.6	1857.3
	s	1.7927	1.8595	1.9172	1.9689	2.0160	2.0596	2.1002	2.1383	2.1744	2.2086	2.2413	2.3028	2.3598
14.096 (212.00)	v	—	30.53	34.68	38.78	42.56	46.94	51.00	55.07	59.13	63.19	67.25	75.37	83.48
	h	—	1192.8	1239.9	1287.1	1334.8	1383.2	1432.3	1482.3	1533.1	1584.8	1637.5	1745.5	1857.3
	s	—	1.8160	1.8743	1.9261	1.9734	2.0170	2.0576	2.0958	2.1319	2.1662	2.1989	2.2603	2.3174
20 (237.90)	v	—	22.36	25.43	28.46	31.47	34.47	37.46	40.45	43.44	46.42	49.41	55.37	61.34
	h	—	1191.6	1239.2	1286.6	1334.4	1382.9	1432.1	1482.1	1533.0	1584.7	1637.4	1745.4	1857.2
	s	—	1.7808	1.8396	1.8918	1.9392	1.9829	2.0235	2.0618	2.0978	2.1321	2.1648	2.2243	2.2834
40 (267.25)	v	—	11.040	12.628	14.168	15.688	17.198	18.702	20.20	21.70	23.20	24.69	27.68	30.86
	h	—	1186.8	1236.5	1284.8	1333.1	1381.9	1431.3	1481.4	1532.4	1584.3	1637.0	1745.1	1857.0
	s	—	1.6994	1.7506	1.8140	1.8619	1.9058	1.9467	1.9650	2.0212	2.0555	2.0883	2.1498	2.2069
60 (292.71)	v	—	7.259	8.357	9.403	10.427	11.441	12.449	13.452	14.454	15.453	16.451	18.446	20.44
	h	—	1181.8	1233.6	1283.0	1331.8	1380.9	1430.5	1480.8	1531.9	1583.3	1636.6	1744.8	1856.7
	s	—	1.6492	1.7135	1.7678	1.8162	1.8605	1.9015	1.9400	1.9762	2.0106	2.0434	2.1049	2.1621
80 (312.03)	v	—	—	6.220	7.020	7.797	8.562	9.322	10.077	10.830	11.582	12.232	13.830	15.325
	h	—	—	1230.7	1281.1	1330.5	1379.9	1429.7	1480.1	1531.3	1583.4	1636.2	1744.5	1856.5
	s	—	—	1.6791	1.7346	1.7836	1.8281	1.8694	1.9079	1.9442	1.9787	2.0115	2.0731	2.1303
100 (337.81)	v	—	—	4.937	5.589	6.218	6.835	7.446	8.052	8.656	9.259	9.860	11.060	12.268
	h	—	—	1227.6	1279.1	1329.1	1378.9	1428.9	1479.5	1530.8	1582.9	1635.7	1744.2	1856.2
	s	—	—	1.6518	1.7085	1.7581	1.8029	1.8443	1.8829	1.9193	1.9538	1.9867	2.0484	2.1056

(Continued)

TABLE F.17 (*Continued*)

Thermodynamic Properties of Superheated Steam

Abs. Press., psi (Sat. Temp.)		Temp. °F												
		200	300	400	500	600	700	800	900	1000	1100	1200	1400	1600
120 (341.25)	v	—	—	4.081	4.636	5.165	5.683	6.195	6.702	7.207	7.710	8.212	9.214	10.213
	h	—	—	1224.4	1277.2	1327.7	1377.8	1428.1	1478.8	1530.2	1582.4	1635.3	1743.9	1856.0
	s	—	—	1.6287	1.6869	1.7370	1.7822	1.8237	1.8625	1.8990	1.9335	1.9663	2.0281	2.0664
140 (353.02)	v	—	—	3.468	3.954	4.413	4.861	5.301	5.738	6.172	6.604	7.035	7.895	8.752
	h	—	—	1221.1	1275.2	1326.4	1376.8	1427.3	1476.2	1529.7	1581.9	1634.9	1743.5	1855.7
	s	—	—	1.6087	1.6683	1.7190	1.7645	1.8063	1.8151	1.8817	1.9163	1.9493	2.0110	2.0683
160 (363.53)	v	—	—	3.008	3.443	3.849	4.244	4.631	5.015	5.396	5.775	6.152	6.906	7.656
	h	—	—	1217.6	1273.1	1325.0	1375.7	1426.4	1477.5	1529.1	1581.4	1634.5	1743.2	1855.5
	s	—	—	1.5908	1.6519	1.7033	1.7491	1.7911	1.8301	1.8667	1.9014	1.9344	1.9962	2.0535
180 (373.06)	v	—	—	2.649	3.044	3.411	3.764	4.110	4.452	4.792	5.129	5.466	6.136	6.804
	h	—	—	1214.0	1271.0	1323.5	1374.7	1425.6	1476.8	1528.6	1581.0	1634.1	1742.9	1855.2
	s	—	—	1.5745	1.6373	1.6894	1.7355	1.7776	1.8167	1.8534	1.8882	1.9212	1.9831	2.0404
200 (381.79)	v	—	—	2.361	2.726	3.040	3.380	3.693	4.002	4.309	4.613	4.917	5.521	6.123
	h	—	—	1210.3	1268.9	1322.1	1373.6	1424.8	1476.2	1528.0	1580.5	1633.7	1742.6	1855.0
	s	—	—	1.5594	1.6240	1.6767	1.7232	1.7655	1.8048	1.8415	1.8763	1.9094	1.9713	2.0287
200 (300.86)	v	—	—	2.125	2.465	2.772	3.066	3.352	3.634	3.913	4.191	4.467	5.017	5.565
	h	—	—	1206.5	1266.7	1320.7	1372.0	1424.0	1475.5	1527.5	1580.0	1633.3	1742.3	1854.7
	s	—	—	1.5453	1.6117	1.6652	1.7120	1.7545	1.7939	1.8308	1.8656	1.8987	1.9607	2.0181
240 (397.37)	v	—	—	1.9276	2.247	2.533	2.804	3.068	3.327	3.584	3.839	4.093	4.597	5.100
	h	—	—	1202.5	1264.5	1319.2	1371.5	1423.2	1474.8	1526.9	1579.6	1632.9	1742.0	1854.5
	s	—	—	1.5219	1.6003	1.6546	1.7017	1.7444	1.7839	1.8209	1.8553	1.8889	1.9510	2.0064
260 (404.42)	v	—	—	—	2.062	2.330	2.552	2.827	3.067	3.305	3.541	3.776	4.242	4.707
	h	—	—	—	1262.3	1317.7	1370.4	1422.3	1474.2	1526.3	1579.1	1632.5	1741.7	1854.2
	s	—	—	—	1.5897	1.6447	1.6923	1.7362	1.7748	1.8118	1.8467	1.8799	1.9420	1.9905
280 (411.08)	v	—	—	—	1.9047	2.156	2.392	2.621	2.845	3.066	3.286	3.504	3.938	4.370
	h	—	—	—	1260.0	1316.2	1360.4	1421.5	1473.5	1525.8	1578.6	1632.1	1741.4	1854.0
	s	—	—	—	1.5796	1.6354	1.6834	1.7365	1.7662	1.8033	1.8383	1.8716	1.9337	1.9912
300 (417.33)	v	—	—	—	1.7675	2.005	2.227	2.442	2.652	2.859	3.065	3.269	3.674	4.078

Pressure values in psia with saturation temperature (°F) in parentheses. For each pressure: v (ft³/lb), h (Btu/lb), s (Btu/lb·°R). Temperatures in °F.

Continuation (350, 400 psia):

P (Tsat)		500	600	700	800	900	1000	1100	1200	1400	1600
350 (431.72)	h	1257.6	1314.7	1368.3	1420.6	1472.8	1525.2	1578.1	1631.7	1741.0	1853.7
	s	1.5701	1.6268	1.6751	1.7184	1.7582	1.7954	1.8305	1.8638	1.9260	1.9835
400 (444.59)	v	1.2851	1.4770	1.6508	1.8161	1.9767	2.134	2.290	2.445	2.798	3.055
	h	1245.1	1304.9	1362.7	1416.4	1469.4	1522.4	1575.8	1629.6	1739.5	1852.5
	s	1.5281	1.5894	1.6398	1.6842	1.7247	1.7623	1.7977	1.8311	1.8936	1.9513

Main table:

P (Tsat)		500	550	600	620	640	660	680	700	800	900	1000	1200	1400	1600
450 (456.28)	v	1.1231	1.2155	1.3005	1.3332	1.3652	1.3967	1.4276	1.4384	1.6074	1.7616	1.8928	2.170	2.443	2.714
	h	1238.4	1272.0	1302.8	1314.6	1326.6	1337.5	1348.8	1359.9	1414.3	1467.7	1521.0	1628.6	1738.7	1853.1
	s	1.5095	1.5437	1.5735	1.5845	1.5951	1.6054	1.6153	1.6250	1.6610	1.7106	1.7486	1.8177	1.8575	1.9381
500 (467.01)	v	0.9927	1.0600	1.1591	1.1883	1.2186	1.2478	1.2763	1.3044	1.4405	1.5715	1.6096	1.9504	2.197	2.442
	h	1231.3	1266.8	1298.6	1310.7	1322.6	1334.2	1345.7	1357.0	1412.1	1466.0	1519.6	1627.6	1737.9	1851.3
	s	1.4919	1.5280	1.5588	1.5701	1.5810	1.5915	1.6016	1.6115	1.6571	1.6962	1.7363	1.8056	1.8683	1.9262
550 (476.94)	v	0.8852	0.9686	1.0431	1.0714	1.0969	1.1259	1.1533	1.1783	1.3068	1.4241	1.5414	1.7704	1.9957	2.219
	h	1223.7	1261.2	1294.3	1306.8	1318.9	1330.8	1342.5	1354.0	1409.9	1464.3	1518.2	1626.6	1737.1	1850.6
	s	1.4751	1.5131	1.5451	1.5568	1.5680	1.5787	1.5890	1.5991	1.6452	1.6868	1.7250	1.7946	1.8575	1.9156
600 (486.21)	v	0.7947	0.8753	0.9463	0.9729	0.9988	1.0241	1.0489	1.0732	1.1899	1.3013	1.4096	1.6208	1.8279	2.033
	h	1215.7	1255.5	1289.9	1302.7	1315.2	1327.4	1339.3	1351.1	1407.7	1462.5	1516.7	1625.5	1736.3	1850.0
	s	1.4596	1.4990	1.5323	1.5443	1.5558	1.5667	1.5773	1.5875	1.6342	1.6762	1.7147	1.7846	1.8476	1.9068
700 (503.10)	v	—	0.7277	0.7934	0.8177	0.8411	0.8639	0.8860	0.9077	1.0108	1.1082	1.2024	1.3853	1.5641	1.7405
	h	—	1243.2	1280.6	1294.3	1307.5	1320.3	1332.8	1345.0	1403.2	1459.0	1515.9	1623.5	1734.8	1848.8
	s	—	1.4722	1.5064	1.5212	1.5323	1.5449	1.5559	1.5665	1.6147	1.6572	1.6962	1.7666	1.8299	1.8881
800 (518.23)	v	—	0.6154	0.6779	0.7006	0.7223	0.7433	0.7635	0.7833	0.8763	0.9623	1.0470	1.2066	1.3662	1.5214
	h	—	1229.8	1270.7	1285.4	1299.4	1312.9	1325.9	1338.6	1398.6	1455.4	1511.0	1621.4	1733.2	1847.5
	s	—	1.4467	1.4863	1.5000	1.5129	1.5250	1.5366	1.5476	1.5972	1.6407	1.6801	1.7510	1.8146	1.8729
900 (531.98)	v	—	0.5264	0.5873	0.6089	0.6294	0.6491	0.6680	0.6863	0.7716	0.8506	0.9262	1.0714	1.2124	1.3509
	h	—	1215.0	1250.1	1275.9	1290.9	1305.1	1318.8	1332.1	1393.9	1451.8	1508.1	1619.3	1731.6	1846.2
	s	—	1.4216	1.4653	1.4800	1.4938	1.5066	1.5187	1.5303	1.5814	1.6257	1.6656	1.7371	1.8009	1.8595

(Continued)

TABLE F.17 (Continued)

Thermodynamic Properties of Superheated Steam

Abs. Press., psi (Sat. Temp.)		Temp. °F													
		500	550	600	620	640	660	680	700	800	900	1000	1200	1400	1600
1000 (544.61)	v	—	0.4533	0.5140	0.5350	0.5546	0.5733	0.5912	0.6084	0.6878	0.7604	0.8294	0.9615	1.0893	1.2146
	h	—	1196.3	1248.8	1265.9	1281.9	1297.0	1311.4	1325.3	1389.2	1448.2	1505.1	1617.3	1730.0	1845.0
	s	—	1.3961	1.4450	1.4610	1.4757	1.4893	1.5021	1.5141	1.5670	1.6121	1.6525	1.7245	1.7886	1.8474
1100 (556.31)	v	—	—	0.4532	0.4738	0.4929	0.5110	0.5281	0.5445	0.6191	0.6866	0.7503	0.8716	0.9885	1.1031
	h	—	—	1236.7	1255.3	1272.4	1288.5	1303.7	1318.3	1384.3	1444.5	1502.2	1615.2	1728.4	1843.8
	s	—	—	1.4251	1.4425	1.4583	1.4728	1.4862	1.4989	1.5535	1.5995	1.6406	1.7130	1.7775	1.8262
1200 (567.22)	v	—	—	0.4016	0.4222	0.4410	0.4586	0.4752	0.4909	0.5617	0.6250	0.6843	0.7967	0.9046	1.0101
	h	—	—	1223.5	1243.9	1262.4	1279.6	1295.7	1311.0	1379.3	1440.7	1499.2	1613.1	1726.9	1842.5
	s	—	—	1.4052	1.4243	1.4413	1.4568	1.4710	1.4843	1.5409	1.5879	1.6293	1.7025	1.7672	1.8263
1400 (587.10)	v	—	—	0.3174	0.3390	0.3580	0.3753	0.3912	0.4062	0.4714	0.5281	0.5805	0.6789	0.7727	0.8640
	k	—	—	1193.0	1218.4	1240.4	1260.3	1278.5	1295.5	1369.1	1433.1	1493.2	1606.9	1723.7	1840.0
	s	—	—	1.3639	1.3877	1.4079	1.4258	1.4419	1.4567	1.5177	1.5666	1.6093	1.6836	1.7489	1.8063
1600 (604.9)	v	—	—	—	0.3733	0.2936	0.3112	0.3271	0.3417	0.4034	0.4553	0.5027	0.5906	0.6728	0.7545
	h	—	—	—	1187.8	1215.2	1238.7	1259.6	1278.7	1358.4	1425.3	1487.0	1604.6	1720.5	1837.5
	s	—	—	—	1.3489	1.3741	1.3952	1.4137	1.4303	1.4964	1.5476	1.5914	1.6669	1.7328	1.7926
1800 (621.03)	v	—	—	—	—	0.2407	0.2597	0.2760	0.2907	0.3502	0.3986	0.4421	0.5218	0.5968	0.6093
	h	—	—	—	—	1185.1	1214.0	1238.5	1260.3	1347.2	1417.4	1480.8	1600.4	1717.3	1835.0
	s	—	—	—	—	1.3377	1.3628	1.3855	1.4044	1.4765	1.5301	1.5752	1.6520	1.7185	1.7766
2000 (635.82)	v	—	—	—	—	0.1936	0.2161	0.2337	0.2489	0.3074	0.3532	0.3935	0.4668	0.5252	0.6011
	h	—	—	—	—	1145.6	1184.9	1214.8	1240.0	1336.5	1409.2	1474.5	1596.1	1714.1	1831.5
	s	—	—	—	—	1.2945	1.3300	1.3564	1.3782	1.4576	1.5139	1.5602	1.6384	1.7065	1.7886
2500 (668.13)	v	—	—	—	—	—	—	0.1484	0.1686	0.2294	0.2710	0.3061	0.3678	0.4244	0.4784
	h	—	—	—	—	—	—	1132.3	1176.8	1303.6	1387.8	1458.4	1585.3	1706.1	1826.2
	s	—	—	—	—	—	—	1.2687	1.3073	1.4127	1.4772	1.5273	1.6088	1.6775	1.7389
3000 (695.36)	v	—	—	—	—	—	—	—	0.0964	0.1760	0.2159	0.2476	0.3018	0.3505	0.3966
	h	—	—	—	—	—	—	—	1060.7	1267.2	1365.0	1441.8	1574.3	1698.0	1819.9
	s	—	—	—	—	—	—	—	1.1966	1.3690	1.4439	1.4984	1.5837	1.6540	1.7163

P (psia)												
3286.2 (706.4)	v	—	—	—	—	—	0.1583	0.1961	0.2288	0.2806	0.3267	0.3703
	h	—	—	—	—	—	1250.5	1355.2	1434.7	1569.8	1694.6	1817.2
	s	—	—	—	—	—	1.3506	1.4309	1.4874	1.5742	1.6452	1.7080
3500	v	0.0806	—	—	—	—	0.1364	0.1762	0.2058	0.2546	0.2977	0.3381
	h	780.5	—	—	—	—	1224.9	1340.7	1424.5	1563.3	1689.8	1813.6
	s	0.9515	—	—	—	—	1.3241	1.4127	1.4723	1.5615	1.6336	1.6968
4000	v	0.0287	—	—	—	—	0.1062	0.1462	0.1743	0.2192	0.2581	0.2943
	h	763.8	—	—	—	—	1174.8	1314.4	1406.8	1562.1	1681.7	1807.2
	s	0.9347	—	—	—	—	1.2757	1.3827	1.4482	1.5417	1.6154	1.6795
4500	v	0.0276	—	—	—	—	0.0798	0.1226	0.1500	0.1917	0.2273	0.2602
	h	753.5	—	—	—	—	1113.9	1286.5	1388.4	1540.8	1673.5	1800.9
	s	0.9235	—	—	—	—	1.2204	1.3529	1.4253	1.5235	1.5990	1.6040
5000	v	0.0268	—	—	—	—	0.0593	0.1036	0.1302	0.1696	0.2027	0.2329
	h	746.4	—	—	—	—	1047.1	1256.5	1369.5	1629.5	1665.2	1794.5
	s	0.9152	—	—	—	—	1.1622	1.3231	1.4034	1.5066	1.5839	1.6499
5500	v	0.0262	—	—	—	—	0.0463	0.0880	0.1143	0.1516	0.1825	0.2106
	h	741.3	—	—	—	—	985.0	1224.1	1349.3	1518.2	1657.0	1788.1
	s	0.9090	—	—	—	—	1.1093	1.2930	1.3821	1.4908	1.5699	1.6309

Source: Abridged from Keenan, J.H. and Keyes, F.G., *Thermodynamic Properties of Steam*, John Wiley & Sons, New York, 1937.

TABLE F.18

Enthalpy of Compressed Water

p (t Sat.)	0				580 (467.13)				1000 (544.75)			
T	v	u	h	s	v	u	h	s	v	u	h	s
Sat.					0.019748	447.70	449.53	0.64904	0.021591	538.39	542.38	0.743
32	0.016022	−0.01	−0.01	−0.00003	0.015994	0.00	1.49	0.00000	0.015967	0.03	2.99	0.000
50	0.016024	18.06	18.06	0.3687	0.015998	18.02	19.50	0.03599	0.015972	17.99	20.94	0.035
100	0.016130	68.05	68.05	0.12963	0.016106	67.87	69.36	0.12932	0.016082	67.70	70.68	0.129
150	0.016343	117.95	117.95	0.21504	0.016318	117.66	119.17	0.21457	0.016293	117.38	120.40	0.214
200	0.016635	168.05	168.05	0.29402	0.016608	167.65	169.19	0.29341	0.016580	167.28	170.32	0.292
250	0.017003	218.32	218.32	0.36777	0.016972	217.99	219.56	0.36702	0.016941	217.47	220.61	0.366
300	0.017453	269.61	269.61	0.43732	0.017416	268.92	270.53	0.43641	0.017379	268.24	271.46	0.435
350	0.018000	321.59	321.59	0.50359	0.017954	320.71	322.37	0.50249	0.017909	319.83	323.15	0.501
400	0.018668	374.85	374.85	0.56740	0.018608	373.68	375.40	0.56604	0.018550	372.55	375.98	0.564
450	0.019503	429.96	429.96	0.62970	0.019420	428.40	430.19	0.62798	0.019340	426.89	430.47	0.626
500	0.02060	488.1	488.1	0.6919	0.02048	485.9	487.8	0.6896	0.02036	483.5	487.5	0.68
510	0.02087	500.3	500.3	0.7046	0.02073	497.9	499.8	0.7021	0.02060	495.6	499 4	0.695
520	0.02116	512.7	512.7	0.7173	0.02190	530.1	512.0	0.7146	0.02086	507.6	511.5	0.712
530	0.02148	525.5	525.5	0.7303	0.02130	522.6	524.5	0.7273	0.02114	519.9	523.8	0.724
540	0.02182	538.6	538.6	0.7434	0.02162	535.3	537.3	0.7402	0.02144	532.4	536.3	0.737
550	0.02221	552.1	552.1	0.7569	0.02198	548.4	550.5	0.7532	0.02177	545.1	549 2	0.749
560	0.02265	566.1	566.1	0.7707	0.02237	562.0	564.0	0.7666	0.02213	558.3	562.4	0.763
570	0.02315	580.8	580.8	0.7851	0.02281	576.0	578.1	0.7804	0.02253	571.8	576.0	0.776
580					0.02332	590.8	592 9	0.7946	0.02298	585.9	590.1	0.789
590					0.02392	606.4	608.6	0.8096	0.02349	600.6	604 o	0.894
600									0.02409	616.2	620.6	0.818
610									0.02482	632.9	637.5	0.834

p (t Sat.)

t	1500 (596.39) v	u	h	s	2000 (636.00) v	u	h	s	2500 (668.31) v	u	h	s
Sat.	0.023461	604.97	611.48	0.80824	0.025649	662.40	671.89	0.86227	0.028605	717.66	730.89	0.9130
32	0.015939	0.05	4.47	0.00007	0.015912	0.06	5.95	0.00008	0.015885	0.08	7.43	0.0000
50	0.015946	17.95	22.38	0.03584	0.015920	17.91	23.81	0.03575	0.015895	17.88	25.23	0.0356
100	0.016058	67.53	71.99	0.12870	.016034	67.37	73.30	0.12839	0.016010	67.20	74.61	0.1280
150	0.016268	117.10	121.62	0.21364	0.016244	116.83	122.84	0.21318	0.016220	116.56	124.07	0.2127
200	0.016554	166.87	171.46	0.29221	0.016527	166.49	172.60	0.29162	0.016501	166.11	173.75	0.2910
250	0.016910	216.96	221.65	0.36554	0.016880	216.46	222.70	0.36482	0.016851	215.96	223.75	0.3641
300	0.017343	267.58	272.39	0.43463	0.017308	266.93	273.33	0.43376	0.017274	266.29	274.28	0.4329
350	0.017865	318.98	323.94	0.50034	0.017822	318.15	324.74	0.49929	0.017780	317.33	325.56	0.4982
400	0.018493	371.45	376.59	0.56343	0.018439	370.38	377.21	0.56216	0.018386	369.34	377.84	0.5609
450	0.019264	425.44	430.79	0.62470	0.019191	424.04	431.14	0.62313	0.019120	422.68	431.52	0.6216
500	0.02024	481.8	487.4	0.6853	0.02014	479.8	487.3	0.6832	0.02004	478.0	487.3	0.6810
510	0.02048	493.4	499.1	0.6974	0.02036	491.4	498.9	0.6953	0.02025	489.4	498.8	0.6930
520	0.02072	505.3	511.0	0.7096	0.02060	503.1	510.7	0.7073	0.02048	501.0	510.4	0.7051
530	0.02099	517.3	523.1	0.7219	0.02085	514.9	522.6	0.7195	0.02072	512.6	522.2	0.7171
540	0.02127	529.6	535.5	0.7343	0.02112	527.0	534.8	0.7317	0.02098	524.5	534.2	0.7292
550	0.02158	542.1	548.1	0.7469	0.02141	539.2	547.2	0.7440	0.02125	536.6	546.4	0.7413
560	0.02191	554.9	561.0	0.7596	0.02172	551.8	559.8	0.7565	0.02154	548.9	558.8	0.7536
570	0.02228	568.0	574.2	0.7725	0.02206	564.6	572.8	0.7691	0.02186	561.4	571.5	0.7659
580	0.02269	581.6	587.9	0.7857	0.02243	577.8	586.1	0.7820	0.02221	574.3	584.5	0.7785
590	0.02314	595.7	602.1	0.7993	0.02284	591.3	599.8	0.7951	0.02258	587.4	597.9	0.7913
600	0.02366	610.4	616.9	0.8134	0.02330	605.4	614.0	0.8066	0.02300	601.0	611.6	0.8043
610	0.02426	625.8	432.6	0.8281	0.02382	620.0	628.8	0.8225	0.02346	615.0	625.9	0.8177
620	0.02498	642.5	649.4	0.8437	0.02443	635.4	644.5	0.8371	0.02399	629.6	640.7	0.8315
630	0.02590	660.8	668.0	0.8609	0.02514	651.9	661.2	0.8525	0.02459	644.9	656.3	0.8459
640					0.02603	669.8	679.4	0.8691	0.02530	661.2	672.9	0.8610
650					0.02724	690.3	700.4	0.8881	0.02616	678.7	690.8	0.8773
660									0.02729	698.4	711.0	0.8954
670									0.02895	722.1	735.3	0.9/72

TABLE F.19

Correlation for Superheated Steam Properties

$C_1 = 80{,}870/T^2$

$C_2 = (-2641.62/T) \times 10^{C_1}$

$C_3 = 1.89 + C_2$

$C_4 = C_3(P^2/T^2)$

$C_5 = 2 + (372{,}420/T^2)$

$C_6 = C_3C_2$

$C_7 = 1.89 + C_6$

$C_8 = 0.21878T - 126{,}970/T$

$C_9 = 2C_6C_7 - (C_3/T)(126{,}970)$

$C_{10} = 82.546 - 162{,}460/T$

$C_{11} = 2C_{10}C_7 - (C_3/T)(162{,}460)$

$v = \{[(C_8C_4C_3 + C_{10})(C_4/P) + 1]C_3 + 4.55504\,(T/P)\}0.016018$

$H = 775.596 + 0.63296T + 0.000162467T^2 + 47.3635 \log T$
$\quad\quad + 0.043557\{C_7P + 0.5C_4[C_{11} + C_3(C_{10} + C_9C_4)]\}$

$S = 1/T\{[(C_8C_3 - 2C_9)C_3C_4/2 - C_{11}]C_4/2 + (C_3 - C_7)P\}$
$\quad\quad \times (-0.0241983) - 0.355579 - 11.4276/T + 0.00018052T$
$\quad\quad - 0.253801 \log P + 0.809691 \log T$

where

  P is the pressure, atm.

  T is the temperature, K.

  v is the specific volume, ft^3/lb.

  H is the enthalpy, Btu/lb.

  S is the entropy, $Btu/lb°F$.

TABLE F.20

Coefficients to Estimate Properties of Dry Saturated Steam with Equation[a]

$$Y = Ax + B/x + Cx^{1/2} + D \ln x + Ex^2 + Fx^3 + G$$

Property	A	B	C	D	E	F	G
Temperature, °F	-0.17724	3.83986	11.48345	31.1311	8.762969×10^{-5}	-2.78794×10^{-8}	86.594
Liquid specific volume, ft³/lb	-5.280126×10^{-7}	2.99461×10^{-5}	1.521874×10^{-4}	6.62512×10^{-5}	8.408856×10^{-10}	1.86401×10^{-14}	0.01596
Vapor specific volume, ft³/lb							
1–200 psia	-0.48799	304.717614	9.8299035	-16.455274	9.474745×10^{-4}	-1.363366×10^{-8}	19.53953
200–1,500 psia	2.662×10^{-3}	457.5802	-0.176959	0.826862	-4.601876×10^{-7}	6.3181×10^{-11}	-2.3928
Liquid enthalpy, Btu/lb	-0.15115567	3.671404	11.622558	30.832667	8.74117×10^{-5}	-2.62306×10^{-8}	54.55
Vaporization enthalpy, Btu/lb	0.008676153	-1.3049844	-8.2137368	-16.37649	-4.3043×10^{-5}	9.763×10^{-9}	1,045.81
Vapor enthalpy, Btu/lb	-0.14129	2.258225	3.4014802	14.438078	4.222624×10^{-5}	-1.569916×10^{-8}	1,100.5
Liquid entropy, Btu/lb °R	-1.67772×10^{-4}	4.272688×10^{-3}	0.01048048	0.05801509	9.101291×10^{-8}	-2.7592×10^{-11}	0.11801
Vaporization entropy, Btu/lb °R	3.454439×10^{-5}	-2.75287×10^{-3}	-7.33044×10^{-3}	-0.14263733	-3.49366×10^{-8}	7.433711×10^{-12}	1.85565
Vapor entropy, Btu/lb°R	-1.476933×10^{-4}	1.2617946×10^{-3}	3.44201×10^{-3}	-0.06494128	6.89138×10^{-8}	-2.4941×10^{-11}	1.97364
Liquid internal energy, Btu/lb	-0.1549439	3.662121	11.632628	30.82137	8.76248×10^{-5}	-2.646533×10^{-8}	54.56
Vapor internal energy, Btu/lb	-0.0993951	1.93961	2.428354	10.9818864	2.737201×10^{-5}	-1.057475×10^{-8}	1,040.03

[a] y = property, x = pressure, psia.

TABLE F.21

Saturation Line, Specific Heat Capacity, and Transport Properties

t, °F	t, °C	P, lb ft/in.²	C_{pf}, Btu/lb °F	$\mu_f \times 10^6$, lb/ft s	$\nu_f \times 10^6$, ft²/s	$\lambda_f \times 10^3$, Btu/ft h °F	$(Pr)_f$	C_{pg}, Btu/lb °F	$\mu_g \times 10^6$, lb/ft s	$\nu_g \times 10^6$, ft²/s	$\lambda_g \times 10^3$, Btu/ft h °F	$(Pr)_g$
32	0.0	0.0886	1.006	1180.0	18.9	329	12.9	0.442	5.91	19500	10.0	0.94
40	4.4	0.1217	1.004	1027.0	16.5	333	11.1	0.443	6.02	14700	10.5	0.91
60	15.6	0.2562	1.000	753.0	12.1	345	7.86	0.447	6.24	7530	10.9	0.92
80	26.7	0.5069	0.998	576.0	9.26	354	5.85	0.447	6.47	4100	11.3	0.92
100	37.8	0.949	0.998	457.0	7.37	363	4.52	0.449	6.71	2350	11.7	0.93
120	49.9	1.693	0.999	372.0	6.03	371	3.61	0.452	6.95	1410	12.1	0.94
140	60.0	2.889	1.000	311.0	5.07	378	2.96	0.458	7.20	886	12.4	0.96
160	71.1	4.741	1.001	264.0	4.33	383	2.48	0.465	7.45	576	13.0	0.96
180	82.2	7.511	1.003	229.0	3.78	388	2.13	0.474	7.70	387	13.5	0.97
200	93.3	11.53	1.006	201.0	3.34	392	1.86	0.484	7.96	268	14.0	0.99
220	104.4	17.19	1.009	179.0	3.00	394	1.65	0.495	8.22	190	14.6	1.00
240	115.6	24.97	1.013	160.0	2.71	396	1.47	0.508	8.50	139	15.2	1.02
260	126.7	35.42	1.018	145.0	2.48	397	1.34	0.522	8.77	103	15.8	1.04
280	137.8	49.20	1.024	133.0	2.29	397	1.23	0.538	9.05	78.2	16.5	1.06
300	148.9	67.00	1.030	122.0	2.13	397	1.14	0.556	9.32	60.3	17.3	1.08
320	160.0	89.64	1.038	113.0	2.00	395	1.07	0.577	9.58	47.1	18.1	1.10
340	171.1	118.00	1.047	105.0	1.88	393	1.01	0.600	9.85	37.3	18.9	1.13
360	182.2	153.00	1.057	98.6	1.79	390	0.96	0.627	10.1	29.9	19.9	1.15
380	193.3	195.7	1.069	92.7	1.70	387	0.92	0.658	10.4	24.2	21.0	1.17
400	204.4	247.3	1.082	87.5	1.63	382	0.89	0.692	10.6	19.8	22.1	1.19

420	215.6	308.8	1.097	82.9	1.57	377	0.87	0.731	10.9	16.3	23.4	1.23
440	226.7	381.6	1.115	78.8	1.52	371	0.85	0.774	11.2	13.6	24.9	1.25
460	237.8	466.9	1.135	75.2	1.47	364	0.84	0.823	11.5	11.4	26.5	1.29
480	248.9	566.1	1.158	71.9	1.44	357	0.84	0.885	11.7	9.60	28.4	1.31
500	260.0	680.8	1.186	68.9	1.41	349	0.84	0.951	12.1	8.14	30.5	1.36
520	271.1	812.4	1.229	66.2	1.38	340	0.86	1.038	12.4	6.94	32.9	1.41
540	282.2	962.6	1.275	63.7	1.37	330	0.88	1.147	12.8	5.95	35.8	1.48
560	293.3	1133.2	1.338	61.5	1.36	319	0.92	1.286	13.2	5.11	39.2	1.56
580	304.4	1326.1	1.420	59.8	1.36	308	0.99	1.472	13.6	4.38	43.3	1.66
600	315.6	1543.3	1.520	58.0	1.37	296	1.07	1.735	14.4	3.85	48.4	1.86
620	326.7	1787.1	1.659	55.7	1.37	283	1.17	2.153	15.3	3.37	54.9	2.16
640	337.8	2060.3	1.880	52.9	1.37	269	1.33	2.832	16.4	2.95	63.6	2.63
660	348.9	2366.0	2.310	49.5	1.37	254	1.62	3.943	17.9	2.58	76.1	3.34
680	360.0	2708.3	3.466	45.2	1.37	231	2.44	5.676	20.2	2.25	97.0	4.26

TABLE F.22

Surface Tension of Water

Temp., °F	lb ft/ft × 10³	Temp., °F	lb ft/ft × 10³
32	5.184	350	2.942
40	5.141	400	2.512
60	5.003	450	2.071
80	4.914	500	1.624
100	4.794	550	1.178
150	4.473	600	0.744
200	4.124	650	0.340
250	3.752	700	0.018
300	3.357	—	—

TABLE F.23

Specific Heat at a Constant Pressure of Steam and Water (Btu/lbm °F)

Temp., °F	Pressure, psia											
	1	2	5	10	20	50	100	200	500	1000	2000	5000
1500	0.559	0.559	0.559	0.559	0.559	0.560	0.561	0.563	0.569	0.580	0.601	0.668
1400	0.551	0.551	0.551	0.551	0.551	0.552	0.553	0.555	0.563	0.575	0.600	0.681
1300	0.543	0.543	0.543	0.543	0.543	0.544	0.545	0.548	0.556	0.570	0.600	0.702
1200	0.533	0.533	0.533	0.533	0.534	0.535	0.536	0.540	0.550	0.567	0.603	0.740
1100	0.524	0.542	0.524	0.524	0.525	0.526	0.528	0.532	0.544	0.564	0.612	0.814
1000	0.515	0.515	0.515	0.515	0.516	0.518	0.519	0.524	0.539	0.566	0.633	0.970
900	0.506	0.506	0.506	0.506	0.507	0.509	0.512	0.518	0.537	0.576	0.683	1.382
800	0.497	0.497	0.497	0.497	0.498	0.501	0.505	0.513	0.544	0.605	0.800	2.420
700	0.488	0.488	0.488	0.489	0.490	0.494	0.500	0.513	0.563	0.681	1.181	1.897[b]
600	0.479	0.480	0.480	0.481	0.483	0.489	0.499	0.522	0.621	0.888	1.453	1.253
500	0.472	0.472	0.473	0.475	0.478	0.489	0.508	0.554	0.773	1.181	1.157	1.106
400	0.464	0.465	0.467	0.470	0.476	0.497	0.536	0.636	1.077	1.072	1.063	1.041
300	0.458	0.459	0.463	0.469	0.482	0.524	1.029	1.028	1.027	1.024	1.019	1.006
250	0.456	0.458	0.463	0.471	0.489	1.015	1.014	1.014	1.013	1.011	1.007	0.996
200	0.453	0.455	0.463	0.475	1.005	1.005	1.005	1.004	1.003	1.002	0.998	0.989
150	0.451	0.455	0.866	1.001	1.000	1.000	1.000	1.000	0.998	0.997	0.993	0.984
100	0.998	0.998	0.998	0.998	0.998	0.998	0.998	0.997	0.996	0.994	0.990	0.980
50	1.002	1.002	1.002	1.002	1.002	1.002	1.001	1.001	0.999	0.996	0.989	0.972
32	1.007	1.007	1.007	1.007	1.007	1.007	1.006	1.006	1.003	0.999	0.990	0.969

[a] Horizontal bars indicate phase change.
[b] Critical point (P = 3206.2 psia; T = 705.4°F).

TABLE F.24

Viscosity of Steam and Water (lbm/h ft)

| Temp., °F | \multicolumn{12}{c}{Pressure, psia} | | | | | | | | | | | |
	1	2	5	10	20	50	100	200	500	1000	2000	5000
1500	0.0996	0.0996	0.0996	0.0996	0.0996	0.0996	0.0996	0.0996	0.1008	0.1008	0.1019	0.1066
1400	0.0938	0.0938	0.0938	0.0938	0.0938	0.0938	0.0952	0.0952	0.0952	0.0961	0.0973	0.1019
1300	0.0892	0.0982	0.0892	0.0892	0.0892	0.0892	0.0892	0.0892	0.0892	0.0903	0.0915	0.0973
1200	0.0834	0.0834	0.0834	0.0834	0.0834	0.0834	0.0834	0.0834	0.0846	0.0846	0.0867	0.0926
1100	0.0776	0.0776	0.0776	0.0776	0.0776	0.0776	0.0776	0.0776	0.0788	0.0799	0.0811	0.0892
1000	0.0730	0.0730	0.0730	0.0730	0.0730	0.0730	0.0730	0.0730	0.0730	0.0741	0.0764	0.0857
900	0.0672	0.0672	0.0672	0.0672	0.0672	0.0672	0.0672	0.0672	0.0683	0.0683	0.0707	0.0846
800	0.0614	0.0614	0.0614	0.0614	0.0614	0.0614	0.0614	0.0614	0.0625	0.0637	0.0660	0.0973
700	0.0556	0.0556	0.0556	0.0556	0.0556	0.0556	0.0568	0.0568	0.0568	0.0579	0.0625	0.171[b]
600	0.0510	0.0510	0.0510	0.0510	0.0510	0.0510	0.0510	0.0510	0.0510	0.0510	0.210	0.221
500	0.0452	0.0452	0.0452	0.0452	0.0452	0.0452	0.0452	0.0440	0.0440	0.250	0.255	0.268
400	0.0394	0.0394	0.0394	0.0394	0.0394	0.0394	0.0394	0.0382	0.317	0.320	0.323	0.335
300	0.0336	0.0336	0.0336	0.0336	0.0336	0.0336	0.441	0.442	0.444	0.445	0.448	0.460
250	0.0313	0.0313	0.0313	0.0313	0.0313	0.551	0.551	0.551	0.552	0.554	0.558	0.569
200	0.0290	0.0290	0.0290	0.0290	0.725	0.725	0.725	0.726	0.729	0.729	0.732	0.741
150	0.0255	0.0255	1.032	1.032	1.032	1.032	1.032	1.032	1.033	1.034	1.037	1.044
100	1.645	1.645	1.645	1.645	1.645	1.645	1.645	1.645	1.645	1.646	1.646	1.648
50	3.144	3.144	3.144	3.144	3.144	3.144	3.144	3.142	3.141	3.139	3.134	3.119
32	4.240	4.240	4.240	4.240	4.240	4.240	4.240	4.239	4.236	4.231	4.222	4.192

[a] Horizontal bars indicate phase change.
[b] Critical point (P = 3206.2 psia; T = 705.4°F).

TABLE F.25

Thermal Conductivity of Steam and Water [(Btu/h ft °F) × 10^3]

Temp. (°F)	Pressure (psia)											
	1	2	5	10	20	50	100	200	500	1000	2000	5000
1500	63.7	63.7	63.7	63.7	63.7	63.8	64.0	64.3	65.4	67.1	70.7	82.0
1400	59.2	59.2	59.2	59.2	59.3	59.4	59.6	59.9	60.9	62.7	66.3	78.2
1300	54.8	54.8	54.8	54.8	54.8	54.9	55.1	55.5	56.5	58.3	62.0	74.6
1200	50.4	50.4	50.4	50.4	50.4	50.5	50.7	51.0	52.1	53.9	57.8	71.6
1100	46.0	46.0	46.0	46.0	46.1	46.2	46.3	46.7	47.8	49.6	53.7	69.8
1000	41.7	41.7	41.8	41.8	41.8	41.9	42.1	42.4	43.5	45.5	50.0	70.7
900	37.6	37.6	37.6	37.6	37.6	37.7	37.9	38.3	39.4	41.5	46.8	80.2
800	33.6	33.6	33.6	33.6	33 b	33.7	33.9	34.3	35.5	37.9	44.9	129.6
700	29.7	29.7	29.7	29.7	29.8	29.9	30.1	30.4	31.8	35.0	47.5	262.8ᵇ
600	26.0	26.0	26.1	26.1	26.1	26.2	26.4	26.9	28.7	34.1	301.9	333.7
500	22.6	22.6	22.6	22.6	22.7	22.8	23.0	23.6	26.9	350.8	357.4	373.8
400	19.4	19.4	19.4	19.4	19.5	19.6	20.0	21.3	383.0	384.9	388.5	398.6
300	16.5	16.5	16.5	16.5	16.6	16.9	396.9	397.2	398.0	399.2	402.0	409.9
250	15.1	15.1	15.1	15.2	15.3	396.9	397.0	397.3	398.1	399.4	402.1	409.7
200	13.8	13.8	13.9	14.0	391.6	391.6	391.8	392.1	393.0	394.4	397.2	404.9
150	12.7	12.7	380.5	380.5	380.6	380.7	380.8	381.1	382.1	383.7	386.7	394.7
100	363.3	363.3	363.3	363.3	363.3	363.4	363.6	363.9	365.0	366.6	369.8	378.3
50	339.1	339.1	339.1	339.1	339.2	339.3	339.4	339.8	340.8	342.5	345.7	354.6
32	328.6	328.6	328.6	328.6	328.6	328.7	328.9	329.2	330.3	331.9	335.1	344.1

ᵃ Horizontal bars indicate phase change.
ᵇ Critical point (P = 3206.2 psia; T = 705.4°F).

Reference

1. J.H. Keenan, F.G. Keyes, *Thermodynamic Properties of Steam*, John Wiley & Sons, Inc., New York, 1937.

Appendix G: Quiz on Boilers and HRSGs with Answers

(The answers to all of these questions can be found in the book. However, e-mail me for clarifications if required. My e-mail is v_ganapathy@yahoo.com.)

1. If boiler efficiency for a typical natural gas fired boiler is 83% on higher heating value basis, what is it approximately on lower heating value basis?

   a. 73%   b. 83%   c. 92%

2. If NO_x in a natural gas–fired boiler is 50 ppmv (3% oxygen dry), what is it on lb/MM Btu (HHV) basis?

   a. 0.06   b. 0.10   c. 0.20

3. In a 100,000 lb/h packaged boiler, 1 in. WC of additional gas pressure drop is worth about how many kilowatts of fan power consumption?

   a. 5   b. 20   c. 50

4. If boiler water concentration in a boiler drum is 1000 ppm and steam purity is 1 ppm, what is the percent steam quality?

   a. 99.9   b. 99   c. 99.99

5. Boilers of the same capacity are located at different sites, whose ambient conditions and elevation are as follows. Which case requires the biggest fan?

   a. 80°F and sea level   b. 100°F and 3000 ft   c. 10°F and 7000 ft

6. In a boiler plant, if the conductivity of the condensate, makeup, and feed water is 800, 40, and 150 mmho/cm, respectively, what is the percent condensate returns in the feed water?

   a. 5   b. 50   c. 15

7. A 20° F change in exit gas temperature of an oil-fired boiler changes boiler efficiency by approximately what percent?

   a. 1   b. 0.5   c. 2.0

8. Approximate airflow (acfm) required in a packaged boiler firing 100 MM Btu/h (HHV) of natural gas is

   a. 19,000   b. 30,000   c. 12,000

9. The steam pressure drop in a boiler superheater is 50 psi when generating 600 psig, 650°F steam. What is it likely to be at 400 psig, 600°F with the same flow?

   a. 70   b. 30   c. 50

10. Which is the worst-case scenario for an economizer from the viewpoint of sulfuric acid condensation? Assume that the oil-fired boiler flue gas contains 12% water vapor and 0.03% SO_2.

    a. Flue gas at 680°F and feed water at 200°F

    b. Flue gas at 320°F and feed water at 275°F

11. If vol% of oxygen (dry) in a natural gas–fired boiler is 2.0%, what is the excess air used?

    a. 15   b. 5   c. 10

12. If boiler casing heat loss is 0.2% at 100% load, what is it at 25% load, assuming that wind velocity and ambient temperature are unchanged?

    a. 1.0   b. 2.0   c. 0.8

13. Plant management decides to change the tube inner diameter of an existing super-heater from 1.7 in. to 1.5 in. The steam-side pressure drop for the same steam conditions will go up by what percent?

    a. 87   b. 65   c. 29

14. The heat transfer coefficient in a finned tube bundle is higher than in a bare tube exchanger for the same gas velocity, temperature, tube size, and geometry.

    a. True   b. False

15. In a fire tube waste heat boiler, a small diameter tube has a higher tube-side heat transfer coefficient and higher heat flux than a larger tube for the same gas velocity.

    a. True   b. False

16. Superheated steam temperature from a boiler firing oil will be higher than when firing natural gas at the same steam generation rate (assuming steam temperature is uncontrolled).

    a. True   b. False

17. More flue gas is generated in a boiler while firing oil than while firing natural gas at the same excess air and steam generation.

    a. True   b. False

18. The maximum possible fuel (natural gas/distillate oil) that can be fired in an HRSG with exhaust gas flow = 200,000 lb/h and 14% oxygen wet in MM Btu/h (LHV) is

    a. 100   b. 150   c. 50

19. Surface area gives a good indication of whether a boiler or HRSG design is adequate or not.

    a. True   b. False

20. An uncooled soot blower lance is located in a boiler convection bank at 1700°F gas temperature. Its temperature will be

    a. >1700°F   b. 1700°F   c. <1700°F

21. (a) A fire tube waste heat boiler using small-diameter tubes will be longer than the design using larger-diameter tubes for the same duty and gas pressure drop.

    a. True   b. False

    (b) A fire tube waste heat boiler using small-diameter tubes requires less surface area than the design using larger-diameter tubes for the same duty and gas pressure drop.

    a. True   b. False

22. For the same mass flow and gas temperature drop, a flue gas containing 16% water vapor will transfer more energy than a gas stream having 5% water vapor.

    a. True   b. False

23. Design of tubular air heaters in steam generators can be improved if finned tubes are used instead of plain tubes.

    a. True   b. False   c. Depends on the fuel used

24. In a cross-flow heat transfer situation, an inline arrangement of plain tubes is better than a staggered one.

    a. True   b. False

25. For the same casing insulation thickness and ambient conditions, aluminum casing will run hotter than carbon steel.

    a. True   b. False

26. If additional steam is required in a cogeneration plant, supplementary firing the HRSG rather than using a packaged boiler will be more prudent.

    a. True   b. False

27. Required thickness of a boiler tube subjected to external pressure will be less than when the same tube is subjected to the same internal pressure at the same temperature.

    a. True   b. False

28. Which is a better choice for fin density for a superheater in an HRSG?

    a. 5 fins/in.   b. 2 fins/in.

29. The exit gas temperature in a single-pressure unfired HRSG generating steam at 600 psig, 700°F can be less than 300°F. (Assume exhaust gas at 950°F and feed water at 230°F.)

    a. True   b. False

30. A boiler designed for 1000 psig, 800°F steam can be operated at the same steam flow at 300 psig without modifications.

    a. True   b. False

31. A gas turbine HRSG economizer is likely to steam at which ambient temperature in unfired mode?

    a. 40°F   b. 90°F

32. More energy can be transferred to a boiler evaporator if the circulation ratio is higher.

    a. True   b. False

33. Heat flux will be higher in a packaged boiler furnace for which fuel? Assume same steam generation.

    a. Fuel oil   b. Natural gas

34. For the same excess air and exit gas temperature, an oil-fired boiler will have a higher efficiency on HHV basis than a gas-fired boiler.

    a. True   b. False

35. For the same mass flow per tube and length of tube, superheated steam at 600 psig, 800°F will have a higher pressure drop than 150 psig saturated steam.

    a. True   b. False

36. Gas-side fouling increases the tube wall temperature in a waste heat boiler.

    a. True   b. False   c. Depends on whether it is a fire tube or water tube boiler

37. The feed pump requires more power to generate a given amount of steam at a given pressure and temperature in a once-through HRSG than in a natural circulation HRSG.

    a. True   b. False

38. The volumetric heat release rate is more important in a gas-fired packaged boiler than the area heat release rate.

    a. True   b. False

39. Large margins on flow and head should not generally be used while selecting the fan for a packaged boiler.

    a. True   b. False

40. If an economizer with counter-flow arrangement is experiencing low-temperature corrosion problems, then re-piping it with a parallel-flow arrangement can fix the problem.

    a. True   b. False

41. Exit gas temperature from a single-pressure HRSG having a superheater, evaporator, and economizer increases as steam generation increases.

    a. True   b. False

42. It is better to preheat condensate or feed water using extraction steam from the steam turbine rather than use the energy in the HRSG exhaust gases.

    a. True   b. False

43. Steam for deaeration should preferably be taken from the boiler outlet rather than from an extraction point in the steam turbine.

    a. True   b. False

44. The maldistribution of steam flow through superheater tubes will be the worst at a boiler load of

    a. 20%   b. 50%   c. 100%

45. Which fuel generates the maximum amount of carbon dioxide per MM Btu fired?

    a. Oil   b. Natural gas   c. Coal

46. Is it possible to predict the off-design performance of an HRSG without knowing its mechanical constructional features?

    a. Yes   b. No

47. Can we have more surface area in an HRSG and yet transfer less duty?

    a. Yes   b. No

48. Can we use finned tubes for the evaporator or superheater of a gas-fired packaged boiler?

    a. Yes   b. No

49. What happens to the pinch and approach points of the evaporator in an HRSG as we increase the supplementary firing rate?

    a. Both increase   b. Both decrease   c. Pinch point increases while approach point decreases   d. They are unchanged

50. In a packaged boiler, the furnace performance and circulation are more critical in oil firing than in gas firing.

    a. True   b. False

51. Can a superheater be located between the evaporator and the economizer in a packaged boiler?

a. Yes   b. No

52. Good steam-separating devices cannot prevent carryover of silica from boiler water into steam at high pressures.

a. True   b. False

53. Superheated steam for use in turbines should have better steam purity than saturated steam.

a. True   b. False

54. Feed water used for attemperation in a desuperheater for steam temperature control should preferably have low to zero solids.

a. True   b. False

55. Tube-side heat flux will be higher in a plain tube evaporator than in a finned tube evaporator for the same gas- and steam-side conditions.

a. True   b. False

56. In a waste heat boiler containing hydrogen chloride gas, a low steam temperature (say, 700°F versus 850°F) is preferred.

a. True   b. False

57. A higher steam pressure requires a higher steam temperature to minimize wetness in steam after expansion in a steam turbine.

a. True   b. False

58. An ammonia–water mixture has a varying boiling point and hence is a better fluid for energy recovery from waste flue gases than steam.

a. True   b. False

59. The cross section of a 100,000 lb/h packaged boiler will be much smaller than that of an unfired gas turbine HRSG generating the same amount of steam.

a. True   b. False

60. Gas conditions being the same, as steam pressure increases, the steam generation in an unfired HRSG

a. Increases   b. Decreases   c. Is unchanged

61. The cross section of a forced circulation HRSG and its surface area will be much different from a natural circulation HRSG for the same duty and pressure drop.

a. True   b. False   c. Can't say

62. A fire tube waste heat boiler generally responds faster to load changes than an equivalent water tube design.

a. True   b. False

63. The amount of deaeration steam is impacted by the conductivity of boiler feed water.

a. True   b. False

64. In a boiler or HRSG evaporator, the allowable steam quality to avoid DNB conditions decreases as the heat flux increases.

a. True   b. False

65. A natural circulation HRSG using vertical evaporator tubes can handle higher heat flux than a forced circulation or once-through unit using horizontal tubes.

    a. True   b. False

66. A gas turbine plant has two options: a supplementary-fired HRSG and an unfired HRSG. The cross section of the supplementary-fired HRSG generating twice the amount of steam as the unfired HRSG should be much larger.

    a. True   b. False

Think about It!

1. Why is multiple-pressure steam generation often required in HRSGs but not in a packaged boiler?

2. Explain how surface areas can be different in steam generators (or HRSGs) and yet the duty transferred is the same.

3. Why is supplementary firing very efficient in HRSGs?

4. Why is an economizer preferred to an air heater in oil- and gas-fired packaged boilers? Give at least two reasons.

5. Why is steaming in the economizer often a concern in HRSGs and not in packaged boilers?

6. Why can we achieve a low exit gas temperature in a packaged boiler at any steam pressure, whereas it is difficult in a single-pressure unfired HRSG?

7. Why is the superheated steam temperature generally lower with oil firing than with gas firing in a packaged boiler?

8. Why is a low fin density, say, 2 fins/in., preferred in an HRSG superheater over, say, 5 fins/in.?

9. Why does raising the gas temperature at the economizer alone not help minimize low-temperature corrosion problems?

10. Compute typical operating costs of fuel and electricity for various boilers and HRSGs in your plant and suggest how to lower these costs.

11. Is a supplementary-fired HRSG a better choice than an unfired HRSG in a combined cycle plant?

12. Why do we not worry about pinch and approach points in a packaged boiler, whereas they are very important in an HRSG?

13. What are the advantages of a convective superheater in a packaged boiler over a radiant design?

14. What are the various factors to be considered while modifying an existing packaged boiler to meet lower emissions of NO_x and CO?

15. In a packaged boiler, why is interstage attemperation for steam temperature control generally preferred to attemperation at the superheater exit?

16. A single-pressure unfired HRSG generates 600 psig steam at 750°F using 230°F feed water with an exit gas temperature of 380°F. To lower the exit gas temperature, is it more prudent to add a condensate heater rather than increase the surface area of the evaporator significantly?

17. Explain why rules of thumb relating surface areas with steam generation can be misleading.

18. An economizer has been removed from a packaged boiler for maintenance. Can the plant generate the same amount of steam as before? What are the concerns?

Answers to Quiz

c, a, a, a, b, c, b, b, a, a, c, c, a, b, a, b, a, b, b, c, a, a, a, b, a, a, a, b, b, b, b, a, b, a, a, b, b, a, b, b, b, b, b, b, a, a, a, a, a, a, a, a, a, a, b, a, a, a, a, b, b, b, a, a, a, b

Conversion Factors

Metric to American, Metric to Metric	American to Metric, American to American

Area

$1\ mm^2 = 0.00155\ in.^2 = 0.00001076\ ft^2$

$1\ cm^2\ = 0.155\ in.^2 = 0.001076\ ft^2$

$1\ m^2\ \ = 1550\ in.^2 = 10.76\ ft^2$

$1\ in.^2 = 645.2\ mm^2 = 6.452\ cm^2$

$\quad = 0.0006452\ m^2$

$1\ ft^2\ = 92{,}903\ mm^2 = 929.03\ cm^2$

$\quad = 0.0929\ m^2$

$1\ acre = 43{,}560\ ft^2$

$1\ circular\ mil = 0.7854\ square\ mil$

$\quad = 5.067 \times 10^{-10}\ m^2 = 7.854 \times 10^{-6}\ in.^2$

Density and specific gravity

$1\ g/cm^3 = 0.03613\ lb/in.^3 = 62.43\ lb/ft^3$

$\quad = 1000\ kg/m^3 = 1\ kg/liter$

$\quad = 62.43\ lb/ft^3 = 8.345\ lb/U.S.\ gal$

$1\ \mu g/m^3 = 136\ grains/ft^3$

$\quad$ (for particulate pollution)

$1\ kg/m^3 = 0.06243\ lb/ft^3$

$1\ lb/in.^3 = 27.68\ g/cm^3 = 27{,}680\ kg/m^3$

$1\ lb/ft^3 = 0.0160\ g/cm^3 = 16.02\ kg/m^3$

$\quad = 0.0160\ kg/liter$

Specific gravity relative to water

$\quad$ SGW of $1.00 = 62.43\ lb/ft^3$ at 4°C or 39.2 °F[a]

Specific gravity relative to dry air

$\quad$ SGA of $1.00 = 0.0765\ lb/ft^{3,b}$

$\quad = 1.225\ kg/m^3$

$1\ lb/U.S\ gal = 7.481\ lb/ft^3 = 0.1198\ kg/liter$

$1\ g/ft^3 = 35.3 \times 10^6\ \mu g/m^3$

$1\ lb/1000\ ft^3 = 16 \times 10^6\ \mu g/m^3$

Energy, heat, and work

$1\ cal\ \ = 0.003968\ Btu$

$1\ kcal = 3.968\ Btu = 1000\ cal = 4186\ J$

$\quad = 0.004186\ MJ$

$1\ J = 0.000948\ Btu = 0.239\ cal - 1\ W\ sec$

$\quad = 1\ N\ m = 10^7\ erg = 10^7\ dyn\ cm$

$1\ W\ h = 660.6\ cal$

$3413\ Btu = 1\ kWh$

$1\ Btu = 0.2929\ Whr$

$\quad = 252.0\ cal = 0.252\ kcal$

$\quad = 778\ ft\ lb$

$\quad = 1055\ J = 0.001055\ MJ$

$1\ ft\ lb\ \ = 0.1383\ kg\ m = 1.356\ J$

$1\ hp\ hr\ = 1.98 \times 10^6\ ft\ lb$

$1\ therm = 1.00 \times 10^5\ Btu$

$1\ BHP\ (boiler\ horsepower) = 33{,}475\ Btu/hr$

$\quad = 8439\ kcal/hr = 9.81\ kW$

Heat content and specific heat

$1\ cal/g = 1.80\ Btu/lb = 4187\ J/kg$

$1\ cal/cm^3\ = 112.4\ Btu/ft^3$

$1\ kcal/m^3 = 0.1124\ Btu/ft^3 = 4187\ J/m^3$

$1\ cal/g\ °C = 1\ Btu/lb\ °F = 4187\ J/kg\ K$

$1\ Btu/lb = 0.5556\ cal/g = 2326\ J/kg$

$1\ Btu/ft^3 = 0.00890\ cal/cm^3$

$\quad = 8.899\ kcal/m^3 = 0.0373\ MJ/m^3$

$1\ Btu/U.S.\ gal = 0.666\ kcal/liter$

$1\ Btu/lb\ °F\ \ \ = 1\ cal/g\ °C = 4187\ J/kg\ K$

(Continued)

Metric to American, Metric to Metric	American to Metric, American to American

Heat flow, power

1 N·m/s = 1 W = 1 J/s
  = 0.001341 hp = 0.7376 ft lb/sec
1 kcal/h = 1.162 J/s = 1.162 W
  = 3.966 Btu/hr
1 kW = 1000 J/s = 3413 Btu/hr
  = 1.341 hp

1 hp = 33 000 ft lb/min = 550 ft lb/sec
  = 745.7 W = 745.7 J/s
  = 641.4 kcal/h
1 Btu/hr = 0.2522 kcal/h
  = 0.0003931 hp
  = 0.2931 W = 0.2931 J/s

Heat flux and heat transfer coefficient

1 cal/cm^2·s = 3.687 Btu/ft^2 sec
  = 41.87 kW/m^2
1 cal/cm^2·h = 1.082 W/ft^2 = 11.65 W/m^2
1 kW/m^2 = 317.2 Btu/ft^2 hr
1 kW/m^2°C = 176.2 Btu/ft^2 hr °F

1 Btu/ft^2 sec = 0.2713 cal/cm^{2}s
1 Btu/ft^2 hr = 0.003 153 kW/m^2
  = 2.713 kcal/m^2 h
1 kW/ft^2 = 924.2 cal/cm^2 h
1 Btu/ft^2 hr °F = 4.89 kcal/m^2 h °C

Length

1 mm = 0.10 cm = 0.03937 in.
  = 0.003281 ft
1 m = 100 cm = 1000 mm = 39.37 in.
  = 3.281 ft.
1 km = 0.6214 mile

1 in. = 25.4 mm = 2.54 cm = 0.0254 m
1 ft = 304.8 mm = 30.48 cm = 0.3048 m
1 mile = 5280 ft
1 μm = 10^{-6} m
1 Angstrom unit = 1 Å = 10^{-10} m

Pressure

1 N·m^2 = 0.001 kPa = 1.00 Pa
1 mm H$_2$O = 0.0098 kPa
1 mm Hg = 0.1333 kPa = 13.60 mm H$_2$O
  = 1 torr = 0.01933 psi
1 kg/cm^2 = 98.07 kPa = 10,000 kg/m^2
  = 10,000 mm H$_2$O = 394.1 in. H$_2$O
  = 735.6 mm Hg = 28.96 in. Hg
  = 227.6 oz/in.^{2} = 14.22 psi
  = 0.9807 bar
1 bar = 100.0 kPa = 1.020 kg/cm^2
  = 10,200 mm H$_2$O = 401.9 in. H$_2$O
  = 750.1 mm Hg = 29.53 in. Hg
  = 232.1 oz/in.^{2} = 14.50 psi
  = 100,000N/m^2
1 g/cm^2 = 0.014 22 psi
  = 0.2276 oz/in.^{2}
  = 0.3937 in. H$_2$O
(For rough calculations,
1 bar = 1 atm = 1 kg/cm^2
  = 10 m H$_2$O = 100 kPa)

1 in. H$_2$O = 0.2488 kPa = 25.40 mm H$_2$O
  = 1.866 mm Hg
  = 0.00254 kg/cm^2 = 2.54 g/cm^2
1 in. Hg = 3.386 kPa = 25.40 mm Hg
  = 345.3 mm H$_2$O = 13.61 in. H$_2$O
  = 7.858 oz/in.^{2} = 0.491 psi
  = 25.4 torr
1 psi = 6.895 kPa = 6895 N/m^2
  = 703.1 mm H$_2$O = 27.71 in. H$_2$O
  = 51.72 mm Hg = 2.036 in. Hg
  = 16.00 oz/in.^{2}
  = 0.0703 kg/cm^2 = 70.31 g/cm^2
  = 0.068 97 bar
1 oz. / in.^{2} = 0.4309 kPa
  = 43.94 mm H$_2$O = 1.732 in. H$_2$O
  = 3.232 mm Hg
  = 0.004 39 kg/cm^2 = 4.394 g/cm^2
1 atm^c = 101.3 kPa = 101,325 N/m^2
  = 10,330 mm H^{2}O = 407.3 in. H$_2$O
  = 760.0 mm Hg = 29.92 in. Hg
  = 235.1 oz/in.^{2} = 14.70 psi
  = 1.033 kg/cm^2
  = 1.013 bar

(Continued)

Metric to American, Metric to Metric	American to Metric, American to American
Temperature	
$°C = 5/9 \, (°F - 32)$	
$°F = (9/5°C) + 32$	
$K = °C + 273.15$	
$°R = °F + 459.67$	
Thermal conductivity	
1 W/m K = 0.5778 Btu ft/ft^2 hr °F	1 Btu ft/ft^2 hr °F = 1.730 W/m K
= 6.934 Btu in./ft^2 hr °F	= 1.488 kcal/m h K
1 cal cm/cm$^2 \cdot$s$\cdot$°C = 241.9 Btu ft/ft^2 hr °F	1 Btu in./ft^2 hr °F = 0.1442 W/m K
= 2903 Btu in./ft^2 hr °F	1 Btu ft/ft^2 hr °F = 0.004139 cal cm/cm^2 s °C
= 418.7 W/m K	1 Btu in./ft^2 hr °F = 0.0003445 cal cm/cm^2 s °C
Thermal diffusivity	
1 m^2/s = 38 760 ft^2/hr	1 ft^2/hr = 0.0000258 m^2/s = 0.0929 m^2/h
1 m^2/h = 10.77 ft^2/hr	
Velocity	
1 cm/s = 0.3937 in./sec = 0.032 81 ft/sec	1 in./sec = 25.4 mm/s = 0.0254 m/s
= 10.00 mm/s = 1.969 ft/min	= 0.0568 mph
Velocity	
1 m/s = 39.37 in./sec = 3.281 ft/sec	1 ft/sec = 304.8 mm/s = 0.3048 m/s
= 196.9 ft/min = 2.237 mph	= 0.6818 mph
= 3.600 km/h = 1.944 knot	1 ft/min = 5.08 mm/s = 0.00508 m/s
	= 0.0183 km/h
	1 mph = 0.4470 m/s = 1.609 km/h
	= 1.467 ft/sec
	1 knot = 0.5144 m/s
	1 rpm = 0.1047 radian/sec
Viscosity, absolute, μ	
0.1 Pa · s = 1 dyne s/cm^2 = 360 kg/h m	1 lb$_m$/hr·ft = 0.000008634 lb$_f$ sec/ft^2
= 1 poise = 100 centipoise (cP)	= 0.413 cP = 0.000413 Pa · s
= 242.1 lb$_m$/hr ft	1 lb$_f$ sec/ft^2 = 115,800 lb$_m$/hr ft
= 0.002089 lb/$_f$ sec/ft^2	= 47,880 cP
1 kg/h · m = 0.672 lb/hr/ft = 0.00278 g/s	= 47.88 Pa s
cm	1 reyn = 1 lb$_f$ sec/in.^{2}
= 0.00000581 lb$_f$ sec/ft^2	= 6.890 × 10^6 cP
μ of water^d = 1.124 cP	μ of air^d = 0.0180 cP
= 2.72 lb$_m$/hr ft	= 0.0436 lb/hr ft
= 2.349 × 10^{-5} lb sec/ft^2	= 3.763 × 10^{-7} lb sec/ft
Viscosity, kinematic, v	
1 cm$^{2/s}$ = 0.0001 m^2/s	1 ft^2/sec = 3600 ft^2/hr = 92,900 cS
= 1 stokes = 100 centistokes (cS)	= 0.0929 m^2/s
= 0.001076 ft^2/sec	1 ft^2/h = 0.000278 ft^2/sec = 25.8 cS
= 3.874 ft^2/hr	= 0.0000258 m^2/s

(*Continued*)

Metric to American, Metric to Metric	American to Metric, American to American

$1 \, m^2/s = 3600 \, m^2/h$
$= 38{,}736 \, ft^2/hr = 10.76 \, ft^2/sec$

v of water^d = 1.130 centistokes v of air^d = 14.69 cS
  = 32 SSU $= 1.581 \times 10^{-4} \, ft^2/sec$
  $= 1.216 \times 10^{-5} \, ft^2/sec$

Volume

$1 \, cm^3$ (cc) $= 0.000 \, 001 \, 00 \, m^3$ $1 \, in.^3 = 16.39 \, cm^3 = 0.0001639 \, m^3$
  $= 0.0610 \, in.^3 = 0.0338$ U.S. $= 0.01639$ liter
  fluid $1 \, ft^3$ $= 1728 \, in.^3 = 7.481$ U.S. gal
oz. $= 6.229$ Br gal
1 liter $(dm^3) = 0.0010 \, m^3 - 1000 \, cm^3$ $= 28{,}320 \, cm^3 = 0.02832 \, m^3 = 28.32$
  $= 61.02 \, in.^3 = 0.03531 \, ft^3$ liters
  $= 0.2642$ U.S. gal $= 62.427$ lb of 39.4°F (4°C) water
                                                     $= 62.344$ lb of 60°F (15.6°C) water

$1 \, m^3 = 1000$ liter $= 1 \times 10^6 \, cm^3$ 1 U.S. gal $= 3785 \, cm^3 = 0.003785 \, m^3$
  $= 61{,}020 \, in.^3 = 35.31 \, ft^3$ $= 3.785$ liters $= 231.0 \, in.^3$
  $= 220.0$ Br gal $= 0.8327$ Br gal $= 0.1337 \, ft^3$
  $= 6.290$ bbl $= sp \, gr \times 8.335$ lb
  $= 264.2$ U.S. gal $= 8.335$ lb of water
  $= 1.308 \, yd^3$ $= 1/42$ barrel (oil)
                                                    1 Br gal $= 277.4 \, in.^3$
                                                      $= 0.004 \, 546 \, m^3 = 4.546$ liters
                                                      $= 1.201$ U.S. gal
                                                    1 bbl, oil $= 9702 \, in.^3 = 5.615 \, ft^3$
                                                      $= 0.1590 \, m^3 = 159.0$ liters
                                                      $= 42.00$ U.S. gal
                                                      $= 34.97$ Br gal

Volume flow rate

$1 \, cm^3/s = 1 \times 10^{-6} \, m^3/8$ 1 gpm (gal/min) $= 60.0$ gph (gal/hr)
1 liter/s $= 1 \times 10^{-3} \, m^3/s$ $= 0.01667$ gps (gal/sec)
$1 \, m^3/h = 4.403$ U.S. gpm (gal/min) $= 0.00223$ cfs (ft^3/sec)
  $= 0.5887 \, ft^3/min$ $= 0.1337$ cfm (ft^3/min)
                                                       $= 0.8326$ Br gpm
                                                       $= 0.227 \, m^3/h$
                                                       $= 1.429$ bbl/hr
                                                       $= 34.29$ bbl/day
                                                     1 gph (gal/hr) $= 0.00105$ liter/s
                                                       $= 0.000037 \, 1$ cfs (ft^3/sec)
                                                     1 cfm (ft^3/min) $= 6.18$ Br gpm
                                                       $= 0.000471 \, m^3/s$
                                                     1 cfs (ft^3/sec) $= 448.8$ gpm
                                                       $= 22{,}250$ Br gph

(Continued)

Metric to American, Metric to Metric	American to Metric, American to American
Weight, force, mass	
1 g = 0.03527 oz avdp mass	1 oz avdp mass = 28.35 g = 0.02835 kg
1 kg mass = 1000 g mass	1 lb avdp mass = 453.6 g = 0.4536 kg
= 35.27 oz. avdp mass	= 4.536×10^6 μg
= 2.205 lb avdp mass	1 lb avdp force = 0.4536 kg force
1 kg force = 1000 g force = 9.807 N	= 4.448 N
= 2.205 lb avdp force	1 lb = 7000 grains
1 metric ton = 1000 kg = 2205 lb	1 short ton = 2000 lb = 907.2 kg
	1 long ton = 2240 lb = 1015.9 kg

[a] 62.35 lb/ft³ at 60°F, 15.6°C; 8.335 lb/U.S. gal.

[b] 0.0763 lb/ft³ for moist air.

[c] Normal atmosphere = 760 torr (mm Hg at 0°C)— not a "technical atmosphere," which is 736 torr or 1 kg/cm². Subtract about 0.5 psi for each 1000 ft above sea level.

[d] Viscosity at STP.

Glossary

acfh	Actual cubic feet per hour.
acfm	Actual cubic feet per minute, a term used to indicate the flow rate of gases, at any condition of temperature and pressure.
°API	A scale adopted by American Petroleum Institute to indicate the specific gravity of a liquid. Water has an API gravity of 10°API and No. 2 fuel oil, about 35°API.
ABMA	American Boiler Manufacturers Association.
ASME	American Society of Mechanical Engineers.
ASR	Actual steam rate, a term used to indicate the actual steam consumption of steam turbines in lb/kWh.
BHP	Brake horsepower, a term used for power consumption or rating of turbomachinery. This does not include the efficiency of the drive.
Btu	British thermal unit, a term for measuring heat.
CFD	Computational fluid dynamics
CO	Carbon monoxide
CO_2	Carbon dioxide
cP	Centipoise, a unit for measurement of absolute viscosity.
CR	Circulation ratio, a term used to indicate the ratio by weight of a mixture of steam and water to that of steam in the mixture. A CR of 4 means that 1 lb of steam–water mixture has $\frac{1}{4}$ lb of steam and the remainder water.
dB	Decibel, a unit for measuring noise or sound pressure levels.
dBA	Decibel, scale A; a unit for measuring sound pressure levels corrected for frequency characteristics of the human ear.
DNB	Departure from nucleate boiling.
FGR	Flue gas recirculation.
fps, fpm, fph	Feet per second, minute, and hour; units for measuring the velocity of fluids.
gpm, gph	Volumetric flow rate in gallons per minute or hour.
HHV	Higher heating value or gross heating value of fuels.
HRSG	Heat recovery steam generator.
ID	Inner diameter of tube or pipe.
in. WC	A unit to measure pressure of gas streams; inches of water column.
kW	Kilowatt, a unit of measurement of power.
LHV	Lower heating value or net heating value of a fuel.
LMP	Larson—Miller parameter.
LMTD	Log-mean temperature difference.
ln	Logarithm to base e; natural logarithm.
log	Logarithm to base 10.
M Ib/h	Thousands of pounds per hour
MM Btu	Millions of British thermal units.
MW	Molecular weight.
NOx	Oxides of nitrogen.
NTU	Number of transfer units; a term used in heat exchanger design.

OD	Outer diameter of tube or pipe.
ppm	Parts per million by weight or volume.
psia	Pounds per square inch absolute, a term for indicating pressure.
psig	Pounds per square inch gauge, a term for measuring pressure.
RH	Relative humidity.
SBV, SBW	Steam by volume and by weight in a steam–water mixture, terms used by boiler designers.
scfm, scfh	Standard cubic feet per minute or hour, units for flow of gases at standard conditions of temperature and pressure, namely at 70°F and 29.92 in.Hg, or 14.696 psia. Sometimes 60°F and 14.696 psia is also used. The ratio of scfm at 70°F to scfm at 60°F is 1.019.
SCR	Selective catalytic reduction.
SSU	Seconds, Saybolt Universal; a unit of kinematic viscosity of fluids.
SVP	Saturated vapor pressure, pressure of water vapor in a mixture of gases.
UHC	Unburned hydrocarbon.
VOC	Volatile organic compound.

Nomenclature

(Wherever applicable, all the three systems of units are shown)

A	air for combustion, kg/GJ (lb/MM Btu)
σ	Steffan–Boltzman constant, $W/m^2\,K^4$
$\in$	Emissivity of casing, effectiveness of heat exchanger
μ	Viscosity, subscripts g and l stand for gas and liquid, kg/ms (kg/m h) (lb/ft h)
β	Volumetric expansion coefficient, 1/K (1/°F)
ΔP	Pressure drop, kPa (kg/cm²) (psi)
ΔT	Temperature difference between tubes and fluid, K (°F)
ς	Damping factor
ψ	Damping criterion
λ	Wave length, m (ft)
A_f, A_o, A_t	Area of fin, obstruction area, and total, (m²/m) (ft²/ft)
A_{ef}	Effective area, (m²/m) (ft²/ft)
b	Fin thickness, (mm) (in.)
C	Amount of water condensed, kg/h (lb/h), also C_{min}/C_{max}
C_{min}, C_{max}	WC_{min}, WC_{max} (product of mass flow and specific heat of fluid, minimum and maximum)
C	Factor $C[(C_p/\mu)^{0.4}k^{0.6}]$ defined in equation for inside heat transfer coefficient in Appendix B
C_p	Gas specific heat, (kJ/kg K) (kcal/kg °C) (Btu/lb °F)
C_1-C_6	Escoa constants for the calculation of heat transfer coefficient and pressure drop in finned tubes
d_i, d_o, d_h	Tube inner, outer diameter, effective diameter, m (in.)
	E = Young's modulus of elasticity, $N/m^2\left(186 \text{ to } 200\times10^9\,N/m^2 \text{ for carbon steel}\right)$
E	Efficiency of HRSG, also excess air
f_n, f_e, f_a	Natural frequency of vibration, vortex shedding frequency, acoustic frequency, 1/s
ff	fouling factor, subscripts i and o stand for inside and outside (m² K/W) (m²h °C/kcal) (ft²h °F/Btu)
F_g	Factor relating gas properties, $[(C_p^{0.33}k^{0.67}/\mu^{0.32})]$
Fr	Froude number
h_1, h_2	Enthalpy of gas at inlet, exit of condenser, (kJ/kg) (kcal/kg) (Btu/lb)
h	Fin height, mm (in.)
h_{fg}, h_f, h_g	Enthalpy of vaporization, feed water, vapor, (kJ/kg), (kcal/kg), (Btu/lb)
h_c, h_n, h_o, h_i	Heat transfer coefficient, convective (condensing), nonluminous, outside, inside, (W/m² K) (kcal/m²h °C) (Btu/ft²h °F)
H	Height of stack, m (ft)
G	Gas mass velocity, kg/m²s (lb/ft²h)
I	Moment of inertia, m⁴
I_0, I_1, K_0, K_1	Bessel functions

g	Acceleration due to gravity, m/s^2 (ft/s^2)
ΔH, H_{fg}	Latent heat, kJ/kg (kcal/kg) (Btu/lb)
k_l	Thermal conductivity of liquid, (W/m K) kcal/m h °C (Btu/ft h °F)
K_m	Thermal conductivity of metal, (W/m K) kcal/m h °C (Btu/ft h °F)
L, L_e	Thickness of insulation, equivalent thickness, mm (in.), effective length, m
LHV, HHV	Lower and higher heating values of fuel, kJ/kg (kcal/kg) (Btu/lb)
LMTD	Log-mean temperature difference, °C (°F)
m	Mass of tube, kg/m (lb/ft)
M	Moisture in air, kg/kg
MW	Molecular weight
n	Mode of vibration
n	Fins/m (fins/in.)
N_d, N_w	Number of tubes deep, wide
NTU	Number of transfer units
Nu	Nusselt number
Pr	Prandtl number
P	Pressure, kPa (kg/cm^2) (psi)
R_1–R_5	Thermal resistances, m^2 K/W (m^2 h °C/kcal) (ft^2 h °F/Btu)
Re	Reynolds number
q	Heat loss per unit area, (kW/m^2) (kcal/m^2 h) (Btu/ft^2 h)
Q	Duty or heat loss, (kW) (kcal/h) (Btu/h)
ρ	Density, subscripts g, l, and v stand for gas, liquid, and vapor, kg/m^3 (lb/ft^3)
S	Strouhal number, fin spacing
S_T, S_L	Transverse and longitudinal spacing, mm (in.)
t_c, t_a, t_g, t_s, t_{w1}, t_{w2}	Casing temperature, ambient temperature, average gas temperature, saturation temperature, inner wall and outer wall temperatures, °C (°F)
T	Absolute temperature in K (°R), subscript dp refers to dew point
U	Overall heat transfer coefficient, subscripts I and o stand for inside and outside, W/m^2 K (kcal/m^2 h °C) (Btu/ft^2 h °F)
UA	Product of U and surface area A, subscripts d and p stand for design and performance, (kW/K) (kcal/h °C) (Btu/h °F)
v	Specific volume, subscripts s, l refer to steam, liquid, m^3/kg (ft^3/lb)
V	Velocity, subscripts h, v, g, and s stand for horizontal, vertical, gas, and sonic, respectively, m/s (ft/s)
VH	Velocity head, kPa (kg/cm^2) (psi)
w	Width of gas path, m (ft)
w	Flow per tube, kg/m (lb/ft)
wc	water column
ws	Fin serration, mm (in.)
W	Flow of fluid, subscripts f, g, i, l, o, and s stand for feed water, gas, inside, liquid, outside, and steam, kg/h (lb/h)
WC	Product of mass flow and specific heat, subscripts min and max stand for minimum and maximum, kW/K (kcal/h °C) (Btu/h °F)

Index

A

Acid dew point temperature, 38–39
Air
  combustion, 2–12
  density, 94, 305, 351
  heat transfer coefficient inside tubes,
    253–257, 380
  moisture, 3
  relative humidity, 1–2
Air heater, 250–251, 362–363
  vs. economizer, 90–91
  impact on NOx, 90
  off-design performance, 252–258
  sizing calculation, 252
  tube wall temperature calculations, 256
Allowable stress values, 135
Ambient temperature,effect on air density,
  305, 351
American Boiler Manufacturers
  Association (ABMA)
  boiler water limits, 335–336
  feed water guidelines, 336
American Society of Mechanical
  Engineers (ASME)
  allowable stress values, 135
  boiler and feed water chemistry,
    335, 337–338
  boiler efficiency calculations, 21–24
  HRSG efficiency, 17, 300
  tube thickness for fire tube boilers, 236–238
  tube thickness for water tube boilers,
    133–135
Ammonia injection grid, 96, 195
Approach point, 266
  design, 267–272
  economizer steaming problem, 267, 281–284
  impact on HRSG size and cost of operation,
    271–272
  off-design, 267, 272
  significance, 197–198, 266–267, 270–272
  suggestions for selecting, 267
Area heat release rate, 67
A-type boiler, 104–105
Auxiliary firing
  advantages of fired HRSG, 19, 205–206, 273
  burner duty estimation, 18, 21
  efficiency, 17, 21, 90–91, 188, 204–205, 325

  exhaust gas analysis, 18–21
  fresh air firing, 299–300

B

Bath heaters, 323–327
Beam length
  fire tube boiler, 157–163, 408
  non-luminous heat transfer, 230, 232, 407–412
  tube bundle, 408
Blow down
  calculation, 335–338
  critical flow estimation, 340–342
Boiler
  A, D, and O types, 104–107
  air heater, 250, 362–363
  basic equations for energy transfer, 358–361
  blow down lines, flow estimation in,
    340–342
  calculations for validating field data,
    199–204, 284–286
  circulation calculations, 73, 77–82
  classification, 103–107
  custom design, 92, 94, 97, 99
  drum coil heater
    advantage, 324
    bath heaters, 325–327
    natural convection heat transfer
      coefficient, 324–325
    using steam, 324
  drum/vessel discharge rate estimation,
    340–343
  dry back, 159
  economizers, 112–113, 361–362
  efficiency
    condensate heater, 109–111
    direct method, 21–22
    excess air *vs.* efficiency, 26
    field data, 109
    glycol heat recovery system, 89, 108
    heat loss method, 22–23
    indirect method, 22
    lower heating value, 12, 22
    low feed water temperature, 109
    performance *vs.* load, 111–112
    simplified formulae, 23–25
  elevated drum design, 160
  emission limits and impact on design, 97

evaporator, 144, 214–215, 369–370

exit gas and fluid inlet temperatures, 370–374

fan calculations
  fan capacity, 354
  fan head, 355
  fan power consumption, 355
  fan selection, 353–354
  flue gas density, 351–352
  pressure drop in ducts, 352–353
  stack effect, 355–356

FGR impact on superheater, 124

finned tubes in convection section, 98

fire tube, 157, 230, 362

flow instability in two-phase circuits, 343–346

forced circulation design, 52, 89, 189

furnace performance, 54, 57

gas and oil firing, 22–23, 26, 32, 36, 41, 44, 46, 87, 91–92, 97–98, 108–109, 112, 122, 157

heat exchangers, 362

horse power, 160

HRSG, 265, 374–375

insulation calculations
  hot casing design, 322–323
  thickness, 320–321

LMTD, 360, 363

load *vs.* performance characteristics, 146–149

low BTU fuel performance, 27–28

low temperature corrosion, 35–38

noise and vibration problems in heat transfer equipment, 327–333

NOx and CO control, 193

number of transfer units method, 366–369

off-design performance evaluation, 146–149, 238, 274–279, 364–366

quiz on, 477–182

significance of surface areas, 113, 429

stack wall temperature, 317–319

steam drum (*see* Steam drum calculations)

steam temperature control methods, 125–128

superheaters, 124–125, 131, 361–362, 364–366

transient calculations, 347–348
  drum level fluctuations, 350–351
  evaporator analysis, 348–350
  superheater heat up time, 165, 346–347

tube geometry, 114, 260

uninsulated duct wall temperature, 317–319

water chemistry, 155–156

water tube boilers, 87–91, 361–362

Boiler horse power (BHP), 160

Boiling heat transfer coefficient
  inside tubes, 67–68
  outside tubes, 231
  Roshenow correlation, 231

Burner selection, 13–14

C

Carbon monoxide (CO)
  catalyst in HRSG, 17, 193–195
  conversion calculations, 29
  efficiency loss due to, 23

Cement plant, 217, 221–223

Cheng cycle, 187

Circulation
  calculations for CR, 74, 77
  forced, 52, 89, 189
  guidelines for design of circulation system, 85–86
  ratio, 69
  Thom's chart
    acceleration loss, 76–77
    friction loss, 75–76
    gravity loss, 76

Circulation ratio (CR), 69

Circulation systems, 72, 89, 167, 169, 189

Classification
  steam generators, 103–107
  waste heat boilers, 165–166

Claus process, 178

Cogeneration systems, efficiency of, 208

Combined cycle, efficiency, 206–207

Combustion calculations
  air flue gas quantities, 3
  air moisture effect, 3
  boiler efficiency
    direct method, 21–22
    excess air *vs.* efficiency, 26
    heat loss method, 22–23
    indirect method, 22
    simplified formulae, 23–25
  burner selection, 13–14
  combustion constants, 3–7
  combustion temperature
    effect of FGR, 15–16, 91
    examples for gas,oil, 15
    NOx formation *vs.* flame temperature, 14
  detailed calculations, 2–10
  dry and wet flue gas analysis, 3, 8
  emission conversion calculations, 14, 29–32
  excess air evaluation, 8–10
  firing low heating value fuels, 27–28

gas turbine exhaust
   duct burner, 17–18
      firing temperatures, 18, 21
      fuel quantity used, 19–21
      oxygen consumption and energy input,
         17–19
   heating values, 12–13
   moisture in air, 1–3
   mole method, 3
   multiple fuel firing, 29–30
   ppmvd of NOx to mg/Nm^3 conversion,
      32–33
   simplified calculations, 10–12
   SO_3 and acid dew point, 35–39
   SO_3 conversion, 33, 39, 97
   turbine exhaust emissions, 34
   water formation, 8
Combustion control, 112
Combustion temperature
   adiabatic, 15, 59
   effect of FGR, 15–16
   impact on NOx, 14, 33
Common economizer concept, 303–304
Condensing economizer
   arrangement of, 310
   condensation heat transfer calculations,
      312–315
   estimation of duty, 311–312
   finned tubes, 315–316
   plain tubes, 314
   schematic, 316
   selection of materials, 309–310
Convective heat transfer coefficient
   ESCOA correlations, 416–417
   formulae in all system of units, 378, 456
   inline and staggered arrangements with
      finned tubes, 417, 435
   inline and staggered arrangements with
      plain tubes, 252, 395–396, 400
   inside tubes, 377–382
   outside tubes, 393–396
   tube entrance effect, 379–380
   variation around tube periphery, 397–398
Corrosion
   acid dew point correlation, 38–39
   air heater minimum wall temperature,
      250, 256–257
   economizer minimum tube wall
      temperature, 149–151
   effect of feed water temperature,
      109–111, 151
   methods to minimize, 151–154
   water dew point, 310

Counter flow and parallel flow superheaters,
      427–428
Critical heat flux (CHF)
   charts, 84
   correlation for tube bundle, 243
   correlations for boiling inside tubes, 82
   correlations for boiling outside tubes,
      242–244
Custom design boiler
   advantages, 97
   dual pressure design, 99–100
   lowering operating cost by changing
      transverse pitch, 100–103
   novel ideas, 97–101
   performance, 97–101, 138, 146

D

Damping factor, 331
Deaeration steam calculations, 111, 335
Density
   effect of ambient temperature, 306, 351
   effect of elevation, 352
   fan sizing, 351–355
Departure from nucleate boiling (DNB)
   correlations, 82–83, 242–244
   Groeneveld tables, 83
Design methods
   air heater sizing and performance, 252–259
   condensing economizer sizing, 309–316
   economizer sizing and performance,
      143, 203, 238
   evaporator sizing and performance,
      144, 203
   fire tube boiler sizing and performance,
      160–163, 230–239
   furnace sizing, 47, 53–58, 411
   steam generator design and performance,
      138–149, 370–373
   superheater sizing and performance,
      131, 202–203, 364
   unfired boiler furnace performance
      sizing, 409
   waste heat boiler performance,
      199–205, 232, 244
Desuperheating steam
   methods of, 126, 128
   sizing exchanger for cooling steam,
      325–326
   spray water calculations, 143–145, 148
   using air to cool steam, 126
   using spray water attemperator, 125–126
   using sweet water condenser, 125–126

Dew point
    acid vapors, 38–39
    water vapor, 310
Direct radiation
    effect of load, 66
    effect of tube geometry, 65–66
    from furnace, 66
Discharge rate estimation from boiler drum,
        340–343
Distillation condensation
        subsystem (DCSS), 224
Drainable superheaters, 126
Drum
    coil exchanger design, 154.324
    hold up time estimation, 339–340
    level fluctuation in transient mode, 350–351
    separation velocities in horizontal and
        vertical directions, 334
    steam pressure change due to load, 99, 349
Drum coil heater
    advantage, 324
    bath heater, 325–327
    natural convection heat transfer coefficient,
        324–325
    using steam, 323–324
Drum exchanger, 127
Dry and wet flue gas analysis, 3, 8
Dry back boilers, 159
D-type boiler, 104, 106–107
Dual pressure design, 99

E

Economizer
    vs. air heater, 35–36
    avoid steaming, 272, 283
    condensing (*see* Condensing economizer)
    design of finned tube, 387
    design of plain tube, 361–362
    horizontal gas flow, horizontal tube, 388–389
    multiple streams, 385, 387, 389
    off design performance evaluation,
        143, 238, 274–279
Effective projected radiant surface (EPRS), 67
Efficiency
    boiler, 21–26
    heat loss method, 22–23
    improving, 109–110
    relation between HHV and LHV, 13
    serrated fins, 418
    simplified formulae, 23
    solid fins, 418
    unfired and fired HRSG, 204, 206

Elevated drum steam generator, 89
Emissions
    conversion calculations for steam generators,
        29–32
    conversion calculations for turbine exhaust,
        34–35
    FGR, 16, 124
    how it affects radiant and convective
        superheaters, 124
    impact on boiler design, 90, 92–97, 193–195
    NOx reduction, 194–195
    SCR, 92, 194
Emissivity of gases, 407
Energy transfer equations, 358–360
Enthalpy
    common gases, 447, 450–451
    compressed water, 468–469
    flue gases, 450
    turbine exhaust gases, 451–452
Evaporator
    bare tube *vs.* finned tube, 99, 414–415
    design, 99, 138, 369
    performance, 138, 144, 276–279
    simplified approach to performance
        evaluation, 199–203, 245–248
    transient calculations, 346–351
Excess air
    determining from dry and wet oxygen
        in flue gas, 8–10
    effect on boiler efficiency, 26
Export steam, impact on boiler
        performance, 208

F

Fan
    calculations
        capacity, 354
        flue gas density, 351–352
        head, 355
        power consumption, 355
        pressure drop in ducts, 352–353
        selection, 353–354
        stack effect, 355–356
    effect of air density, 353
    effect of elevation, 351
    flue gas recirculation, 353
    margins on flow, head, 353
    sizing for fired HRSGs, 299
Ferrule
    purpose, 180
    tube sheet temperature with and without,
        248–250

Field performance of installed HRSG, 286
Film temperature calculations, 227, 318
Finned exchanger, 127
Finned tubes
    advantages over plain tubes, 414
    chart for optimizing fin geometry, 438–439
    in condensing economizers, 315–316
    effectiveness, 418–419
    effect of fin geometry
        bundle weight, 433–435
        heat flux inside tubes, 427
        heat transfer, 432–437
        tube wall temperature, 433
    effect of fin thickness and thermal
        conductivity, 439–440
    effect of gas side fouling, 437
    effect of tube side fouling, 436–437
    effect of tube size on heat transfer, 435
    efficiency, 418–419
    fin tip temperature estimation, 68–69, 420
    flue gas pressure drop, 419, 424
    in gas fired package boiler, 98–99, 415
    gas side heat transfer coefficient, 417, 430
    heat flux inside tubes, 425, 433
    heat transfer coefficient
        convective, 416–417
        effect of fin geometry, 432–437
        fin efficiency and effectiveness, 418–419
        fin selection, 431–432
        nonluminous, 419
    heat transfer correlations, 417
    inline *vs.* staggered arrangement, 435
    solid and serrated, 413–414, 416–417
    superheater, 425–429
    superheater fin selection, 426
    tube life calculations, 426–427
    variations in surface area, 438, 445
    what you should know about, 445
    why no fins in air heater, 440–445
Fire tube boiler
    boiler horse power, 160
    design of, 230
        addition of economizer, 238–241
        convective heat transfer coefficient, 230
        critical heat flux, 242–244
        gas analysis effect, 172, 241–242
        nonluminous coefficient, 230, 408
        performance evaluation, 245–248
        simulation, 239, 241
        tube bundle diameter estimation, 245
        tube sheet temperature, 248–250
        tube sheet thickness, 248
        tube size effect, 244–245
    dry back, 158–159
    gas pressure drop, 233
    grooved tubes, 157–158
    off-design performance, 238
    reformed gas boiler, 165–166
    simplified performance calculations,
        160–163
    sizing and performance calculations,
        160–163, 230, 238
    steam drum, 174, 176
    superheater, 157–158, 173
    vs. water tube boiler, 168, 173–175
    wet back, 158–159
Firing low Btu fuels, 27–28
Flow accelerated corrosion, 195–197
Flow diagram for performance evaluation,
    139, 200
Flow in parallel paths, 129–130
Flow instability in two-phase circuits, 343–346
Flue gas analysis
    before and after condensation of water, 319
    before and after turbine exhaust firing, 21
    combustion calculations, 3, 8
    composition and gas pressure, 168–172
    density, 351–352
    dry and wet basis, 3, 8
    effect of gas pressure on heat transfer, 401
    effect on heat transfer, 242
    fuel oil combustion, 11, 30
    molecular weight, 8, 30, 351
    natural gas combustion, 2, 11, 19, 311
    water dew point, 310
Flue gas recirculation (FGR), 15–16, 91, 94
Fluid heaters and film temperature, 227
    fuel degradation, 227
    properties of thermic fluid, 229
Forced circulation boiler (FC-OSSG), 116
Formulae in all system of units
    condensation heat transfer, 361
    density of flue gas, 455
    heat transfer inside tubes, 378, 455
    pressure drop inside tubes, 389, 455
    pressure drop outside tubes, 399, 456
    two-phase flow losses, 456
    velocity inside tubes, 361, 390, 455
Fouling factor
    cement plants, 221
    due to scale, 221
    tube wall temperature in finned/plain tubes,
        151, 433
    variation with time, 221–222
Fresh air firing, 299–300
Friction factor, 390, 399

Froude number, 82, 176
Furnace
    boiler circulation
        cross-flow boiler, 73
        downcomers, 71
        elevated drum package boiler, 71
        forced, 72, 74, 189, 191
        natural, 70–71, 191
    boiling heat transfer coefficient, 67–68
    boiling process, 69–70
    circulation calculations, 77–82
    circulation system design guidelines, 85–86
    critical heat flux and allowable steam
            quality, 82–84
    duty calculation, 57–60
    effective projected radiant surface, 67
    effect of fuel, 47, 55
    effect of steam pressure, 47
    exit gas temperature evaluation, 53–60
    external radiation, 61–65
        distribution of, 65–66
    fin tip temperature estimation, 68–69
    flow stratification in horizontal tubes, 82
    furnace duty estimation
        adiabatic combustion temperature, 59
        Boltzman number, 57
        coefficient of thermal radiation, 58
        combination firing, 60
        heat flux, 60
        temperature profiles, 58
    Gurvich method, 57, 59
    heat flux, 61, 64–65
    heat release rates, 46
    radiation distribution to tube banks, 61–64
    radiation to superheater, 66
    refractory lined furnace, 41–42
    Thom's two-phase loss factors, 75–77
    unfired performance, 407, 409
    view factor, 61–63
    water cooled design, 44–45
    water cooled furnace
        front,rear and side walls, 43
        membrane walls, 42
        O-ring header design, 44–46
        start up rates, 43
        steam–water mixture circulation, 44

G

Gas analysis
    effect on properties, 453
    from industrial effluents, 171, 183
Gas bypass system, 165, 167

Gas mass velocity, 101, 131, 314, 330, 393–394,
            399, 402, 416, 430–431, 493
Gas pressure drop
    how to evaluate operating cost, 101, 198
    inside tubes, 233
    over finned tubes, 419–420
    over plain tubes, 399–401
Gas turbine exhaust combustion calculation
    duct burner,HRSG, 17–18
    firing temperatures, 18, 21
    fuel quantity, 19–21
    oxygen and energy input, 17–19
Gas turbine HRSGs
    Cheng cycle system, 187
    cogeneration system, 184, 186
    combined cycle plant *vs.* cogeneration
            plant, 185
    dual pressure natural circulation, 191–192
    economizer, 188, 233
    forced circulation, 188–189
    furnace fired, 191–192
    inclined evaporator tubes, 188, 190
    intercooler, 191, 193
    natural circulation, 188–189
    once through, 188, 190
    performance evaluation, 199–203
    simulation (*see* Simulation)
    unfired *vs.* fired, 187
Gurvich method, 57, 59

H

Heat flux
    critical, 82–83, 242, 244
    film temperature, 227–228
    in furnaces, 61, 64, 66, 75, 79
    inside finned tubes, 74–75, 427–428
    inside plain tubes, 74–75, 426–432
    outside tubes, 242–244
Heating values, 11–13
Heat loss
    in boilers, 22–23
    in ducts, stacks, 319
    effect of load in boilers, 24
    equation, 320, 323
    method, 22–23
    through insulation, 320
Heat pipes, 251–252
Heat recovery
    cement plants, 217, 221–223
    hydrogen plants, 166, 182–184, 242
    kalina cycle, 226–227
    solar energy, 184–185

in sulfuric acid plants, 167, 173, 181–182
in sulfur plants, 176, 178–181
Heat recovery steam generator (HRSG)
  accuracy of simulation, 279, 281
  ambient temperature effect, 305–306
  applications of simulation, 265–266
  approach points, 197–198, 266–267, 270–272
  CO catalyst, 194–195
  cogeneration plant application, 301–303
  combined cycle plant (fired), 206–208
  complex HRSG configuration, 264, 308
  design of, 198, 263, 308
  duct burner, 17–18
  efficiency of, 204–205, 300–301
  emission control in, 194–196
  evaporator, 272–273
  exhaust gas flow and temperature, 306–308
  field data evaluation, 284–286
  fired performance, 206–208, 279, 281
  firing temperature, 18, 21
  flow accelerated corrosion, 195–197
  forced circulation, 189–191
  fresh air firing, 299–300
  furnace fired, 17, 191–192
  gas–steam temperature profiles, 267–269
  geometric data of, 279–280
  heating surfaces, 217–219
  improving efficiency, 303
  natural circulation, 191–194
  NOx catalyst, 194
  NTU method, 264, 274, 277
  off-design performance, 199–203, 274–279,
    286, 288
  once through, 188–189
  optimization, 303–305
  optimum utilization of boilers, 205–206
  performance and field data
    calculations, 203, 267, 274–279
    economizer performance, 203, 274
    evaporator performance, 203, 274
    gas pressure drop, 203
    logic diagram for performance
      evaluation, 199–200
    superheater performance, 202–203, 275
    thermal performance data, 201
    tube geometry data, 199–200
    unfired HRSG, 199
  physical design, 263–264, 286–287
  pinch point, 197, 266–267, 270–272
  quiz on, 477–482
  simulation, 263
  single-multiple pressure, 191–192, 286–294
  split superheater design, 208–213, 295–299

steaming economizer
  approach point selection, 267
  minimization methods, 283
  performance of, 284
supplementary firing, 273–274
temperature profiles, 194, 196, 267
unfired performance, 189, 199, 279–280, 374
with and without export steam, 208, 211,
  214–217
Heat recovery vapor generator (HRVG), 224
Heat release rates
  area, 46
  furnace size, 47
  steam pressure
    distribution of energy, 47–48
    high-pressure steam generator, 47, 49
    low-pressure steam generator, 47, 51
    medium-pressure steam generator, 47, 50
  volumetric, 46
Heat transfer coefficient
  angle of attack, 396
  in boiling, 67–68, 231–234
  condensation, 453–454
  convective, 360, 377, 393, 416–417
  effect of fin geometry
    evaporator, 433
    Fishenden and Saunder's equation, 397
    Grimson's correlation, 394–396
    inline *vs.* staggered arrangement,
      435–436
    tube length and spacing, 432
    tube side fouling, 436–437
    weight reduction, 433–435
  effect of longitudinal spacing, 412
  examples, 382–383, 396, 421–425
  fin efficiency and effectiveness, 418–419
  fin selection, 431–432
  gas pressure drop in finned tubes, 419–420
  gas-side, 429–431
  inside tubes, 380
  natural convection, 324
  nonluminous, 407
  outside finned tubes, 416
  outside plain tubes, 393
  overall, 359–362
  simplified correlations, 377, 394
  superheaters
    high fin density, 425–426
    high tube wall temperatures, 426–429
  tube entrance effect, 379–380
  tube wall and fin tip temperatures,
    420–421
  weight of finned tubes, 421

Heat transfer equipment, noise and vibration
    problems in
  acoustic frequency, 328
  damping criterion, 331–332
  fluid elastic instability, 332–333
  natural frequency of vibration of tubes, 327–328
  vortex shedding frequency, 328–331
Higher heating value
  efficiency based on, 22–24
  fuel data, 12
Holdup time in steam drum, 339
Horizontal gas flow, horizontal tube
    economizer, 388–389
Horizontal tube superheater, 119, 385–386
Hot casing design, 322–323
Hottel's charts, 410
Hydrochloric acid dew point, 38
Hydrogen plant
  fire tube boiler, 166, 183
  flue gas boiler, 184
  gas properties, 182, 242, 450
  reformed gas boiler, 184
  steam reforming process, 182–183

I

Important formulae in all systems of units,
    455–456
Incineration plant boiler, 165–167
Inlet gas temperature and analysis, 168, 170
Inline arrangement
  finned tubes, 417, 435
  plain tubes, 397–398
  staggered tube banks, 393, 395, 417
Instability in two phase flow, 343–346
Insulation
  effect of wind velocity, 320
  heat loss calculation, 321–322
  performance, 320
Inverted loop superheater, 137–138, 388–389

K

Kalina cycle
  advantages of, 223–225
    condensing pressure, 225
    energy recovery potential, 226–227
    low inlet gas temperature heat sources, 224
  ammonia water mixture, 223
  cement plant, 223
  distillation condensation sub system, 224
  gas–fluid temperature profile, 223–224
  3 MW demonstration plant, 225–226

L

Large capacity, high pressure, forced
    circulation steam generator, 89
Large package boilers, 87–89
Larsen Miller parameters for estimating
    superheater life, 426–429
Log-mean temperature difference (LMTD),
    360, 363
  correction factors, 360
Longitudinal spacing
  effect on gas pressure drop, 403–404
  effect on nonluminous heat transfer
    coefficient, 403, 412
Lower heating value (LHV), 12–13
  boiler efficiency, 22–24, 108
  fuel data, 12
Low NOx burner, 13, 92–93
Low temperature corrosion in boilers, 35–38,
    151–154

M

Materials
  condensing service, 35, 38, 310
  stress values, 135
  superheater, 125, 427
Modular steam generator designs, 88
Moisture in air, 1–2
Molecular weight of gases, 3, 351–352
Mole method of combustion calculations, 3
Multiple pressure HRSG, 189, 192, 286–294

N

Natural circulation, 52, 69–71, 74, 77, 81, 104, 118,
    167, 188–189, 191, 193, 217
Natural convection heat transfer, 324
Natural frequency of vibration, 327
Nitric oxide (NO)
  ammonia injection grid, 96
  burners, 13–14, 17, 92–93, 160
  conversion calculations, 29–30
  effect of air temperature, 91
  effect of FGR, 15–16
Noise and vibration problems, *see* Heat transfer
    equipment
Nonluminous heat transfer
  absorptivity, 407
  attenuation factor, 409
  coefficient, 419
  correction factor, 409–410
  effect of beam length, 419

effect of longitudinal spacing, 403
emissivity, 409–410
finned tube bundle, 408
Hottel's charts, 410
inside tubes, 408
large tube spacing, 412
plain tube bundle, 408
radiation exchange, 407
waste heat boiler furnace, 411
Number of transfer units (NTU) method, 142,
 202, 366, 371
 air heater, 257
 economizer, 143
 effectiveness factor, 367
 evaporator, 238, 264
 inlet gas and inlet fluid temperature, 367
 theory, 366–367
Nusselt number, 377, 393

O

Off-design calculations
 conventional method, 366
 NTU method, 143, 202, 366
 simulation method, 263, 277
 steam generator, 131–135, 143
 waste heat boiler, 197, 200
Once-through steam generator (OTSG), 104
O-type package boiler, 98–99
Outside fouling factor, 437
Outside heat transfer coefficient, *see* Heat
 transfer coefficient
Oxygen
 for combustion of fuels, 17
 determining excess air, 8
 reduction due to firing in HRSGs, 19–21
 relating FGR with, 16

P

Package boilers, 87–89
Pinch point
 effect of inlet gas temperature, 168, 271
 optimizing, 197–198
 relation to surface area, 266
 selection of, 267
Plain tube bundles, 393–396
Plain tube *vs.* finned tube evaporator, 414–415
Prandtl number, 377, 393
Pressure
 inside tubes, 134
 outside tubes, 236

Pressure drop
 in ducts, stacks, 352–353, 356
 formulae in all system of units, 455
 inside tubes, 389–391
 outside finned tubes, 419
 outside plain tubes, 399
 tube entrance and exit, 134
Pressurized hot air water preheater, 191, 193
Properties
 estimation of, 447
 flue gas, 448–449
 steam, 457–475
 water, 457–475
Properties of gases and steam
 coefficients, dry saturated steam, 471
 enthalpy, 447, 450–451, 468–469
 Fg factor, 447, 450
 flue gas mixture properties calculation, 447
 saturated water, 453–454
 saturation line, specific heat capacity, and
  transport properties, 472–473
 specific heats, 447–449, 474
 superheated steam properties, 470
 surface tension of water, 474
 thermal and transport properties, 447
 thermal conductivity, 447–448, 450, 476
 thermodynamic properties
  dry saturated steam, pressure table,
   457–459
  dry saturated steam, temperature table,
   460–462
  superheated steam, 463–467
 turbine exhaust gases, 451–452
 viscosity, 447–448, 450, 475

S

Scale
 conductivity, 234–235
 effect on fouling factor, 235
 tube wall temperature rise, 235–236
Selective catalytic reduction (SCR)
 in HRSGs, 194–195
 in steam generators, 92–94, 96
Simulation
 accuracy of, 263, 279
 applications of, 265–266
 comparison with physical design, 277–279
 HRSG temperature profiles, 267–269
 optimizing HRSG arrangement, 303–305
 optimizing temperature profiles, 303
 predicting off-design performance, 274–279
 single/dual pressure, 286–294

Specifications, waste heat boilers, 258–259
Specific heat
    air, flue gas, 448
    gas mixture, 447
    steam and water, 474
    various gases, 447–448
    water, 474
Split superheater design, 208–213, 295–299
Stack
    draft loss, 355
    drop in gas temperature, 317
    wall temperature calculation without
        insulation, 317–319
Staggered arrangement
    finned tubes, 416–417
    plain tubes, 393–394
Standard boiler, 100–101
Steam
    exchanger in drum, 126–127
    heat transfer coefficient, 377–379
    properties, 457–469
    purity, 155–157
    temperature control, 125–128
Steam assisted gravity drainage (SAGD), 115
Steam drum calculations
    blow down calculations, 335, 340
        ABMA boiler water limits, 335–336
        ASME boiler and feed water chemistry,
            335, 337–338
    hold up calculations, 339–340
    horizontal steam velocity, 333
    normal and low water levels,
        339–340
    vertical steam velocity, 334
Steam generator
    air heater, absence of, 90–91
    boiler classification, 103–107
    boiler design, changes in, 90
    boiler efficiency, 22–24
        condensate heater, 109–111
        economizer, 112–113
        field data, 109
        glycol heat recovery system, 108
        performance *vs.* load, 111–112
    boiler performance, 142–146
    corrosion minimization, 154
    custom designed boilers, 97
    design and performance
        data required for, 138, 140–142
        flow diagram, 138–139
        part load performance, 146–149
        tube geometry data, 138

dual pressure design, 99–100
emissions
    ammonium sulfate, 97
    catalysts, 92
    chemical reactions, NOx reduction, 96
    flue gas recirculation, 93–94
    flue gas temperature profile, 94, 96
    low NOx burner, 92–93
    NOx and CO levels *vs.* excess air, 92
    SCR system, ammonia injection grid,
        94, 96, 195
    sulfur containing fuels, 97
fire tube boilers
    boiler horse power, 160
    dry back, 158–159
    grooved tubes, 157–158
    sizing and performance calculations,
        139, 160–163, 230–232
    superheater, 157–158
    wet back, 158–159
flow in parallel paths, 129–130
furnace design (*see* Furnace)
with import and export steam, 129
inverted loop superheaters, 137–138
large package boilers, 87–89
low temperature corrosion problem, 151–154
novel ideas in design, 97
oil sands application
    boiling process, 117
    evaporator technology, 115
    forced circulation boiler, 115–116
    heat flux, 117
    steam-assisted gravity drainage, 115
    steam quality, 116
    water quality variations
O type package boiler, 98–99
standard *vs.* custom-designed boilers,
    100–103
superheaters
    arrangement of, 119–121, 123–125, 387–388
    FGR effect, 124
    horizontal tube, 118–119
    inverted loop, 118, 120
    location of, 118
    off-design performance of, 131–135
    radiant *vs.* convective, 119–123
    steam temperature control, 125–128
    with tube failure problems, 135–137
    vertical tube, 118, 120
tube wall temperature estimation, 132–133
water chemistry, carry over, steam purity,
    155–157

water cooled furnaces, 88, 90
why economizer does not steam, 112
Steaming economizer
  approach point selection, 267
  how to minimize steaming, 283–284
  performance of, 284
Steam pressure
  effect on boiler furnace size, 104
  effect on HRSG temperature profiles, 270
Stream
  arrangement of boiler tubes, 385
  significance of, 384
Strouhal number, 328–331
Sulfuric acid dew point
  air heating/superheating, 182
  feed water temperature, 182
  gas analysis, 173, 181
  heat recovery, 181–182
  membrane wall casing, 182
  scheme, 181
Sulfur plant
  catalytic reactors, 181
  ceramic ferrules, 179–180
  Claus process, 178
  condenser, 178–179
  fire tube boiler, 176, 179
  flue gas, 181
  heat flux, 180
  heat recovery, 178–181
  reaction furnace, 179
Superheater
  arrangement of, 119–121, 123–125, 387–388
  convective, 121–124
  design, 118, 364–366
  direct radiation at full and part loads, 67
  gravity loss, 137
  high fin density, 425–426
  high tube wall temperatures, 426–429
  horizontal gas flow, 387, 389
  horizontal tube, 118–119, 385–386
  inverted loop, 118, 120, 388–389
  location of, 118
  mal-distribution of flow at low loads, 104
  off-design performance of,
      131–135, 364–366
  radiant, 118, 122
  radiant *vs.* convective, 119–123
  remaining life estimation, 114, 426–427
  special design, 115
  split design in HRSG, 208–209
  steam temperature control, 125–128
  with tube failure problems, 135–137

tube wall temperature estimation, 132–133
  vertical tube, 118, 120, 385–386
Supplementary firing, HRSG, 273–274
Surface area
  effect of fin geometry, 113, 433
  estimation of, 264, 282, 363, 371
  finned tubes, 438, 445
  plain tubes, 245, 419
  significance of, 113, 429
Swell effect, 350

T

Temperature profiles
  approach point, 266
  in HRSGs, 265
  optimization of, 185, 303
  pinch point, 266
  SCR location, 194
  steam pressure effect, 267
Thermal conductivity
  estimation for a flue gas mixture, 450
  flue gases, 447, 450
  steam and water, 476
  various gases, 447–448
Thermodynamic properties
  dry saturated steam pressure table,
      457–459
  dry saturated steam temperature table,
      460–462
  superheated steam, 463–467
Thom's method, 75–77
Transient calculations in boilers, 347–348
  drum level fluctuations, 350–351
  effect of steam demand on drum pressure,
      348, 350
  evaporator heat up time, 348–350
  superheater heat up time, 348–350
Tube side fouling, 436–437
Tube side heat transfer coefficients
  C values
    for compressed water, 375, 379
    for dry air, flue gases, 377–378
    for saturated and superheated steam,
        377–379
  equations, 377–378
  gas pressure effect, 380–381
  heat transfer at tube entrance, 379–380
  streams, 384–389
Tube size
  effect on heat transfer, 244–245, 401, 433
  reduction in bundle weight, 244, 433–435

Tube thickness
  external pressure, 174, 236
  internal pressure, 133–134, 174
  stress values, 134–135
Tube wall temperature
  in air heater at full and part load, 257
  in economizer, 149–151
  effect of fouling factors, 234–236, 437
  in fire tube boiler, 163
  in superheater, 125–126, 132
  superheater tube life estimation, 426–428
Tubular air heaters, 250–251
  arrangement of heat pipe, 252
  heat transfer coefficient, 253–257, 362
  part load performance, 257

U

Unfired furnace performance, 407, 409
Unfired HRSG performance, 188–189, 374

V

Velocity
  in fire tube boiler, 244–245
  formulae in all system of units, 455
  gas mass, 393, 399, 429–431
  horizontal flow in drum, 334
  inlet and exit pressure losses,
      233, 255, 425
  vertical flow in drum, 334
  water, 382
Vertical tube superheater, 385–386
Vibration
  acoustic, 327
  baffles to increase acoustic
      frequency, 327
  damping factor, 331
  fluid elastic, 332
  natural frequency, 327
  vortex shedding frequency, 328
View factor
  distribution of radiation around furnace
      tubes, 61–62
  estimation of, 61, 63
Viscosity
  flue gases, 447, 450
  steam and water, 475
  various gases, 447–448
Volumetric heat release rate (VHRR), 67
Von Karman vortices, 327

W

Waste gas analysis, 171–172
Waste heat boiler
  air heaters, 250–251
  cement plant waste heat recovery, 217,
      221–223
  circulation systems, 72, 167, 169, 173–177, 192
  classification of, 165–166
  design of fire tube boilers
    addition of economizer, 238–241
    boiling heat transfer coefficient, 231–238
    convective heat transfer coefficient, 230
    critical heat flux, 242–244
    nonluminous heat transfer coefficient,
        230, 412
    performance evaluation, 245–248
    simulation, 239, 241
    tube bundle diameter estimation, 245
    tube sheet temperature, 248–250
    tube size effect, 244–245
  design of tubular air heaters
    heat transfer coefficient, 253–257
    part load performance, 257
  flue gas composition and pressure,
      168–172
  fluid heaters and film temperature, 227
    properties of thermic fluid, 229
  furnace section, 165–168, 410–411
  gas bypass system, 165, 167
  gas turbine plant, 191
  heat pipes, 251–252
  heat recovery
    in hydrogen plants, 166, 182–184, 242
    solar energy, 184–185
    in sulfuric acid plant, 167, 173, 181–182
    in sulfur plant, 176, 178–181
  HRSG (*see* Heat recovery steam
      generator (HRSG))
  hydrogen plant, 174, 182–183
  incineration plant, 165, 167
  Kalina cycle
    advantages of, 224–227
    ammonia–water mixture, 223
    distillation condensation sub system, 224
  reformed gas application, 165–166
  shot cleaning system, 167, 170
  specifying, 258–259
  steel mill kiln, 165, 169
  thermal design, 259–261
  water tube *vs.* fire tube boilers, 168, 173–175

Water-cooled furnace, 43, 88, 90, 409
    advantages of, 43–46
    front rear and side walls, 43
    membrane walls, 42
    O-ring header design, 44–46
    reradiation, 44
    start-up rates, 43
    steam water mixture circulation, 44
Water dew point
    description, 310
    natural gas products of combustion, 311
    recoverable energy through condensation,
        311–313
    sensible and latent energy recovery,
        312–313
Water level fluctuations during transients, 350
Water tube boiler, 361–362; *see also* Boiler

vs. fire tube boilers, 168, 173–175
furnace section, 53, 57, 165, 168
sizing and performance
    cement plant, 217, 221–223
    combined cycle plants and fired HRSGs,
        206–208
    efficiency, 204–205
    field data, 189, 199–204
    heating surfaces, 217–219
    optimum utilization, 205–206
    pinch and approach points, 197–198, 266
    split superheaters, 208–213
    steam generator performance, 142, 192,
        205–206, 370
    with and without export steam, 208, 211,
        214–217
Wet back boilers, 159

Printed in the United States
by Baker & Taylor Publisher Services